Surface-Mount Technology for PC Boards

Second Edition

by

Glenn R. Blackwell, P.E.
and
James K. Hollomon, Jr.

THOMSON

DELMAR LEARNING ™

Australia Canada Mexico Singapore Spain United Kingdom United States

THOMSON

DELMAR LEARNING

Surface-Mount Technology for PC Boards, Second Edition

Glenn R. Blackwell, P.E. and James K. Hollomon, Jr.

Vice President, Technology and Trades SBU:
Alar Elken

Editorial Director:
Sandy Clark

Senior Acquisitions Editor:
Stephen Helba

Development:
Dawn Daugherty

Marketing Director:
David Garza

Channel Manager:
Dennis Williams

Marketing Coordinator:
Stacey Wiktorek

Production Director:
Mary Ellen Black

Production Editor:
Barbara L. Diaz

Library of Congress Cataloging-in-Publication Data:

Hollomon, James K.
   Surface-mount technology for PC boards / by James K. Hollomon and Glenn Blackwell.
      p. cm.
   Includes bibliographical references and index.
      ISBN 1-4180-0011-6 (alk. paper)
   1. Printed circuits—Design and construction—Textbooks.
2. Surface mount technology—Textbooks. I. Blackwell, Glenn R. II. Title.

TK7868.P7H65 2006
621.3815'31–dc22 2005048522

NOTICE TO THE READER

Publisher does not warrant or guarantee any of the products described herein or perform any independent analysis in connection with any of the product information contained herein. Publisher does not assume, and expressly disclaims, any obligation to obtain and include information other than that provided to it by the manufacturer.

The reader is expressly warned to consider and adopt all safety precautions that might be indicated by the activities herein and to avoid all potential hazards. By following the instructions contained herein, the reader willingly assumes all risks in connection with such instructions.

The publisher makes no representation or warranties of any kind, including but not limited to, the warranties of fitness for particular purpose or merchantability, nor are any such representations implied with respect to the material set forth herein, and the publisher takes no responsibility with respect to such material. The publisher shall not be liable for any special, consequential, or exemplary damages resulting, in whole or part, from the readers' use of, or reliance upon, this material.

Contents

Preface

Surface-Mount Technology, or *SMT*, is defined most simply as the second-level interconnection of components using planar mounting. Thus, SMT includes numerous circuit and packaging technologies that do not use lead-through-hole assembly techniques. Widely used SMT component-packaging formats include the *chip carrier, small-outline IC (SOIC), small-outline transistor (SOT), chip capacitor, chip resistor, flip chip, ball grid arrays (BGAs)* and similar packaging methods, which are all planar mounting techniques and thus SMT.

SMT drives the electronics industry by making possible the smaller and more powerful devices demanded by users. As variations such as chip scale packages (CSPs) continue to be developed, combined with the continuing reduction in silicon features at the die level, this trend can only continue.

This book was written to provide a two-way bridge between the manufacturing technology and the printed wiring board (PWB) layout considerations for SMT. In it, we present standards for design and discuss why these standards exist. We look beyond legalistic adherence to design rules, and we study their basis and the application-specific issues that might force departure from the standards. This book is intended for the working engineer or manager, the student, or the interested layman who would like to learn to deal effectively with the many trade-offs required to produce high manufacturing yields, low test costs, and manufacturable designs using SMT. The chapter organization of the text is:

1. An Introduction to Surface-Mount Technology (SMT)

2. Surface-Mount Components (SMCs)

3. SMT Manufacturing Methods

4. System Design Considerations for SMT

5. Printed-Wiring Layout Using SMCs

6. Assembly-Level Packaging and Interconnections

7. Quality Assurance in SMT

8. SMT and Design for Manufacturability

9. Hybrid Circuits and Multi-Chip Modules (MCMs)

New to this second edition is information on newer IC packaging technologies such as BGAs and CSPs, updates of computer-aided design (CAD) software information, and introductory information on the requirements for lead-free solder that are now being put into place across the globe. In addition, we have updated many of the examples and figures.

Textbooks have been slow to appear since the first uses of SMT in Japan in the mid-1970s, and they continue to be scarce. The availability of an SMT-knowledgeable technologist (the human resource side of the SMT revolution) is generally the most significant aid to surface-mounting progress. It is our hope that this book makes the benefits of SMT more obtainable for you and those who may work with you.

The authors have both worked and taught in the field of SMT and PCB design. We hope to bring others the knowledge that will allow them to enjoy this field as much as we have.

James K. Hollomon, Jr.
Glenn R. Blackwell, P.E.

Acknowledgments

The authors and Thomson Delmar Learning wish to acknowledge and thank the members of the second edition review panel for their suggestions and comments during the development process. Thanks go to:

William Murray
Broome Community College
Binghamton, NY

David Hartle
State University of New York
Canton, NY

Harry M. Smith
Mt. San Antonio College
Walnut, CA

chapter 1

An Introduction to Surface-Mount Technology (SMT)

Objective

The object of this chapter is to introduce the reader to surface-mount technology (SMT) and surface-mount components (SMCs). After reading and understanding this chapter, the reader should be able to discuss:

- some of the advantages and disadvantages of SMT use compared to through-hole components (THTs)
- background material necessary to understand topics in the remainder of the book
- the basics of when SMT or THT technologies make the most sense
- issues surrounding the move to the use of lead-free solder

Introduction

SMT, Application Specific Integrated Circuits (ASICs), Field-Programmable Gate Arrays (FPGAs), computer integrated manufacturing (CIM), and advanced computer technologies are all impacting the way we work and live our lives. Surface-mount technology usage is growing rapidly. In 1985, SMT accounted for a small fraction of the U.S. components market. In 2002, SMT had an 85 percent share. In many cases, SMT components are now less costly than their through-the-hole packaged counterparts. No small number of parts are now available only in SMT packaging. Such explosive growth is an indication that SMT has a great deal to offer the circuit designer, and requires that the design team consider SMT in each new and revised

1

design. While there is still a place for, and usage of, through-hole technology that will continue for some time, SMT is the primary technology of electronic design and assembly.

This first chapter will introduce some of the advantages and some of the disadvantages of the use of surface-mount components compared to the more traditional through-hole components (THCs). It will also offer some historical perspective to the initial rationales for the introduction and use of SMT in electronic design and manufacturing.

1.1   What SMT Can Do for Circuitry

At the time this second edition is being written, there is no longer be a question of, "Should our designs incorporate surface-mount technology components?" The appropriate questions now may be:

- How do I incorporate SMT into a design?
- What are the newer component packaging technologies I need to be aware of?
- What are the newer assembly technologies I need to be aware of?
- What are the newer overriding issues I need to be aware of, such as the march to lead-free assemblies as mandated by the European Commission?

We will begin to discuss these and other issues in this chapter.

What are the benefits that are driving SMT's continued growth? Understanding the rewards that come from mastery of a subject is vital to the learning experience. So, we'll begin our discussion of SMT design with a study of what it has to offer.

As we introduce a multitude of acronyms to the reader, there are two that are used interchangeably and with some confusion as to their actual meanings. Printed circuit boards—PCBs—technically boards that include both point-to-point interconnects, commonly as copper traces, and printed components, for example thick-film resistors, on a common substrate. Printed wiring boards—PWBs—are substrates that include only interconnect structures such as copper traces. Nevertheless, PCB is the most common acronym for the substrate structure with interconnects on which will be mounted through hole and surface mount components. For that reason we will generally use PCB as the acronym in this book. For discussions where it is necessary to differentiate between the characteristics of PCBs and those of PWBs, we will do so.

1.1.1   Size and Weight Improvements

If you were to ask any group of electronics engineers for the major advantages of SMT, it's a safe bet that "size and weight reductions" would head the list, as would advantages in high-speed designs. So, let's start our discussion of advantages with a look at some positive aspects of circuit shrinking with the use of SMT.

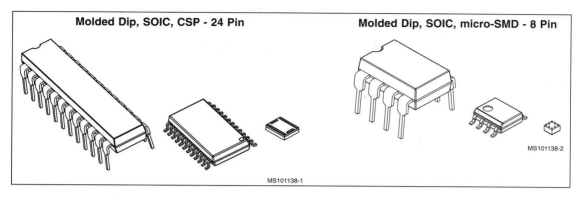

Figure 1-1 SMT and THT IC comparison. More packages will be discussed in Chapter 2.

Real-Estate Reductions with SMT

Surface-mount technology is a real-estate miser for several reasons. Most obvious is the size comparison between SMCs and typical THT components. Figure 1-1 shows a comparison of equivalent SMT and THT packages. For readers who are new to SMT components, Figure 1-1 also provides a study tool for some package identification. More SMT packages will be identified in Chapter 2 and in all major IC and passive component manufacturers' websites and data books.

While the smaller size of SMCs is a major factor in space savings, the ability to populate both sides of a circuit board often yields far greater reductions than an A/B comparison of through-hole and SMT component sizes would suggest. While the leads of THT components require clearance for each lead on both the top and the bottom of the PCB as well as on all inner layers, SMT components leave all layers clear except for the mounting layer. Additional space savings result from the elimination of through holes and their large annular rings at each component lead connection. Annular rings (the copper rings around any through hole) occupy a good deal of real estate on dense boards.

The net result is that SMT may deliver a 70 percent or greater improvement in density. Savings of 35 percent to 50 percent are routine. This alone would make SMT an attractive technology. Most electronics designers are struggling with space efficiency in some form. Either we need to pack a given circuit into a smaller box, or we need to pack more circuitry into a given box. In applications where size is not important, weight or cost may be.

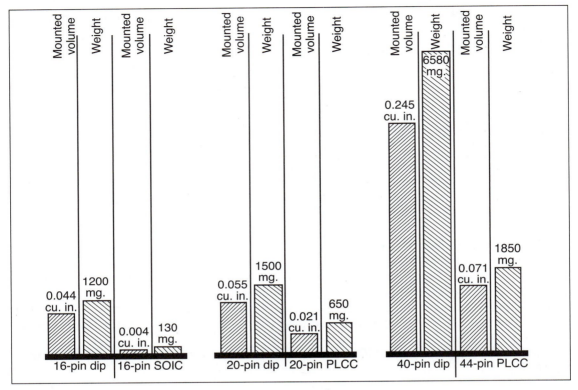

Figure 1-2   SMC and THT weights and volumes compared.

Weight Savings of SMT

The lower weight of SMCs is of particular interest in aerospace and portable equipment applications. Lower mass may also give SMT a distinct advantage in high vibration and shock-load environments. Figure 1-2 compares the volume and weight of equivalent SMC and THT devices.

1.1.2   Better Performance

Looking again at Figure 1-1, it is apparent that SMCs have far smaller leads than THTs. In fact, as will be seen in Chapter 2, chip passives have no leads at all. Small outline ICs (SOICs) and chip carrier packages have miniature or nonexistent external leads, and their internal lead paths are far shorter than a Dual In-line Package (DIP) of equal pin count. Shorter lead paths boost SMT performance over THT designs in several areas.

Circuit Propagation Delays

Each additional millimeter of lead adds parasitic inductance, capacitance, and resistance to the signal path. Research by IBM,[1] Intel, and others indicated early on that, in high-performance integrated circuits and computers, as much as 60 percent of the signal delay was attributed to THT interconnect parasitics. Circuit performance improvements of over 20 percent have been documented through changeovers to SMT.[2]

Noise Immunity

Component leads act as antennas that receive and contaminate the circuit with environmental radio-frequency noise. Compared to through-hole technology, SMT affords substantial miniaturization, particularly when fine-pitch technology is used. Miniaturization means shorter interconnections, and thus, less coupling of radio-frequency interference (RFI) and electromagnetic interference (EMI). The substantial reductions in lead length afforded by surface mounting give SMT a major edge in external noise immunity.

Crosstalk

Circuit noise is particularly troublesome in any circuits where a number of lines are switched simultaneously with very short rise times. Early research by Digital Equipment Corp. and Unisys has shown that induced noise spikes on unswitched lines are primarily produced by package inductance. Therefore, SMT has a distinct advantage over THT components in the reduction of coupled noise.[3]

1.1.3   Manufacturability Advantages of SMT

Most engineers are aware of the size and performance advantages of SMT. In addition, surface-mount boards are often more manufacturable than through-hole assemblies. Let's look at some of the factors that give an SMT design the edge over a traditional THT implementation.

Manufacturing Floor Space

For automated assembly, THT factories generally require one of several separate automated machines for each unique package style. A typical automated THT factory might have a DIP inserter, an axial component inserter with its associated sequencers (one sequencer for each forty to sixty axial part numbers to be inserted in one PWB assembly setup), a radial component inserter, and a pin inserter. Still, it's not unusual for as many as 50 percent of the assembly's THT components to require hand insertion. Typically, through-hole factories dedicate substantial floor area to hand-assembly stations.

In contrast, it takes only one flexible automated placement machine to handle the assembly of ICs, transistors, diodes, capacitors, resistors, trimmers, coils, small transformers, crystals, and other components in appropriate surface-mount packages. This placement machine is often smaller than any of the previously cited through-hole machines, although high-volume lines will typically have several placement machines, one dedicated to small components, and one to large components. Thus, in applications not requiring automated assembly of mixtures of THTs, surface-mount manufacturing generally saves floor space.

Table 1-1 Examples of SMC vs. THT Warehousing Space

Component	Through-Hole Type	SMT Type
Resistors	$^1/_4$-watt axials on tape and reel; 5 cubic feet	$^1/_8$-watt 1206 on EIA RS 481 tape; 0.35 cubic feet.
Capacitors	Radials taped per EIA RS-468; 22.4 cubic feet.	1206 Chip Capacitors on EIA RS 481 tape; 0.35 cubic feet.
ICs	16-pin DIP in plastic sticks; 18.20 cubic feet.	16-pin SOICs in plastic sticks; 2.92 cubic feet.

Additional floor space savings accrue from the warehousing of SMCs and Work in Process (WIP). Table 1-1 compares the dramatic differences in storage space requirements for reels and sticks carrying SMCs and THTs. The comparison shown is for storage of 100,000 each of the various components in their typical shipping containers.

Even more savings of chip storage are possible with the use of bulk packaging of chip resistors and capacitors. AVX Corporation data sheets state that the bulk case packaging of chips takes approximately $^1/_{20}$ the volume of tape and reel.

Where new facilities and equipment would be required for a project, many electronics manufacturers have switched from through-hole to SMT operation because the savings in building and automation costs compelled them to do so. Others have elected to change over to increase production in a given space, thus saving the cost of having to erect new facilities.

Low-Volume High-Mix Assembly

Hand assembly may be even easier with *standard pitch* SMT than with through hole, and assembly rates are typically higher. Thus, SMT is not a volume manufacturing technology limited only to products built in automated factories. In fact, we often design circuits using SMT for no other reason than to facilitate low-volume hand assembly. Typically, hand assembly is done with the use of a vacuum pick carried on an x-y arm, with reel, tube, or tray parts holders. An example of a hand-assembly station is shown in Figure 1-3.

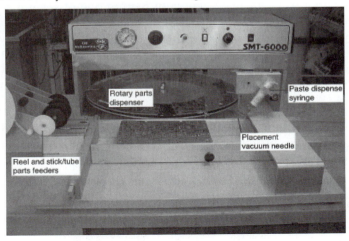

Figure 1-3 An example of an SMT hand-assembly station.

First-Pass Yield

Process quality varies greatly from factory to factory, and even for different products in the same factory. A good indication of quality is First Pass Yield, that percentage of completed products that conform to specifications *without requiring any rework* as they leave the manufacturing line and successfully pass final test.

A key determinant of process quality is the degree of process control on the manufacturing line. Few SMT versus THT comparisons involve identical manufacturing environments. Thus, it is difficult to prove that SMT offers inherently better yields than THT. However, a great deal of statistical data is at hand to show that, with equally stringent process controls, SMTs first-pass yields can at meet or exceed those of THT.

As an example, let us review process quality from one early SMT factory where a comparison is meaningful; that is, (1) we can compare products that are identical except that one is surface mount and the other is through hole, (2) the products are built in the same factory with similar process controls, and (3) both versions have been produced in sufficient volumes that a strong bank of statistical data on yields has been evaluated. In their early conversion from THT to SMT, Cincinnati Microwave, Inc. found SMT yields of 93 percent to 97 percent good assemblies entering final test, while through-hole yields for the identical circuit schematic ran between 86 percent and 87 percent.[4]

CIM Adaptability

The electronic manufacturing community is under considerable competitive pressure to maximize manufacturing cost efficiency and quality. Both of these goals are driving us toward higher levels of automation. Computer integrated manufacturing (CIM) is the target of many top electronics manufacturers today, but few have hit the bull's eye as of this writing.

Although we think of our industry as the essence of high technology, the fact is that the metals industry often leads electronics in adopting innovative manufacturing methods. This is partially because the automation for THT assembly was primarily designed for manual setup and as standalone islands of automation. Also, many of the tasks necessary to complete a THT assembly are still very difficult to automate, and off-the-shelf automation does not exist to address them.

In contrast to through-hole manufacturing, a great deal of factory-automation-compatible SMT equipment is available today. There are numerous suppliers of high flexibility machines capable of self-reconfiguration per the instructions of a host computer. The dramatic storage space reductions offered by SMT components (as shown in Table 1-1) open the door to warehousing a full complement of components for a family of board part numbers on one assembly machine. The electronic computer-aided design (ECAD) design software can provide the bill of materials (BOM) list for each design. Components can then be selected under computer control for each board part number, the placement machine can also receive its necessary placement location information from conversions of the ECAD software file data, and finished boards can be tested and stored or sent for shipping through the use of smart-cart handling of work in progress. It is, therefore, not surprising that the majority of electronics factories with full CIM implementation are SMT factories.

When it was a relatively new technology, there were hurdles that had to be cleared before CIM was applied to an SMT factory. However, software is now available for design rule checking, test engineering, test fixture development, and shop feedback to CAE/CAD systems.

1.1.4 Reliability Increases with SMT

SMT is more reliable than THT, as substantiated by numerous companies both in their labs and on their productions lines. Discussed next are the three major factors which, in various combinations, will contribute to a reliability edge for SMT over analogous THT designs.

Reduction in Number of Solder Joints

Figure 1-4 shows how surface mounting components, such as chip capacitors and chip resistors, reduces the number of interconnects. Since every solder connection is a potential failure point, the 50 percent reduction in interconnects associated with the surface mounting of these components can have a marked impact on capacitor and resistor reliability.

We should mention that the degree to which this benefit applies is dependent on several factors. First, the impact of solder-joint count reduction is directly proportional to the ratio of chip resistors and capacitors to other components in the system (i.e., those components which, in through-hole versions, do not have leads internally attached by soldering). Second, this benefit relies on quality solder joints, proper design, and on the management of thermally induced stresses. We will discuss these topics in more detail later.

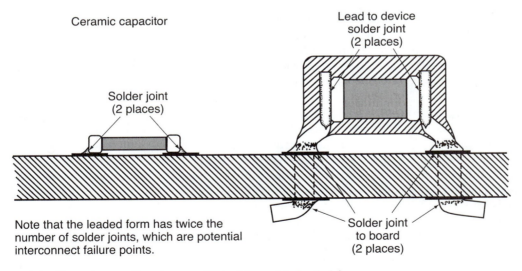

Figure 1-4 Ceramic capacitors in monolithic chip and in leaded form.

Noise Immunity

As stated earlier, each component lead and PWB trace on a circuit serves as a tiny radio antenna. Radio-frequency interference (RFI), electromagnetic interference (EMI), and electromagnetic pulse (EMP) can sneak into circuitry through these antennae, wreaking havoc with the specified performance of the circuit. By drastically reducing the antenna area of a circuit, SMT makes designs much more noise tolerant. In Chapter 6 (Section 6.4.6), we consider the effects of radiation and EMP on circuitry. A more thorough discussion of the effects of lead and line length on circuit performance will occur in Chapter 5.

Component Mass

Assemblies exposed to vibration and shock-induced stress benefit from the previously discussed order-of-magnitude lower mass of SMCs. This is a primary factor fueling the avionics and automotive communities' intense use of SMT.

1.1.5   Cost Reductions Through Surface Mounting

Cost is always a factor in choosing directions for electronics designs. In some industry segments, it runs third place behind performance and reliability, but more often, it is of critical importance. As SMT component volumes have increased, more commercial designs are selecting SMT because its use reduces system cost. IBM's early PS/2® line of PCs is a perfect example of a cost-driven selection of SMT. A Bourns advertisement of that era for trimmer resistors noted: "To survive in the marketplace, more and more products need the cost savings, space efficiency, and high performance of surface mounting."[5]

We will cover the details of the influence of design on cost in Chapter 4, Section 2. But now, we'll look at the key factors which make SMT a penny pincher.

Board-Level Savings

A printed wiring board for an SMT circuit is usually considerably less expensive than a board for the same THT circuit. Board cost savings result from three factors.

First, SMT boards are typically 35 to 70 percent smaller than similar circuits using THTs. Thus, there is considerably less board to buy. Second, SMT boards do not require holes at every component lead for insertion of leaded components. Via holes are required only where electrical communication between layers or sides of the board must be established. So, in some designs, SMT allows a reduction in hole count, which has a direct relation to board cost.

Third, multilayer boards can often be routed with fewer layers using SMT. This is true because the reduction in through holes, mentioned above, leaves more room open for inner layer traces. Also, where via holes are required, they can be less than one-half the diameter of through holes on through-hole boards where clearance for the component lead must be provided. IBM reported on a particularly high-density application. Using pin grid arrays (PGAs) and THTs, twenty-eight layers were required. However, the same circuit was routed in SMT in six signal layers.[6] Since multilayer board costs usually escalate by 7 to 15 percent per layer, savings in multilayer board costs can be substantial when switching from THT to SMC constructions.

System-Level Savings

At a system level, several options are available to the designer who wishes to economize using SMT. With a median 50 percent reduction in real estate, it becomes possible to put the functions of two THT boards on one SMT board of the same size. In a multiboard system, halving the number of boards required reduces the cost of backplanes and board-level interconnects the cost of boards, and the cost of board-mounting hardware and enclosures.

In single-board systems, system cost may be reduced by downsizing the board and the housing for the board. Another approach is to keep the housing the same size and add more functions through surface mounting. This may permit giving customers greater value added and can justify a higher selling price, which equates to a more profitable position for the manufacturer. Adding functions to aging products may revitalize a tired line, saving substantial new product development costs and allowing the continued amortization of previous investments.

Cost of THT lead prep, sequencing, and insertion is often higher than the cost of placing the SMT version of the same component. This is true both in automated and in-hand assembly. Whenever comparing costs, it is vital to look at cost-of-ownership on the board, not just the raw cost of the parts involved.

What About Volume?

While some electronics companies that have not converted to SMT feel that high-volume production is necessary to justify the conversion to SMT, the reality is that virtually all ranges of volume can efficiently use SMT. There is SMT production equipment available for all ranges of volumes, from prototype volumes of ten to the thousands of assemblies per day produced by companies like Motorola and Delphi Electronics. Assembly equipment manufacturers like Fuji, Universal, Panasonic, Siplace and Assembleon (nee Philips) can provide machines that will place up to 60,000 components per hour. Conversely, suppliers such as MannCorp, AIC Technologies, and Advanced Production Systems can provide systems that allow manual placement of components, with vacuum tools and vision-assistance, at whatever rate an individual operator can support, as well as low- and medium-volume automated equipment.

Plant Investment Savings

As we discussed in Section 1.1.3, for similar product volumes, automated SMT production requires less equipment and floor space investment than automated THT assembly manufacturing. Where brick-and-mortar investment is required for a new product, the savings in front-end capitalization may be a major factor in system costing.

1.2   Applications Where Mixed SMT and THT Technologies Make Sense

To date, no single circuit-assembly technology satisfies all applications. Some needs are best served by SMT. Other applications need custom large-scale silicon integration, hybrids, and some may require THT components. Before we begin a detailed analysis of SMT design, let's review application issues that might speak for some other technological approach.

1.2.1   Component-Driven Disadvantages for SMT

Price of Components

A few components cost more in surface-mount packaging than in through-hole style, while most SMCs enjoy a price advantage over their THT analogs. Note that, even in applications where SMT components are somewhat more expensive than through-hole components, the other surface-mount cost savings discussed earlier may still make SMT the most economical choice. Whether SMT component costs are a burden or an advantage is entirely a function of the mix of component styles and the weight of the noncomponent savings for a given application.

Availability of Components

Most component types are widely available in SMT styles. Some components, such as certain large coils, transformers, power resistors, high-capacitance or high-voltage capacitors, high-power transistors, large rectifiers, and other large components, are of limited availability as SMCs, although the design team must check the latest components, since more and more are becoming available in SMT. In applications where most components must be through hole, the slight real-estate gain offered by changing a small percentage of the population to SMT may not be justifiable. Conversely, many new ICs are only available in SMT packages.

1.2.2   "Support Concerns" as a Delaying Factor

Even in situations where SMT has a clear cost, size, and performance advantage, there are additional considerations to weigh when determining if a change to surface mount is a sound business decision. Next, we'll look at several issues that may lead some management to consider delaying the entry into SMT, even though the advantages at the product level are clear.

Field Repair of Assemblies

SMT assemblies are relatively simple to repair and rework in the manufacturing arena. Components are generally removed and replaced in special workstations using heated air or hot bar tools to desolder and resolder. Figure 1-5 shows examples of SMT repair tools. Repair

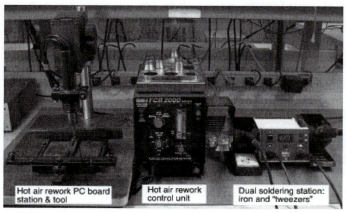

Figure 1-5   A typical SMT repair station with hot air and iron tools.

technology for standard parts is straightforward, and the necessary workstations are not expensive. However, workstations aren't free. Also, transporting them for field service would be difficult.

For companies with field-repair strategies for embedded electronics, repairing SMT might mean a considerable investment in training and equipment. Getting SMT repair gear to work sites could be a burden. In such cases, it might be wise to develop the new strategies for repair before changing manufacturing technologies. Instead of field repair, management might opt for a game plan involving the isolation of faults to the board level and the field exchange of boards.

Through-Hole Installed Base

Having a large installed base in the field makes some companies think twice about committing to SMT. To support such an installed base, some through-hole capability must be maintained, or new SMT designs of the same form, fit, and function as the existing THT boards may be developed.

1.2.3   Amortization of Existing Through-Hole Equipment

Announcing to the directors that the new million-dollar-plus automated insertion line is obsolete before it's fully operational can have a profound and undesirable effect on a career. The fear of having to make such an announcement has certainly delayed development of SMT in some companies.

In reality, changing to SMT is very seldom an overnight, 100 percent proposition. Many firms have started into SMT by redesigning their circuit boards to use SMT passives on the bottom side of the board, which allows continuing use of wave solder technology. Others have added a few pieces of SMT-specific equipment to an existing through-hole factory.

Managers must decide between the pain of asking for new capital and the pain of explaining a loss of competitive edge and market share if they don't ask for new capital. Such an avoidance/avoidance conflict may needlessly delay an important technological move. A proactive measured development of new technology without rendering existing strengths obsolete should be examined closely if you feel that your insertion investment is too new to retire.

1.3   SMT Manufacturing Styles and Their Effect on Benefits

There are currently two definitions of SMT manufacturing styles in common use. One definition divides the spectrum of SMT manufacturing into three types. Called Type 1, Type 2, and Type 3 SMT, these classifications were developed by Jim Hall of Dynapert HTC in 1983 to help explain the soldering technology of SMT.[7] In recent years, the IPC has defined six levels and six types of assemblies in their publications noted in Section 1.3.2.

Each definition of electronic manufacturing styles is distinct in the soldering approach(es) it requires. Since the manufacturing style chosen for a design has an impact on the mix of SMT benefits achieved by the design, we'll discuss manufacturing style from a benefits/limitations point of view in this chapter. Later, we'll discuss design for manufacturing relative to each of the assembly styles. First we will present an overview of Types 1, 2, and 3 SMT as well as the IPC type and level definitions. Note in Figure 1-6 Intel's succinct definitions of Types 1, 2, and 3. These are denoted by the component technologies used and the soldering processes required.

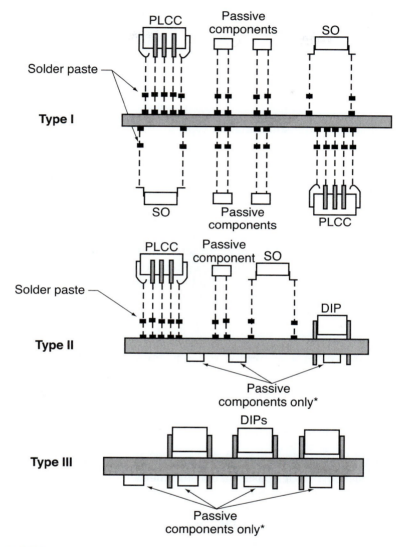

*NOTE: Intel does not recommend active devices be immersed in solder wave.

Figure 1-6   Examples of Types 1, 2, and 3. *(Courtesy Intel Corp.)*

The manufacturing processes whose terms are used in this and the next section will be described in more detail in Chapter 3. Please refer to that chapter and to the Glossary if you are unfamiliar with the terms used.

1.3.1   Types 1, 2, and 3 SMT

1.3.1.1   Type 1 SMT

Type 1 SMT uses SMCs only. No through-hole mounting of components is used in Type 1. Type 1 assembly is characterized by the use of only reflow soldering. Type 1 manufacturing allows the use of all styles of surface-mounted components.

Advantages

Since Type 1 SMT involves reflow soldering, all types of SMCs may be used, including those intended exclusively for reflow soldering.

Type 1 SMT is relatively straightforward from a manufacturing process viewpoint. Single-side-populated Type 1 boards require only one reflow soldering pass. Double-side-populated Type 1 boards are generally soldered one side at a pass. Typically the "bottom" side with small components would be reflowed first, then the top side with the larger component. During the second reflow cycle, the bottom side components are held on the board by the surface tension of the molten solder. Therefore, Type 1 boards do not require adhesives.

Finally, for product lines requiring a new automated-equipment investment, Type 1 manufacturing requires no investment in insertion equipment or a wave soldering machine. Since several different insertion machines, each dedicated to one single package style, are required to automate the typical insertion line for THT components, Type 1 SMT may bring substantial savings in both capital equipment and floor space.

Limitations

Type 1 SMT allows no use of THTs, except by special lead prepping where:

- the THT leads are cut square to make "butt joints" on matching SM pads, or
- the THT leads are bent out in a gull wing pattern to be placed on matching SM pads.

Even where all components for a design are available in SMC types, the designer may prefer to use a THT component in some instances. Types 2 and 3 do not place this restriction on the designer.

Like all manufacturing styles that use reflow soldering, Type 1 SMT also requires knowledge of solder pastes, a solder-paste dispensing method, and reflow soldering ovens—technologies foreign to many through-hole assembly shops. However, survival in the twenty-first century will require the acquisition of this knowledge at some point. Type 2 SMT requires the same new processes as well as existing THT processes, but the simpler Type 3 avoids both reflow soldering and the use of solder pastes.

Any application that requires that several boards be sandwiched closely together in a minimum housing volume should profit from a Type 1 approach. As Figure 1-7 shows, by sticking exclusively to SMCs with their low profiles, boards may be mounted much more closely than THT heights would allow.

Figure 1-7   A controls PCB using Type 1 SMT.

1.3.1.2   Type 2 SMT

Type 2 SMT is defined as a surface-mount assembly method where both SMCs and THTs are mixed together on at least one side of the board. As this definition implies, Type 2 boards may be populated on one or both sides, with bottom-side SMCs typically confined to passive, diodes, and transistors.

Type 2 SMT is distinct from Type 1 and Type 3 in that two separate soldering approaches are used. Where SMCs and THTs are integrated on the same side of the board, the SMCs are generally reflowed and the THTs are wave soldered. There are three exceptions to wave soldering in Type 2 assemblies:

- where hand soldering is used in lieu of one or both automated soldering processes,
- when laser or other point-soldering technique is used, and
- when using the pin-in-paste process, described later.

Type 2 SMT covers the widest variation of assembly styles. Boards may be populated on one or both sides. As long as at least one side combines surface attachment with insertion mounting, the assembly is Type 2.

Advantages

Where component availability is a barrier to converting a THT assembly to Type 1 SMT, Type 2 methods may provide a convenient path around the obstacle. Type 3 SMT also provides a way to integrate THTs and SMCs in one assembly. However, Type 3 usage constrains the designer to using only immersion-solderable SMCs. Type 2 has the same constraints on the wave-solder side of the assembly. Generally, leaded and leadless chip carriers (ICs), power transistors, aluminum electrolytic capacitors, connectors, and switches should not be passed through molten solder. Type 2 excels in applications where the designer wishes to take advantage of SMCs which are not immersion solderable, and where he also wishes to use some THTs for cost or availability reasons.

It's also worthwhile to note that certain THTs, such as Single In-line Packages (SIPs) or Zig-zag In-line Packages (ZIPs) packages for resistor networks and multichip memory modules, are very real-estate efficient in single-board systems where board stacking height is not a concern. So, maximum use of available board square area is another factor that may suggest Type 2 usage.

Limitations

The advantages of Type 2 SMT are high density, component availability, and component cost shopping flexibility. In short, Type 2 SMT seems almost too good to be true. The catch is the overall complexity of the manufacturing processes required, which will be discussed in detail in Chapter 3. Briefly, the dual soldering process requirement (wave + reflow soldering) can complicate manufacturing, add process steps, and push manufacturing costs up while depressing first-pass yield. Test access may prove a problem. There are design rules to minimize the impact of these problems. We'll discuss these in detail when we cover manufacturability and design for test. For now, we will just note that the benefits of Type 2 SMT come at a price.

Best Applications

Type 2 SMT is a component-driven technology. It should be chosen when the benefits of SMT are very attractive but component availability or component cost would shut the door on an all-SMT design and Type 3 design would be unsatisfactory.

1.3.1.3   Type 3 SMT

Surface-mount technology, as a major commercial force, originated in Japan in the early 1970s. In the mid-1970s, Panasonic was granted patents on SMT pad designs (not enforced to allow broad SMT usage). Early Japanese surface mounting, and the bulk of surface mounting in Japan as late as 1990, was Type 3. Type 3 SMT uses SMT passives and small active components on the bottom side of the board.

Type 3 utilizes a single-pass wave soldering approach and no reflow soldering. In Type 3 assembly, all SMCs are glued to the solder side of a board. THTs are inserted on the classical component side. The board is then wave soldered, attaching components on both sides of the assembly in a single automated soldering operation.

Advantages

For a shop familiar with through-hole manufacturing, Type 3 presents a more manageable learning assignment than Types 1 or 2. There is no need to develop any solder-paste expertise, reflow soldering technology, or new solder-materials knowledge. However, in no way is learning to produce quality, high-yield, Type 3 assemblies a cake walk. The Type 3 process book has fewer chapters than the other types, but each chapter is plump, and it must be fully digested if good results are to be obtained.

Type 3 SMT may well satisfy the need to adopt surface-mount technology while protecting an investment in insertion automation. Type 3 SMT integrates into an existing THT automated line with less impact than either Type 2 or Type 1. You need only find space in your line for a placement machine, along with its associated adhesive dispense and curing ancillaries. No additional soldering steps are required. (Note that standard single-wave soldering systems will not give quality soldering results on SMCs. SMT adoption dictates that through-hole wave-soldering machines be replaced with dual-wave soldering machines that have been proven to provide successful SMT wave soldering.)

Component engineers find Type 3 assemblies a boon. The types of SMCs most commonly used in this approach—chip capacitors, resistors, and SOTs—are the most widely available SMCs. They are also reasonably priced. SOICs (wave-solderable glued-on packages) are typically near parity with or less expensive than DIPs. Chip capacitors are usually less expensive, and chip resistors are usually close to parity or less expensive than their leaded cousins. Be certain that the IC manufacturer of your parts certifies that the ICs can survive a trip through a wave-solder bath. This will be particularly important if the change to lead-free solders results in higher soldering temperatures.

The authors must note the strongest caution on designing ICs for bottom side wave soldering. Some IC manufacturers, such as Intel, state that none of their ICs should be subjected to wave soldering. If the change to lead-free solder (introduced in Section 1.5) and soldering results in higher soldering temperatures, more IC manufacturers may state a similar prohibition.

Limitations

Type 3 SMT constrains the designer in several areas, whereas Types 1 and 2 give great flexibility. If these constraints are understood, the design team can make an informed decision on the use of Type 3.

The first significant constraint is the requirement that any SMCs selected for Type 3 assembly must be immersion solderable. The second is that PWB designers are more rigidly constrained in the component location of SMCs for immersion soldering than they are for reflow soldering. Components may not be located so close that solder bridges from one termination to another. Also, the component layout must take into account possible eddy currents, or shadowing in the solder wave, which might cause solder misses on terminations. These constraints are well understood. We will deal with them later in our discussion of Type 3 design and in circuit board layout. However, the constraints place limits on the benefits the designer can reap from Type 3 SMT.

Best Applications

Type 3 is best used by the Printed Circuit Assembly (PCA) assembler who has no prior experience with SMT and needs to both begin to understand the design process using SMT components, the SMT pick-and-place equipment, and/or needs to amortize the existing THT equipment. Good applications for Type 3 SMT are determined by the mix of components required for the circuit. The top candidates have no very high pin-count devices (>40 pins), and have a large number of discretes. If a THT circuit has a large contingent of capacitors and resistors such that ≥60 percent of its components can be converted to immersion-solderable SMCs, the chances are good the circuit will be a cost-effective Type 3 changeover candidate.

Note in Figure 1-8 the through-hole relays on the top of the board and SMT chip components glued on to the bottom of the board. All components are then soldered using a wave solder machine.

Figure 1-8   Top (with relays) and bottom (with chip capacitors) of a relay board using Type 3 SMT assembly.

1.3.2 IPC Levels and Types

The IPC levels and types define many more levels of complexity than do the Types 1, 2, and 3. For the designer new to SMT, this can be confusing. The reader should understand that the IPC classifications attempt to define both levels of "producibility" as well as board types based on the complexity of the substrate itself. The board types are further defined in these documents:

- For rigid printed circuit boards, IPC-2222 and IPC-6012, "Design and Qualification and Performance Specification for Rigid Printed Boards"
- For flexible printed boards, IPC-2223 and IPC-6013, "Qualification and Performance Specifications For Flexible Printed Boards"
- For overall PCB component mounting documentation, IPC-CM-770E, "Guidelines for Printed Board Component Mounting"

The reader must note that all documents and standards noted in this book may be revised on a regular basis. This book presents the information as it was available at the time of writing, but the reader is advised to determine if updates and/or revisions have occurred before s/he uses the information as gospel.

The reader should also note that each recent update of these documents has changed the PCB classifications and their definitions. It is one author's opinion that IPC is attempting to chase a moving target with the definitions and they will never be static. The reader should use these definitions without casting them in stone. Many design and manufacturing engineers still use the Types 1-3 terminology and further define the component technologies in use in their own product.

The following discussion uses many terms from the IPC documents. If the reader is unfamiliar with any of them, they are defined in the Glossary or can be found in the Index.

This section will only serve as an introduction to rigid board documents, and the reader that needs more detail must refer to the IPC documents already noted and to IPC CM-770E, "Guidelines for Printed Board Component Mounting," dated January, 2004. The producibility level definitions are taken from CM-770E, and are:

- Level A: THT component mounting only
- Level B: SMT component mounting only
- Level C: simplistic THT and SMT components intermixed
- Level X: complex intermixed assembly, THT, SMT, fine-pitch technology (FPT) and ball grid arrays (BGAs)
- Level Y: complex intermixed technology, THT, SMT, ultra fine pitch (UFP), and chip scale technologies
- Level Z: complex intermixed technology, THT, UFP, chip on board (COB), flip chip, and tape automated bonding (TAB) technologies.

Each producibility level can then be divided by the technology of the board type. For rigid boards:

- Type 1: single-sided printed circuit board (PCB)
- Type 2: double-sided PCB
- Type 3: multilayer PCB without blind or buried vias
- Type 4: multilayer PCB with blind and/or buried vias
- Type 5: multilayer metal-core PCB without blind or buried vias
- Type 6: multilayer metal-core PCB with blind and/or buried vias

IPC CM-770E also provides definitions that apply to flexible boards and to boards using High Density Interconnect (HDI) technology. Please see CM-770E for these definitions. They will be briefly discussed later in this book.

There are many combinations of producibility levels and board types. Some of these examples, as shown in Figure 1-9 and Figure 1-10, along with their required soldering technology(s) are:

- 1A: THT on one side, all components inserted from top side, wave soldering
- 1B: SMT on one side, all components place on top side, reflow soldering
- 1C: THT and SMT on one side only, reflow for SMT and wave for THT
- 2B: SMT on top and bottom, reflow soldering
- 2C: THT on top side, SMT glued on bottom side, wave soldering only
- 1X: 'complex' SMT on top, THT (if present) on top, reflow soldering of SMT, wave soldering if THT is present
- 2X: SMT fine pitch, BGA on top and bottom (glued), THT (if present) on top, reflow soldering of SMT, wave soldering if THT is present
- 2Y, 2Z, SMT UFP, COB, flip chip, TAB, on top and bottom (glued), THT (if present) on top, reflow soldering of SMT, wave soldering if THT is present

Manufacturing for examples of these types will be discussed later.

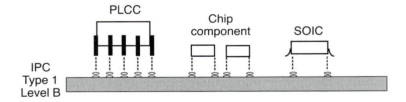

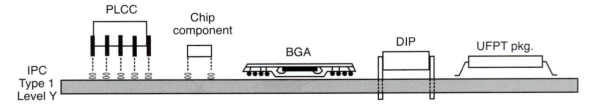

Figure 1-9   Examples of IPC single-sided Type 1 PCAs.

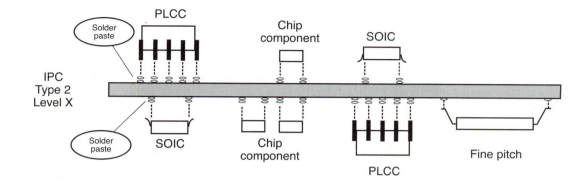

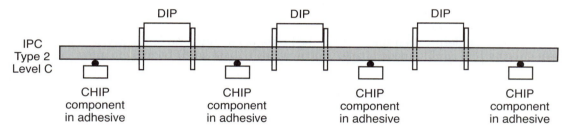

Figure 1-10   Examples of IPC double-sided Type 2 PCAs.

1.3.2.1   IPC Type 1 Level A

IPC 1A, with all THT on top side and wave soldering, is the oldest and most basic type of assembly in the IPC definitions. Assembly would consist of selecting the bare PCB, inserting all THT parts, wave soldering, then cleaning if necessary.

1.3.2.2   IPC Type 1 Level B

The simplest SMT assembly type, 1B assembly would consist of selecting the bare PCB, depositing solder paste with a dispenser or with a stencil printer (commonly called a screen printer, even though screens are rarely used), picking and placing SMT components, reflow soldering, then cleaning if necessary.

1.3.2.3   IPC Type 1 Level C

THT and SMT on one side only; this is not the simplest introduction to mixed SMT and THT assemblies since it requires both reflow and wave soldering. Generally, 1C assembly would consist of applying solder paste to the PCB, placing SMT components, reflow soldering, then inserting THT components and wave soldering.

1.3.2.4 IPC Type 2 Level B

For double-sided SMT, the steps described for IPC 1B are repeated for each side. Normally, large ICs are confined to the top side. Smaller bottom-side parts are placed in paste and reflowed first. The board is turned over, then large and small top-side parts are placed in paste. When the second reflow pass occurs, the bottom-side solder does again melt, but the surface tension of the molten metal solder will hold all the bottom-side parts in place while the top-side solder paste reflows and creates solder joints for the top-side parts. Alternatively, a high-temperature solder can be used for bottom-side parts and a low-temperature solder used for top-side parts.

1.3.2.5 IPC Type 2 Level C

Typically the easiest "introduction" to SMT for shops that are familiar with THT assembly, IPC 2C process steps would include:

- Board inverted, glue deposited at the location of each bottom-side SMT part
- SMT parts placed in glue, then glue cured either by IR or heat
- THT parts inserted from top side
- Wave soldering then solders both the bottom-side glued-on SMT parts and the top-side THT parts.

Other Type 2 manufacturing includes some variation of the steps noted above. They will be described in more detail in Chapter 3.

1.4 Lead-Free Issues

Beginning in 2006, certain countries and regions around the globe have requirements in place that will ban the use of lead in electronic interconnects. While additional issues regarding the use of lead-free solder will be discussed later in the book, this section will introduce the topic.

There are three primary environmental issues with the use of lead-bearing solder in electronics assembly. One is the control of the lead-bearing solder during its manufacturing and during its use during the assembly process. The second is contamination of cleaning water used in the cleaning process following the soldering processes. Contamination relating to these first two issues is closely regulated in most developed countries.

The third issue is the disposal of electronic products at their end-of-life, when most consumer electronic devices are typically consigned to the local landfill since they are not designed to be economically repaired or recycled. Concerns about lead leaching from the solder into ground water have resulted in the bans discussed below.

At this writing, electronic and electrical waste is the fastest growing category in the municipal waste stream.[8] A few local U.S. jurisdictions have already banned electronic devices from their landfills (wasteage.com), and others are beginning to offer recycling of electronic devices. As of this writing, four states have banned electronics (or "e-waste") from landfills: California, Massachusetts, Maine, and Minnesota, while South Dakota's largest landfill has also banned electronics. The European Union (EU) has presented a Restriction of Hazardous Substances (RoHS) directive which will be proposed in July, 2006, as EU legislation to ban the use of lead in electronic devices starting in 2006, with a phase-in over several years. The EU has also presented a Waste in Electrical and Electronic Equipment (WEEE) directive with a similar implementation timetable. Japan has implemented lead-free deadlines, and other parts of Asia are considering a similar ban. All this is being implemented without complete background information. As Dowds writes, "One study using the EPA's test method has shown that lead does not appreciably leach into water, while lead-free alloys containing silver do. Whatever the truth, it is too late to stop the industry's momentum toward lead-free." Nguyen, et. al, write "Market forces, trade restrictions, and customer perceptions rather than environmental realities have driven the lead-free movement. However, it cannot be turned around."[9] In contrast to these comments, two studies—one from the University of Florida's Solid and Hazardous Waste Engineering Program, the other from the California EPA's Department of Toxic Substances—determined that discarded cell phones should be managed as hazardous waste[10] due to both their batteries and the materials they contain.

The EU RoHS directive proposes bans on five substances: lead, mercury, hexavalent chromium, polybrominated biphenyls (PBB), and polybrominated diphenyl ethers (PBDE), effective July, 1, 2006, for consumer and many commercial devices. A phase-in of the ban applies to certain commercial and industrial devices. In addition to the coverage of the five substances, the directive also requires material content information be available to satisfy legal and regulatory requirements. Suppliers should be expected to have materials declarations available to users by July, 2005, to allow time for the users to verify their products will meet the new requirements in time to allow their products to be sold after July, 2006. It is important to note that these directives are still being defined and refined as this book is being written. In January, 2005, the EU released the "third and final" draft consultation on both RoHS and WEEE to assist with the understanding and interpretation of both directives. However, the guide is purely informative and has no legal authority. It does define several issues. Exemptions may be extended to encompass products where electrical or electronic components are not needed to fulfill the products primary function—e.g., musical greeting cards may not need to comply with the RoHS directive and resulting legislation. This guide has also further defined the term "homogeneous" as it applies to the directive. The latest definition seems to indicate that each component of different materials must comply; i.e., an IC consists of many homogeneous materials, such as the copper leadframe and the gold bonding wires, and therefore the latest guide states that each separate material in the IC must individually comply with the directive. Good luck to all involved.

The European phase-in beginning in July, 2006, of requirements for lead-free soldering materials (technically lead-free electronic package attachment techniques, as conductive adhesives are one possible option) requires:[11]

- Components that do not use lead as a component material in the tinning of leads and terminations
- Circuit boards that do not use lead as a component material in the tinning of traces and pads
- Solder material or component attachment material that does not contain lead

While this may seem admirable, there is no drop-in replacement for Sn/Pb solder from either an economical or from a manufacturing standpoint.

Lead-free components, lead-free solders, and other lead-free assembly materials are already working their way into the supply chain. The switchover requires documentation trails that in many cases do not currently exist. IC manufacturers have tinned their component terminations for SnPb solder for years, but they never had to document and "prove" what materials they were using. They must now do that, then have that documentation available to users and in a format for the users to pass along to regulatory agencies. Many component manufacturers are using or are considering using pure Sn for plating their terminations. While this meets the regulatory needs and pure tin is an acceptable conductor, tin produces "whiskers" when it undergoes the reflow process. This production is poorly understood, and the whiskers can grow long enough to potentially cause short circuits under components.

To gain some control over the documentation issues, and particularly to assure that components with lead-free terminations are designated separately from components with lead in their terminations, The National Electronics Manufacturing Initiative (NEMI) supports the JEDEC Standard JESD97: "Marking, Symbols, and Labels for Identification of Lead (Pb) Free Assemblies, Components, and Devices." This standard attempts to develop a standard vocabulary for lead-free transitions and recommends separate part numbers for lead-free devices.

At this point, there are a number of issues surrounding any changeover to lead-free component attachment.

- Can the change be made using existing processes?
- If process changes are required, what are they, and what is their overall impact on the design and assembly processes?
- What are acceptable lead-free solders?
- What are other non-solder attachment techniques that do not incorporate lead?

Many of these issues, as well as others, are addressed on IPC's lead-free website, http://www.leadfree.ipc.org. In their 2002 article, Nguyen, et al., proposed a "Structured Approach to Lead-Free IC Assembly Transitioning."[12] It contains many good points for the transitioning of ICs to meet lead-free standards.

Different areas of the world and areas of manufacturing are progressing to lead-free at a variety of rates. Some manufacturers, like Motorola[13] have demonstrated lead-free implementation in certain devices. Some of these manufacturers have implemented lead-free without making significant process changes. Others feel that the total implementation of lead-free will require major process changes.

A change to lead-free solders may affect circuit board design since different solders have different characteristics that can affect pad design. One of those issues is that certain tin (Sn) alloys, and as mentioned earlier particularly pure tin used on component terminations, have a much greater tendency to create "whiskers" of solder during the reflow process. This is being addressed by NEMI's Tin Whisker User Group and others. At this point, several of the electronic manufacturing groups that are working hard to establish a lead-free solder have recommended variations of an Sn/Cu/Ag alloy. One issue with this approach is that there are two major overlapping patents, held by Iowa State University and Senju Matsushita, that cover several variations of the alloy. The relative coverage of the two patents has not been addressed in the courts. Another issue is that any of the Sn/Cu/Ag alloys have a melting temperature 20 to 30°C above that of Sn/Pb eutectic solder.[14] If one assumes that the Sn/Cu/Ag alloy is the final replacement for Sn/Pb, there are two possible approaches to implementation. Some authors[15] feel that implementation can occur with no change in process temperature profiles, as most current oven profiles run considerably above the melting point of Sn/Pb. Maintaining the same profiles would most likely require the use of enhanced Sn/Cu/Ag solder alloys, with the addition of bismuth and/or indium. The enhancement results in a lower melting-temperature alloy that has also enhanced wetting ability during the reflow phase of the oven temperature profile.

Others project an approach using one of the basic Sn/Cu/Ag alloys, with its inherent need for higher reflow temperatures, typically projected to be above 235°C. At this writing, few integrated circuit manufacturers have certified their products for reflow at the projected elevated temperatures. Additionally, the glass transition temperature of the standard FR-4 board material is approximately125°C, and the effects of higher reflow temperatures on the substrates must also be evaluated. Studies comparing eutectic Sn/Pb and Sn/Cu/Ag have been done by Boeing, IDEALS, and the NCMS 1 and 2 projects.

One piece of good news is presented by Roth,[16] whose July, 2004, article reports tests that demonstrate that Sn/Ag/Cu solder compositions can be inspected by x-ray systems with the same degree of success as eutectic Sn/Pb solders. Comparing x-ray images of solder joints made with Sn/Pb solder and Sn/Ag/Cu solder, he states, "In this study . . . none of the solder joint types, including BGA, through-hole, and QFP, showed any significant change in appearance of the x-ray image." He does go on to note that x-ray inspection of joints such as those made by silver-filled adhesives do need further study and continuing work on the development of appropriate standards.

Therefore, circuit board designers will need to determine:

- The solder materials to be used in the manufacture of their design
- New soldering issues that the new materials will bring with them
- Pad criteria that may be different with new materials
- Substrate materials that may be appropriate for higher reflow temperatures
- Compatibility of components and other materials with the higher reflow temperatures

Again, at this writing there is still a great deal of work to be done in all areas of circuit board design, component temperature issues, and selection of appropriate soldering alloys as the trend towards lead-free continues. Many articles have been written on the subject, as a search of the Internet and a search of the Engineering Index (Compendex) will show, and will continue to be written as the European deadline approaches and as other deadlines are put into place. Some suppliers are trying to lead the way with information and process capability analyses. For example, Cookson Electronics and Practical Components offer several lead-free process kits. More information is available on Cookson's lead-free website, http://www.shared-intelligence.com/leadfree/index.html.

Industry organizations that are working on the lead-free efforts include the IPC, NEMI, the Lead-Free Project, and others. See these organizations' websites (listed in section 1.5) along with IPCs lead-free website, for up-to-date information. Because of the interest in the conversion to lead-free soldering and assemblies and the uncertainty of any final standards as of the date of this writing, lead-free references are listed at the end of this section.

The WEEE directive brings a different set of issues to bear on electronics producers. This directive requires that European Union member states enact national legislation placing financial responsibility for "collection, treatment, reuse, recovery, and recycling of waste from electrical and electronic equipment" on equipment producers. This will require take-back operations on the part of suppliers for all electronic goods covered by the directive.

In addition to the take-back requirement, the WEEE directive also requires high levels of reuse and/or recycling. This will require design for disassembly (requiring a new acronym, DFD?) to make recovery easier.

As noted earlier in this chapter, the reader must note that all documents and standards referred to in this book may be revised on a regular basis. This book presents the information as it was available at the time of writing, but the reader is advised to determine if updates and/or revisions have occurred before s/he uses the information as gospel.

Lead-Free Reference Sites

IPC: http://www.leadfree.org

SMT in Focus: http://www.smtinfocus.com/leadfree.html

Cookson Electronics: http://www.shard-intelligence.com/leadfree/index.htm.

Electronics Components, Assemblies, and Materials Association: http://www.ec-central.org/LEAD-FREE/index.htm

RoHS Key Elements: http://www.pb-free.info/laymans_terms.htm

Lead Free Magazine: http://www.leadfreemagazine.com

Tin Technology and Waste: http://www.tintechnology.biz/soldertec/soldertec.aspx

1.5   Organizations

Following is a list of organizations that publish information, documents, standards, etc., that relate to PCB design, PCA assembly, SMT components, wiring, testing of both components, boards, and assemblies, safety, etc.

AES—Audio Engineering Society; http://www.aes.org/
AIA—Aerospace Industries Association; http://www.aia-aerospace.org/
ANSI—American National Standards Institute; http://www.ansi.org/
ASTM—American Society for Testing and Materials; http://www.astm.org/
ATIS—Alliance for Telecommunications Industry Solutions; http://www.atis.org/
CEA—The Consumer Electronics Association; http://www.ce.org/
CENELEC—European Committee for Electrotechnical Standardization; http://www.cenelec.be/
CSA—Canadian Standards Association; http://www.csa.ca/

CTIA—Cellular Telecommunications and Internet Association; http://www.wow-com.com/

DISA—Defense Information Systems Agency Center for Standards; http://www.disa.mil

ECA—The Electronic Components, Assemblies, and Materials Association; http://www.ec-central.org/

ECMA—Standardizing Information and Communication Systems; http://www.ecma-international.org/

ESDA—Electrostatic Discharge Association; http://www.esda.org/

ETSI—European Telecommunications Standards Institute; http://www.etsi.fr/

GEIA—The Government Electronics and Information Technology Association; http://www.geia.org/

IEC—International Electrotechnical Commission; http://www.iec.ch/

IEEE—Institute of Electrical and Electronics Engineers; http://www.ieee.org/

IPC—Institute for Interconnecting and Packaging Electronic Circuits; http://www.ipc.org/

JEDEC—The JEDEC Solid-State Technology Association; http://www.jedec.org/

LFSP—Lead-Free Solder Project; http://eerc.ra.utk.edu/ccpct/lfsp-projectinfo/

NEMA—National Electrical Manufacturers Association; http://www.nema.org/

NEMI—National Electronics Manufacturing Initiative; http://www.nemi.org

NIST—National Institute for Standards and Technology; http://www.nist.gov/

SCC—Standards Council of Canada; http://www.scc.ca/

SCTE—Society of Cable Telecommunications Engineers; http://www.scte.org/

SEMI—Semiconductor Equipment and Materials International; http://wps2a.semi.org/

SIA—Semiconductor Industries Association; http://www.sia.org

TIA—The Telecommunications Industry Association; http://www.tiaonline.org/

UL—Underwriters Laboratories; http://www.ul.com/

USNC of IEC—U.S. National Committee of the International Electrotechnical Commission; http://public.ansi.org/ansionline/portal/search

VESA—Video Electronics Standards Association; http://www.vesa.org/

VITA—VMEbus Trade Association; http://www.vita.org/

1.6   Review

After reading and understanding this chapter, you should be able to answer these questions and note the reference location for the information in the chapter.

1. What are the two physical characteristics of SMT packages that allow for PCB real-estate reduction?
2. SMT components have shorter internal and external leads than THT components. Explain the advantages this provides in high-speed circuit performance.
3. For passive components stored in a factory, estimate the space savings for 50,000 components with the use of SMT compared to THT components.
4. Define "first-pass yield."
5. List three reasons why the reliability of SMT-based assemblies is typically higher than the reliability of THT-based assemblies.

6. List three reasons in favor of the continuing use of THT components in PCB assemblies.
7. List three reasons in favor of the use of SMT components in PCB assemblies.
8. Describe the component mix in a Hall Type 1 PCB assembly, then do the same for Type 2 and Type 3 assemblies.
9. For the assembly types noted in the previous question, list the IPC type and style of board/assembly that would most closely equal a Type 1, a Type 2, and a Type 3 assembly.
10. What are the IPC documents that apply to rigid printed circuit boards?
11. What are the two European Union (EU) directives that are intended to reduce both electronic waste and lead in electronic assemblies?
12. Do you think the EU directives will be successful in their goals?

1.7 References

1. Kolias, John T., "Packaging Impact on System Performance," *IEPS Surface-Mount Technology Compendium of Technical Articles Presented at the 1st, 2nd, and 3rd Annual Conference,* International Electronics Packaging Society, Inc., Glen Ellyn, IL, 1984, pg. 363.

2. Hutchins, Dr. Charles, "VLSI Prompts Designers to Ponder Packaging," *Electronic Design,* Hasbrouck Heights, NJ, December 22, 1983.

3. Moore, Robert and Weaver, Bruce, "Electrical and Mechanical Considerations in Design of LCC for Higher Performance Applications," *IEPS SMT Compendium of Technical Articles from 1st, 2nd, and 3rd Annual Conference,* International Electronics Packaging Society, Inc., Glen Ellyn, IL, 1984, pg. 529.

4. Kuk, Donald, Advanced Technology Development Manager, Cincinnati Microwave, Cincinnati, OH; Telephone interview with Hollomon, 19 November 1986.

5. Bourns, Inc., Advertisement, *Electronic Component News,* July 1987, pg. 13.

6. Caswell, Greg, et al., *Surface-Mount Technology.* International Society for Hybrid Microelectronics, (ISHM), Reston, VA, 1984, p. 6.

7. Hall, James W., "Soldering Techniques for Manufacturing Surface Mounted Circuits," *International Journal for Hybrid Microelectronics,* Volume 6, Number 1 (currently available from ISHM) Reston, VA, October 1983.

8. Dowds, S., "Current Trends in Lead-Free Solder." *SMT Magazine,* v. 18 #1, January 2004, p. 50.

9. Nguyen, L., Walberg, R., Lin, Z., Koh, T., Bong, Y. Y., Chua, M. C., Chuah, S., Yeoh, J. J., "A Structured Approach to Lead-Free IC Assembly Transitioning." Proceedings of SEMICON West 2002, Semiconductor Equipment Manufacturing Initiative, San Jose.

10. Levs, M. L., "Calling on Cell Phones." Waste Age, www.wasteage.com/mag., June 1, 2004.

11. Crawford, J., "One Step Closer to Lead-Free Standards." *Circuits Assembly,* v. 15 #4, July, 2004, p. 18.

12. Nguyen, L., Walberg, R., Lin, Z., Koh, T., Bong, Y. Y., Chua, M. C., Chuah, S., Yeoh, J. J., "A Structured Approach to Lead-Free IC Assembly Transitioning." Proceedings of SEMICON West 2002, Semiconductor Equipment Manufacturing Initiative, San Jose.

13. Lasky, R., Jensen, T., "Best Practices in Implementing Pb-Free Assembly." *Global SMT and Packaging,* v. 3 #8, Nov. 2003.

14. Dimock, F., "Effects of High-Temperature Requirements for Lead-Free Solder." *Circuits Assembly,* v. 15 #4, July, 2004, pp. 20–22.

15. Hwang, J., "Step 3: Solder Materials." *SMT Magazine,* v. 18 #3, March 2004.

16. Roth, H., "X-Ray Inspection of Lead-Free Solder Joints." *SMT Magazine,* v. 18 #7, July 2004.

chapter 2

Surface-Mount Components (SMCs)

Objectives

After reading and understanding this chapter, the reader should be able to:
- understand and be able to identify the basic types of SMT components, both active and passive
- understand basic component selection issues
- understand component-related issues that will later be shown to impact PCB design
- be able to find appropriate standards and references that relate to SMT components
- understand additional issues surrounding the use of lead-free solder

Introduction

The starting point of design with SMT is understanding the components. Figure 2-1 shows examples of some of the active components available at this writing.

This book was written for the working engineer who must be included in the design team of products that incorporate the use of SMT components. In the first edition, published in 1989, detailed component data and land pattern (footprint) recommendations were included in an appendix in an attempt to serve the engineer's needs. In the ensuing years, as we have watched the proliferation of SMT device packages, we have recognized the limitations of this strategy. In our 1995 update, we added considerable component information and continue to do so for passive components. However, for this second edition, the number of available IC packages has grown so large we cannot include them all. Please be cautioned to check IPC, EIA, JEITA, and JEDEC standards and—above all—consult your component vendors for package outline and

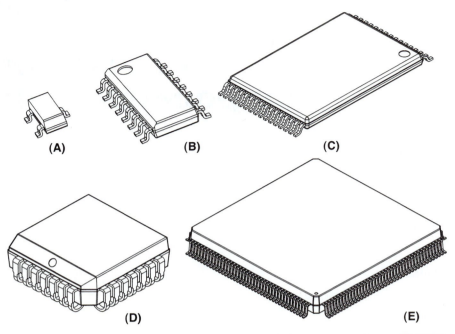

Figure 2-1 A variety of SMT IC packages. (A) SOT23—diode/transistor package. (B) SOIC—small outline IC package. (C) TSOP—thin small outline package (fine pitch). (D) PLCC—plastic leaded chip carrier. (E) QFD—quad flat pack (fine pitch). *(Courtesy National Semiconductor Corp.)*

footprint information to ensure that you are working with the most up-to-date and relevant package and footprint data, and whether that package and footprint are dimensioned in inches or millimeters. If you are truly new to SMT and PCB design, the "footprint" is the pattern of pads on the board that is designed to accommodate all the terminations for a given component. A footprint for a chip resistor will consist of two rectangular pads where the footprint for a 44-pin SOIC will have four sets of eleven rectangular pads that correspond to the package leads, as shown in Figure 2-2. Footprints are discussed in more detail later.

Tin Whiskers

Comments will be made in this chapter about tinning of component terminations. Typically we will talk about SnPb, nickel, and other materials *that may be superseded with lead-free materials as industry agrees on the appropriateness of replacement materials. With the exception of pure tin, which is one of the replacements, we will not mention any other potential replacement materials.*

One topic that is becoming of more concern with the move to lead-free requirements is the growth of "tin whiskers"[1] which can occur on the tin-plated leads of any component. These whiskers seem to occur in any pure tin application with an applied voltage and result in the spontaneous and poorly understood growth[2] of tiny, needle-like protrusions that break loose and can then potentially short circuit conductive features on a circuit assembly. Implicated in

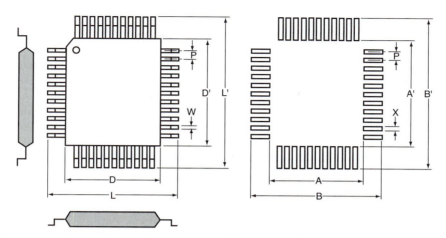

Figure 2.2   Package and footprint for a 44-lead quad flat pack. *(Courtesy National Semiconductor Corp.)*

everything from relay failures to the failure of the Galaxy 4 satellite whose failure shut down 40 million pagers in 1998, tin whisker growth occurs more rapidly in pure tin than it does in tin-lead solders. Pure tin, with its accompanying risk of whisker growth,[3] is what virtually all component manufacturers are tinning their component leads with to meet lead-free requirements. With all this growth in the use of pure tin, the tin-whisker failure mode is re-emerging. In the past, electrically conductive tin whiskers have been responsible for the loss of billions of dollars worth of satellites, missiles, and other equipment. Tin whiskers can develop under typical operating conditions on any product type that uses lead-free pure tin coatings. Driven by the accelerating movement to lead-free products, tin whiskers pose major safety, reliability, and potential liability threats to all makers and users of high-reliability electronics and associated hardware. The reason(s) for the growth of tin is not well understood, and so far there does not appear to be sufficient control of tin whiskering in high-reliability systems. Many authors believe that decisive action must be taken soon to fund development and implementation of a strategic action plan to devise short-term stopgap procedures and medium-term investigation of whisker-mitigation techniques.

Manufacturing

In this chapter, we will discuss component selection and specifications for an SMT project. In component selection, one important factor to consider, in addition to electrical characteristics, is the manufacturing equipment capabilities. An equally important factor is choosing the proper component for circuit performance, but this book is not intended to cover that aspect of design. The placement system used to assemble boards may restrict both the types of components that can be automatically handled as well as the type of component carriers in which the parts must be purchased. Teamwork from all departments should be used in component selection to avoid costly surprises on the production and test floor.

Metric Dimensioning

It is important to note that, as in so many industries, there is a definite move to metric dimensioning by the U.S. SMT component manufacturers. Already universal in Asian and European manufacturers, this means that PCB designers must be vigilant in determining the controlling dimension for all parts to be used in a design. While inch-dimensioned and metric-dimensioned parts may be similar in size and lead pitch, the inaccuracies generated by some PCB design software must be determined. Virtually all new parts are metric-dimensioned, and for this reason metric dimensions will be used in this book with inch dimensions being used by exception and will be so noted.

For up-to-date information on component packages, always refer to the Electronic Industries Association standard, EIA-PDP-100, JEDEC registered listings, Standard Mechanical Outlines for Electronic Parts, and your component vendor's data. See the references listed in Section 2-6.

2.1   SMT Passives

Let's begin our study of surface-mounted components with *passives*. Passives are circuit elements (components) that do not change their basic character when an electrical signal is applied. Passives include such elements as capacitors, resistors, inductors, etc.

The reader should note that the dimensioning of passive components is a mixed bag of inch and metric. For example, chip capacitor and resistor sizes are commonly referred to in the U.S. as 1206, 0805, 0402, etc., where the length and width of the packages are denoted in thousandths of an inch based on the two pairs of numbers in the package size. A 1206 chip is 0.120" × 0.060," and an 0402 chip is 0.040" × 0.020." The similar metric size to the 1206 is a 3216, which would be 3.2 mm × 1.6 mm, equivalent to 0.125" × 0.063." Other examples include those shown below.

Chip device inch and metric notations

Inch Notation	Metric Notation of Similar Size
0402 (0.040" × 0.020" = 1.016 mm × 0.508 mm)	1005 (1.0 mm × 0.5 mm = 0.039" × 0.197")
0603 (0.060" × 0.030" = 1.524 mm × 0.762 mm)	1608 (1.6 mm × 0.8 mm = 0.063" × 0.32")
0805 (0.080" × 0.050" = 2.032 mm × 1.27 mm)	2012 (2.0 mm × 1.2 mm = 0.0787" × 0.047")
1008 (0.100" × 0.060" = 2.54 mm × 20.032 mm)	2520 (2.5 mm × 2.0 mm = 0.098" × 0.0787")
1206 (0.120" × 0.060" = 3.048 mm × 2.032 mm)	3216 (3.2 mm × 1.6 mm = 0.126" × 0.063")
1812 (0.180" × 0.120" = 4.572 mm × 3.048 mm)	4532 (4.5 mm × 3.2 mm = 0.177" × 0.126")

Similar inch/metric size comparisons can be drawn for other passive components as well. The designer must know which is the "controlling" dimensional measure for each component used in the design—inch or millimeter. In an attempt to add to the confusion, some American chip manufacturers continue to use the inch designation, e.g., 1206, but actually manufacture and dimension their parts in millimeters! Hopefully this will soon pass.

Terminations of passive devices should be solder coated with solder that is close to the 63/37 tin/lead eutectic *or* with pure tin for lead-free designs. The reader is cautioned that the lead-free requirements may change. Also, pure tin is subject to "whiskers" as discussed earlier in this section. Check with suppliers. The higher tin contents may exhibit improved solderability after a long storage period. However, tin/lead contents matching those of the solder paste and substrate metallization tin/lead ratios promote high reflow soldering yields. Solder coatings may be either hot-dip applied or electroplated. Hot-dipped terminations must be free from lumps, icicles, or other solder surface irregularities and must not show any dewetted or pinholed areas. Plated terminations must be reflowed after plating to fuse the solder to the base metal. The tin/lead finish should be at least 7.5 microns (0.0003 inch) thick.

2.1.1   Capacitors

Capacitor packages have few variations in through-hole designs. The design engineer may have to do some homework occasionally to find a part with suitable voltage ratings, capacitance, dissipation factor, etc. But, for most through-the-hole uses, capacitor selection is a nearly rote activity.

Surface-mount package constraints place practical limits on the capacitance and voltage ratings available. SMT processing may further constrain component selection since the component may be required to survive a trip through a reflow soldering furnace or even direct immersion in a molten solder bath. Fortunately, there is a rapidly growing collection of parts that answer most if not all of these needs. In the following, we'll analyze each of the major SMT capacitor families available today.

We have not attempted to catalog every available SMT component. Even if we had, new offerings appear daily. Only common parts are covered herein. We mention this in order to caution you that you should not be deterred from a search for a specific SMT component that meets a particular requirement merely because we have not mentioned it in this work.

Ceramic Chip

The *ceramic chip* is electrically the same animal you have used to in leaded ceramic capacitors. It is the building block of the ceramic leaded capacitor, only no leads have been soldered to its end terminations and it has not been encapsulated, as shown in Figure 2-3. Beyond those

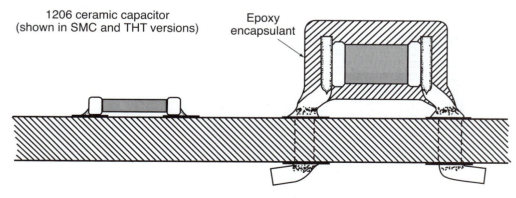

Figure 2-3   Comparison of leaded and SMT chip capacitors.

fundamental differences, SMT ceramic capacitors have plated end terminations designed to yield quality results in the required soldering process.

Ceramic chip capacitors are the workhorse capacitors of SMT. They go by various names, such as chip cap, multilayer ceramic chip (MLC or MLCC for chip capacitor), and ceramic cap. Figure 2-4 shows the construction details of an MLC cap. As explained in the next few paragraphs, the MLC/MLCC designation is descriptive of the component's construction.

MLC chips are built up of multiple laminated layers of a partially metallized green ceramic material (Figure 2-4). Each layer's metallization extends all the way to one end of the device, but stops just short of the other end. The layers are arranged such that the terminated end alternates. In green form, the layers are then laminated by heat and pressure. The green monolith is fired for up to 16 hours at around 1200°C. End terminations are then added to the fired monolith and it is again fired at around 800°C for about 15 minutes. A nickel barrier layer may then be plated or sputtered over the base termination for protection in the soldering process. Some parts are made with a palladium-based termination. Unless process engineering develops a foolproof method for soldering of such parts, they are best avoided. They may be slightly cheaper than capacitors using a precious-metal system, but the savings in component cost will almost certainly be offset by higher processing expenses brought on by the poor solderability of the palladium metallization. As noted for other parts, chip capacitors designed for lead-free soldering commonly have a pure tin finish.

The reader is cautioned about chip caps that are terminated with thick film mixtures of silver, palladium, and glass grit. Silver readily goes into solutions with tin, the major ingredient in most electronics solders. This silver/tin solubility poses the danger that the silver in the termination will enter into solution with the tin in the solder leaving the end termination damaged and destroying its cohesive bond to the capacitor (the metallization going into solution with the solder is a phenomenon called *leaching*).[4] Nickel barriers, with an over-coating of tin/lead for solderability, are the most common approach to solving this problem from the component side

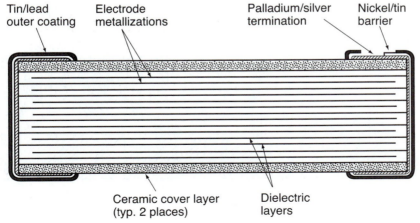

Figure 2-4   A cutaway view of an MLC chip capacitor.

of the equation. For lead-free applications, some manufacturers are providing a pure Sn final layer. The procurement member(s) of the team must verify that any MLCCs being purchased for the design have the nickel barrier.

Leaching is a time/temperature-based problem, as shown in Table 2-1. Note that other termination materials may follow different curves. As long as time-at-temperature is kept within safe limits, as indicated by the component vendor, leaching should not be a problem. However, remember that this time/temperature consideration must include any rework or repair that the component may undergo in its lifetime.

Since leaching was identified as an SMT problem years ago, the tendency has been toward specifying the use of nickel or tin barriers on all parts to avoid any possible problems. Virtually all SMT capacitors have some type of leach-resistant barrier on their terminations, whether leaded like tantalum caps or leadless like multi-layer ceramic (MLC) caps. The supplier of capacitors in a particular design is the best source of termination information.

Carrier Packaging—Chip capacitors have been used in the hybrid industry for many years. They may be supplied in bulk containers and fed to automated placement equipment by a bulk vibratory feeder. The most widely accepted packaging is tape and reel. Figure 2-5 shows the basic dimensions of SMT tape and reels (Table 2-2 and Table 2-3). For more details, please refer to the manufacturers' specifications.

Table 2-1   Leach Resistance of Typical MLC Capacitors.

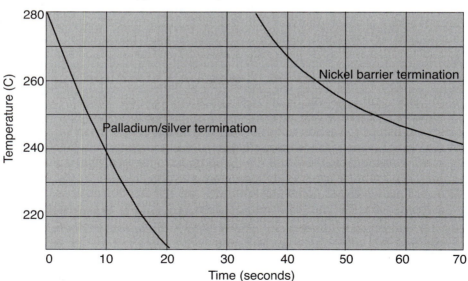

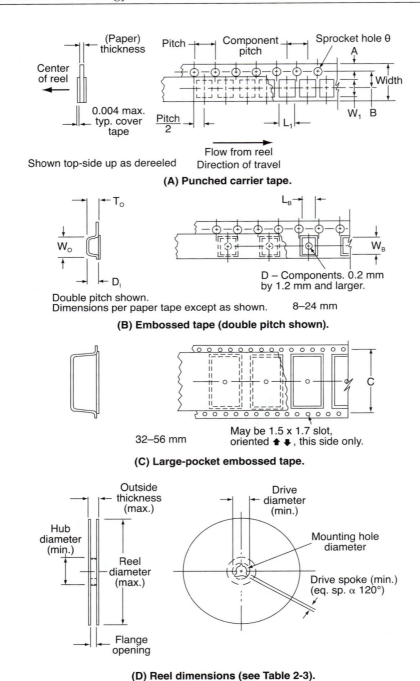

Figure 2-5   Standard tape-and-reel packaging dimensions. *(See Table 2-2)*

Table 2-2   Standard SMT Tape-and-Reel Packaging Dimensions. *(See also EIA RS-481)*

Width	Pitch	Component Pitch	Thickness (see Note 2)	Sprocket Hole φ	A	B	C	D	D_I Embossed Only	L_B Embossed Only	L_I Punched Only	T_O Embossed Only	W_B Embossed Only	W_I Punched Only	W_O Embossed Only	Z (see Note 3)
8	4	4	1.1 (max)	1.5	1.75	3.5	—	1.0 (min)	Note 1	Note 1	Note 2	2.4 (max)	Note 1	Note 2	4.2 (max)	25
12	4	4.8	—	1.5	1.75	5.5	—	1.5 (min)	Note 1	Note 1	—	4.5 (max)	Note 1	—	8.2 (max)	30
16	4	4, 8, 12	—	1.5	1.75	7.5	—	1.5 (min)	Note 1	Note 1	—	6.5 (max)	Note 1	—	12.1 (max)	40
24	4	12, 16, 20, 24	—	1.5	1.75	11.5	—	1.5 (min)	Note 1	Note 1	—	6.5 (max)	Note 1	—	20.1 (max)	50
32	4	16, 24, 32	—	1.5	1.75	14.25	28.5	2.0 (min)	Note 1	Note 1	—	10.0 (max)	Note 1	—	25.6 (max)	50
44	4	24, 32, 40	—	1.5	1.75	20.25	40.5	2.0 (min)	Note 1	Note 1	—	10.0 (max)	Note 1	—	37.6 (max)	50
56	4	48	—	1.5	1.75	26.25	52.5	2.0 (min)	Note 1	Note 1	—	10.0 (max)	Note 1	—	49.6 (max)	50

Notes:

1. D_I, L_B, W_B and W_O are determined by the component size. Dimensions are selected to yield a cavity with a 0.05 mm (min) to 0.50 mm (max) clearance on each axis and to prevent a component rotation of more than 10° within the cavity.

2. Material thickness L_I and W_I are determined by the component size. Dimensions are selected per Note 1.

3. Tape with components must withstand bending around a mandrel of radius Z without damage to tape, cover seal, or components.

4. All dimensions are in millimeters.

5. Some small chip components are available in reels with 2 mm pitch. However, some tape feeders may not index properly when used with 2 mm pitch tape.

Table 2-3   Tape Reel Dimensions.

Tape	Reel Diameter (max)	Flange Opening	Outside Thickness	Hub Diameter	Mounting Hole	Optional	
						Drive Diameter	*Drive Spoke*
8	330	8.4	14.4	50	13.0	20.2	1.5
12	330	12.4	18.4	50	13.0	20.2	1.5
16	330	16.4	22.4	50	13.0	20.2	1.5
24	330	24.4	30.4	50	13.0	20.2	1.5
32	330	32.4	38.4	50	13.0	20.2	1.5
44	330	44.4	50.4	50	13.0	20.2	1.5
56	330	56.4	62.4	50	13.0	20.2	1.5

Note: Miniaturized parts in 0603, 0402, and 0201 sizes are being widely used in high-density applications. Larger sizes than those in this table are available from vendors. *0201, 0402, and 1812 sizes are typically specified only for reflow soldering. Other sizes may be reflowing or wave soldered.

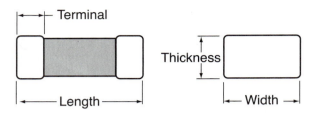

NOTE: See dimensions in Table 2-2.

Figure 2-6   Chip capacitor standard sizes.

Standards—For chip capacitors, the typical industry-standard sizes are shown in Figure 2-6 and Table 2-4.

In the table, please note that the dimensions are given in metric with inch equivalents in parentheses, and the EIA "size designation" follows the inch dimension.

Design Considerations—End-termination specifications should be written based on the required assembly processing requirements. For reflow soldering, components with end terminations as shown in Figure 2-7 are recommended for maximum soldering yield. However, components with a greater tin content may provide a longer shelf life. With the advent of lead-free solders and soldering, some chip manufacturers are making the outer coating of the termination of pure tin (Sn). For wave soldering, end terminations with a tin content matching that of the solder in the wave-machine pot are recommended.

Table 2-4   Chip Capacitors: Examples of EIA Standard Sizes.

EIA Size Designation	Dimensions (mm/in)			
	Length	*Width*	*Thickness (max)*	*Terminal*
0201*	0.60 ± 0.03 (0.024 ± 0.001)	0.30 ± 0.03 (0.011 ± 0.001)	0.33 (0.013)	0.15 ± 0.05 (0.006 ± 0.002)
0402*	1.0 ± 0.10 (0.040 ± 0.004)	0.50 ± 0.10 (0.020 ± 0.004)	0.56 (0.022)	0.25 ± 0.15 (0.010 ± 0.006)
0603	1.60 ± 0.15 (0.063 ± 0.006)	0.81 ± 0.15 (0.032 ± 0.006)	0.86 (0.034)	0.35 ± 0.15 (0.014 ± 0.006)
0805	2.01 ± 0.20 (0.079 ± 0.008)	1.25 ± 0.20 (0.049 ± 0.008)	1.4 (0.055)	0.50 ± 0.25 (0.020 ± 0.010)
1206	3.2 ± 0.20 (0.126 ± 0.008)	1.6 ± 0.20 (0.063 ± 0.008)	Varies on WVDC	0.5 ± 0.25 (0.020 ± 0.010)
1210	3.20 ± 0.20 (0.126 ± 0.008)	2.5 ± 0.20 (0.098 ± 0.008)	Varies on WVDC	0.50 ± 0.25 (0.020 ± 0.010)
1812	4.5 ± 0.30 (0.177 ± 0.012)	3.20 ± 0.20 (0.126 ± 0.008)	Varies on WVDC	0.61 ± 0.36 (0.024 ± 0.014)

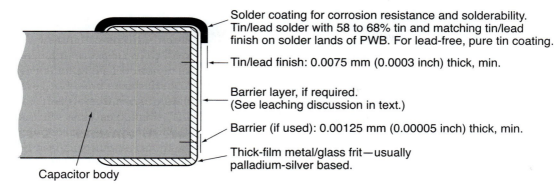

Figure 2-7 Chip capacitor end terminations.

Because of their leadless design, chip capacitors have a lower equivalent series resistance (ESR) and equivalent series inductance (ESL) than their THT cousins. Therefore, they are a boon in high-speed circuitry design. However, like THT capacitors, MLCCs have a variety of dielectrics that affect their stability with temperature and applied voltage changes. Types Z5U are typically the least stable, with Y5R being a medium stable material, and NPO/COG dielectric being the most stable with changes in temperature and applied voltage. Examples of variations in capacitor stability with applied changes are shown in Figure 2-8. These are *not* universal graphs and will vary with the type of capacitor (low-voltage, high-voltage, etc.), the type of dielectric, DC versus AC applied voltage, the manufacturer, etc.

To minimize capacitance changes with temperature and/or applied voltage, designers should specify low dielectric-constant, temperature-compensating Type NPO/COG. The other

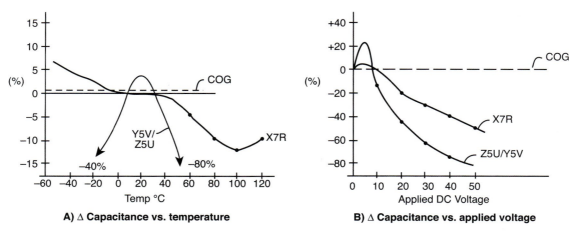

Figure 2-8 Variations in capacitor stability with temperature and applied DC voltage.

dielectric types use higher dielectric constant materials allowing larger values of capacitance in smaller volumes but with the penalty that temperature and voltage stability is considerably worse than with NPO/COG dielectrics. The limitation of the NPO/COG dielectric is that it is limited to lower values of capacitance.

Tantalum and Aluminum Electrolytic Capacitors

Tantalum capacitors are typically specified when a stable high-value capacitor is needed. They are the most stable of the electrolytic capacitors. SMT electrolytics with aluminum electrolyte materials should be specified in the same manner as they are in THT designs with information from manufacturers' data sheets.

The dimensions for four standard sizes of this package, recognized by the EIA, are given in Table 2-5. The package outline is illustrated in Figure 2-9. Four additional EIA sizes form an "extended range" for high capacitors or voltage requirements. Their outline is shown in Figure 2-10 and dimensions are listed in Table 2-6. See the transistor specifications (discussed later in the chapter) for comments on end-termination treatments for use on component procurement drawings. Other package formats are available, but often present difficulties in PWB assembly.

Construction—While construction details for tantalum capacitors vary from manufacturer to manufacturer, these traits are common. A solid tantalum dielectric material is employed, from which these capacitors draw their name. End-termination leads are welded to the plates of the capacitor, and a protective encapsulation of epoxy resin plastic may be molded around the capacitor body such that the end terminations are left exposed for attachment to a PWB.

Table 2-5 Typical Tantalum "Brick" Capacitors. *(See Figure 2-9)*

EIA RS-228 Size	Size Code/Standard Capacitive Range (mm/in)			
	A Case	*B Case*	*C Case*	*D Case*
Length	3.0–3.4 (0.118–0.134)	3.3–3.7 (0.130–0.146)	5.7–6.3 (0.224–0.248)	6.8–7.6 (0.268–0.299)
Width	1.4–1.8 (0.050–0.070)	2.6–3.0 (0.102–0.118)	2.9–3.5 (0.114–0.138)	4.0–4.6 (0.157–0.181)
Height	1.4–1.8 (0.055–0.070)	1.7–2.1 (0.067–0.083)	2.2–2.8 (0.087–0.110)	2.5–3.1 (0.098–0.122)
Termination Length	0.5–1.1 (0.020–0.043)	0.5–1.1 (0.020–0.043)	0.5–1.1 (0.020–0.043)	0.5–1.1 (0.020–0.043)
Termination Width	1.1–1.3 (0.043–0.051)	2.1–2.3 (0.083–0.090)	2.1–2.3 (0.083–0.090)	2.3–2.5 (0.090–0.098)
Termination Height	0.7 Min (0.028)	0.7 Min (0.028)	1.0 Min (0.040)	1.0 Min (0.040)

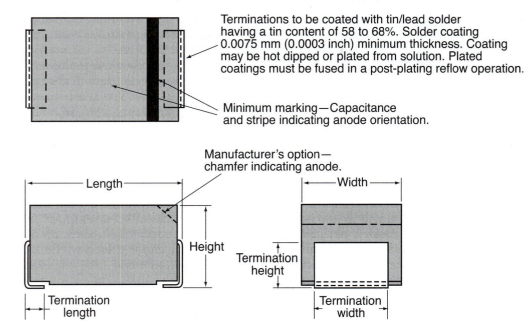

Terminations to be coated with tin/lead solder having a tin content of 58 to 68%. Solder coating 0.0075 mm (0.0003 inch) minimum thickness. Coating may be hot dipped or plated from solution. Plated coatings must be fused in a post-plating reflow operation.

Minimum marking—Capacitance and stripe indicating anode orientation.

Manufacturer's option—chamfer indicating anode.

Figure 2-9   Typical tantalum "brick" capacitors. *(Dimensions are given in Table 2-5)*

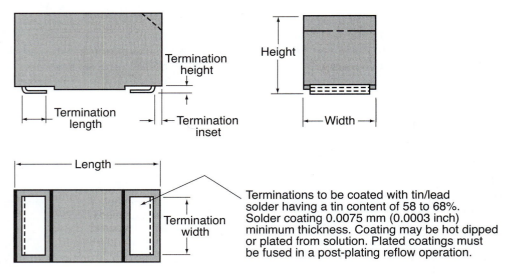

Terminations to be coated with tin/lead solder having a tin content of 58 to 68%. Solder coating 0.0075 mm (0.0003 inch) minimum thickness. Coating may be hot dipped or plated from solution. Plated coatings must be fused in a post-plating reflow operation.

Figure 2-10   Extended-range tantalum "brick" capacitors. *(Dimensions given in Table 2-6)*

Table 2-6   Extended-Range Tantalum "Brick" Capacitors. *(See Figure 2-10)*

EIA RS-228 Size	Size Code/Extended Capacitive Range (mm/in)			
	3518 Case	3527 Case	7227 Case	7257 Case
Length	3.3–3.7 (0.130–0.146)	3.3–3.7 (0.130–0.146)	6.9–7.5 (0.272–0.295)	6.9–7.5 (0.272–0.295)
Width	1.6–2.0 (0.063–0.079)	2.4–3.0 (0.095–0.118)	2.4–3.0 (0.095–0.118)	5.2–6.2 (0.205–0.244)
Height	1.7–2.1 (0.067–0.083)	1.7–2.1 (0.067–0.083)	2.5–3.1 (0.098–0.122)	3.0–3.7 (0.118–0.146)
Termination Length	0.6–1.0 (0.024–0.040)	0.6–1.0 (0.024–0.040)	0.8–1.2 (0.031–0.047)	0.8–1.2 (0.031–0.047)
Termination Width	1.6–1.8 (0.063–0.071)	2.4–2.6 (0.095–0.102)	2.4–2.6 (0.095–0.102)	5.4–5.8 (0.213–0.228)
Termination Height	0.7 (0.028)	0.7 (0.028)	1.0 (0.040)	1.2 (0.047)
Termination Inset	0.4–0.6 (0.016–0.024)	0.4–0.6 (0.016–0.024)	0.6–0.8 (0.024–0.032)	0.6–0.8 (0.024–0.032)

Carrier Packaging—Like chip capacitors, tantalum capacitors are available in tape-and-reel packaging. This is the preferred handling method. In small quantities, tantalum bricks may also be purchased in plastic sticks or in bulk. Bulk packaging is not recommended, however, except for very small quantities where hand placement will be the assembly method. This is because tantalum capacitors are polarized components, and typical polarizing markings vary from being difficult to being impossible to automatically orient. Also, we should note that many automatic placement systems do not have standard accessories that permit them to deal with tantalums in plastic sticks. Therefore, tape-and-reel packaging is the preferred handling method for these parts. Table 2-7 shows tape sizes for standard tantalum bricks.

Table 2-7   Tape-and-Reel Sizes for Tantalum Bricks.

Standard Brick Size	Tape Specifications		Extended Brick Size	Tape Specifications	
	Width	Pitch		Width	Pitch
A Case	8 mm	4 mm	3518	8 mm	4 mm
B Case	8 mm	4 mm	3527	8 mm	4 mm
C Case	12 mm	8 mm	7227	12 mm	8 mm
D Case	12 mm	8 mm	7257	12 mm	8 mm

2.1.2   Resistors

Leadless chip resistors are, as of this writing, the most widely used component in the SMT stable. They are to SMT what the ¼-watt carbon or metal-film axial resistor is to insertion-mount processing. The wide variety of available SMT resistor packages, tolerances, and voltage ratings means that the design team must verify that their selected components meet all circuit needs as well as compatibility with the placement machines.

Chip resistors are like chip capacitors in that the absence of leads reduces parasitic inductance and allows a faster, more noise-tolerant operation. Like THT resistors, they are available in a wide variety of tolerance ratings.

Chip Resistors

Commercial chip resistors are typically offered in ±5% and ±1% tolerances. Standard resistance values, as defined by the EIA, are listed in Table 2-8.

The reader is cautioned that the power dissipation rating of chip resistors is not the same for a given size—it depends both on size and the technology of the resistor. For exampole, at this writing, Yageo America rates its 9C1206 series of 0805 thick film resistors at 0.125 watt, as does Panasonic for its thick film series, while Susumu Co., Ltd., rates its RR series of 0805 thin film resistors at 0.1 watt.

Table 2-8   Standard Resistor Values.

Series E24 (±5%)		Series E96 (±1%)							
10X	33X	10.0X	13.3X	17.8X	23.7X	31.6X	42.2X	56.2X	75.0X
11X	36X	10.2X	13.7X	18.2X	24.3X	32.4X	43.2X	57.6X	76.8X
12X	39X	10.5X	14.0X	18.7X	24.9X	33.2X	44.2X	59.0X	78.7X
13X	43X	10.7X	14.3X	19.1X	25.5X	34.0X	45.3X	60.4X	80.6X
15X	47X	11.0X	14.7X	19.6X	26.1X	34.8X	46.4X	61.9X	82.5X
16X	51X	11.3X	15.0X	20.0X	26.7X	35.7X	47.5X	63.4X	84.5X
18X	56X	11.5X	15.4X	20.5X	27.4X	36.5X	48.7X	64.9X	86.6X
20X	62X	11.8X	15.8X	21.0X	28.0X	37.4X	49.9X	66.5X	88.7X
22X	68X	12.1X	16.2X	21.5X	28.7X	38.3X	51.1X	68.1X	90.9X
24X	75X	12.4X	16.5X	22.1X	29.4X	39.2X	52.3X	79.8X	93.1X
27X	82X	12.7X	16.9X	22.6X	30.1X	40.2X	53.6X	71.5X	95.3X
30X	91X	13.0X	17.4X	23.2X	30.9X	41.2X	54.9X	73.2X	97.5X

Note: X in the table is a power of 10 multiplier yielding the following ranges: E24 ±5%, 10 ohms to 2.2 megohms; E96 ±1%, 49.9 ohms to 2.2 megohms.

Construction—Chip resistors are generally built in a batch process. Many individual resistors are fabricated on a single ceramic substrate. The process derives from thick-film hybrid manufacturing methods. First, conductive metal terminations are screened on the top surface of the substrate where individual resistor ends will be. These terminations are then fired. Next, conductive ink, generally based on ruthenium oxide, is screened between the terminations and fired. The resistors are then laser trimmed to tolerance. The substrate is cut into individual resistors which are terminated with a palladium-silver metallization. A nickel barrier is usually applied over the base termination to prevent silver leaching, and a final solderable coating, generally tin/lead or pure tin, is plated or dip-applied over the barrier. It is important to note that chip resistors must be mounted with their silk-screened side up for two reasons. Number one, the component identification is on that side and should be readable, and number two, the most important reason, the resistor will not be able to dissipate heat well to ambient air if it is not mounted right-side up. If mounted upside down, the I^2R heat will be trapped between the ceramic body and the top surface of the circuit board, causing the resistor to overheat and fail before it reaches its rated heat dissipation.

Figure 2-11 shows the standard construction of a chip resistor and details the critical dimensions for standard sizes. Table 2-9 lists common dimension values.

Within the U.S. market, the 1206 resistor has been the dominant size, but there is a trend toward greater use of the 0805 for high-density designs, as well as 0603, 0402, and 0201 sizes (nominally 0.060" by 0.030" to 0.020" by 0.010").

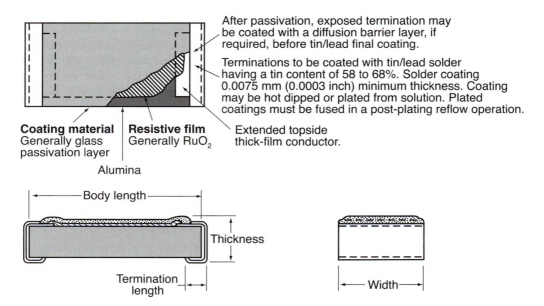

Figure 2-11   Diagram of a typical chip resistor.

Table 2-9 Chip Resistor Dimensions. *(Additional sizes similar to chip caps in Table 2-4)*

Size Code	Dimensions (mm/inch)			
	Body Length	*Width*	*Thickness*	*Termination Length*
RC0805	1.8–2.2 (0.070–0.087)	1.0–1.4 (0.040–0.055)	0.3–0.7 (0.012–0.028)	0.3–0.6 (0.012–0.024)
RC1206	3.0–3.4 (0.118–0.134)	1.4–1.8 (0.055–0.070)	0.4–0.7 (0.016–0.028)	0.4–0.7 (0.016–0.028)
RC1210	3.0–3.4 (0.118–0.134)	2.3–2.7 (0.090–0.106)	0.4–0.7 (0.016–0.028)	0.4–0.7 (0.016–0.028)

Standards—In 1989, the EIA completed amendments to EIA/IS 30. The new specification, EIA/IS 30A, adds considerable data on solderability and leach resistance and it tightens physical tolerances.[5] Refer also to EIA 575 for standard SMT resistors and EIA 576 for precision parts. These standards, as well as others, will be modified or superseded to reflect lead-free termination coatings.

Carrier Packaging—As discussed in the section on chip capacitors, tape and reel is the preferred handling medium for chip resistors. The 1206, 0805, and 0603 sizes fit in 8 mm tape on 4 mm pitch, while 0402 and 0201 sizes fit on 8 mm tape on 2 mm pitch.

Specifications—Resistor parameters that must be specified are resistance, resistive tolerance, power rating, mechanical size, TCR in PPM, 10,000-hour stability, and end-termination material and format. To prevent tombstoning and other solder defects, end terminations should be leach resistant and of equal length on both ends of the part. It is also important that the termination project a minimum of 0.25 mm (0.010 inch) in from the end of the part. Components with little or no top and bottom terminations (terminated on the ends only) are much more prone to tombstoning. End terminations should be uniform in thickness. Any lumps or globs on the terminations will likely cause significant process problems during the assembly and soldering operations.

For wave-soldered consumer products that will not be reworked (low enough in cost and/or high enough in yield to allow scrapping of those assemblies that fail final test), eliminating the nickel barriers will cut cost. However, where this cost savings is used, care must be taken with soldering time/temperature profiling to avoid leaching and/or silver contamination in the wave-machine solder pot.

Cylindrical Resistors

Cylindrical resistors are commonly called MELFs (metal-electrode face-bonded). MELFs are typically supplied in two sizes. The MLL-34 is a $^1/_8$-watt resistor, and the MLL-41 handles $^1/_4$-watt at 70°C. There is a larger package that is both called and rated as "$^1/_4$ watt." As with chips, MELF wattage capacities must be derated per the manufacturer's suggested curve if they are to be operated in high-temperature ambient conditions.

The MELF is not widely used in the U.S. market, but is very common in Asian SMT. An understanding of regional preferences is useful in deciding when to specify chips and when

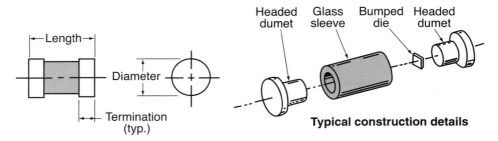

Figure 2-12   MELF component construction.

MELFs are a better choice. The Asian electronics market is biased toward consumer-electronics manufacturing. While there are numerous ultrahigh-tech applications, such as space instrumentation and super-computers in Japan, consumer electronics is a major force in determining the market share of the various component styles there.

Construction—In general, MELF resistors are constructed in similar fashion to metal or carbon-film axials except that no axial leads are attached. Some manufacturers attach a metal cap to the component body and others apply metallizations similar to those on chip capacitors or chip resistors.

Standards—No specific EIA standard exists for MELF construction. Most available parts conform to JEDEC DO-35 form, but without leads, as in Figure 2-12. This basic form factor is used for cylindrical components of all types. Capacitors, resistors, inductors, and diodes are available in this form factor. Table 2-10 lists the dimensions for each case style.

Carrier Packaging—For low- to moderate-volume production, tape and reel is the preferred handling method. Certainly, it is favored for MELF diodes and other polarized components. However, some high-volume placement equipment is available with bulk hopper-style feeders for nonpolar MELFs of a given body size. This handling approach minimizes component cost and has proven very reliable.

Table 2-10   MELF Component Directions.

Case Style	Dimensions (mm/inch)		
	Length	*Diameter*	*Termination*
MLL-34	3.3–3.7	1.5–1.7	0.29–0.55
(SOD-80)	(0.130–0.146)	(0.059–0.067)	(0.011–0.022)
MLL-41	4.8–5.2	2.44–2.54	0.35–0.51
(SOD-87)	(0.189–0.205)	(0.096–0.100)	(0.014–0.020)

Note: Additional case sizes 0805, 1206, 1406, and 2309 have bee added since the first edition.

Specifications—MELF resistors are specified in the same manner as chip resistors. Particular attention should be paid to both the end terminations, per the specifications for capacitors, and the termination bond integrity. Also, MELFs, more than rectangular components, require process-engineering support. The process engineer should be involved in the decision to use MELFs. The engineer should understand the importance of the advantages of MELFs since their use may require the investment of considerable process-engineering resources. In addition, the process engineer should be involved from the beginning of the selection process so that last-minute process development does not become a stumbling block to production startup.

Trimmer Potentiometers

The good news here is that there is a wealth of different trimming potentiometers available in a variety of SMT packages. The bad news is that there is still a proliferation of available styles with the accompanying proliferation of footprints, and some parts might have difficulty meeting any specification that a rational group would develop. The watchword is caveat emptor. Be sure you understand how the potentiometer is constructed and how your process and field use will affect it.

Construction—SMT trimmer construction details obtained from different manufacturers vary greatly. Many are built on 96 percent alumina substrates with attachment metallizations similar to those of chip resistors. Others are constructed on epoxy-glass substrates. Organic substrate types employ a wide range of metallizations and resistive inks. Trimmers may be plastic encapsulated, cermet body, or open frame.

We can make one safe generalization regarding construction. Whatever the materials and construction details, worthwhile trimmers must be built to survive the soldering, cleaning and rework operations of the manufacturing environment. For sealed units, this implies protection from contamination that might enter the part during soldering or cleaning and suggests that the seals around the adjustment-shaft entry are areas of critical concern. Where applications permit, an open construction may provide a low-cost part without such a convenient hiding place for contaminants and corrosives. The decision between use of sealed and open constructions involves considering what contaminants might remain in an open part after processing, and then assessing the protection available for the open part from environmental contaminants while in service or storage.

Design and Certification Issues—Due to the proliferation of SMT trimmer layouts, designers must devote time to becoming SMT trimmer literate or use the mixed-technology approach and stick with leaded trimmers which are through-hole mounted. For in-house certification, consider humidity testing per MIL-STD-202, Method 103 for 100 hours with a minimum insulation resistance of 10 megohms. Leak testing in 85°C perfluorinated hydrocarbon fluid may be used to test the integrity of sealed units.

Carrier Packaging—Tape and reel packaging, where available and compatible with placement equipment feeders, is the preferred carrier package. However, many trimmers do not fit the tape thickness limits of EIA-RS-481A tape standards. Therefore, the tape may not fit dereeler feeders designed to handle this standard. Check your carrier format options and involve the manufacturing-engineering department in the decision.

Specifications—Beyond the benchmarks and cautions stated earlier, specify the resistance range, resistance tolerance, number of turns, physical description, TCR, absolute minimum resistance (where the desired part should approach zero resistance), insulation resistance, dielectric strength, resistance stability (over 10,000 hours or after high-temperature exposure, whichever is more appropriate), temperature range, load life, processing-environment survival characteristics, environmental requirements, and mechanical life.[6]

Resistor Networks

As space savers, SMT resistor networks are easy winners when compared to through-hole or SMT discrete resistors. But SMT resistor networks usually come in a poor second in board area consumed when compared to THT SIP or ZIP packages. Let's not rule against SMT networks too hastily, though. Where boards must be sandwiched together closely or the component side must be kept low profile for packaging considerations, SMT networks should be considered. Also, the small board-area premium required by the SMT part may be more than offset by processing considerations where a design would be pure surface mount were it not for a few SIP resistor networks.

Construction—Construction varies among manufacturers. In general, networks are fabricated using thick-film resistive techniques on 96 percent alumina substrates. Networks are available in packages similar to leadless and leaded chip carriers and in packages resembling gull-wing and J-leaded SOICs. Of course, similarity to IC packaging goes no further than the outside form factors—if it goes that far. There is no die inside the package. Ceramic substrates are often protected by encapsulation instead of lid attachment. Also, some network packages are not as tall as the IC packages they mimic. And, a number of manufacturers supply pseudo SOIC parts. These are wider than standard and more narrow than wide-body SOs and are called Small-Outline Medium (SOM) by some vendors. Be forewarned, vendors may make scant mention of the fact that their SOM does not fit IPC SOIC land-pattern dimensions.

Standards—Resistor network standards are progressing as EIA project number PN 1905. Currently, pseudo standardization is achieved by following, or sort of following, either the LCC, PLCC, SOIC, or SOJ outline. In dual-in-line formats, the SOJ is less popular than its gull-wing counterpart. However, it is gaining some momentum because it maximizes resistor network substrate area for a given slice of PWB real estate. Also, J-lead solder joints are easier to visually inspect than gull wings despite the common perception to the opposite. See more on lead-form inspectability in Chapter 7.

Carrier Packaging—Tape and reel and plastic tubes, or "sticks," are popular carrier packages. Sticks are appropriate for small quantities and repeated material changes on the production line. Tape and reel, if available, is better for long production runs. Of course, the choice is fundamentally dependent on the handling capabilities of the placement machine.

Specifications—Surface-mount resistor networks are specified in a fashion similar to through-hole networks. The package style, number of pins, electrical configuration (isolated, bussed, or dual terminator), and resistance specifications are typically called out. As with all SMT devices, the solderability of the lead finish is of critical importance and bears far more attention than you might give for a through-hole device. For leaded SMT networks, see comments on lead finish given in Section 2.2.2 under "Specifications" of transistors.

2.1.3   Inductors and Coils

Tuning Coils, Transformers

Construction—Surface-mountable inductors and coils are available in open designs or shielded designs (to reduce magnetic coupling between adjacent components). The winding cores are usually either ceramic or ferrite. The internal joints are generally soldered with high-tempera-ture alloys that are impervious to process environments. Protected types are postmolded with an epoxy resin or other appropriate molding compound. Bare beryllium-copper end termina-tions are available on some parts, but solder-coated terminations with 60 to 63 percent tin (near eutectic tin/lead solder) should be specified to ensure acceptable process yields.

Standards—Standards activity is underway in the EIA under its P-3.8 committee for inductors and coils. As of this writing, there is no issued standard for molded parts. As a workaround, most users choose parts conforming to the package outlines for molded brick-style tantalum capacitors (Figure 2-5 and Figure 2-6). However, dimensions for inductor bricks may follow their own directions.

Open-style inductors, tuning coils, and transformers vary widely in package outline. A careful analysis of the available offerings and a well-defined package on the specification con-trol drawing are the order of the day with these components.

Carrier Packaging—Parts are generally available in bulk or in Tape & Reel shipping packages. Coils and transformers are often problem parts for vibratory feeders, so automated handling usually requires that taped components be specified.

Specifications—Specifications for SMT coils should include all the electrical considerations typi-cal of through-hole components, plus a close control of the physical package and attention to solderability and soldering process resistance. For inductors with actual metal-tab terminations rather than fired-on metallizations, see the comments on lead finish given in Section 2.2.2 (Transistor Specifications). For fired-on terminations, refer to the comments given under Multilayer Ceramic Chip Capacitors (Section 2.1.1).

2.2   SMT Actives

Since the 1990s, electronic products have been significantly reduced in size. The most significant factor in this reduction has been the continuing reduction in the size of surface mount compo-nents (SMCs) and especially SMT ICs. The change from leaded DIP components to lead SMCs was the first great wave in this continuing reduction. The second wave includes packages such as the ball grid array (BGA) and the chip scale package (CSP). SMT seminars and papers have brought forward a great deal of data on the physical dimensions and processing methods for

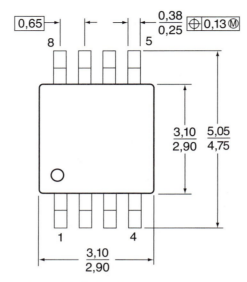

Figure 2-13   Example of lead pitch of SMT ICs, dimensioned center-to-center of adjacent pins. This is a 0.65 mm (0.040") pitch 8-pin device, as noted with the 0.65 mm measurement shown between pins 7 and 8 *(Courtesy Texas Instruments)*

the wide variety of SMCs now available. Data, such as package thermal characteristics and package-related propagation delays, are available from most semiconductor manufacturers to aid in power and high-speed design.

As we discuss active devices, the term "pitch" will be used. Pitch is the center-to-center lead distance measurement between two adjacent leads, as shown in Figure 2-13.

Another issue which affects all semiconductor devices is their temperature ratings. While we will discuss this further in Chapter 5 and Chapter 6, the standard temperature ratings for devices include commercial, 0–70°C, industrial, –25 to +125°C, and military or harsh, –55 to +125°C. Variations on these ratings exist. There may be times when a part is not rated for the thermal environment the product is expected to operate in. For some parts, it is possible to uprate them for use in wider temperature extremes. Regardless of any uprating, designers must do all thermal modeling and/or testing necessary to be certain that no part exceeds its absolute maximum ratings as defined by the part manufacturer.

2.2.1   Diodes (2- & 3-Lead Packages)

Construction—Diodes are commonly packaged in MLL-34 or MLL-41 MELF bodies, 2-lead gull-wing SOD-123, or in plastic-encapsulated parts based on the 3-lead SOT-23 outline. Dual diodes are also available in SOT-23 and SOT-143 packages. Construction details for the MELF are sim-

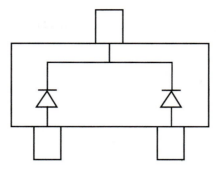

Figure 2-14   Diodes in SOT package.

ilar to a DO-35 diode except that the leads are replaced with face-bonded metallizations for sur-face mounting. SOT-packaged construction is the same as for SOT transistors. There is a trend toward dual diodes in SOT-23 packages and toward diode arrays in SOIC packages. SOT-23s are available with two diodes, with a common-anode, common-cathode, and series connection, or as a single diode, with one lead of the SOT unused. See Figure 2-14 for an example of diodes in an SOT package. The SOD-123 is also available with terminations formed under the body in the fashion of the "brick" style tantalum.

Standards—The mini MELF MLL-34 (SOD-80) and MELF MLL-41 follow roughly after JEDEC DO-35 and specifically after the EIAJ standards for MLL designations. The SOT-23 outline is controlled by EIA TO-236. The SOT-143 package closely follows the body outline of the TO-236, but it has four pins and slightly different dimensions from the SOT-23.

Carrier Packaging—For most applications, the preferred carrier is 8 mm tape on 4 mm pitch.

Specifications—The major difference in specifying diodes in SMT packages versus leaded diodes is in the resistance to soldering. Where nonleaded termination systems have a precious-metal base layer, a barrier layer may be specified to prevent diffusion of the precious metal during sol-dering, as stated earlier in the discussion on capacitors. For MELFs with bonded end caps or leaded device specification, follow the advice given next under Specification of Transistor Terminations.

2.2.2   Transistors (3-Pin Packages)

Construction—SMT small signal transistors are typically packaged in the SOT-23 format. Higher-current devices are packaged in a wide variety of power packages including SOT-89, DPAK (TO-252) and on up to packages such as the SOT-227 package, capable of in excess of 500 watts dissipation with proper heat sinking. Like SMT ICs, power package designs continue to proliferate, so any designer should check the latest data sheets for components most suited to the design.

Plastic-case transistors in SOT packages are constructed like TO-92 leaded transistors. The die is eutectic or conductive epoxy attached to the die flag on the lead frame. The die bonding pads are interconnected to the leads of the lead frame by wire bonding, and then the package is postmolded with epoxy novalac or a similar encapsulant. Finally, the individual devices are trimmed away from the lead frame runner, lead formed, tested, sorted, and packaged for shipment.

Standards—The SOT-23 format is controlled by EIA TO-236 within the U.S. and Europe. SOT-89 specifications are per TO-243. Note that the SOT-23 outline is a rough envelope. These parts are also covered by EIAJ specifications which vary from the EIA outlines. So caution is the word when substituting vendors, and purchase decisions best not be made on price alone!

Carrier Packaging—The SOT-23 format is packaged in 8 mm tape with 4 mm pitch, while the SOT-89 is packaged in 12 mm tape with an 8 mm pitch. For packaging of other lead-formed transistors, consult the vendor.

SOT-23 Specifications—As shown in Figure 2-15, there are three board standoff heights available within the TO-236 outline. The industry trend is toward a medium-profile part with 0.08 mm to 0.13 mm (0.003 inch to 0.005 inch) standoff since this part works in both reflow and wave soldering and provides sufficient clearance for cleaning. Note also that some diodes and some 3-lead ICs also use the SOT-23 package.

As discussed under passive devices, terminations of active devices should be solder coated with solder that is close to the 63/37 tin/lead eutectic or with pure tin for lead-free designs. The higher tin contents may exhibit improved solderability after a long storage period. However, tin/lead contents matching those of the solder paste and substrate metallization tin/lead ratios promote high reflow soldering yields. Solder coatings may be either hot-dip applied or electroplated. Hot-dipped terminations must be free from lumps, icicles, or other solder surface irregularities, and must not show any dewetted or pinholed areas. Plated terminations must be reflowed after plating to fuse the solder to the base metal. The tin/lead finish should be at least 7.5 microns (0.0003 inch) thick. *Note:* This termination treatment applies to all leaded SMCs and, therefore, this discussion will not be repeated under each category.

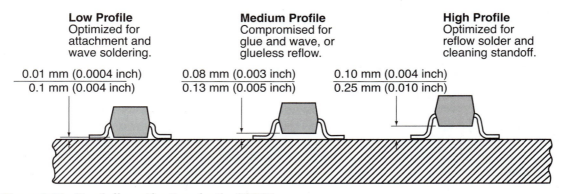

Figure 2-15   Standoff specifications for the SOT-23.

Power Packages—There are numerous power transistor packages available, each with its own unique board footprint. The most common are the SOT-223 and the TO-252AA, also known as a Motorola DPAK. Both of these packages have heat sink tabs that are designed to mount to a corresponding copper pad on the surface of the PCB. The designer must note carefully the data sheet specifications with regard to heat dissipation and maximum current. For example, the datasheet for International Rectifier (http://www.ir.com) DPAK devices notes that the maximum current is 7 amps, but the maximum current with the heat sink tab attached to a 1" × 1" copper pad is 2.4 amps. Maximum dissipation is likewise dependent on the thermal abilities of the PCB pad and any associated thermal *vias* or other thermal dissipative devices.

2.2.3   Transistors (4-Pin Packages)

The SOT-143 format may be used to package unijunctions and gated devices requiring four leads, and can also be used for RF switching transistors where the extra lead improves high-frequency performance. Another case style, the SOT-223, provides a more robust package with an enlarged lead on one side and three smaller pins on the opposite side. The enlarged lead makes this package useful for high-dissipation semiconductors.

Construction—Construction of the 4-lead SOT-143 is similar to the SOT-23 package discussed earlier.

Standards—The SOT-143 format basically follows the outline of TO-236, except that it has a fourth extra-large lead. Virtually every dimension differs slightly from the SOT-23 package, however. Power transistors are frequently packaged in SOT-223 or SOT-252AA (D-PAK) packages. The SOT-223 is a true 4-lead package in which the TO-252AA has its middle lead visible but cut short and will not extend to a pad on the board.

Carrier Packaging—The recommended carrier package for SOT-23 is 8 mm tape with 4 mm pitch. Larger devices use different carrier size and pitch, and the manufacturer must be consulted.

Specifications—See comments under "Transistor Specifications" above.

2.2.4   SMT Integrated Circuit Introduction

SMT packages have taken the IC market by storm. They are generally called "chip carriers" where chip refers to the IC silicon die contained at the center of the package. The primary defining characteristics that separate the various SMT IC packages include:

- material, plastic or ceramic
- leaded or leadless, where leadless chips have soldering terminations flush with the sides and/or bottom of the carrier body or inset into the sides and/or bottom of the carrier body
- lead shape if leaded

Leadless carriers are used with ceramic packages, and some low lead-count fine and ultra fine pitch SMCs also use this style, such as Carsem's MLP "Micro Leadframe Package" which has no lead projection from the sides of the quad pack and 0.02 mm projection on the bottom of

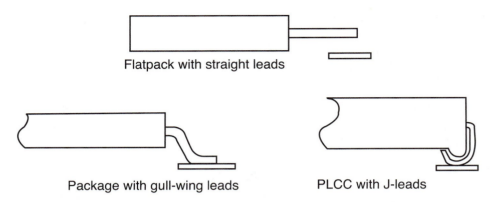

Flatpack with straight leads

Package with gull-wing leads

PLCC with J-leads

Figure 2-16   SMT lead styles. The flatpack may have its leads exiting the body at the center, as shown, or at the bottom of the body.

the package, and National Semiconductor's LLP "Leadless Leadframe Package" with lead counts to 56 and 0.050 mm nominal lead projection on the bottom

Leaded carriers typically have one of three leadshapes: flatpack, J-lead, or gull wing, shown in Figure 2-16.

Note that the flatpack may need to have its leads formed at the assembly facility prior to placement, where the gullwing and J leads are formed as one of the steps that occur during IC manufacturing and are shipped in their final form in the appropriate carrier.

In examining the proliferation of packages, the number available is mind boggling. As an example, here is a listing of *some* of the SMT IC packages listed by Texas Instruments (TI) and National Semiconductor Corp. (NSC) at the time of this writing:

TI and NSC Package Terminology
BGA: Ball Grid Array
CBGA: Ceramic Ball Grid Array
CDIP or **CERDIP:** Glass-Sealed Ceramic Dual In-Line Package
CERPACK: Ceramic Flatpack
CERQUAD: Ceramic Quad Flatpack
CFP: Both formed and unformed CFP
CPGA: Ceramic Pin Grid Array
DFP: Dual Flat Package
FBGA: Fine Pitch Ball Grid Array
FC/CSP: Flip Chip/Chip Scale Package
HLQFP: Thermally Enhanced Low-Profile QFP
HQFP: Thermally Enhanced Quad Flat Package
HSOP: Thermally Enhanced Small-Outline Package
HTQFP: Thermally Enhanced Thin Quad Flat Pack

HTSSOP: Thermally Enhanced Thin Shrink Small-Outline Package
HVQFP: Thermally Enhanced Very Thin Quad Flat Package
JLCC: J-Leaded Ceramic or Metal Chip Carrier
LBGA: Laminate Ball Grid Array
LCCC: Leadless Ceramic Chip Carrier
LCSP: Laminate Chip Scale Package
LGA: Land Grid Array
LQFP: Low Profile Quad Flat Pack
Micro SMD: Micro Surface-Mount Device
Mini SOIC: Mini Small Outline Integrated Circuit
PQFP: Plastic Quad Flat Package
QFP: Quad Flat Package
SOIC: Small Outline Integrated Circuit
SOJ: J-Leaded Small-Outline Package
SOP: Small-Outline Package (Japan)
SSOP: Shrink Small-Outline Package
TFP: Triple Flat Pack
SOT: Small Outline Transistor
TQFP: Thin Quad Flat Package
TSSOP: Thin Shrink Small-Outline Package
TVFLGA: Thin Very Fine Land Grid Array
TVSOP: Very Thin Small-Outline Package
VQFP: Very Thin Quad Flat Package

*Additional types marked by TI as "DO NOT USE," but some are still active by other IC manufacturers!

DIMM*: Dual-In-Line Memory Module
HSSOP*: Thermally Enhanced Shrink Small-Outline Package
LPCC*: Leadless Plastic Chip Carrier
MCM*: Multi-Chip Module
MQFP*: Metal Quad Flat Package
OPTO*: Light Sensor Package
PLCC*: Plastic Leaded Chip Carrier
PPGA*: Plastic Pin Grid Array
SIMM*: Single-In-Line Memory Module
SODIMM*: Small Outline Dual-In-Line Memory Module
TSOP*: Thin Small-Outline Package
VSOP*: Very Small-Outline Package

Additional package types, or the same package types with other designations, are available from other IC manufacturers, so this list is far from totally inclusive and is shown only to give the reader some feel for the number of IC packages available.

Note that the section marked "do not use" in the TI designations does not necessarily mean the packages are not being produced by TI. Many are currently in production. The "do not use" designation means that the manufacturer has projected an end of life for that package, and it should therefore not be used in new designs. The design team is cautioned to be alert for these sort of notations on any component being designed into a new product. Other manufacturers may, however, have some or all of these packages as active components without the end-of-life designation.

The authors must also note again that it is impossible to cover all the available SMT IC packages in this book, not only because of the number of them, but also because more will become available between the time of this writing and the time of publication. Again, readers are encouraged to seek out manufacturers' data sheets for the latest information. Also note that many of the ICs do incorporate specifics with regard to circuit board design. These issues will be discussed in detail in Chapter 5.

2.2.5   SO ICs

2.2.5.1   SO ICs: 4- to 8-Pin SOT-Type Packages

ICs are available in packages that use the same basic body size as the SOT transistor but add leads to make 4- to 8-lead devices as seen in the example shown in Figure 2-17. These very compact ICs are used by a variety of manufacturers for analog, digital, and microprocessor components. As one example, MicroChip packages one of their basic PIC microcontrollers in an 8-pin mini-SO package. As with most IC packages, different manfacturers use different package

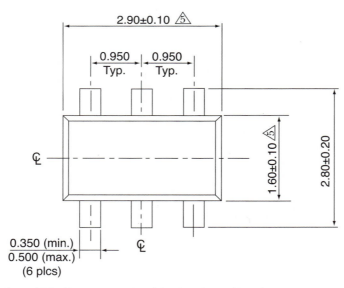

Figure 2-17   SOT-packaged IC. *(Courtesy National Semiconductor Corp.)*

terminology for these packages. As examples, Maxim and National uses the designation SC70 for a millmeter-dimensioned 6-pin SOT23 package, while National uses the designation MF06A for essentially the same package with inch dimensions.

2.2.5.2   SO ICs: 8- to 28-Pin Standard Pitch (0.050" or 1.27 mm Dual-In-Line Packages)

Surface-mount ICs have been packaged in DIL flatpacks for years. Hermetic flatpacks are available, and this package was popular for some time in military and hi-rel design. However, the flatpack package is expensive. It requires handling for lead trimming and forming operations, and its leads are very fragile and easily damaged. Because of these disadvantages, flatpack production has been on a downward trend since the mid-1980s.

The 50-mil (1.27 mm) pitch SO family of ICs is the primary direct alternative to DIP ICs. Their small size (about one-eighth of the material of a DIP package) and their substantial volume usage make them very price competitive with DIP packages. In a recent study, we found merchant-marketed ICs in SO at parity with, and up to 10 percent cheaper than, DIPs. The SO package has a number of variations and applications by IC manufacturers. Standard SO dimensions are shown in Figure 2-18.

Construction—The SOIC and its wide-bodied cousin the SOW (W for wide, also known as SOL for Large) are manufactured like DIPs. Dies are generally attached to a bonding flag on the lead frame and are interconnected electrically to the lead frame by wire bonding. Alternatively, dies are inner-lead bonded to TAB tape which is then outer-lead bonded to the lead frame. In either case, the assembly is then postmolded with a protective epoxy covering.

Standards—JEDEC has standardized a 0.150-inch (3.8 mm) nominal-width body SOIC in 8-, 14-, and 16-pin sizes, and a 0.300-inch (7.6 mm) nominal-width body in 16, 20, 24 and 28 pins, the SOW or SOL.

There is also a standard for a nominal 0.300-inch body-width part having J-leads, although this part is little used at this writing. The standard covers 14- to 28-pin devices in 2-pin

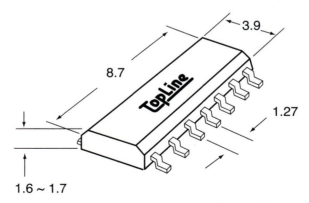

Figure 2-18   SOIC package and standard outline dimensions. *(Courtesy Topline, Inc.)*

increments. This package has been used for memories, where its J leads allow very close order spacing for high-density designs. Center pins may be omitted in selected part numbers to facilitate busing of a wide power trace. Most memory applications now use fine pitch or BGA devices.

Carrier Packaging—For short production runs, plastic tubes (also known as "sticks") are the packaging format of choice. For longer runs where suitable tape feeders and taped components are available, reeled components are recommended. One standard plastic tube carries 48 16-pin SOICs. A 16 mm tape with 8 mm pitch holds at least 2500 (see Table 2-11) of the same device. (Many customers hesitate to fill the reels to the capacity because they cannot consume the full reel in a single production run.) Therefore, a placement machine with one tube at a feed station would have to be reloaded 52 times for every 1 reload required with the tape. Table 2-11 details reel sizes for various SO family parts.

Specifications—There are some areas where caution is called for in specifying SOICs. First, be aware that there are several standards competing in the world. EIAJ parts, for instance, come in a variety of body widths. To make matters worse, older JEDEC parts are controlled by an inch-based lead pitch of 0.050 inch (1.27 mm). Some, but not all, Japanese parts are controlled by a metric pitch of 1.25 mm (0.04921 inch). In an 8-pin device, this is of no concern. However, for 28-pin devices, the incremental error because of pitch discrepancy is a substantial 0.010 inch, enough to make bridging between leads likely if a board is laid out for inch parts and metric parts are used, or vice versa.

There are also a number of components built in what the manufacturers call an SO package that do not fit standard pads for either the JEDEC SO or SOL devices. Resistor networks and crystals are among the prime offenders. A word to the wise is for the user to check the manufacturer's outline drawing. Nonstandard parts will generally work just fine as long as you have laid out your PWB for the part you are getting.

The basic comments regarding the specification of lead finish for transistors will apply equally to SO packages.

Table 2-11   13-inch (330 mm) Tape-and-Reel Sizes for SO Family ICs.

Package Type	Tape Size		Reel Width	Typical Component Count Per Reel*
	Width	*Pitch*		
SO-8	12 mm	8 mm	18.4 mm	2500
SO-14	16 mm	8 mm	22.4 mm	2500
SO-16	16 mm	8 mm	22.4 mm	2500
SOL-16	16 mm	12 mm	22.4 mm	1000
SOL-20	24 mm	12 mm	30.4 mm	1000
SOL-24	24 mm	12 mm	30.4 mm	1000
SOL-28	24 mm	12 mm	30.4 mm	1000

*Based on partially filled reels; may be more at user request.

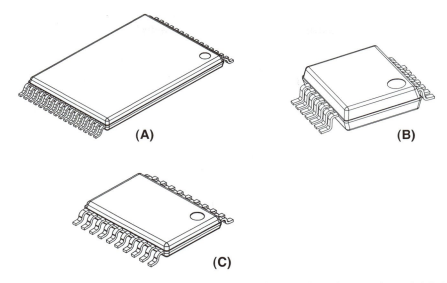

Figure 2-19 Varieties of the fine pitch SO package. (A) Thin small-outline package (TSOP). (B) Shrink small-outline package (SSOP). (C) Thin shrink small-outline package (TSSOP). *(Courtesy National Semiconductor Corp.)*

2.2.5.3 Fine Pitch SO ICs (14- to 68-Pin Dual-In-Line Packages)

Fine pitch SO packages started as identical to 1.27 mm pitch devices except using 0.65 mm (25 mil) pitch packaging. They rapidly proliferated to a variety of types in a variety of pitch sizes, including the thin small-outline package (TSOP), the shrink small-outline package (SSOP), the thin shrink small-outline package (TSSOP), and the very small-outline package (VSOP) among others. Variations on this package are shown in Figure 2-19. Some of these packages will have pitch dimensions smaller than 0.65 mm, e.g., 0.5 mm is very common, and these are frequently referred to as "ultra fine pitch" devices.

Internal wire-bonding lead lengths have limited the dual packages to 68 pins. Packages with pin counts higher than 68 use all four sides for their pins or are packaged in one of the array packages.

2.2.6 Quad ICs (20- to 304-Lead Packages)

There is such a diversity of package alternatives in this category that some explanation is in order. If nothing else, this exercise may ease the pain of dealing with the package proliferation. At best, it may help you make better package selection decisions.

The SOIC solved many problems inherent in the original SMT flatpack format. However, the SO series was not ideal for high lead-count devices, and the growth in high lead-count devices has become exponential. Since the dual-in-line SO package is both space inefficient and very difficult to wirebond without crossovers (potential short circuits) in packages of 20 pins and up, four-sided devices and grid array devices were developed. Figure 2-20 shows a comparison with a DIP. Early four-sided chip carriers for IC packaging included the plastic leaded chip carrier (PLCC), Figure 2-21. The PLCC is still in use by some manufacturers. The most common quad part currently in use includes variations of the quad flat pack (QFP). One example is shown in Figure 2-22. While there are some quad packs shown in some literature with lead counts up to 304, much of the industry has maximized available leads at 208 due to internal wire bonding issues.

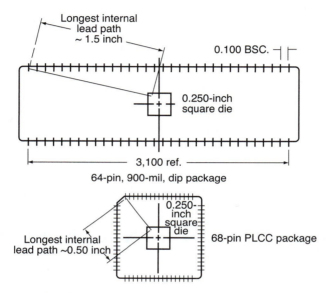

Figure 2-20   DIP and quad IC packaging compared.

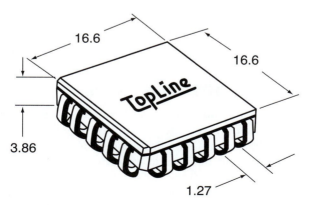

Figure 2-21   The 20-lead 1.27 mm (0.050") pitch PLCC package. *(Courtesy Topline, Inc.)*

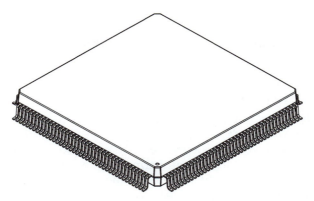

Figure 2-22   Example of the 100-lead fine pitch thin quad flat pack (TQFP) package. *(Courtesy National Semiconductor Corp.)*

To allow chip carrier use on organic PWBs and reduce packaging cost, a number of package formats were developed using leaded construction, plastic postmolding, and varying lead forms. The plastic leaded chip carrier (PLCC), developed around 1980, became the dominant style in the U.S. market. PLCCs provided well-protected leads to form flexural elements, thus relieving expansion stresses that might otherwise damage solder joints. This package met the needs of a large segment of the PWB assembly market through the 1980s but is losing market share to packages that mount lower to the board and have a higher lead count. Due to its design, the PLCC is not suitable for lead counts greater than 84, partly due to its standard 50-mil pitch. Texas Instruments, for one, considers the PLCC a package not to be used in new designs, and they are not designing any new parts in that package.

JEDEC Type-A PLCCs are built using dies attached to lead frame and wire bonding or TAB interconnect to the lead frame and postmolding in the same manner that SOICs or plastic DIPs are made. A small percentage of PLCCs are premolded parts designed to be socket attached to board assemblies. Premolded parts may also be reflow attached. JEDEC postmolded specifications cover only 28-, 44-, 52-, and 68-pin parts. Square premolded JEDEC PLCC specifications detail eight sizes—from 20 to 124 pins—but most 20- and 84-pin PLCCs are actually postmolded. This is because the original JEDEC Type-A standard left to interpretation many details of part construction. There are also rectangular premolded PLCCs in five styles—from 18 to 32 pins.

For designers that either must conform to certain MIL-spec requirements or for whom heat is a major issue, ceramic packages are available. They will be discussed first, with plastic packages to follow.

2.2.6.1   Ceramic ICs (20- to 124-Pin Packages)

Ceramic packages are generally expensive and are used primarily to meet MIL-spec requirements. For that reason, they will only be covered in passing in this book. The reader who needs detailed ceramic part information is directed to the manufacturers' data sheets and other texts specifically aimed at military components and design. We will note there are these variations of ceramic IC carriers:

- Leaded ceramic chip carriers
- Leadless ceramic chip carriers
- Preleaded ceramic chip carriers
- Postleaded ceramic chip carriers

Ceramic package styles are defined in JEDEC Publication 95.

Ceramic chip carriers were early contenders as quad packages, but ceramic carriers are much more expensive to produce than postmolded plastic devices of equal pin count. And, because there are no leads to absorb stresses of thermal expansion, the CLLCC (ceramic lead-less chip carrier) package demands special attention in the matching of the thermal coefficient of expansion (TCE) between the ceramic package body and the substrate. Common board materials, like epoxy/glass, expand at about three times the rate of the alumina ceramic of the chip-carrier body and, therefore, makes a poor substrate for CLLCCs. Most materials, such as ceramics that match alumina's TCE, also match or exceed its cost (about three times FR-4). Leaded ceramic packages also suffer from cost and other issues that severely limit their use. Readers who need the hermaticity of ceramic carriers are advised to consult the IC manufacturer's application information.

Construction—A family of ceramic packages was developed as shown in Figure 2-23. There are four main 50-mil CLLCC families registered with JEDEC for square parts. Types A, B, C, and D differ somewhat in construction details, but they share the following points in common. All are

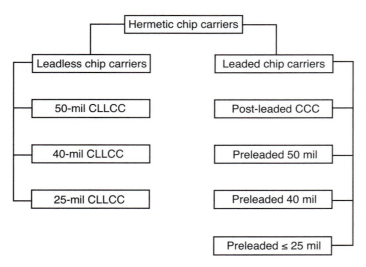

Figure 2-23   The hermetic chip carrier family tree.

built on a ceramic substrate (base carrier) having metallizations for die attach and interconnection to the outside world. Each is covered using a lid, which is generally glass-to-metal sealed for hermeticity.

Type A is designed for lid-down mounting in a socket, placing its heat-dissipating side up for air cooling (particularly forced air) rather than conductive heat removal through the board. Its cavity-down orientation makes it well suited to heat-sink mounting on its top side. However, it does not provide for solder connection to boards. It is not widely used today.

Type B and Type C are designed for lid-up mounting via reflow soldering to TCE-compatible materials, or by socketing on noncompatible substrates. The lids on both packages extend above the CCC body. The Type-B lid is metal (such as gold-plated Kovar), and Type C's lid is ceramic. Type C dominates today's CCC marketplace.

Type D is similar to Type B except that its metal lid is recessed so that its top surface is flush. The package is designed for cavity-down mounting making it like Type A, suitable for air-cooled applications. However, unlike Type A, this package allows the option of reflow or socket attachment to board assemblies.

True preleaded CCCs are manufactured with a thin metal-foil lead frame enclosed by two ceramic body halves. The body is joined around the lead frame with hermetic glass-to-metal seals in CERDIP fashion. The specifications for such parts are covered in JEDEC MS-044. They have J-lead forms. However, the specification provides only for cavity-up 68- and 84-pin parts.

To meet the need for other lead counts and cavity-down orientations, manufacturers may post-attach lead frames to MS-003, MS-004, or MS-005 parts. This produces a chip carrier modeled after side-brazed ceramic DIPs. The leads are brazed or thermocompression-bonded to the top surface of the carrier. In some cases, the leads are attached to side castellations. Since these carriers are supplied to the user with leads in place, it is customary to speak of them as preleaded even though leads are attached as a secondary operation. Figure 2-24 shows the typical lead configurations used. We can see in Figure 2-24 that side-attached leads provide less of a flexural element for a given component standoff. Thus, top-attached leads are preferred to maintain the lowest possible component profile for a needed flexural capability in the leads.

Figure 2-24 also provides a convenient point to evaluate the relative strength of solder joints produced by various lead forms. The chart in Figure 2-24B shows the results of pull and shear tests of J-lead, gull-wing, and butt lead joints. Clearly, both the J-lead and gull-wing formats provide higher initial tensile and shear stress resistance than the butt lead. While joint strength may have some relationship to long-term reliability, the equation for this relationship has not been demonstrated. Also, joint reliability is probably impacted by a large collection of variables outside the lead form. The chart in Figure 2-24C comes closer to the point of comparing the joint reliability of the J-leads versus the I or butt leads. The chart shows the results of extended thermal cycling tests of J- and butt-leaded parts.

Where preleaded CCCs are unavailable, users may add a lead. Postleaded CCCs are built per standard CLLCC Types B, C, and D. After fabrication and lidding, the leads are soldered on by the component vendor or by the end user. Additionally, post-lead vendors can attach wire leads to the CLLCC castellations using a high-temperature solder so that the lead attachment

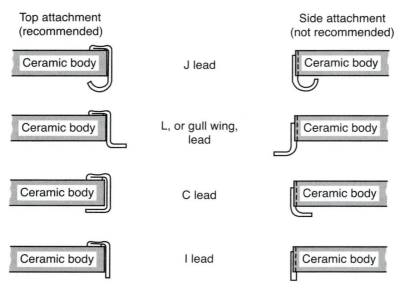

NOTE: While directed at CLLCC lead frames, these comments, except those regarding side and top attachments, apply equally to other devices with these lead forms.

(A) The lead configurations.

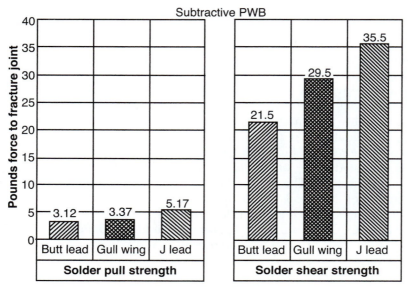

(B) Pull and shear test results.

Figure 2-24 Typical lead configurations for CLDCCs. *(Continues on next page)*

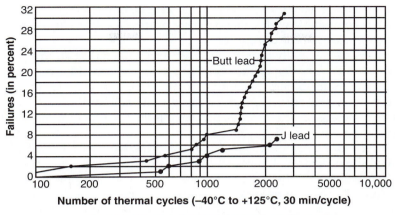

(C) Results of extended thermal cycling tests.

Figure 2-24 *(Continued)* Typical lead configurations for CLDCCs.

remains unaffected by subsequent soldering during PWB assembly or rework. In general, leads are of the two types shown in Figure 2-25.

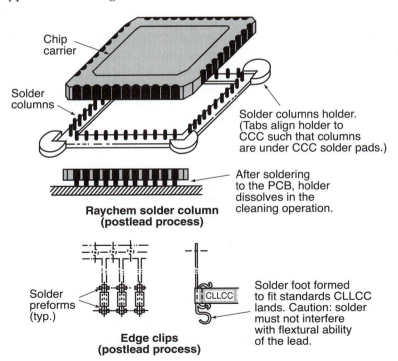

Figure 2-25 Lead types for postleaded chip carriers.

Specifications—In specifying leaded or leadless chip carriers, pay close attention to the lead finish to ensure that it is compatible with your soldering process. Leaded chip carriers should have lead finishes as detailed earlier for transistors.

Leadless carriers are generally constructed with a trimetal termination of gold, nickel, and a refractory metal, such as tungsten or molybdenum. Gold is fond of forming extremely brittle intermetallics with the tin in solders. Such intermetallics are a sure source of failure for solder joints which, under stress, will crack right along the intermetallic and solder junction. This problem is corrected by selecting tin-free solders, by keeping the time at solder-reflow temperatures short, or by overcoating the gold.

Another concern is the solderability of parts. Aggressive fluxes would provide an easy answer for through-hole devices. However, leadless chip carriers allow very little cleaning space between their bellies and the board. Therefore, it is not good practice to use fluxes that might harm the circuit if the cleaning process does not do a perfect job. To ensure (and test) solderability, it is a common practice to dip-solder coat the CLLCC terminations.

CLLCCs are available with (relatively) low-cost glass-frit sealed lids or with more expensive solder-sealed lids. Solder attachment allows a lower temperature sealing than the CERDIP-style glass-frit seal, and solder seals meet all the process requirements of MIL-M-38510. Therefore, solder-sealed parts are required for some military and hi-rel applications.[7]

So, for CLLCCs, engineers must determine how the parts will be soldered, who will solder-dip the leads, what the end use dictates in the seal selection, and whether heat will be primarily dissipated into the PWB or into the air (cavity up or cavity down, respectively).

Standards—Within the U.S. market, JEDEC's JC-11 committee handles standards for component outlines. JEDEC, a branch of the EIA, can be contacted at:

Engineering Standards Manager
2500 Wilson Blvd.
Arlington, VA 22201
(703) 907-7500

EIA Japan (EIAJ) activity is also of interest and is important to assess when selecting Japanese suppliers for components.

2.2.6.2   Plastic Quad ICs (20- to 208-Pin Packages)

Plastic quad packages have taken the IC market by storm. In the previous listing of Texas Instruments' and National Semiconductor Corp.'s available packages, no fewer than ten of them were variations on plastic quad packages. Plastic quad flatpacks (QFPs and PQFPs), like the 144-lead QFP shown in Figure 2-26, are widely used for IC packages with 20 leads and over, up to 208 leads. There are a number of packages in this family on a variety of inch and metric dimensions, including 50 mils (0.050"), 25 mils (0.025"), 20 mils, 1.0 mm (nominally 0.040") shown in Figure 2-27, 0.65 mm (nominally 0.025"), 0.5 mm (nominally 0.020") and 0.4 mm (nominally 0.016") lead pitches. Other pitches may be used by IC manufacturers. Regardless of the pitch of the package leads, it is important that the IC manufacturer maintain close control over the coplanarity parameter of the package. Shown in Figure 2-28, coplanarity is important to

Figure 2-26   QFP package.

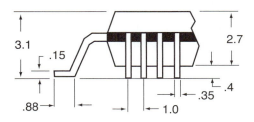

Figure 2-27   Dimensioned drawing of a standard 1.0 mm pitch QFP package.

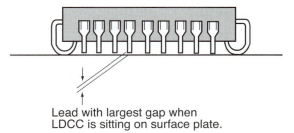

Lead with largest gap when
LDCC is sitting on surface plate.

Figure 2-28   Coplanarity variance is defined as the total deviation of leads from a single plane.

minimize the risk of open solder joints developing during the solder process. It must be matched by flatness of the substrate pads meeting the same criteria, which for fine pitch devices and all BGAs is 4 mils, or 0.004".

Carrier Packaging—With leaded carriers, coplanarity of leads, as defined by the sketch of Figure 2-24, is of critical importance in maintaining high manufacturing yields and in keeping solder defects in check when using SMT. Carrier packaging for other SMT components serves primarily to maintain the parts in an oriented position for delivery to a placement machine. With leaded

Table 2-12   13-inch (330 mm) Tape-and-Reel Sizes for Chip Carriers.

Package Type*	Tape Size		Reel Width	Typical Component Count Per Reel**
	Width	*Pitch*		
PLCC-20	16 mm	12 mm	22.4 mm	1000
PLCC-28	24 mm	16 mm	30.4 mm	750
PLCC-32	24 mm	16 mm	30.4 mm	750
PLCC-44	32 mm	24 mm	43.4 mm	500
PLCC-52	32 mm	24 mm	43.4 mm	500
PLCC-68	44 mm	32 mm	55.4 mm	250
PLCC-84	44 mm	32 mm	55.4 mm	250

*PLCCs shown. CLLCCs and LDCCs are the same.
**Based on partially filled reels. This number can be increased to suit user needs.

chip carriers, packaging must both maintain the pin-1 orientation and protect the component leads from bending or damage during the shipping and handling operations. Leadless chip carriers of similar sizes must be protected from smashing together on edge, so tubes are rarely used for leadless parts. Rather, carrier tapes and matrix trays are popular for these parts. Carrier tape specifications for PLCCs and CLLCCs are shown in Table 2-12.

The older JEDEC PQFP has tabs (also known as bumpers) protruding from its four corners. These devices were also called bumpered quad flat packs, or BQFPs, as shown in Figure 2-29. The tabs extend beyond the perimeter of the leads so that a device banging around in a tape carrier cavity is afforded some protection from lead damage. These tabs make it possible to ship PQFPs in plastic tubes as well as tape-and-reel carriers, however issues with the bumper have precluded its inclusion in newer packages.

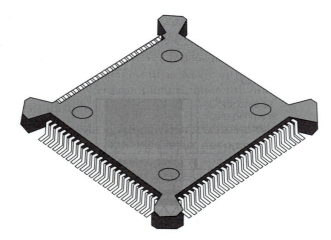

Figure 2-29   BQFP package *(Courtesy Topline, Inc.)*

Fine-Pitch Packages—Packages having over 84 pins become problematic with standard pitch (1.27 mm, 0.050") due to internal wire bonding issues. They also become inefficient in their use of circuit board space. For these reasons, fine pitch (0.80 mm, 0.65 mm, 0.025") and ultrafine pitch (0.5 mm and below, 0.020" and below) packages are widely available in lead counts up to 304 with a variety of applicable JEDEC standards. Packages can be rectangular or square in both EIA and EIAJ registrations. An example is shown in Figure 2-30.

The reader must be careful to check the manufacturer's data sheets, as the same size package may have varying lead pitchs and lead counts, and the circuit board footprint must be chosen accordingly. Similarly, packages with identical lead counts from the same manufacturer may have different lead pitches, again requiring careful selection of the correct footprint. Standoff, the vertical distance from the bottom of the lead plane and the bottom of the package also varies among packages and among manufacturers. Prior to designing any chip components under a fine pitch device, the standoff dimension must be verified.

Handling fine pitch packages is normally accomplished through the use of tape-and-reel or matrix tray (also known as waffle packs) carriers. Most modern placment machines have matrix tray handlers available that allow automatic changing of empty trays. Due to the thin and narrow leads on fine-pitch gull-wing packages, lead damage is always a risk, and the user should assure proper handling of package carriers as well as consider the implementation of coplanarity verification tools on the placement machine. These may be camera or laser-based devices. Designs with fine-pitch packages normally require a placment machine with a vision system.

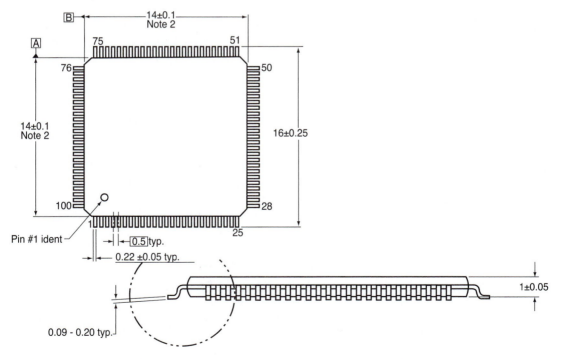

Figure 2-30 Dimensioned fine-pitch quad package. *(Courtesy National Semiconductor Corp.)*

2.2.7   ICs: Tape Automated Bonding (TAB) Packages

For uses in which the standard quad packs are too thick, there are variations of the tape auto-mated bonding (TAB) package. In Europe, Siemens developed a variation of the TAB procedure, called Micropack™. Micropack is a TAB direct-on-board technique that uses a three-layer copper polyimide tape with outer terminations typically on 0.508-mm (0.020") centers. Three-layer tape can be used for very close center inner-lead bonding on the chip. Wire bonds on the IC die are generally set to a 0.254 mm (0.010") minimum pitch to allow room for the bonding tool and to prevent bond wire shorting. A 6-mil pitch (0.152 mm) is about the limit for current bonding technology. Three-layer tape has been successfully used in production for bonds on 0.102 mm (0.004") centers, and 0.051 mm (0.002") is the current limit. This technology can handle produc-tion devices with over 300 leads and, in studies, has handled 600 I/Os. Special Micropack-style designs provide excellent thermal paths for very high thermal dissipation.

Disadvantages of the TAB on-board approach are the cost of three-layer tape, the require-ments for specialized equipment for full parametric test and burn-in of components before placement, and a lack of protection for the die.

In the United States, National Semiconductor developed the TapePak™ approach for SMT use of TAB. TapePak uses a 0.071 mm (0.0028") single-layer copper-tape "lead frame" to form interconnections to a die. The device is postmolded with a protective package around the die. A protective plastic ring is simultaneously molded outside the attachment leads of the device. The leads for attachment to the circuit are on 20- or 10-mil (0.508-0.254 mm) centers, but the pro-tective ring supports an outer perimeter of fanned-out leads on 1.27 mm (0.050") centers allow-ing package handling with current burn-in and test technology. The disadvantages of TapePak are the limits of lead spacing and bond-pad centers on the die which are imposed by single-layer copper tape, and the requirement for specialized excising stations to remove the outer pro-tective ring and lead from the component before placement on pick-and-place machines. The disadvantages apparently outweigh the advantages since National no longer shows TapePak devices on their website.

There is still some interest in TAB, although it is very limited. Hitachi Cable entered the TAB packaging arena in 1997 with their TAB system known as Hitachi Compliant Chip Packaging (HCCP) based on Tessera's µBGA package. Tessera also uses a variant of TAB in their CSP packages. It remains to be seen how much of an impact TAB will have in the future.

All of these technologies have found growing uses. In space-constrained applications such as notebook computers and hand-held electronics, TAB, Flip Chip, and COB are methods of choice. These applications have spawned rapid growth in equipment, sources of known good die, service infrastructure, and standards for these advanced manufacturing technologies.

2.2.8   ICs: Ball Grid Array (BGA) Introduction

This section is excerpted, with permission from National Semiconductor Corp. Application Note AN1146, "Ball Grid Arrays." It will present an overview of BGA packages.

Leaders in the consumer electronics industry will be determined by their ability to deliver increasingly miniaturized products at lower costs. The Ball Grid Array (BGA) package achieves these objectives by providing increased functionality for the same package size while being compatible with existing SMT infrastructure. Some of the other benefits of using BGA packages over similar lead count packages include:

- Efficient use of board space
- Improved thermal and electrical performance. BGAs can offer power and ground planes for low inductances and controlled impedance traces for signals
- Improved surface mount yields compared to similar fine pitch leaded packages
- Reduced package thickness
- Potentially lower cost of ownership compared to leaded packages by virtue of their reworkability

This application note provides general information about Plastic Ball Grid Array (PBGA) packages and its variants—the TE-PBGA (Thermally Enhanced BGA), EBGA (Enhanced BGA) and TSBGA (Tape Super BGA). Information on FBGA (Fine Pitch BGA) and LBGA (Low Profile BGA) packages can be found in National Semiconductor's Laminate CSP application note (AN 1125) (Figure 2-31).

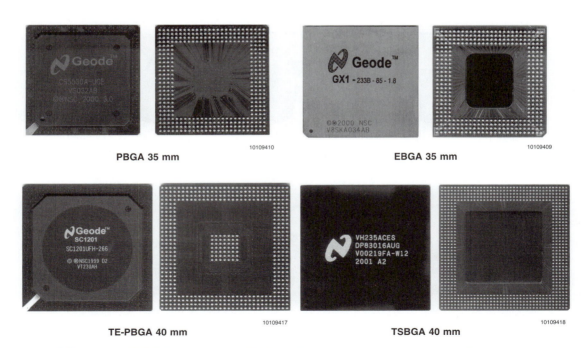

PBGA 35 mm   10109410   **EBGA 35 mm**   10109409

TE-PBGA 40 mm   10109417   **TSBGA 40 mm**   10109418

Figure 2-31 PBGA, TE-PBGA, EBGA and TSBGA. *(Courtesy National Semiconductor Corp.)*

Package Overview

PBGA (PLASTIC BGA) CONSTRUCTION

The PBGA (Plastic Ball Grid Array) package is a cavity-up package based on a PCB substrate fabricated of Bismaleimide Triazine (BT) or FR5 epoxy/glass laminate (Figure 2-32). The BT/FR5 core is available in several thicknesses with rolled copper cladding on each side. The final plated copper thickness is typically 25–30 μm.

Solder mask is applied on both sides over the copper pattern to ensure that all the substrate vias are completely tented. Four-layer substrates are available for applications requiring power or ground planes (they also provide additional routing flexibility). For thermal applications, the inner layers can be clad with thicker (2 oz.) copper (~70 μm).

The IC is attached on the top side of the substrate using die attach. The chip is then gold wire-bonded to bondfingers on the substrate. Traces from the bondfingers transfer the signals to vias that then carry them to the bottom of the substrate and finally to circular solder pads on the same side.

The bottom side solder pads are laid out in a square or rectangular grid format with a pitch recommended by JEDEC registration standards (MO-151) for PBGAs. The part is then over-molded to completely encapsulate the chip, wires, and substrate bondfingers.

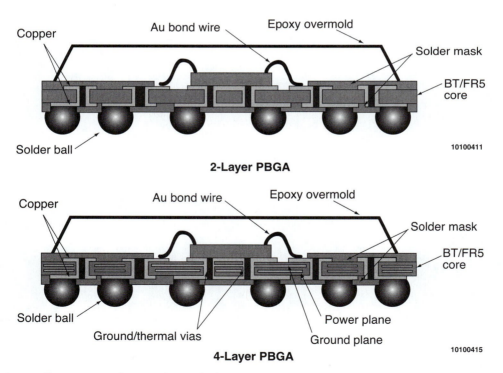

Figure 2-32 Cross-sectional view of 2- and 4-layer PBGA. *(Courtesy National Semiconductor Corp.)*

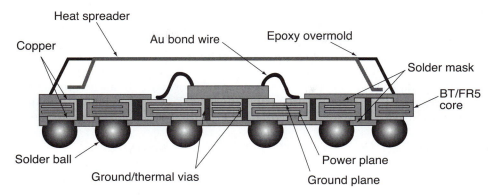

Figure 2-33   Cross-sectional view of 4-layer TE-PBGA. *(Courtesy National Semiconductor Corp.)*

TE-PBGA (THERMALLY ENHANCED BGA) CONSTRUCTION

The TE-PBGA (Thermally Enhanced Plastic Ball Grid Array) package is a variant of the PBGA package for enhanced thermal dissipation, Figure 2-33. A drop-in heat slug is added to a 4-Layer PBGA with 2 oz. (70 μm) copper on the inner layers. This provides a much better thermal path to the top surface of the package. The heat slug can be grounded to provide an EMI shield for the package. Thermal vias are provided under the die and are typically connected to the package ground plane thereby conducting the heat to the PCB ground plane.

EBGA (ENHANCED BGA) CONSTRUCTION

The EBGA (Enhanced Ball Grid Array) package is a cavity-down package configured to provide enhanced thermal and electrical performance, Figure 2-34. The thermal advantage of this package is realized by attaching the die to the bottom of a heat spreader or heat slug that also forms the top surface of this package. Since the heat spreader is on top of the package and is exposed to airflow, the thermal resistance is very low.

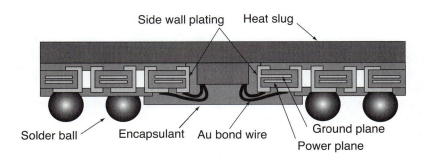

Figure 2-34   Cross-sectional view of EBGA. *(Courtesy National Semiconductor Corp.)*

The heat spreader or heat slug is laminated to a printed circuit board (PCB) substrate fabricated of BT or FR5 epoxy/glass laminate. The IC die is bonded to the heat spreader using a die attach adhesive. The die is then gold wirebonded to bondfingers on the substrate. Traces from the bondfingers transfer the signals to solder pads. Unlike typical PBGAs, vias in an EBGA are primarily used to connect the solder pads to inner layers (typically the power and ground planes).

The solder pads are in a square or rectangular grid format with a pitch recommended by the JEDEC registration standard (MO-151) for the EBGAs. The package is encapsulated to completely cover the chip, wires, and substrate bondfingers.

2.2.8.1   ICs: Ball Grid Array (BGA) Package Variations

The variety of packages introduced in Section 2.2.8 come with many variations in size, ball pitch, ball coverage, and number of interconnects. This section will cover some of the available package variations.

Size of the overall BGA is directly related to ball pitch, ball coverage, and number of interconnects. Ball pitch ranges from 1.27 mm pitch for 'standard' plastic and ceramic BGAs down to 0.40 mm. Fine-pitch BGAs (FBGAs), similar to large CSPs, may have ball pitch down to 0.8 mm (0.032") which can still be handled by most standard placement systems.

Ball pitch of 1.27 mm (0.050") is advantageous to the manufacturing segment of the design team. This will be covered further in Chapter 3, but the 1.27 mm pitch plastic devices can be placed with standard placment machines and are self-centering on their pads during the reflow process as long as 50 percent of each ball is on its respective pad. Smaller-pitch devices are useful in compact designs, and in RF designs where IC manufacturers minimize size to minimize parasitic capacitance and inductance. Kobayashi et al.[8] report on parasitic reductions in μBGA packages.

With the use of standard pitch for many packages, BGAs combine the desired traits of using standard paste printing, placement, and reflow equipment with a vastly increased number of interconnects. As will be discussed in more detail in Chapter 5 on PCB design, one necessary criteria for the use of BGAs in a design is to maintain flat PCB pads and minimize any coplanarity issues on the pads themselves The main downside of designs with BGAs is the need for x-ray inspection systems and specialized hot-air rework tools. The x-ray systems are both expensive and relatively slow compared to other inspection systems, but there is no other way to examine the quality of the solder joints under the BGA package.

Ball coverage, as seen on the bottom of the package, generally falls into three categories:

- Full coverage, that is the bottom is completely covered with a uniform pattern of pads/balls
- Peripheral coverage, that is where there are three to six rows of pads/balls around the periphery of the bottom face of the chip
- Peripheral coverage with center coverage, where the peripheral coverage is augmented with a square pattern of pads/balls in the center of the package directly under the silicon die

Examples of the various types of coverage are shown in Figure 2-35. The peripheral + center coverage pattern is used to enhance the transfer of heat from the die to the substrate.

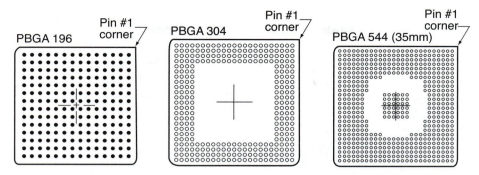

Figure 2-35 Examples of full-coverage, peripheral ball coverage, and peripheral + center ball coverage BGAs. *(Courtesy Intel Corp.)*

The number of interconnects varies greatly among manufacturers, but the authors have seen counts reported on data sheets of 64—>2000 interconnects. Routing of 2000—interconnect devices boggles the mind but can be done.

As would be expected, the BGA has its drawbacks. In addition to the need for x-ray systems to inspect BGA solder joints, Yee, et al.,[9] report on intermetallic formations that threaten BGA joint reliability when material compositions are different between ball materials and pad plating materials. BGAs also suffer from moisture-related problems. All plastic-bodied IC packaging materials suffer from absorption of moisture. SO and most QFP packages use the same molding compound that DIP packages use, and it absorbs moisture, as will be discussed further in Section 2.2.13. However, BGA packaging uses a bismalide triazine (BT) resin that is even more moisture absorbant than other compounds. This means that BGA users must be even more careful with storage and with the length of time BGA packages are open prior to reflow.

2.2.9 ICs: Chip Scale Packages (CSPs)

The variety of Chip Scale Packages (CSPs) is exploding, indicating the interest in the package style by both IC manufacturers and users. There are over fifty variations of the CSP on the market at this writing. The CSP package incorporates an IC silicon die in a package that is nominally no greater than 20 percent larger than the bare die itself in the X and Y dimensions, with a ball pitch of 1 mm or less. The are some manufacturers, such as National Semiconductor, who have "wafer-level" CSP packages that are nominally the same X-Y dimensions as the die. In their application note AN-1112, National describes the fabrication of their wafer-level micro SMD: "The micro SMD manufacturing process steps include standard wafer fabrication process, wafer re-passivation, deposition of solder bumps on I/O pads, backgrinding (for thin version), application of protective encapsulation coating, testing using wafer sort platform, laser marking, singulation and packing in Tape and Reel." Chip bumping and other aspects of micro SMD-type packages will be discussed in Section 2.2.10, Flip Chips.

As IC fabrication lithography processes create smaller features, the package size may remain constant, changing the die-to-package ratio so the 20 percent guideline may not be

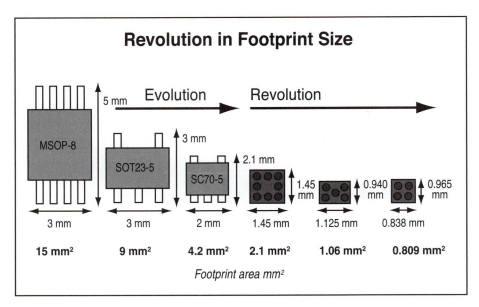

Figure 2-36 CSP package examples. *(Courtesy Intel Corp.)*

accurate for some packages. Another purpose for keeping the package size the same is that the manufacturer can offer variations on the part, such as larger or smaller memory capacity, on the same board footprint.

IC manufacturers may call the package a CSP, a Micro SMD or a μBGA, but they are all CSPs. The CSP commonly has BGA-type terminations, although other terminations are possible. Figure 2-36 shows the possible reduction in package size with the use of the CSP package.

That CSPs use BGA-style interconnects contributes both to the small size of the package as well as its self-centering aspects during reflow soldering. Most CSPs use this style of interconnect, as shown in Figure 2-37. This figure shows Intel's μBGA, a true CSP. Larger CSPs are fre-

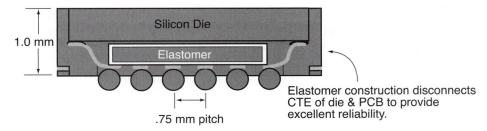

Figure 2-37 CSP package cutaway. *(Courtesy Intel Corp.)*

quently known as FBGAs, or fine-pitch BGAs, as mentioned in the previous section. It is sometimes difficult to determine exactly where a true CSP ends and a larger BGA package begins since many BGA packages also have a ball pitch of 1.0 mm and smaller.

Note particularly the elastomer interposer between the IC die and the interconnect balls. This allows the die and the substrate/PCB to expand and contract at different rates with minimal risk of cracking the die.

Like many small packages, new varieties of CSPs are regularly introduced. Some of these include packages that stack more than one silicon die inside one CSP.[10] This, along with the variety already in existence, precludes us from trying to show "typical" dimensions and footprints. The reader who chooses to use a particular IC manufacturer's CSP device must determine the characteristics of that device from the appropriate manufacturer.

2.2.10 ICs: Flip Chip Packages

Flip chip packages basically take the bare silicon die of a particular semiconductor device, create "bumps" at the die interconnect pad locations on the top of the die (which then allow the die to be "flipped" over with the interconnect bumps facing down), and placed on corresponding pads on the substrate. The advantage is that the flip chip is the smallest "package" available—no larger than the die itself in the X and Y dimensions, which then brings parasitic L and C to their lowest level possible. The disadvantages are that it has ultra-fine pitch interconnects, commonly 0.25 mm, and if used on FR-4 substrates, must be underfilled with an epoxy to prevent cracking of the interconnects during thermal excursions. The flip chip does require special handling and squaring of the part during the pick-and-place process.

First developed by IBM for use in their mainframe computers in the 1960s and then for automotive applications by Delco Electronics in the 1970s, flip chips are now used in a variety of applications that include automotive, watches, cell phones, pagers, PDAs, and microprocessor peripherals. Estimates are that 5 to 8 percent of die are bumped by semiconductor manufacturers and others. Bumping services are offered by a number of companies worldwide:

APAK, Taiwan
APTOS, USA
Chipbond, Taiwan
CS2, Belgium
Flip Chip Technologies, USA
Focus Interconnect, USA
Fujitsu, Japan
IC Interconnect, USA
Microfab Technology, Singapore
TLMI, USA
Unitive Electronics, USA

IC manufacturers may call their flip chips by other terms. National Semiconductor's micro SMD is essentially a flip chip, although National calls it a "wafer-level SMD," with solder-

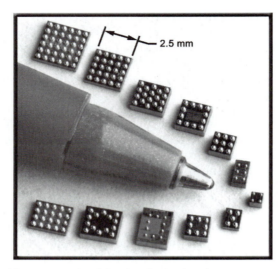

Figure 2-38   Micro SMD. *(Courtesy National Semiconductor Corp.)*

bumped pads as shown in Figure 2-38. The bump size and pitch varies among individual devices. One issue with any vendor choosing to bump bare die is that of known good die (KGD) discussed further in Section 2.2.12.

There are a number of bumping technologies currently in use. It is important that a potential user understand the technologies in order to make an informed decision about the one most appropriate for a given design. Bumping precedes sawing the silicon wafer into individual die. The bump serves a number of purposes:

- to provide the electrical interconnect from the die to the substrate interconnects
- to provide the mechanical interconnect from the die to the substrate interconnects
- to provide a thermal path for heat generated in the die during operations
- to act as a very short lead to provide some relief of mechanical strain as die and substrate expand and contract during thermal excursions, although underfill is usually required for additional strain relief

There are several ways of creating the bump at each die interconnect point. Typically, the bump is formed over an appropriate under bump metallization (UBM) that allows good electrical and mechanical contact. This is necessary due to the aluminum that is used on the basic die interconnect points (bond pads) which rapidly oxidizes if unprotected. UBM technologies vary depending on the bump technology but are generically similar to that shown in Figure 2-39. Lippold[11] and Patterson, et al.[12] provide a good discussion on bumped die.

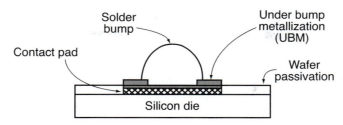

Figure 2-39   Generic UBM and bump.

Bumping techniques include:

• Adhesive bump following electroless nickel under-bump metallurgy (UBM)
• Solder ball bump, also known as stud bump
• Evaporated bump
• Electroplated bump
• Electroless nickel UBM with printed solder bump

Adhesive bumps are formed by stencilling thermoset or adhesive to form bumps on the under-bump metallization. The adhesive is then cured, and assembly to the substrate makes use of the additional layer of conductive adhesive.

Solder ball bumps are formed by depositing solder paste on the UBM with stencil or syringe-deposition techniques or through the use of evaporation or electroplating. A resulting bump is shown in Figure 2-40.

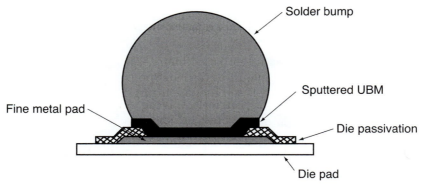

Figure 2-40   Detail of UBM with solder paste bump.

Bumped die flip chip assembly operations with solderable bumps include:

1. Component placement using flip chip mounting/placement equipment
2. The placement step involves application of flux to the solder bumps either using a spray fluxing arrangement or a flux-dip station on the pick-and-place machine
3. An alternative method using no-flow or flux underfill may also be used for assembly
4. Flux dip should involve wetting of at least $^1/_3$ of the total bump height with flux
5. Standard reflow (convection preferred) to form solder joint interconnections
6. Cleaning step (depending on type of flux used)
7. Underfill application using typical underfill equipment

Figure 2-41 illustrates the process steps involved. Adhesive bumps would involve only adhesive deposition and curing. Figure 2-42 shows a closeup of a flip chip with underfill applied to minimize the thermal stress on the interconnect bumps.

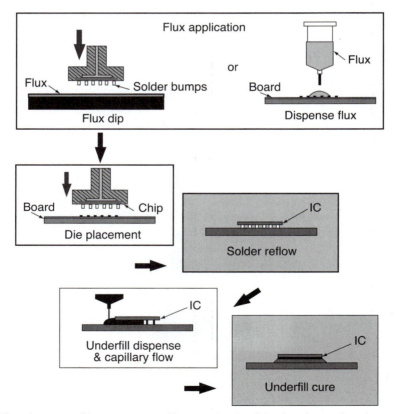

Figure 2-41   Flip chip assembly operations. *(Courtesy National Semiconductor Corp.)*

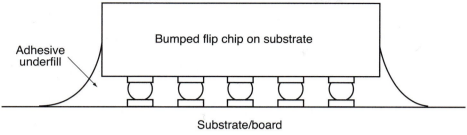

Figure 2-42   Flip chip on board with underfill.

Underfill is dispensed around the periphery of the flip chip, then capillary action carries it under the chip and around the interconnect bumps. To calculate the amount underfill material that should be dispensed, one needs to calculate the total volume between the flip chip and the substrate and subtract the volume occupied by the interconnect bumps or balls. As shown in Figure 2-43, this calculation would include:

- V_B, the volume between the flip chip and the substrate, $V_B = L_C \times W_C \times C_S$, where:
  - L_C = length of the chip
  - W_C = width of the chip
  - C_S = chip standoff
- V_I, the volume of the interconnect bumps or balls
- V_F, the volume of the fillet around the package, and
- V_U the total volume of the underfill

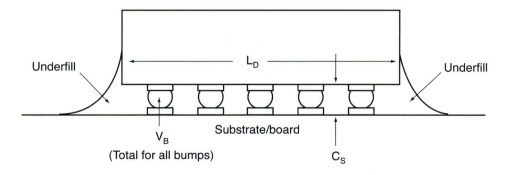

Figure 2-43   Flip chip underfill calculations.

Then $V_U = V_B - V_I + V_F$. Recognize that the fillet volume serves several purposes. It is necessary to have a fillet since it is impossible to calculate an exact underfill volume. The fillet then serves as extra material if the under-chip volume was underestimated and also serves as an overflow if the under-chip volume was overestimated.

In one study on flip chip underfill and flux residue using lead-free solder,[13] the authors present data on seventeen different flux systems with two underfill materials in a lead-free assembly process. They found joint quality ranging from poor to good over the range of their work. Of the seventeen fluxes studied, five gave better results than the other twelve.

EIA/IPC standards covering the implementation of flip chips include:

J-STD-012: "Implementation of Flip Chip and Chip Scale Technology"
J-STD-026: "Semiconductor Design Standard for Flip Chip Applications"
J-STD-027: "Mechanical Outline Standard for Flip Chip and Chip Size Configurations"
J-STD-028: "Performance Standard for Construction of Flip Chip and Chip Scale Bumps"
J-STD-029: "Test Methods for Flip Chip Scale Products"
J-STD-030: "Qualification and Performance of Flip Chip Underfill Materials"
IPC 7078: "Sectional Requiremens for Flip Chip Component Mounting (Direct Chip Attach)"

2.2.11   ICs: Chip on Board, COB

Chip on board, or COB, is the use of the bare silicon die directly on a substrate and has been in use since the 1970s. Typically used for devices like watches and calculators, it involves testing the bare die to overall specification, also known as known good die testing which is discussed further in Section 2.2.12, placing the bare die on the substrate with the bond pads facing up and wire bonding the bare die pads to matching contact pads on the substrate. Once wire bonding is complete, the die and wires are normally encapsulated with a black epoxy that is dispensed over the die + wire area on the substrate. Because of the manner of deposition and its appearance, the protective coating is frequently know as a "glob top."

2.2.12   ICs: Known Good Die (KGD)

As noted in the flip chip section, there are only a few vendors available outside of semiconductor manufacturers who will "bump" die for flip chips. Part of the reason for this is the same issue briefly mentioned in chip on board (COB), the issue of known good die (KGD). Semiconductor manufacturers test at several points in their overall manufacturing process, including at the end of wafer processing and at the end of the overall packaging process, e.g., when a QFP has been encapsulated and separated from its lead frame carrier. The testing performed at the wafer level is not intended to be a complete functionality test to specification. Therefore, any user of bare die, or any user contracting with a vendor for a flip chip bumping process, needs to understand the issues around KGD.

EIA JESD 49, "Procurement Standard for Known Good Die (KGD)," addresses many of the issues a user should be aware of. In this standard, the EIA warns users that the levels of quality and reliability they have come to expect of packaged devices are unlikely to be met with bare die. Therefore, the user and the supplier need to work together to assure that:

- Specifications for final performance are agreed upon in advance
- Specifications for final quality tests and expected quality levels are agreed upon in advance
- The supplier has a die testing program and capability in house, with specifications for percent fault coverage, stress tests, UL, ISO or MIL-SPEC certifications, etc.
- Recordkeeping standards meet the needs of both the supplier and the user

KGD is less of a problem issue today than it was, but it can still be a stumbling block in the implementation of COB or flip chip devices. Therefore, any potential user should strongly consider working with experienced suppliers.

2.2.13   ICs: Moisture Absorption of IC Packages

As briefly mentioned under BGAs, moisture absorption is a problem with IC packaging materials. The mechanism has been described in many documents by which any moisture absorbed creates a major problem during the reflow soldering process.[14] During this process, the moisture may expand and crack the package, typically under the die flag as shown in Figure 2-44.

Figure 2-44(A) shows moisture entering the package as a result of the porosity of the molding compound to moisture. Saturation level and speed of moisture entry depend on the the porosity of the compound material, the RH of the environment as well as temperature, and exposure time. In addition to moisture, there are other mechanisms at work to increase the probablity of crack formation, including the relative thermal coefficient of expansion between

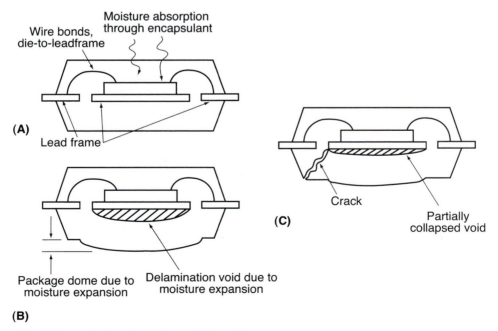

Figure 2-44   Moisture-induced cracking of IC package.

the molding compound, the leadframe material (typically Cu or alloy 42), the die itself, and the silver-bearing die-attach epoxy.

Figure 2-44(B) shows the results of the moisture-laden air expanding during the reflow heating process. The bond strength is typically weakest between the bottom of the die flag and the surrounding encapsulation. At this point, if the pressure of the expanding gas is stronger than the bond strength, the two materials will delaminate and a pressure dome will form on the bottom of the package.

Figure 2-44(C) shows the result of the collapsed void as the temperature decreases. The formation and relaxation of the dome creates a crack in the body.

As a result of these events, failure of the functionality of the IC may occur at the time of crack formation or soon after as a result of degradation of the package and/or aluminum pad corrosion inside the device, or it may cause intermittent functionality due to damage to the internal gold wire ball bonds.

Because of both the serious nature of this failure and the fact that all ICs are susceptible to it, a number of studies have been conducted starting in the mid 1980s. Early work included that of Fukuzawa et al.,[15] Steiner & Suhl,[16] and Alger et al.[17] Studies have shown that packages with no die or a very small die will not exhibit cracking. Therefore, die size is an important characteristic in determining moisture sensitivity. As mentioned in the BGA section, BGAs use a different molding compound than other ICs, and this BT compound absorbs more moisture than the compounds used for other SMT or THT IC packages. Work on this issue has been done by Yip, et al.,[18] Ahn, et al.,[19] and Giacobbe.[20]

Both the IPC and EIA have standards relating to moisture sensitivity of IC packaging, proper storage, and labeling with regard to moisture sensitivity of plastic packages.[21] Working together they have developed J-STD-020, "Moisture/Reflow Sensitivity Classification for Plastic Integrated Circuit Surface-Mount Devices," as well as others. It is instructive to note that all the moisture problems and standards are only an issue for SMT ICs, not for DIP devices. This is because only the leads of DIP devices are directly heated by wave soldering, and DIP bodies rarely exceed 150°C. SMT ICs, however, must undergo reflow soldering and normally reach peak temperatures in excess of 200°C. Obviously, any DIP package with modified leads that will undergo reflow soldering will have the same concerns regarding moisture absorption.

Standards that apply to the issues in this section include:

- IPC/JEDEC J-STD-020: "Moisture/Reflow Sensitivity Classification for Plastic Integrated Circuit (IC) SMDs"
- IPC/JEDED JSTD-033: "Standard for Handling, Packing, Shipping and Use of Moisture-Sensitive SMDs"
- IPD/JEDEC J-STD-0354: "Acoustic Microscopy for Non-Hermetic Encapsulated Electronic Components"
- IPC-9501: "PWB Assembly Process Simulation for Evaluation of Electronic Components (Preconditioning IC Components)"
- IPC-9502: "PWB Assembly Soldering Process Guideline for Electronic Components"
- IPC-9503: "Moisture Sensitivity Classification for Non-IC Components"
- IPC-9504: "Assembly Process Simulation for Evaluation of Non-IC Components (Preconditioning Non-IC Components)"

- JEDEC JESD22-A103: "High-Temperature Storage Life"
- JEDEC JEP113: "Symbol and Labels for Moisture-Sensitive Devices"
- JEITA (formerly the EIAJ) Technical Report EDR-4701: "Handling Guidance for Semiconductor Devices"

There are also testing standards for manufacturers, such as EIA JESD 22-A120, "Test Method For the Measurement of Moisture Diffusivity and Water Solubility in Organic Materials Used in Integrated Circuits." Readers who have need for this type of information are encouraged to search the EIA and JEDEC standards.

Packages that have absorbed a certain amount of moisture should not be reflowed. Normal factory packaging of moisture-sensitive components includes both a humidity indicator and dessicant bages inside a moisture barrier bag that has a factory seal date. The standards do allow opening the bag for an inspection time of no greater than $1/2$ hour at no greater than 30°C/60%RH. To allow for use of packages that have received exposure to moisture or for which either the time indicated on the bag has been exceeded or for which the humidity indicator shows an exposure to over 20% RH, industry standards provide for a "two-bake" procedure, one at lower temperature and one at a higher temperature. Once the exposure limits listed on the shipping package have been reached, a package can be baked at:

- 125°C (±5°C), 48 hours, oven RH <50%, or
- 40°C (±5°C), 192 hours (8 days) minimum, oven RH <5%

The high-temperature bake is problematic for tube or tape-and-reel carrier devices since these carriers will not withstand the high temperature and the devices must be removed from their carriers prior to baking. Conversely, low-temperature bake can be done in the carriers, but the time involved is considerable.

To put perspective on the issue and provide guidelines for users, the moisture sensitivity classification levels in J-STD-020 allow the IC manufacturer to classify each package based on its allowable floor time once removed from the hermetic shipping package. Table 2-13 shows the eight classification levels and their allowable floor time.

Table 2-13   J-STD-020 Moisture.

Classification Level	Floor Life Once Removed from Bag	Moisture Exposure: °C/RH	Moisture Exposure: Time (hours)
1	Unlimited at ≤85% RH	30°/85%	168 (7 days)
2, 2a	1 year (level 2) or 4 weeks (level 2a) @ ≤30°/60% RH	30°/60%	168 (7 days)
3	1 week @ ≤30°C/60% RH	30°/60%	168 + MET
4	72 hours @ ≤30°C/60%RH	30°/60%	72 + MET
5, 5a	48 (level 5) or 24 (level 5a) hours, as specified on label, @ ≤30°C/60% RH	30°/60%	Time on label + MET
6	Mandatory bake: bake before use and after bake must be used by time limit on label	30°/60%	Time on label

The MET notation in the exposure time limits is the Manufacturer's Exposure Time. This factor allows compensation for the time the IC manufacturer needs to process components after encapsulation and before bag seal and shipping with an included dessicant. It also includes an allowance for open bag time and repacking with dessicant at a distributor's location.

The exposure times in Table 2-13 are for the specific temperature and RH conditions shown. A level 4 device must be baked before reflow if it has been exposed to 60% RH at 30°C for 72 hours or more. There are correction tables for other times, temperatures, and RHs. These are best obtained from the IC manufacturer.

Also note that the JEDEC labelling requirements for moisture sensitivity in JEP-113B state that level 1 devices do not require special storage conditions provided "They are reflow soldered at a peak body temperature which does not exceed 235°C." If the conversion to lead-free requires higher reflow soldering temperatures, this entire topic will require intense study. Vaccaro, et al.,[22] examine the impact of the proposed lead-free reflow temperatures on moisture-sensitive plastic ICs.

2.2.14 ICs: Thermal Issues

Basic thermal issues exist with all ICs regardless of their package types. Heat is transferred from the silicon die itself to the case, from there to the heat sink if there is one, and ultimately to the air. While there is a subset of electronics intended for liquid cooling, we will use air for our cooling medium. Heat transfer issues are discussed in some detail in Chapter 5 on PCB design since many SMCs transfer all or part of their heat load to the PCB. In this section, we will note that designers must be aware of both the maximum allowable junction temperature T_J of any heat-producing ICs that will be used and the ambient environment in which the IC will be operated.

If operation is necessary outside the normal range of the chip, the design team must consider uprating the IC. Uprating is defined as "a process to assess the capability of a part to meet the functionality and performance requirements in which the part is used outside the manufacturer's specification range."[23] The main need for a design team to consider the uprating of a device is that fewer parts seem to be available with wide temperature specifications. Teams must consider the effects on the parts when they are uprated as well as the proper techniques for performing the uprating process.[24]

2.3 Connectors and Sockets

There are numerous and varied connector offerings now in the SMT arena, and the family is expanding rapidly. Many connectors (particularly true of the pure SMT types) are low, or zero, insertion/extraction force designs for use in minimizing the forces applied to surface-mount solder joints during use. If moderate-to-high insertion/extraction forces are anticipated (such as in designs involving large pin or tin-plated parts or in designs requiring repeated insertions and removals), it is best to choose an interconnect device providing some form of mechanical fastening to the board. Typical mechanical hold-down schemes include screw connection, heat staking of bosses, and snap-in bosses. THT connectors can also be used with the reflow soldering process using the pin-in-paste (PIP) technique that will be described in Chapter 3.

Construction—Construction of SMT connectors and sockets differs from inserted interconnect construction in several ways. First, since the solder connections are on the component side of the board, many SMT connectors employ gull-wing lead forms to provide for visual inspection of solder joints. Gull-wing mounting requires more board real estate than typical insertion styles. Second, special plastics are often used as dielectrics for SMT connectors because the parts must withstand the full heat of the soldering-process environment. Additionally, if the PIP process is to be used, the connectors will be designed to have the body stand off from the PCB surface.

Standards—Activity to provide standards is underway in a subcommittee of EIA P-5.1.

Carrier Packaging—Carrier packaging varies dependent on the assembly methods and equipment to be used and on the manufacturer's available offerings. The most common approaches are specialized carrier tapes for high-volume robotic applications, matrix trays or tubes for medium-volume automation, and bulk bagging for small runs where hand assembly will be used.

Specifications—Beyond the considerations involved in specifying through-hole PWB-mounted connectors, SMT brings a concern about the solderability of lead finish, the survivability in soldering and rework/repair processes, and the mechanical fastening needs.

2.4   Switches

Construction—Switches are not yet fully standardized for SMT. The variety of jobs that switches must address makes standardizing to a limited number of designs difficult if not impossible. Most SMT switches are of the small slide or rotary type. A few push-button types are also available. Outwardly, many of the SMT switches currently available look like gull-winged versions of through-hole designs. However, their materials and sealing may be quite different in order to suit the relatively hostile process environment that SMT components may encounter.

Standards—Consult EIA P-13 for the current status of work underway in P-13.1 on rotary switches, P-13.2 for slide, rocker, and toggle switches, and P-13.3 for the push-button and keyboard switches.

Carrier Packaging—Generally, switches are supplied in bulk or in plastic tube, but some are available in specialized tapes. Selection of the carrier package should be determined by the availability from the switch suppliers and the requirements of assembly automation.

Specifications—As with all leaded SMCs, surface finish of the leads should be controlled. Refer to the discussion of transistor specifications (Section 2.2.2) for details. Also, the specifications should detail the board-attachment method to use for stress relief, if any, and the carrier packaging needed.

2.5   Other Components

The following is just a sampling of the support components available in SMT packaging today. Except for very large and high-power devices, virtually all component types are now offered in SMT style.

2.5.1   Crystals

An early objection to surface-mount packaging for crystals was that their size and bulk did not blend with 2-pin SMC packaging. As a solution to this, a variety of larger device packages have been adapted by various crystal manufacturers. Crystals are available in chip carrier and SO packages and also in metal cans with the two wire leads formed in gull-wing fashion for surface mounting.

2.5.2   Relays

The first SMT relays on the market, gull-wing lead-formed DIP relays, are still widely available today. The alternatives include plastic rectangular packages with the leads trimmed and formed in "C" fashion around each end and some variations on standard surface-mount IC packages.

2.5.3   Jumpers/Zero-Ohm Resistors

Construction—Jumpers are basically shorting bars constructed like chip resistors but with conductive inks instead of resistive inks fired on the top surface. Chip jumpers are used to avoid double-sided, or multilayer, signal traces on boards where only a few tracks cannot be routed without crossovers. Also available and regularly used for the same purposes are zero-ohm resistor. The zero-ohm resistor is also commonly used to allow variations in electrical design with no variation in PCB design, e.g., an op-amp circuit can have a variety of gains or even filter implementations with no change in the PCB layout if zero-ohm resistors are used in some implementations.

Standards—Chip jumpers and zero-ohm resistors are available in standard SMT chip resistor packages, while some jumpers are available in CR21 and CR32 sizes. Like regular resistors, this can vary from 0201 to 1210 and larger sizes.

Carrier Packaging—Carrier tape is the preferred shipping format, except where specialized placement equipment dictates some other choice.

Specifications—Jumpers are specified by component outline, termination material, resistance to the soldering process, and the maximum voltage/wattage rating. A CR21-size jumper will handle roughly one order of magnitude more power than the equivalent-size chip resistor, or 1 ampere at 70°C, while the CR32 handles twice that amount.

2.5.4   Variable Capacitors

A variety of caseless and enclosed variable capacitors are available for surface mounting. Construction types include Teflon™ dielectric and air trimmers. Care must be taken in selection in order to consider the entire manufacturing process and choose components that will still be trim caps after assembly. See comments in the earlier discussion regarding trimmer resistors.

2.5.5   Thermistors

Chip thermistors are available with form factors roughly following those of chip capacitors. However, there are many sizes of thermistors outside the five recommended by the EIA for MLC chips. Refer to the discussions of the chip-cap specification process (Section 2.1.1) for guidance in specifying the end terminations for surface mounting thermistors. Surface-mount thermistor chip wraparound terminations for PC board mounting are shown on the right, and hybrid wire bondable terminations are on the left. For volume uses, 8- or 12-mm carrier tape is the preferred packaging format. Where tape is not available from the component supplier but desirable for manufacturing, service bureaus exist to reel parts.

2.5.6   Inductors

Construction—SMT inductors are generally wire wound around a ceramic or ferrite core. Some are left bare, some are conformally coated, and some are encapsulated, for protection and to improve automated handling characteristics. A molded part with a flat top is important for pick-and-place equipment, and many inductors are available in EIA-standard packages from 1nH to 10,000µH.

Coils are available in both unshielded or shielded form. Shielded designs reduce the magnetic coupling between components. Internal soldered connections should be welded or high-temperature soldered to withstand SMT assembly-process environments.

Standards—Two case sizes have been proposed by the SMTA components committee and are specified in ANSI/IPC SM-782. However, our 1987 survey of component availability indicated less than an enthusiastic support of any standard from suppliers. Therefore, it is wise to allow time for shopping when purchasing inductors.

Carrier Packaging—The two standard types listed above come in 8-mm/4-mm pitch, and 12-mm/8-mm pitch tape. However, those companies that are battling to resist the standard are just as determined to package parts in innovative ways as they are committed to creativity in package size. Again, check early with prospective suppliers and be aware of those services that will tape components. Note, taping may add some lead time since the taping house must purchase suitable tape and tooling for some nonstandard components.

Specifications—For parts with end terminations fired onto the body, refer to the chip capacitor specifications (Section 2.1.1) for end termination treatment. For molded parts with metal lead terminals, refer to the specifications of SOT-23 for termination specifications.

2.5.7   Optoelectronics

Construction—Packaging optocouplers for SMT use is a challenge. Typically, the package bodies are too small to provide the space required for the needed 2500 to 4000-volt isolation. Three avenues are being used for SMT as of this writing. Typical DIP parts are sometimes lead formed to gull-wing configuration or are trimmed to butt-lead form for surface mounting. In a more generically SMT approach, Siemens offers a high-body version of the 50-mil SOIC. Motorola plans to second source this package. The critical footprint is the same as a JEDEC SOIC, and the larger body height needed to meet isolation specifications does not interfere with most automated assembly operations. For SMT applications requiring hermeticity, TRW offers a CLLCC variant with two pins each on two sides.

Standards—As of this writing, no standard exists for optocouplers. While many optocouplers are packaged in 4-pin SO packages, users should check dimensions carefully.

Carrier Packaging—The SOIC- and CLLCC-style parts can be supplied in Tape & Reel for large production runs. However, the SOIC's nonstandard height violates EIA tape thickness specifications. This may impact feeding of the tape in some de-reelers. SOICs may also be supplied in plastic sticks for short runs or small lot jobs. Matrix trays are available for small quantities of the CLLCC.

Specifications—For parts with end terminations fired onto the body, refer to the specifications for CLLCCs for end termination treatment. For molded parts with metal-lead terminals, refer to the specifications for SOT-23 (Section 2.2.2) for termination specifications.

2.6   Component Standards and References

There are many organizations that write standards that apply to electronic components, their testing, and their application to PCAs. These include, from the overall listing in Chapter 1:

JEDEC—The JEDEC Solid State Technology Association; http://www.jedec.org
ECA—The Electronic Components, Assemblies, and Materials Association; http://www.ec-central.org/
ESDA—Electrostatic Discharge Association; http://www.esda.org
SEMI—Semiconductor Equipment and Materials International; http://wps2a.semi.org
SIA—Semiconductor Industries Association, http://www.sia.org

2.7   Review

After reading and understanding this chapter, you should be able to answer these questions and note the reference location for the information in the chapter.

1. List three organizations that set forth standards for electronic components and/or the matching footprints on the PCB.
2. Define a "tin whisker."
3. Explain why tin whiskers will be a bigger problem with the move to lead-free solders and component lead tinning.
4. Define an MLCC, then describe its construction.
5. Explain why MLCCs typically provide better circuit performance than their THT counterparts.

6. Explain what a "brick" capacitor is.
7. For a chip capacitor with the inch notation 1206, what is the closest equivalent size metric notation?
8. Give two reasons why an SMT resistor must be mounted with the silk-screened side up.
9. Define "pitch" as used in relation to SMT IC leads.
10. Many SMT diodes are packaged in a 3-lead SOT-23 package. Sketch possible 1- and 2-diode configurations inside a 3-lead package.
11. Define "pitch" with respect to tape-and-reel carriers.
12. What lead pitch dimension do the following types of ICs use: standard pitch, fine pitch, ultra-fine pitch?
13. What is the main driver in the use of ceramic IC packages?
14. Define "bumping" as applied to flip chip packages.
15. Briefly explain why moisture absorption is an issue with SMT ICs and what the proper handling procedures are.
16. Why is moisture absorption a larger concern with BGAs compared to other SMT IC packages?

2.8   References

1. NASA Goddard Space Flight Center Tin Whisker Home Page. Available at http://nepp.nasa.gov/whisker/.

2. Farnell RoHS Directive website, "What Are Tin Whiskers," available at http://uk.farnell.com/static/rohs.

3. Oberndorff, P. J. T. L, Dittes, M., and Petit, L., "Intermetallic Formation in Relation to Tin Whiskers." Available at http://www.infineon.com.

4. "Technical Summary, Multilayer Capacitors." Available at http://www.microcapacitors.com, "Technical notes."

5. Karnezos, M., "Stacked-Die Packaging." *Advanced Packaging,* August, 2004, pp. 41–44.

6. Toleno, B. J., Carson, G. (2004), "Flip Chip Underfill and Flux Residue with Lead Free." Circuits Assembly, v. 15 #6, June, 2004, pp. 24–29.

7. Vaccaro, B. T., Shook, R. L., Gerlach, D. L., "The Impact of Lead-Free Reflow Temperatures on the Moisture Sensitivity Performance of Plastic Surface-Mount Packages." *Journal of SMT,* Oct–Dec, 2003, v. 16 #4, pp. 21–26.

8. Kobayashi, M., Hidetoshi, M., Yasuda, T., Kumakura, T., "Reduced Parasitic Inductances in μBGA Package Using Conductive Plane." Available at Hitachi Cable, http://www.hitachi-cable.co.jp/micro BGA/ubga/designguide/paper/vlsi_pkg.pdf.

9. Yee, S., Chen, L., Zeng, J., Jay, R., "Ternary Intermetallic Compound—A Real Threat to BGA Solder Joint Reliability." *Journal of Surface Mount Technology,* v. 17, #2, Apr–Jun 2004, pp. 29–35.

10. Karnezos, M., "Stacked-Die Packaging." *Advanced Packaging,* August, 2004, pp. 41–44.

11. Lippold, M., "Eutectic Bumped Die Qualification and Processing." National Semiconductor Corp. white paper. Available at http://www.national.com/appinfo/die/0,1826,856,00.html.

12. Patterson, D. S., Elenius, P., and Leal, J. A. (1997). "Wafer Bumping Technologies—A Comparative Analysis of Solder Deposition Processes and Assembly Considerations." INTERPack '97, June 15, 1997, EET Vol. 19-1, Advances in Electronic Packaging—1997, pp. 337–51, ASME, New York, 1997.

13. Toleno, B. J., Carson, G. (2004), "Flip Chip Underfill and Flux Residue with Lead Free." Circuits Assembly, v. 15 #6, June, 2004, pp. 24–29.

14. Suhl, D., "The Mechanism of Plastic Package Cracking in SMT and Two Solutions." IPC-TP-683, IPC Spring Meeting, April, 1988.

15. Fukuzawa, I., et al., "Moisture Resistance Degradation of Plastic LSIs by Reflow Soldering." Proceedings of the 23rd Reliability Physics Symposium, 1985, pp. 192–97.

16. Steiner, T., Suhl, D., "Investigation of Large PLCC Package Cracking During Surface-Mount Exposure." IEEE Transactions on Component, Hybrid, and Manufacturing Technology, v. CHMT-10, #2, June, 1987, pp. 209–16.

17. Alger, C., Huffman, W., Gordon, S., Prough, S., Sandkuhle, R., Yee, K., "Moisture Effects on Susceptibility to Package Cracking in Plastic Surface Mount Components." IPC Technical Review, February, 1988, pp. 21–28.

18. Yip, L., Massingil, T., Naini, H., "Moisture Sensitivity Evaluation of Ball Grid Array Packages." Proceedings of the 1996 Components and Technology Conference, pp. 829–35.

19. Ahn, E-C., Cho, T-J., H, H-S., C, T-G., Oh, S-Y., "Reliability of FBGA Package Using PCB Substrate." Proceedings of SPIE, v. 39 #6, 1999, pp. 599–604.

20. Giacobbe, F. W., "Moisture Absorption by Encapsulated Microchip Packages." Journal of Electronics Manufacturing, v. 10 #3, Sept., 2000, pp. 201–10.

21. Rowland, R., "An Overview of IPC Standards Related to Moisture-Sensitive Components." Journal of SMT, Oct–Dec., 2003, v. 16 #4, pp. 11–15.

22. Vaccaro, B. T., Shook, R. L., Gerlach, D. L., "The Impact of Lead-Free Reflow Temperatures on the Moisture Sensitivity Performance of Plastic Surface-Mount Packages." Journal of SMT, Oct–Dec, 2003, v. 16 #4, pp. 21–26.

23. Condra, L., Hoad, R., Humphrey, D., Brennon, T., Fink, J., Heebrik, J., Wilkinson, C., Narlborough, D., Das, D., Pendse, H., & Pecht, M., "Terminology on the Use of Electronic Parts Outside the Manufacturer's Specified Temperature Ranges." IEEE Transactions on Component and Packaging Technology, v. 22 #3, September, 1999, pp. 355–56.

24. Mishra, R., Keimasi, M., & Das, D., "The Temperature Ratings of Electronic Parts." Electronics Cooliing, v. 10 #1, February 2004, pp. 20–26.

chapter **3**

SMT Manufacturing Methods

Objectives

After reading and understanding this chapter, the reader should be able to:

- Describe a variety of SMT and mixed SMT + THT manufacturing processes
- Describe the steps involved in the various processes
- Describe appropriate decisions involving the selection of solder paste, placement techniques, reflow and wave soldering, cleaning, and testing of SMT assemblies
- Understand and perform calculations supporting the pin-in-paste (PIP) process

Introduction

Manufacturing methods have such an impact on design directions that we are devoting an entire chapter of this book to the manufacturing side of SMT. In order to make appropriate use of the variety of SMT devices in the design of a circuit assembly, designers must understand the assembly process. Each machine involved in the assembly process has potential impacts and limitations for PCB layout, component choices, and final size of the PCA. This chapter will introduce the designers to the SMT assembly process. The engineer truly new to SMT processes may also find that Chapter 7 in Intel's Packaging Handbook provides an short introduction to the processes.[1]

High flexibility, high levels of computer control, and design for manufacturability (DFM) are the engineering watch words that guide the design of modern SMT assembly facilities. Traditional high-volume production lines deliver low manufacturing cost through economy of scale. High-mix assembly lines make maximum use of the flexibility of many modern machines. This pass-through system's economy of scope makes cost-effective manufacturing of a large product mix possible.

Engineers who have recently entered the SMT arena often disagree about fundamental points of design methods, but there is one area where we find amazing unity. Experienced engineers agree that SMT is a manufacturing process-driven technology and therefore requires that design engineers understand these processes.

3.1　An Overview of the SMT Manufacturing Process

Based on the expected process steps as defined in Chapter 1 and components used in the design as defined in Chapter 2, development and manufacturing of an SMT PCB then involves a number of process steps:

- Circuit design, simulation, and verification
- Test development following design for testability (DFT) guidelines
- PCB design, including DFM and DFT concepts
- PCB fabrication and bare-board testing
- Attachment-media deposition and/or solder paste deposition
- Component placement
- Adhesive attachment-media curing, if necessary
- Soldering
- Cleaning
- Test

After these steps, boards passing final test are shipped. Depending on the relative costs involved, boards failing final test may be scrapped or may be disassembled, the noncomplying components/areas removed, reassembled, then retested using essentially the same process steps. Of course, this is a gross oversimplification of the actual manufacturing process flow, but this simple view is very useful in understanding SMT processing.

Fiducials

The automated equipment used for several of the processes to be discussed require that the machine "know" the location of the PCB to a certain level of accuracy. For low-volume machines that are dealing with THT components and standard-pitch SMDs, mechanical indexing is sufficient. In this case the board has been designed into it several indexing, or tooling, holes through which the machines will insert indexing/tooling pins to locate the board, typically to within 0.20 mm (0.008"). However, faster machines and those dealing with fine pitch components and their corresponding board footprints will need more accurate indexing. These machines use camera-based vision systems and accompanying vision software to examine the features on a circuit board and adjust the X-Y axis location accordingly.

The reader should note that indexing per se is not the only reason for vision systems. The PCB fabrication process has inherent inaccuracies, and it is not uncommon to have the features on a 6" × 9" PCB be as much as 0.2 mm off where the CAD data indicates they should be. Add this to the machine's mechanical indexing being as much as 0.2 mm off and the need for knowledge of the exact location of features becomes obvious.

Most machines today rely on fiducials being present in the copper layer of a PCB. Fiducials are vision targets used by the vision systems of the machines to determine the exact location of features. They must be in the copper layer since the machines that use them for indexing will be working with features in the copper layer, primarily the footprints. There are two types of fiducials:

- Global fiducials: used to index the entire board
- Local fiducials: typically used to index high I/O ICs with fine-pitch terminations

While fiducial shapes vary, most vision systems accept round fiducials in the external copper layer. Exact size and preferred shape of the fiducial is best determined from the machine/vision system supplier, although SMEMA Standard 3.1 "Fiducial Mark Standard" calls for a round fiducial. Typically, three global fiducials are used on a board and as many local fiducials as necessary to index parts. Global fiducials are used to set the PCB itself, and local fiducials are for more accurate indexing in the immediate vicinity of a fine- or ultra-fine pitch part. A typical board with fiducials might look like Figure 3-1. Note that a local fiducial can also serve as one of the global fiducials.

There are three global fiducials to allow the vision system to calculate exact PCB feature locations in both X and Y dimensions. Once this is done and the machine has adjusted its data accordingly, two local fiducials are sufficient to allow accurate placement of fine-pitch components. On many boards, one or two local fiducials are also used as global fiducials. Three global fiducials allow the automated systems to determine both the actual location of the board, as well as whether location of features on the board agree with the CAD data. Since the boards are made with a photographic process, there is always the possibility of some shrink or stretch of the feature locations—that is, the overall size of the fabricated board may be slightly smaller or slightly larger than the CAD data. This causes features to become closer or further apart than the CAD data. Since the fiducials are in the copper layer, a vision system can correct the location of solder paste deposition or parts placement for shrink or stretch as a result of the imaging and fabrication processes.

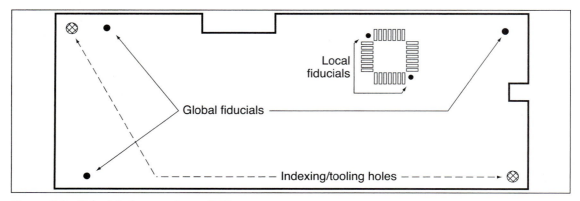

Figure 3-1 Fiducial placement on a PCB.

3.1.1   Attachment Media—Adhesives

Attachment media means the material that will hold the SMC in position after placement and prior to reflow. Depending on the media, it may also form the conductive and mechanical attachment between the pads on the PCB and the termination of the components. Attachment media may be solder paste, conductive adhesives, and non-conductive adhesives. Application of non-conductive adhesives, which typically perform the holding function prior to and during reflow, will be discussed later in this chapter. They may be cured by convective heat or by infrared (IR). Component-holding adhesives are applied under the body of the component. An excellent discussion on adhesives can be found in Prasad.[2]

Conductive adhesives are typically solder-bearing epoxy adhesives that perform both the tasks of creating a conductive connection and of creating a physical attachment. They may be conductive all the way through, or may be z-axis conductive only, dependent on proper z-axis downforce. They may be two-part air-cure, or one-part heat or IR cure. Since IR lamps are typically line-of-sight devices, any IR-cure adhesives must be placed on the PCB in locations where the adhesive location exposes at least part of the adhesive deposit to IR energy.

Attachment media deposition is dependent on the medium being used. However, in any event and regardless of the manner of deposition, the material must be and remain viscous enough to hold components in place until further processing steps permanently attach them. Non-conductive adhesives for component holding may be deposited by stencil, by syringe deposition, or by transfer pins.

3.1.2 Attachment Media—Solder Paste

Solder paste application is the first step in the reflow soldering process:

- Place a bare printed circuit board in a stencil printer or paste dispensing machine.
- Apply solder paste to the PC board by the stencil or dispensing process.
- Place the component on the PCB such that the component terminations are in the corresponding solder paste deposits.
- Reflow the solder paste by heating the PCB + components assembly until the paste reaches its melting temperature (also known as "liquidus") and flows at each termination and pad, then cools, creating a solder joint.

Solder paste is by far the most common attachment medium. While this discussion will confine itself to SnPb solder, please see the section on lead-free solder in Chapter 1 and the discussion on tin whiskers in Chapter 2. The primary reason for not discussing lead-free solder further is that at the time of this writing, virtually all the changeover issues are still in discussion. Therefore, any additional information we might provide here could be obsolete by the time you read this. Should the reader find that a decision has been made on SnPb replacement for your facility, please replace any SnPb guidelines with the appropriate guidelines for the replacement solder!

Also, in this discussion the reader should remember that the final word on the most appropriate chemistry and mechanical characteristics of a particular paste comes from the board itself after the soldering process is complete and inspection for appropriate quality levels is performed. Several industry standards address solder materials, flux materials, and tests both the paste itself and the final inspection results of the solder joint level. Additionally, the solder paste manufacturers have a wealth of knowledge for users.

Solder paste is a mixture of tiny spherical metallic solder particles, flux and its activators, and carrier vehicles. The components used in solder paste create a material that is thixatropic. Thixatropic material becomes less viscous and flows when pressure is applied but regains viscosity after the pressure is removed, preventing unwanted spreading of deposits. SnPb solder is most commonly used in its eutectic composition of 63 percent Sn, 37 percent Pb. The reason for the use of eutectic SnPb is two-fold: as shown in Figure 3-2, it has the lowest melting point of any SnPb composition and it has no "pasty" phase during the transition from solid phase to liquid phase.

Tin is a necessary component in most metal solders because molten tin has the ability to dissolve many other metals and it is a good conductor of electricity. This dissolution capability leads to the formation of the intermetallic layer. This intermetallic layer will be discussed further in Section 3.1.6.

In lead-bearing solder, the primary function of the lead component is to lower the melting temperature of pure tin, as also seen in the phase diagram in Figure 3-2 . The relatively low 183°C melting temperature of eutectic SnPb solder allows safe processing of the components and the common FR-4 substrate material used in most electronic assemblies today. It should be noted here that in the makeup of SnPb solder paste, one characteristic of good quality solder paste is the consistency of the size and the roundness of the spheres when examined under a microscope.

Other metal solder compositions include SnPbBi compounds, where the bismuth lowers

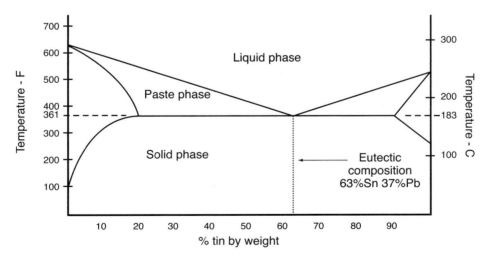

Figure 3-2   Phase diagram of SnPb solder paste.

the melting temperature, 63/36/2 SnPbAg, which reduces leaching of any silver-bearing component termination material into the solder; and 96/4 SnAg, which is a high-temperature solder with a melting point of 221°C. This is useful for tasks such as soldering thermocouples onto the PCB for the thermal profiling process, which will be discussed later.

All metallic solder paste compositions must include flux. The primary purpose of flux is to reduce oxides that form when any metal is exposed to the oxygen in the air (remember oxidation reduction in high school chemistry?). Any oxidation on the metals prevents the molecular action between the solder metals and the metal at the surface of the PCB pad or component termination. Some metals, like copper, create visible oxides rapidly when left uncovered, but even SnPb tinning on pads will oxidize and create poor solder joints if allowed enough time to oxidize (solderability tests can determine safe storage times). Flux reduces oxides of SnPb and of copper, although only of microscopic oxides. Any visible copper oxidation must be mechanically removed before flux can work. The reduction of oxides then allows both good wetting of the surfaces and proper formation of the intermetallic layer.

Wetting describes how well solder flows over the surface of the substrate and termination base metals and is effectively the opposite to the effects of surface tension. Since the surface tension of any liquid, including molten metal, wants to create a formation that occupies the smallest possible volume, it attempts to form a sphere. Much like the water drops formed on the highly polished hood of a car, surface tension of molten solder will attempt to form a solder ball if the surfaces it is resting on cannot be adequately wetted. The degree to which wetting counteracts the effects of surface tension is indicated by the angle at which the edge of the solder paste meets the pad. Poor wetting allows the surface tension to exert control over the form that molten solder takes. The first three examples shown in Figure 3-3 assume that a correct amount of solder paste has been deposited. Too much solder paste will overfill a pad and create an angle that may be misinterpreted as poor wetting.

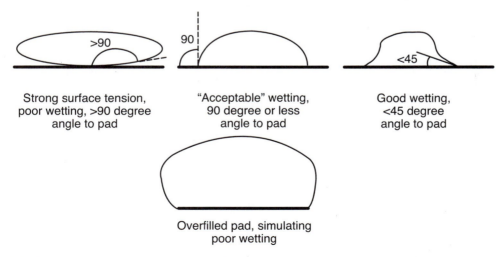

Strong surface tension,
poor wetting, >90 degree
angle to pad

"Acceptable" wetting,
90 degree or less
angle to pad

Good wetting,
<45 degree
angle to pad

Overfilled pad, simulating
poor wetting

Figure 3-3   Wetting versus surface tension and an overfilled pad.

Solder paste combines the component metals, flux for reduction of oxides, other cleaners, and various binders to create an optimal viscosity and stickiness. Common flux chemistries include Rosin Mildly Activated (RMA), the standard flux for years, no-clean fluxes, and water-soluble fluxes.

This may be a good time to note that there are three types of solder "balls" that will enter into our discussions of SMT design and manufacturing:

- The spherical terminations on the bottom of ball grid arrays
- The spherical metal balls that are included in the components of solder paste
- The unwanted metal balls that form on the PCB as a result of design and process problems

Be certain you understand which type of solder ball is being referred to in any SMT discussions, whether in this book or other sources. The unwanted post-reflow balls can be caused by many factors, including:

- Paste deposition volume
- Paste deposition location accuracy
- Component placement accuracy and downforce
- Reflow profile

Good design will help, but manufacturing issues are obviously the prime contributors to the formation of unwanted solder balls.

Fluxes

RMA leaves a yellow, non-conductive residue on the board. Although the residue is initially non-conductive and non-corrosive, RMA flux residue is sticky and will attract and absorb both dirt and moisture which can ultimately create a poorly conductive residue. The residue can also interfere with automated test equipment (ATE) spring-loaded test probes. RMA flux is derived from pine tree sap, then has various activators added, hence the MA term in the name. There is also an RA flux, which is more highly activated, leaves a more corrosive residue, and is rarely used in electronic soldering. RMA flux used to be cleaned with CFC-based cleaners, which are now prohibited by the Montreal Protocol. It can be cleaned with water and saponifiers (*sapo* is Latin for soap), which reacts with oils and other components of the rosin to make them water soluble.

No-clean flux is similar to RMA flux and is modified to minimize the amount of residue remaining on the board. It is intended to not require cleaning off the PCB surface since it has a lower solids content than RMA. This also reduces its aggressiveness as a cleaner, so when using no-clean flux it is necessary to pay more attention to handling and cleanliness of bare PCBs, and to maintain a tight process window. Some manufacturers of small components with low stand-off recommend cleaning of even no clean flux. No-clean fluxes can be rosin or resin-based. Rosin-based no-clean leaves a thin sticky residue (like RMA) that can be cleaned if desired. Resin-based no-clean leaves a thin, hard, residue that is very difficult to remove. Since oxides form in the presence of oxygen, many users of no-clean flux opt for blanketing their wave or reflow oven with nitrogen to drive oxygen from the area of the heating process.

Water-soluble flux is more aggressive in its cleaning capabilities than either RMA or no-clean and is based on the oxidation-removing abilities of organic acids. It must be cleaned off the surface of the board since its residue is still corrosive. Its chemistry is designed to be cleaned with deionized water, which then must be treated before disposal. The surface tension of pure water is higher than that of the solvents or saponifiers used for the removal RMA and no-clean fluxes. Therefore, it should be noted by the design team that low-standoff components may be difficult to clean under. See further information under "Cleaning."

Like all manufacturing decisions, selection of flux should be made in consultation with both the paste/flux supplier as well as the component suppliers. For instance, for certain of their SMT inductors, the muRata data sheet specifically says "Use rosin-based flux . . . Do not use water-soluble flux."[3] Ignorance of this sort of information guarantees problems later on.

With all these discussions of flux, it should be noted that solder paste also contains various other materials, principally binders and viscosity controllers. These will become important during the reflow soldering process since many of them will go through a gaseous state during the soldering process, and the speed with which they go through that state must be controlled. If it happens too fast, the gas expands fast enough to move portions of the solder paste.

A Few Lead-Free Issues

Lead-free solder, as introduced in Chapter 1, brings with it a number of issues. As this is written, the leading contenders are variations on an SnAgCu mix, including some with bismuth (Bi) as a fourth component. Bi is being strongly considered since it contributes two plusses to the process: it lowers the melting point of the solder to a temperature very similar to the melting point of SnPb solder, and it lowers the surface tension which contributes to better wetting—wetting being a recognized shortcoming of SnAgCu solder. However, the lower surface tension needs to be evaluated relative to SnPb tension, and it may force users of double-sided reflow to recalculate the "holding power" of the surface tension when inverted parts are again sent through the reflow oven and are expected to hang on to the bottomside of the board with the surface tension of the molten solder. Also unclear is bismuth's longer-term leaching characteristics, and its availability could be a problem since it is in much smaller supply than any of the other constituent materials being considered for lead-free solders.

As mentioned in Chapter 1, with the European directive to eliminate solder in most electronic assemblies (there are exceptions) by 2006, there are many lead-free references and articles available to designers and users. Among these is the lead-free page at the SMT in Focus site, http://www.smtinfocus.com/leadfree.html, which has many lead-free references listed.

3.1.3   Attachment Media Deposition

Solder paste may be placed by stencil or syringe dispensing. This will now be discussed in more detail, divided by the type of automated soldering method to be used.

Reflow Soldering

SMCs on reflow-processed assemblies are generally placed in preapplied solder paste. Thixatropic solder pastes may be dispensed by screen or stencil printing and by syringe/pump dispensing.

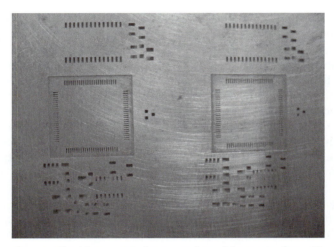

Figure 3-4   Example of a stencil with a stepped pattern for solder paste deposition.

Figure 3-4 shows a stencil of the type used for depositing solder paste with a "stepped" pattern for the QFP patterns. This means that the stencil is thinner in the immediate area of those patterns, resulting in less solder paste being deposited than would have been deposited had the stencil been a uniform thickness. Figure 3-5 shows a syringe deposition system.

Screen printing, analogous to the screen printing of T-shirts with ink, is not widely used. Its primary advantage is that in low-volume high-mix assembly situations, the screen does not need to be changed with a change in circuit board designs. Screens are not adequate for fine-pitch or smaller devices.

Figure 3-5   Syringe deposition system to deposit solder paste on an SMT board.

The vast majority of assembly today uses stencils in the assembly process. The stencil, made of brass, stainless steel, or other materials, has an opening of the appropriate size and shape at each component termination that is to have paste applied and be reflow soldered. While a different stencil is required for each circuit board, the accuracy with which stencils can be fabricated means that the correct quantity of solder paste can be deposited at the correct location on the bare PCB.

Reflow-soldered assemblies may also be produced by placing pre-tinned parts in sticky flux on pre-tinned boards, or by using a combination of viscous liquid flux and solder preforms. These methods are not widely used but have been proven worthwhile in solving certain application challenges. Certain ball-grid arrays can be successfully soldered with the application of only flux on the board.

Wave Soldering

Surface-mounted components on wave-soldered boards are generally placed in adhesive. After being cured, the adhesive holds the SMCs in place as they are passed through a solder bath, which makes electrical connections between the components and board, and which reinforces the mechanical bond supplied by the adhesive. Through-hole components that may also be on the board are held in place by virtue of their leads being inserted in the appropriate PCB through holes and then having the leads cut and clenched in place. Generally, no further attachment or attachment medium is required.

Thixatropic adhesives may be applied by all the methods used with solder pastes. Nonthixatropic adhesives are syringe or pump dispensed or are offset printed.

When considering the use of adhesives to hold parts in place, you should also consider if the parts may need to be reworked or repaired. Rework or repair of glued-on parts will require knowledge of the appropriate rework and repair techniques by the personnel involved.

3.1.4   Component Placement

After deposition of solder paste, placement of SMT components is next. Most current automated placement machines use a vacuum tool, sometimes called a pipette, to pick components out of the component holders, carry the component to the proper location over the PCB, and place the components.

It is important to note that the component is intended to be placed *into* the solder paste deposit, not just set on top of it. In between the pick action and the place action, the component must either be mechanically squared on the pipette (brought into alignment with the X-Y axes of the machine) or pass over a component-recognition camera to allow the head to rotate the pipette and square the component prior to placement. Acceptable placement requires consideration of a number of issues. The most common factors include:

- Equipment:
  - The choice of placement machine
  - The accuracy of the X-Y axes of the machine
  - The accuracy of the placement head and its associated nozzles
  - Overall placement accuracy
  - Vision system(s) needs and accuracy

- Equipment Ancillaries:
  - Feeders, their compatibility with components and component packaging to be used
  - Nozzles, suitable for the components to be placed
  - PCB support to prevent deflection with the downforce of the placement process
- Materials and Documentation:
  - Components, including coplanarity and identification to be viewed by the vision system (if any)
  - Component holders, e.g. tape and reel, stick, bulk, matrix tray
  - PCB artwork shrink and stretch, and the ability of the vision system to compensate
  - PCB flatness
  - Footprint flatness on the PCB, especially for high lead-count devices.
  - Solder paste, as the assembly was designed to use
  - Adhesive, if used, and any curing method necessary
- Operating Areas and Personnel:
  - Temperature
  - Humidity
  - Cleanliness
  - Operator training and knowledge of the processes being used
  - Quality control techniques in place

Current automated SMT placement machines fall into several categories. These include:

- Linear X-Y systems with a single head and one or more vacuum pickup tools and a fixed PCB
- Linear X-Y systems with one or two vertical rotary heads and a fixed PCB
- Fixed rotary turret heads with X-Y table for PCB indexing and feeder indexing

Figure 3-6 shows a single pipette placement head with two sets of squaring jaws to square the component to the machine's X-Y axis prior to placement. This is a low-volume automatic place-

Figure 3-6   Example of a pick-and-place tool of a linear X-Y SMT placement machine with a single head.

ment machine. Medium- and high-volume machines use both vision systems for detecting part rotation and squaring, and may also use multiple pipette heads in either a linear or rotary array.

Examples of manufacturers of low- and medium-volume equipment that typically use the linear X-Y system with a single head include MannCorp, AIC Technologies, and Automated Production Systems (APS). Examples of manufacturers of medium- and high-volume equipment include Philips/Assembleon, Zevatech/Juki, Siemans/Siplace, Universal, Fuji, and Panasonic/Create. Each manufacturer has one or more of the types of placement machines previously discussed. Their respective websites provide detailed information on the operation of each type of machine.

For engineering prototyping and very low-volume assembly or for the assembly of boards with very low population counts, hand assembly is often used. Some low-volume assembly is also handled by assisted placement using a machine to present the correct components and possibly show on a TV screen the placement locations for a hand-assembly operator. One example of an assisted-placement machine is shown in Figure 3-7. Examples of manufacturers of placement machines that are suitable for assisted hand placement include MannCorp and APS. The reader should note that all SMT placement machines are rated in components placed per hour, but that these numbers are normally optimistic relative to day-to-day performance in a produc-

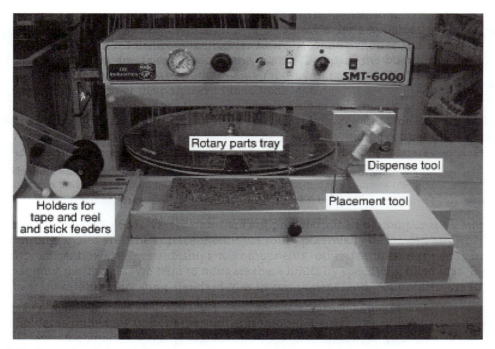

Figure 3-7   Example of a hand-placement system.

tion line. While under ideal conditions most machines will actually place at their stated rate, issues such as these prevent those rates from being realized under normal conditions:

- Slower machines upstream or downstream
- PCB loading and unloading time
- Changeout of feeder reels, tubes, or trays
- Optimal feeder locations and pad locations
- Optimal component mix
- Operator errors

Automated pick-and-place machines select, orient, and place on the board a variety of components under software control. To do this, they must have the board and the component square with the machine's X-Y axes or be able to detect the actual orientation of both and correct for any out-of-square condition. These two tasks can be accomplished in several ways:

- The PCB can be squared to the machine using tooling holes in the board and corresponding tooling posts located on the machine.
- The PCB can have vision targets, called fiducials, designed into the copper layer of the board. Fiducials are viewed by a downward-viewing camera or laser. If the board is out of square, (which means each component footprint on the PCB is also out-of-square by the same amount), the placement tool will rotate each component by that amount prior to placement.
- The head may have a set of mechanical jaws to square each component prior to placement or it may use a set of squaring heads on the machine itself.
- The head may fly the component over an upward-viewing camera which, in conjunction with its associated software, will calculate the amount the component is out of square and cause the placement tool to rotate a corresponding amount prior to placement.
- The placement machine may use a laser or LED-based system to evaluate the orientation of the component and/or to evaluate the co-planarity of IC components.

Placement machines typically lock the board into place with edge clamps that force the board up against the rails above the conveyor. At this point, we have seen that there are several features that must be considered during PCB design:

- Any required fiducials must be designed into the copper layer of the board. The placement machine documentation should be consulted to determine the optimal shape and size of the fiducials. Three fiducials are typically located on each side of the board on which components will be placed. These may be any combination of global and local fiducials.
- Any tooling holes must be located in the design.
- The edges of the board must be kept clear of components and traces ("keep-out" areas) to insure that the edge clamps do not cause damage during the clamping process

Pick-and-place machines are available to fill virtually any application requirement. Machines can handle a manufacturing volume ranging from a thousand components to upwards of a million components per 24-hour day. Some machines handle only a restricted range of similar components while others handle a very wide range of component styles and sizes. Pick-and-place machines also differ greatly in their level of machine intelligence (their ability to sense and adapt to changes in the process environment). Finally, a matter of little surprise, equipment varies widely in price. Automatic placement equipment starts in the $20,000 range and goes upward to the multiple $100Ks. In general, prices increase with the four desirable features: throughput, intelligence, flexibility, and reliability (accuracy, repeatability, and consistent operation).

Placement Troubleshooting

There are a number of failures that are identified in finished PWAs that result from errors during the placement process. These can include:

- missing component(s).
- component out of alignment.
- bridging, if due to component misplacement rather than excess solder paste.
- tombstoning, if only one end of a chip component is placed in solder paste. Related to this is skewing if a component is placed with its end terminations unequally in their respective solder paste deposits.
- solder ball formation if a result of too much down force which forces some of the solder paste off the pad and onto the solder mask. If on the mask, the surface tension of the molten solder may create solder balls.

3.1.5   Attachment Media Curing

In some processes, after component placement, the attachment media (solder paste or adhesive) must be cured before the soldering step. This generally occurs with solder paste if vapor-phase ovens will be used for reflow, and it occurs for all adhesives. It may or may not occur with solder paste if IR ovens are used, and it is never done if convection ovens are used. The solder paste manufacturers' data sheets and the oven manufacturer's documentation should be consulted to determine the necessity for paste curing.

Solder Paste

Solder paste curing is accomplished by heating the assemblies with the components placed in wet paste. Temperatures and soak times vary dependent on materials, but they generally run between 80°C and 150°C for 15 to 60 minutes. Where curing times, temperatures, or assembly materials dictate, curing may be done in a nitrogen or forming-gas atmosphere to prevent excessive oxidation of assembly materials and/or solder.

Curing evaporates the volatiles from solder paste, thereby preventing a rapid boil off that might create defects in certain soldering processes. Higher-temperature curing may also begin flux activation with certain pastes.

Curing has been shown to dramatically reduce solder defects in vapor-phase soldering where steep temperature gradients can cause rapid outgassing of flux volatiles in uncured paste. Escaping gas bubbles have been shown to increase tombstoning, produce solder balls, and leave voids in solder joints.[4]

To be accurate yet brief, we cannot cover specific data on all the solder paste formulations used in surface mounting, especially the newer lead-free varieties which are still undergoing development as this is being written. Pastes may be made from varying combinations of tin, lead, silver, indium, gold, palladium, etc. Rosin, organic acids, halides, and synthetic materials are used as fluxes. The activity levels of the fluxes can be varied from unactivated to highly active. In addition, thixatropic vehicles are made of a wide range of materials in varying viscosities and with distinct properties. To further complicate matters, there are infinite permutations and combinations of these basic materials. To cover them all would require a separate and exhaustive book. Solder paste and process equipment vendors are an excellent source of data on the materials to be used. In Section 3.3, we will introduce more information on paste, fluxes, and lead-free soldering.

Adhesives

Since very few air-activated, or aerobic, adhesives are used in SMT component attachment, curing is a common step used in the wave-soldering assembly process to accelerate crosslinking of the epoxy adhesives. Some adhesives are cured by heat, some by exposure to infrared radiation, and some by a combination of the two.

3.1.6   Soldering

After placement (and after curing where required), the next major process step to consider is soldering. Below, we will briefly cover the common methods used in SMT soldering. Your application and the components on the board will determine the choice of soldering method. Process selection is such an important fact of design that the full SMT team should participate in the selection.

First, members of the design team should recognize one common requirement for any SMT soldering system that will create solid high-quality solder joints: an appropriate thermal profile. As discussed earlier, the soldering paste contains various components that either start as liquid or turn to liquid then become gaseous at soldering temperatures. These components must be given time to evaporate off at a controlled rate or they can create several problems:

- The gases can be entrapped in the solder joint, weakening it
- Rapidly expand/explode and force small quantities of solder paste out of the deposit and onto the solder mask where they will form solder balls

Variations of these problems can occur when using virtually any type of soldering system, including hand soldering, laser soldering, wave soldering, and the several types of reflow soldering.

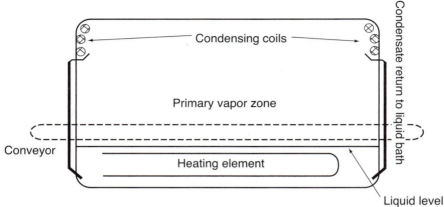

Figure 3-8   Vapor phase soldering (VPS) oven schematic.

Reflow Heating Technologies

Reflow soldering describes the application of sufficient heat to an assembly to melt pre-applied solder. The solder is usually in the form of paste, but it may be in preforms or pre-tinned components and boards. The term *"reflow"* probably stems from soldering by the heating of pre-applied solid solder as opposed to the application of molten solder to assemblies (flow or wave soldering). The solder paste "flowed" once when it was applied, and now it will "reflow" as it is heated to liquidus temperature.

Reflow soldering may be accomplished in several ways. The most common reflow approaches include vapor-phase soldering (VPS)[5,6] shown in Figure 3-8, convective shown in Figure 3-9, nonfocused IR (panel and T-3 lamp), focused IR (laser and spot-focused lamp), laser, and conductive. Each of these methods is characterized by a somewhat different heat-transfer mode:

- Vapor phase uses the latent heat of condensation to heat the assembly. It gives even heating during the reflow stage but is less controlled during preheating. It has the advantage that the reflow environment is inherently inert.

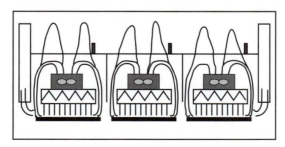

Figure 3-9   Convective/circulating hot air oven schematic.

- Infrared uses line-of-sight radiant heat transfer, although most IR ovens have some convective transfer due to the thermal movement of air. IR tends to heat unevenly since radiant heat is absorbed based on, e.g., the color of the item under the elements.
- Laser is spot heating of individual leads.
- Conductive is forced air heating, and is considered by many to be the most even heating technology.

The previous statements are general comments on these styles of heating and are not meant to preclude an in-depth study of reflow technologies by potential users.

At this stage of development of the reflow processes, full convection ovens are generally considered to be the most reliable in the quality of the formed solder joint.

In reflow soldering, the PC assembly with its applied solder paste is heated along the appropriate thermal profile until the melting point of the metals in the paste (the *liquidus* point) is passed, then it is cooled until the solder metals harden and create a permanent interconnection between the pads on the board and the terminations of the component. Either during this process or earlier during the pad tinning process, the tin in the molten solder paste interacts with the copper on the pads to create an intermetallic layer at the junction of the copper pad and the solder metals.

Reflow Thermal Profile

The traditional thermal profile for a reflow oven is shown in Figure 3-10. It consists of a preheat ramp, a soak or activation plateau, and a rapid reflow bump that is a rapid heating past and above the point of melting the metals in the paste (or the "liquidus" temperature) followed

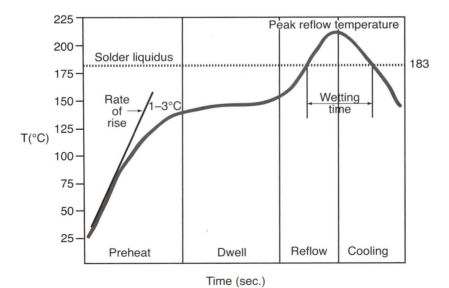

Figure 3-10   Traditional SMT reflow oven ramp-to-dwell ramp-to-peak (RDRP) thermal profile. *(Courtesy Research International)*

immediately by a cooling phase. This is sometimes known as a "ramp-to-dwell—ramp-to-peak" or RDRP profile and was developed when IR heating was the primary reflow oven technology. The level of the dwell or soak time varies by the flux technology. As examples, some manufacturers of OA water-soluble flux suggest a 160°C soak time, which also drives off a number of the solubles in the paste. Manufacturers of polymer-based water-soluble flux may recommend a 140°C soak temperature.

However, as convection heating has shown itself to be the superior heating technology, work done by a number of companies and researchers has shown deficiencies with the RDRP profile. The work of Lee[7] and others has shown that the RDRP profile may not be optimum for RMA-flux pastes and that a profile with a continuous gentle ramp-to-reflow results in higher yields from a convection oven. Shown in Figure 3-11, the ramp-to-reflow (RR) profile has a steady ramp to the liquidus temperature (183°C for eutectic SnPb), a brief dwell at liquidus, then a smaller reflow bump and cooldown. While the total time to the peak of the reflow bump is similar in both profiles and the ramp rate to the reflow peak is similar, the initial ramp of the RDRP profile is considerably faster than the continuous ramp of the RR profile. It should be noted that all reflow profiles should be measured at selected solder joints on the PCB. Whether using the RDRP or the RR profile, the cooling rate should be kept no greater than 4°C/second. A fast cooldown rate not exceeding this will result in a finer grain structure in the solder joint, which results in a stronger and shinier joint. However, cooldown faster than 4°C/second can result in thermal shock and damage to the assembly.

The original RDRP profile was developed when IR ovens were the norm, and it creates significant yield problems due to the relative unevenness of its heating and its relatively steep

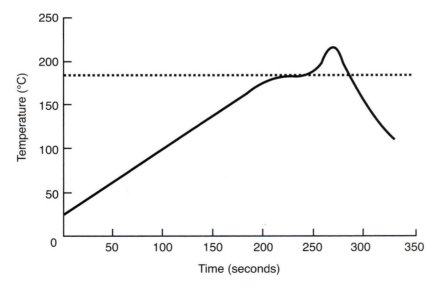

Figure 3-11   The ramp-to-reflow (RR) thermal profile.

positive ramps. In the original IR ovens, the initial heating followed by the lengthy dwell at 140° to 160°C was thought to be required due to the uneven heating of those ovens. The relatively long dwell time allowed the temperature to equalize across the oven and the board and allowed time for the flux to activate and reduce metallic oxides. However, this profile also contributes to several of the problems seen after reflow:

- Escape solder balls: Due to the rapid ramp after dwell in the RDRP profile, the solvents present in solder paste can rapidly heat to their gaseous transition point, at which time the escape of the gases blows small amounts of paste out of the deposit on a given pad. This paste reflows on the solder mask, and due to the lack of solder adherence to the mask, the surface tension of the molten paste causes it to form into a spherical solder ball. This type of escape solder ball can be difficult to remove during the cleaning phase. It was noted in the section on solder paste deposition that stencil design can also affect escape solder balls.
- Tombstoning: Shown in Figure 3-12, tombstoning (also known as the Manhattan effect) occurs during the rapid ramp phase. This will also be discussed under "Reflow Problems." If during travel through a reflow oven the solder paste on the leading edge reflows before the solder paste on the trailing edge, the surface tension of the reflowed solder can actually lift the other end out of its paste.
- Solder wicking: Solder wicking occurs when the component termination (particularly IC leads) heats faster than its corresponding pad due to the additional thermal mass of the pad on the substrate. When the termination reaches reflow temperature before the pad does, the paste in contact with it melts and wicks up the termination, thereby starving the pad and leaving an open condition between the pad and the lead.

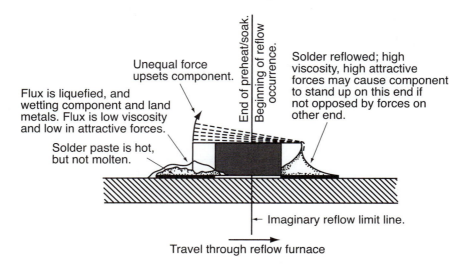

Figure 3-12   Diagram showing how uneven reflow can contribute to tombstone defects.

- Hot slumping: Slumping of paste deposits can occur directly after deposition, in which case the problem is with the solder paste composition and/or the stencil. Slumping of paste during the heating process can result from the rapid initial ramp and the long dwell or soak time.

Lee's work also notes that a target peak reflow temperature of 215°C is preferable to higher temperatures, as is a relatively short time above the liquidus point. In his own work, one of the authors (Blackwell) has found that in over 5 years of working with polymer-based RMA flux paste in a full-convection oven using a profile between the RDRP and RR profiles, no tombstones and no wicking have been observed.

In their confirmation of Lee's work, Dahle and Lasky[8] recommend RR profile characteristics similar to the following:

- The initial ramp at 0.5° to 1.0°C/second to 175°C
- The gradual rise through liquidus (183°C for eutectic SnPb solder paste) at <0.25°C/second for less than 45 seconds
- Reflow ramp to 215°C peak, ± 5°C at 1 to 2°C/second
- Ramp down at 2 to 4°C/second, with total time above solder liquidus of 30 to 90 seconds

While you must determine the best profile for your product, the best beginning profile is that recommended by the paste supplier. Your oven's ability to provide any particular profile is dependent upon its type of control system and the final arbiter of the paste, and the profile is the quality of the solder joints produced. Any deficiencies at the solder joint must be solved considering the paste and flux composition, the oven profile, and the plating characteristics of both the substrate and the component terminations. Each facility and each product is unique and must be treated as such.

Thermal Profiling

To determine the thermal profile that your product is exposed to during the reflow process, thermocouples are attached to appropriate locations on the printed circuit board, typically with high-temperature solder. The time-temperature profile is then recorded in one of a number of ways: to a recorder, in a profiling "mole," on a PC, etc. The quality of the thermal profile depends on appropriate selection of profiling locations and following the correct procedures before and during the profile process.

While it is the heat carried in the air as indicated by the temperature of the air in a convection oven that does the actual heating of the PCB and components, it is important to note that it is the temperatures of features on the board that are of interest in determining the quality of the soldering process, not the temperature of the air in the oven. Therefore, profiling must measure only the temperature of PCB features.

Items that must be taken into account when determining profiling locations include:

- Component with recognized heating issues. An example is BGAs due to difficulty getting heat transferred to inner ball locations. A monitoring thermocouple should be located at the inner balls. Another example of a component with heating issues is any device that uses a large heat sink, whether the heat sink is attached to the body as on a large processor or a power semiconductor device with a large heat sink plane on the circuit board surface.

- Variations in component thermal mass. A low thermal mass component and a large (ideally the largest) thermal mass component on the board must be monitored. Since most commercial profile monitoring systems will monitor at least five thermocouples, "large" thermal mass can be determined by a single thermal mass such as a device with a heat slug or by a device with a set of thermal vias, such as a BGA with centered thermal vias. In any event, choose a mix of components with some that are likely to reflow early and some that are likely to be the last to come up to reflow temperature. In this fashion, you can determine if some components are likely to be overheated and if some do not adequately exceed the liquidus temperature of the solder.

- Variations in PCB thermal mass. When using PCBs with one or more large metallic planes, whether they are internal or on the surface, recognize that they will act as heatsink(s) and absorb enough heat to affect the profile in their immediate area. For example, an area of a PCB with a surface plane for a TO-252/DPAK device will be slower to rise in temperature since it takes more heat input to raise the temperature of this large thermal mass. Components in the immediate area of a localized plane must be monitored.

Process issues that can affect the quality of profiling include:

- Thermocouple attachment. Attachment issues are related to how thermocouples measure temperature. A thermocouple consists of two wires, each of a different metal, that create an electrical EMF at the point where the wires are joined. No power supply is required. However the temperature versus voltage curve that results is nonlinear, and conditioning circuitry is added that both linearizes the voltage generated by the thermocouple wires and amplifies it to a standard value, such as 10 mv/°C. A thermocouple's measurement point is where the two wires join. The thermocouple is attached to the measurement point using a high-temperature solder that will not melt during the solder process. If during the attachment process the solder attaching the wires to the board is inadvertently allowed to run up the wires, the voltage generated (and thereby the temperature measured) is an average over the joined distance. This means that part of the measurement is the surrounding air temperature, and that is not a measurement of interest for profiling. Epoxy is sometimes used to secure the wires but cannot be allowed to cover the measurement point. Since epoxy is a very poor conductor of heat compared to metal, the measurement will be incorrect.

- Thermocouple routing is important. The metal wires in the thermocouple (T/C) absorb heat from the heating process and can affect the local profile. T/C wires must be routed away from any measurement points on the PCB. Since large components are usually mounted on the top of the board (typically for reflow heating), the T/C wires should be routed on the bottom of the PCB away from any measurement points and the most critical components. For a wave solder process, T/C wire routing is on the top of the board so that heat is not absorbed from the waves and the flow of the waves is not affected by the T/C wires.

- If fine uninsulated T/C wires are used to monitor underneath low-profile components or individual balls on a BGA, the wires must be insulated from each other, typically through the use of Kapton tape. If the wires touch at a location other than the measurement point, again the voltage generated is proportional to the average of the temperatures at the touching points.

In general, the profiling process must be controlled as carefully as the manufacturing process that is being monitored. Failure to do so results in temperature data that is faulty and will result in incorrect process decisions.

Reflow Soldering Problems

Tombstoning, mentioned earlier, is one of the chief gremlins in reflow soldering, assuming the proper solder-paste deposits are present. This defect is produced by a combination of the wetting and surface-tension forces in molten solder. It occurs when forces at one end of a component are not reasonably balanced by forces at the other end. The imbalance may be produced by land-geometry variations, component-termination variations, or time/temperature reflow variations. Whatever the source, when the solder-force imbalance is great enough, tombstoning is the predictable result. Figure 3-12 illustrates how uneven reflow can be a factor in tombstoning.

We will consider imbalances from land geometries or component terminations in the PCB design discussion in Chapter 5. Here, we are concerned with recognizing the tombstone phenomenom in the reflow process.

If tombstoning plagues a reflow operation, you may resort to batch (non-conveyorized) reflow to avoid the moving reflow limit-line effect, or you may glue or fixture the offending components to prevent their movement during reflow.

Intermetallic Layer Formation

The intermetallic layer forms between the base Cu metal of pads and terminations/leads and the Sn of SnPb solder. It forms both the mechanical and electrical integrity of the solder joint. It is created at soldering temperatures from the action of molten tin on copper and is optimally 1 to 2 microns thick. The thickness of the layer is controlled primarily by the time above liquidus during the initial tinning and during the wave and/or reflow process. Generally this time should be limited to 45 to 60 seconds.

One of the reasons that tin is a component of virtually all solders is its ability to interact with many other metals when molten. Many electronic components use pure copper or a metallic copper alloy for the leads/terminations, and the pads on the PCB are typically copper. During the process of tinning the leads or during the soldering process, the molten tin in the solder has a solvent reaction with the copper, forming a set of alloys composed of Sn and Cu, the

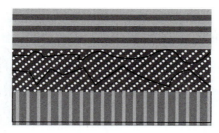

Figure 3-13   The intermetallic layer (IML). *(Courtesy Research International)*

intermetallic layer or IML shown in Figure 3-13. Near the copper pad, the formation is Cu_3Sn, and near the solder itself the formation is Cu_6Sn_5. The formation must take place rapidly at reflow temperatures to assure a good IML structure and a strong solder joint.

The IML is both a blessing and a curse in the soldering process. It is a blessing since it is the IML (the intermediate layer between the wetted copper and the solder) that creates the bond between the copper of the terminations and pad and the SnPb of the solder itself. A properly formed IML creates the strength of this bond. It is a curse because the IML is a crystalline structure that continues to grow in the presence of elevated temperatures. It grows in thickness very little below 125°C, but if the soldering temperatures are too high or held high for too long or if the product is used in the presence of continuous elevated temperatures, the IML continues to grow. If it becomes too thick, its crystalline structure becomes brittle, and the risk of solder joint cracking increases. In truth, the IML continues to grow at all temperatures, but its growth at room temperatures is so slow that it is not an issue. Here again, the higher reflow temperatures required for many of the lead-free solder formulations may present additional brittle joint problems. Suraski[9] discusses IML formation in SnPb solder, while Shin et al.[10] discusses IML formation in high-temperature Sn-Cu solder joints. Rouband, et al.[11] discuss IML growth using four different lead-free solders with plastic BGAs.

Flow Soldering

SMT places new requirements on flow (or wave) soldering, and it generally requires specialized equipment to meet these requirements. SMT flow-soldering systems must solve two opposing problems inherent with the soldering of SMCs on a board's bottom (solder) side. These are voiding (too little solder) due to gas entrapment and eddy currents around the SMC, and bridging (too much solder) between adjacent SMC terminations. Virtually all wave soldering machines designed to provide quality solder joints use a dual-wave-soldering system with a laminar wave and a turbulent wave. This has proven to be the best wave system. A dual-wave-soldering system

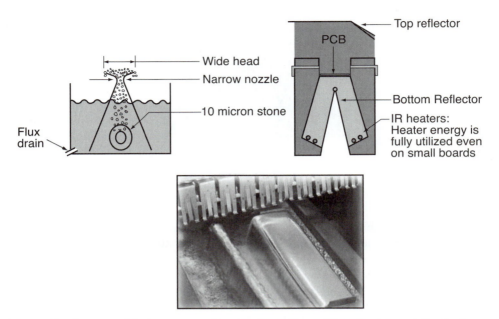

Figure 3-14   Dual-wave-soldering system showing foam fluxer, IR preheater, and the dual-wave: turbulent wave for SMT parts and laminar wave for THT parts and to finish SMT parts. *(Courtesy NovaStar)*

is shown schematically in Figure 3-14. Like most soldering operations, successful use of a wave-solder system requires:

- An appropriate design
- Good equipment
- Proper fluxing procedures prior to the wave since the waves in a wave soldering system do not include flux in the molten solder
- Proper preheat to prevent thermal shocks to the board and the components
- Proper temperatures in the various zones to create an appropriate thermal profile

The important process parameters in a wave-soldering process are concerned with time and temperature relationships. The wave-solder process must be profiled just like the reflow process as discussed previously under the reflow process. The process steps in wave soldering are:

- Preheating: The board and components assembly must be preheated to prevent thermal shock as a result of sudden contact with a 220° to 240°C molten solder wave. Maximum slope should be 2°C/second, and board temperature should reach a minimum of 90°C before the board enters the wave.
- Fluxing: Since the molten solder waves carry no flux, flux is sprayed onto the bottom of the board during the preheat process. Preheat temperatures do not activate the flux, but the rapid heating in the wave does.

- Soldering: Passage through the wave heats the pads, through holes and vias of the boards, and leads/terminations of the components to soldering temperature above the melting point of eutectic solder
- Cooldown: The soldered assembly must cool down below liquidus temperature before moving off the wave-solder machine. This may or may not include forced cooling. If the heat from the carrier causes the assembly temperature to continue to rise after it has left the wave, forced cooling is necessary at the end of the wave machine.

Wave-solder process parameter timing shown in Figure 3-15 includes:

- Preheating and preheat time
- Wetting Time: The time from the first contact of the board and component leads with the molten solder until complete wetting of the metallic parts by molten solder has occurred
- Soldering time: The amount of time from first contact of the board and component leads with the molten solder until the time the solder begins to solidify
- Dwell time: The moment the board and leads spend in actual contact with the wave

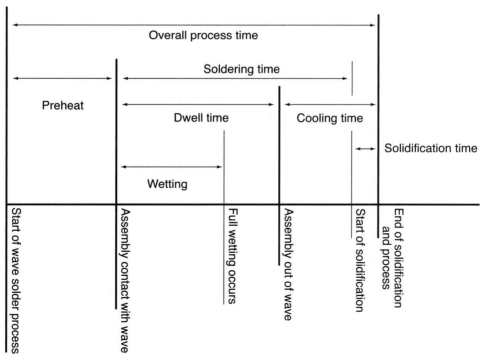

Figure 3-15   Wave solder timing.

- Solidification time: When the assembly is no longer in contact with the wave, the time between when the solder first begins to solidify and when complete solidification has occurred
- Cooling time: The time from when contact with the molten wave is lost until complete solidification has occurred

Not surprisingly, wave soldering invokes its own problems and own set of component orientation rules. Here, one of the driving considerations is in not creating eddy currents that can cause solder skips by blocking the solder from some terminations as the assembly moves through the molten solder. The other primary concern is to position the parts to avoid bridging. Since one defect involves too little solder deposited while the other is due to too much solder, the solution to one problem may work against the other. Rules for PCB layout with the best compromise will be discussed in Chapter 5.

Today's dual-wave-solder machines may obviate the need for some of these rules. However, the production environment, state of repair of equipment, operator training, and assembly vendor selection throughout the entire life of a product are much less under the control of the circuit designer than is the layout of the board. Once these layout rules are well understood, their cost is minimal and needs be paid only once per design. The expense of violating them can be substantial and will be exacted many times per day over the life of the manufacturing cycle.

Lead free soldering issues raise concerns with the wave solder process as they do with reflow solder process. A study of lead free wave soldering was reported by Aim Solder (www.aimsolder.com). In a test to determine the dissolution of copper in the wave solder pot, they compared Sn/Ag3/Cu0.5 (LF218) and Sn/Ag2.5/Cu0.8/Sb0.5 (CASTIN). Over a 30-minute test the Sn/Ag3/Cu0.5 alloy resulted in twice the copper loss than did the Sn/Ag2.5/Cu0.8/Sb0.5 alloy—an average over three test coupons of 1.84 grams compared to 0.86 grams for the antimony-containing alloy.

Aim also compared wave solder pot maintenance issues. The copper dissolution and the fact that the density of either lead-free alloy is lower than that of Sn63/Pb37 solder means that the Cu6Sn5 intermetallics that form in the pot will sink to the bottom rather than float on top as they do with lead-based solder. This by itself means far more difficult and expensive wave solder pot maintenance.

Figure 3-16 shows the result of the Aim testing of the copper dissolution rate for several lead-free solders relative to Sn63/Pb37 solder. Note that while the lead free solder containing antimony has a lower dissolution rate than Sn63/Pb37, the Sn/Cu/Ag lead free solders have a higher dissolution rate. And all the lead free solders will have the solder pot maintenance problems already noted. The final sentence in the Aim study notes "Regardless of the solder alloy or barrier metal chosen, waste and changeover costs will dramatically increase with lead-free wave soldering."

Nihon Superior in Japan has developed a Sn/Cu/Ni alloy that appears to offer advantages compared to the Sn/Cu/Ag solder selected by NEMI. Readers should investigate all these aspects of the conversion to lead free wave soldering before embarking on this hazardous voyage.

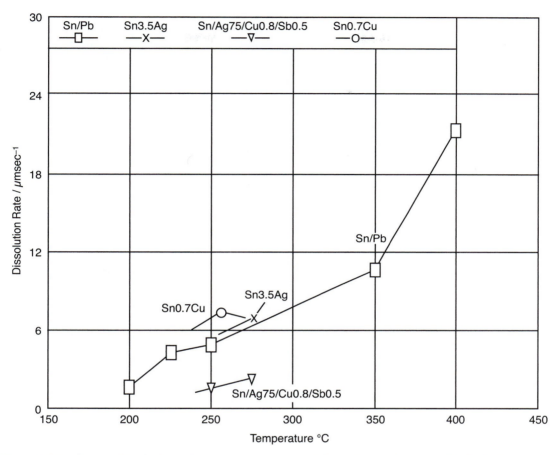

Figure 3-16   Copper dissolution rate versus temperature. *(Used with permission of AIM Solder Inc.)*

3.1.7   Cleaning

Component "standoff" distances provide clearance, allowing the cleaning solvents space in which to flow easily under components and remove harmful contaminants. Many SMCs mount very close to the board. This makes cleaning under the components a special concern in the engineering of the SMT cleaning process. Leadless components are particularly troublesome since they have no leads that can be used to provide standoff. And, leadless chip carriers are the most difficult because they have little standoff, cover a large board area, and have interconnections that form a picket fence around all four sides, which seriously retard cleaning agent flow.

In the early days of SMT, cleaning was a challenge, but the approach was obvious. Everybody knew that you used solvent cleaners to remove the soils from under SMCs with their tiny cleaning clearances. The solvent of choice was almost universally CFC based. Of course, concern over the depletion of the earth's ozone layer by CFCs has changed all this. As the Montreal Accords clicked into full effect, there was considerable confusion and debate over the proper way to clean SMT assemblies; but the one nearly universal given was that the CFC route was not it.

Cleaning Strategies

Electronics manufacturers today have numerous cleaning strategy choices available to them. They can select from:

1. Aqueous cleaning with saponifiers
2. Aqueous cleaning without saponifiers
3. No-clean solder/flux systems
4. Semi-aqueous cleaning
5. Solvent cleaning with ozone-friendly chemistries

Which one is best? We will pass on the chance to take a firm stand and just provide a list of considerations in your selection process.

Aqueous cleaning with saponifiers can be used with RMA and synthetic-activated (SA) fluxes. The bonus of this approach is that a wide range of solder/flux formulations may be selected, active SA fluxes can be used where poor solderability is a problem, and finished work can be of excellent, clean, shiny-solder appearance. The downside is that an increasing number of jurisdictions impose stringent environmental regulations on the contaminants allowed in water dumped into their sewers. In many cases, the water you return to the sewer must be cleaner than the water the locality supplies you from the tap. The result is that very costly waste-water treatment plants or closed-loop water systems may be a requirement. Finally, water molecules are relatively small items, and it would seem they would fit into the tiniest of spacers. However, water has a very high surface tension, and that property of water tends to inhibit its penetrating capacity. The saponifier is added to reduce the surface tension of the water and increase its penetration and, thus, its cleaning action in the tight spaces so prevalent in SMT assemblies. Unfortunately, cleanliness requirements often force us to remove the saponifiers from the assembly, and we cannot do this with saponified water. Deionized water is used and, again, we are up against the surface-tension problem. The answer to the apparent conundrum is in design of assemblies with sufficient component standoff to permit cleaning.

Aqueous cleaning without saponifiers follows the discussion immediately above but adds limitations on the types of flux formulations permitted. If no saponifier is used, the water solubility of the flux residue must be adequate to ensure full removal during the cleaning operation.

No-clean soldering is an attractive method. Where there is no very high-impedance circuitry on a board, and where use environments are reasonably dry, it is often possible to simply forego cleaning. In such cases, a mildly activated rosin flux is used, and the solder formulation is carefully selected for a balance of printing, wetting, fluxing, and low residue characteristics. The flux chemistry should be such that residues are very low in ionic content, and that any ionics that remain after soldering are rendered harmless by the encapsulating effect of the rosin base of the flux. The benefit of the no-clean method is that the cost and difficulty of thorough cleaning are eliminated. Downsides of this approach are that no-clean solders tend to do poorly unless the solderability of all parts and boards is very good, and the residue that remains may interfere with test probing or contaminant probes. Also, the circuit still looks as if it needs cleaning, and this bothers some veteran inspectors. Caution: do not clean a no-clean for the sake of looks. In doing so, you may defeat the encapsulating effect of the flux residue and leave exposed ionic contaminants in a position to harm the circuit materials.

Semi-aqueous cleaning makes use of a solvent-water mixture where the solvent does the cleaning and the water serves as a carrier to flush away the dissolved wastes. Semi-aqueous cleaners generally are closed-loop in design. Contaminants are purged from the cleaning vehicle, and it is recycled through the system. Advantages are very high levels of cleanliness and the freedom to use more active fluxes for difficult-to-solder materials. The penalties are the cost of equipment and chemicals, and the fact that some of the semi-aqueous methods require the use of highly flammable chemicals.

Solvent cleaning with ozone-friendly chemistries is an obvious and much perused alternative to CFC cleaning. The challenge remains finding a chemistry that approaches the cleaning power, low cost, and handling ease of CFCs without bringing its own set of environmental concerns. Some of the early contenders are already on the list of substances identified as ozone depleters.

When Is an SMT Circuit Clean?

Cleanliness testing is often done by measuring the conductance of a 75/25 percent-by-weight solution of isopropanol and water. This method is complicated by the same standoff problems that make the cleaning process difficult. Contaminants hiding under a tightly clinging SMC are as difficult to dissolve for conductometric measurement as they are difficult to remove in a cleaning machine. One solution to this is to use a Surface Insulation Resistance (SIR) test. The SIR test uses defined test patterns that are designed into test boards developed specifically for the SIR test. An example of SIR test patterns is shown in Figure 3-17.

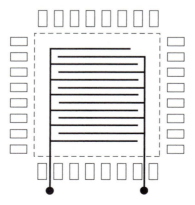

Figure 3-17   SIR test pattern example as used under a quad package.

The SIR pattern on the test board is made up of a pair of interlocking comb patterns that are located underneath components and uncovered by solder mask to test for the effectiveness of cleaning. Patterns under chip components may only have two "teeth" to the comb. The varying sizes of components make it impossible to use one standardized SIR test pattern, but in all cases the pattern should fill as much of the area as possible under a component. Note that SIR tests, therefore, are not done using production boards, but require that SIR test boards be inserted into the production line from time to time. After cleaning is complete, the SIR tests are performed.

SIR values are dependent on ambient test conditions, including relative humidity and temperature in the test chamber, time in the test chamber, the test voltage, and the purity of dionized water used in the test chamber. Test requirements are set in IPC-TM-650.

A destructive sampling program (including the removal of low standoff components to facilitate contaminant extraction) may be required to ensure cleanliness on an SMT line. Turbidity testing involves a light-scattering test to determine how much organic contaminants are precipitated from an isopropanol wash solution when a dilute aqueous acid is added to the wash. Turbidimeters measure the light-scattering angle at $0°$. They are accurate only when measuring a fairly turbid solution. Nephelometers measure light scattering at an angle to the incident beam. For nearly clear solutions, they are much more sensitive than turbidimeters, and thus perform better in SMT testing. Nephelometers are available through scientific instrument distributors.

Ultraviolet spectrometric analysis shows the abietic acid level in an isopropanol wash solution by measuring the sharp peak absorption of abietic acid at 241 nanometers. Abietic acid is the principal component in rosin and gives an accurate measurement of rosin presence down to 1 PPM in the wash solution. Ultraviolet spectrometers are also available through scientific instrument distributors.

3.1.8   Testing

The Challenges of SMT Testing

Inspection and test of SMT boards must accommodate the close lead spacings (1.27 mm/0.050" and less) of SMCs, the tight geometries of SMT assemblies, and the potential of assemblies populated on both sides. Each of these SMT characteristics poses a challenge for visual inspection, in-circuit testing, and repair/rework. SMT also facilitates packaging high pin count ICs. Devices with a large number of I/O pins may complicate testing since many are untestable by in-circuit means.

Close Lead Spacings

Bed-of-nails testing of bare and populated THT boards is a simple and well-understood science. The standard 2.45 mm (0.100 inch) lead centers of THTs lend themselves to probing, and probe cards built on such generous center distances are reasonably priced and rugged.

But what happened when we introduced SMT with 1.27 mm, 1.016 mm, 0.635 mm (0.050", 0.040", 0.025") and finer-pitch devices? Typical of technological advances, we stumbled along for a while, then engineers developed spring-loaded probes for fine pitches. At first, these probes were expensive, but competitive pressures quickly brought them into line with coarser-pitch probes. Today, SMT pitches are readily addressed by probe cards. The cost per point for 1.27 mm (0.050 inch) probe cards is reasonable. Finer-pitch probe cards do carry a cost-and-ruggedness penalty, however. The moral is "Design no finer than is absolutely required."

Dense Assemblies—Self Test and Boundary Scan

Planning is the testability watchword introduced by increasing circuit density. Pay very close attention to device clearances (particularly around large and tall components), and make all test nodes accessible. These concepts will be discussed further in Chapter 5.

Consider the case for testing designed in to the circuitry itself. A little investment here will pay enormous benefits in manufacturing and test costs in a production environment. Built-in self-test can be applied to both analog and digital circuits.[12,13,14] Boundary scan for digital circuits, also following the protocols of IEEE 1149.1 (also known as JTAG), is widely supported both in documentation and in component availability. The best way to start on a boundary scan is to visit the IEEE website, http://www.ieee.org, and search for "IEEE 1149.1." The required test access port (TAP) and boundary architecture are well defined. Virtually all digital IC manufacturers have 1149.1-compliant components that they support. National Semiconductor makes it particularly easy to find, at http://www.national.com/scan, but all digital manufacturers, including Texas Instruments, Altera, Maxim, and others, address the boundary scan issue.

Double-Sided Population

Provided proper design rules are followed, double-sided boards, with only passives and a few transistors on the bottom side (Type 3 technology), are not much of a test problem. However, SMT allows the building of complex Type 1 and 2 assemblies with dense IC populations on both sides, and these can spell trouble for test engineers. Test nodes for top-side components are often inaccessible from the bottom of the board because a bottom-side component obscures them or because they connect through a buried or hidden via or one obscured by a bottom-side component. Solutions to this include built-in testing, active partitioning for functional testing, double-side probing for functional or in-circuit testing, or some combination of these approaches.

Please note that functional testing generally requires access to many or all the circuit nodes in order to reduce test time and provide the most accurate troubleshooting guidance from testing. Therefore, planning to functionally test a board is no carte blanche for covering up test access points.

High Pin Counts

One of SMT's primary benefits is that, by reducing pin spacings and allowing I/O on four sides of a device instead of the DIP's two, SMT facilitates much higher I/O counts than does through-hole packaging. While this is great news for circuit designers looking for ways to replace fifty glue logic chips with one ASIC, it places a burden on test engineers. High pin count custom ICs and microprocessors often cannot be fully tested in-circuit. Functional and/or built-in testing may be a requisite with high pin count devices.

The Answer

Testable and manufacturable SMT boards, using the principles of DFT and DFM, occur by chance about as often as a group of monkeys types out the great books of the western world in random play with typewriters. More often, testable complex assemblies are the result of rigorous design efforts guided from the inception by a multidisciplinary team that includes process, test, manufacturing, and quality engineers. We will discuss specific design-for-testability issues in detail in Chapters 4 through 7. For now, let's review key differences between SMT and THT assemblies in preparation for a study of various manufacturing approaches to building SMT circuits.

3.1.9   Key Differences Between SMT and THT Assembly

The manufacturing of circuit boards using surface-mounted components differs from through-hole manufacturing in several important ways. We have touched on these differences in the preceding chapters, but we will recap them here before beginning a detailed discussion of the manufacturing processes.

Differences in Engineering

Aside from the obvious adaptations to new component packaging and mounting techniques, team engineering is a major change being pushed by SMT. In our earlier discussion of testing, we stated that a team approach is vital to the success of SMT engineering efforts. While team (or simultaneous) engineering may be employed to benefit in just about any manufacturing operation, it is unlikely you will manage SMT's complexity without teamwork.

This was not the case in THT manufacturing. Manufacturing and test engineers in electronics assembly operations have traditionally been largely ignored during the design phase of product development and are then assigned the thankless task of solving whatever riddles design engineering might "throw over the wall" to them.

The new emphasis on complete teamwork is elevating the status of manufacturing, process, test, and quality engineering in the electronics industry. On the downside, early involvement in the design operations removes the hedge a manufacturing group used to enjoy. When manufacturing participates in the design, it cannot refuse to build a new product on the grounds that it is unmanufacturable.

Differences in Component Attachment

The primary difference in SMT component attachment is the lack of the mechanical connection to the board that is supplied by the through-hole-mounted leads of THTs. The SMT solder joint is not only the electrical connection for the device, it is the major or sole mechanical connection as well. Therefore, solder-joint integrity is of critical concern in producing reliable SMT assemblies.

Operating in opposition to the need for high-integrity solder connections, SMT joints are often more difficult to inspect than THT solder connections. The solder joints are smaller, on closer centers, often obscured under the component they mount, and are not inspectable from the board's opposite side. Thus, a very high-yield soldering process, tightly monitored and under excellent control, is a vital part of production SMT manufacturing.

Differences in Soldering Methods

Solder difficulties cannot be resolved by simply turning up the temperature in the solder pot or by passing boards through for a second or third try in the solder wave. Component-lead and board-metallization solderability must be near perfect.

SMT soldering also differs from THT soldering in the control of the amount of solder applied. With limited process attention, wave soldering of inserted components in properly sized through-plated holes results in the proper amount of solder being deposited, and the solder deposit is relatively easy to visually inspect on an THT assembly. In contrast, reflow soldering requires close attention to design of the solder land, the solder print area, the solder deposit thickness, the solder paste consistency, and the printing operation in order to produce joints with just the right amount of solder applied. Further, wave soldering of SMT circuits involves resolving the trade-off between defects produced by too much solder and defects caused by too little solder. Both these defects can and do occur in the same process setup, as shown in Figure 3-18. Thus, SMT soldering, far more than THT soldering, requires an attention to the amount of solder applied.

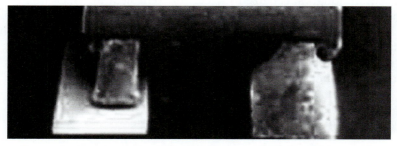

Figure 3-18   A surface mount component with leads that show both insufficient and excessive solder.

Differences in Component Testing

As we stated earlier, SMCs have much closer lead centers than THTs. Also, SMT boards may have components on both sides. These differences complicate the accessibility of SMT assemblies for testing. Matters are made worse by the fact that, unlike THTs, probing on device leads is prohibited in SMT. This is because the probe pressure might press an open lead into contact with a land and give a false pass. Also, misalignment with a lead might damage fragile fine-pitch probes.

SMT boards may also have very high pin count devices, many of which are not testable by in-circuit means. The functional testing of VLSI is just as dependent as in-circuit testing on test node access.

The differences in SMT and THT tests are much more obvious than the way to resolve them. Therefore, test engineering must be part of the initial design of SMT boards. Test access requires space and often conflicts with the miniaturization efforts. Access vs. real-estate-savings trade-offs must not be resolved in a kangaroo court.

Differences in Repair Methods

In their most basic form, SMT and THT repair requires the same general functions. Both involve the taking off of nonconforming parts or circuitry as necessary, the cleanup of the rework area (including removal of solder where components were removed), and the repetition of the original process steps (or similar steps) to replace those components. In some instances, the lack of through-hole clenched leads makes SMT repair simpler than THT repair, but the higher densities of the surface-mount and double-sided component population can add complexity where SMT is concerned, although specialized "tweezers" will heat all leads of two- or four-sided parts simultaneously allowing easy removal. Depending on the heating technology used, the time required to reflow all of the leads of a large SMT IC package may require the shielding of adjacent circuitry from conductive and convective heat.

As with testability, repairability for SMT assemblies must be a goal of the initial design. Repair department personnel should be a part of the design team.

3.2   Manufacturing Styles of SMT

As discussed in Chapter 1, manufacturing style or type (Types 1, 2, and 3), more than any other single factor, determines the process flowchart used to assemble a given SMT product. Armed with our understanding of the SMT process from the previous overview discussion, we will now study how process steps are varied to suit different SMT manufacturing styles.

3.2.1   Type 1 SMT Manufacturing

Type 1 Defined

To review, Type 1 SMT is an exclusive surface-mount method with components mounted on one or both sides. All styles of surface-mount components are used in Type 1 processing. Type 1 components are virtually always reflow soldered. A typical Type 1 process flowchart is shown in Figure 3-19.

Figure 3-19 shows that double-sided populated boards are reflow soldered twice. On the second pass, the board is inverted and the top-side components are placed upside down in the reflow oven. New SMT engineers often stagger at this. They see visions of all the upside-down

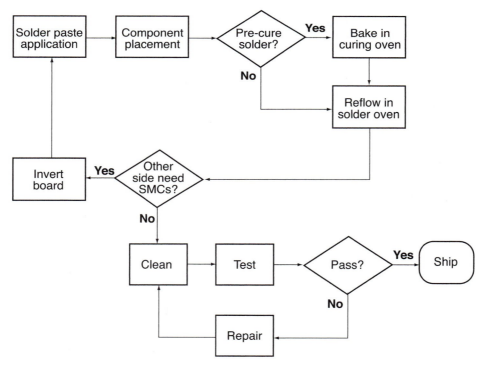

Figure 3-19   A typical Type 1 SMT process flowchart.

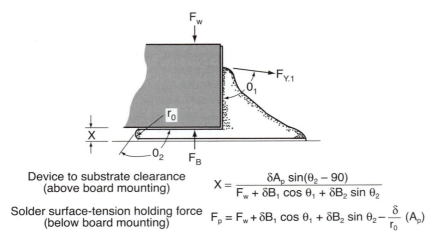

Device to substrate clearance
(above board mounting)

$$X = \frac{\delta A_p \sin(\theta_2 - 90)}{F_w + \delta B_1 \cos\theta_1 + \delta B_2 \sin\theta_2}$$

Solder surface-tension holding force
(below board mounting)

$$F_p = F_w + \delta B_1 \cos\theta_1 + \delta B_2 \sin\theta_2 - \frac{\delta}{r_0}(A_p)$$

Figure 3-20   Factors for calculating surface-tension attraction and stand off height.

components winding up on the floor of the furnace. In fact, with all standard SMCs, solder surface tension will hold upside-down components in place, even though they are reflow soldered on the second pass. The ratio of surface-tension attractive force to component weight determines whether a component will remain on the bottom of a board when held only by molten solder. Surface-tension attractive force is determined by the termination surface area and can be calculated as illustrated in Figure 3-20.[15]

The equation for calculating the device-to-substrate clearance (for above-board mounting) is given in Equation 3-1.

$$X = \frac{\delta A_p \sin(\theta_2 - 90)}{F_w + \delta B_1 \cos\theta_1 + \delta B_2 \sin\theta_2} \qquad \textbf{(Eq. 3-1)}$$

where,

X = Clearance, device to substrate
δ = Surface tension
A_p = Total area of chip-carrier bottom pads
F_w = Weight of chip carrier
B_1 = Wetted perimeter of chip-carrier castellation
B_2 = Wetted perimeter of chip-carrier bottom pads
θ_1, θ_2 = Wetting angle

The equation for the solder surface-tension holding force (for below-board mounting) is given in Equation 3-2.

$$F_p = F_w + \delta B_1 \cos\theta_1 + \delta B_2 \sin\theta_2 - \frac{\delta}{r_0}(A_p) \qquad \textbf{(Eq. 3-2)}$$

where,

F_p = Pull-off force
r_o = Radius of curvature at the bottom pad location

At the point of pull-off, the expression reduces to

$$F_p = F_w + d(B_1 + B_2)$$ **(Eq. 3-3)**

In practical terms, it is very difficult to determine the amount of surface tension, but empirically each lead of a J-lead PLCC adds about 0.08 gram of attractive force in the bottom-side reflow of previously soldered joints. Thus, a 44-pin PLCC needs 24 leads soldered to hang upside down by molten solder (an error margin of 45 percent if all leads are soldered). However, an 84-pin PLCC must have 71 leads attached, leaving a margin of only 15 percent. The obvious conclusion is that it is better to locate large ICs on the last reflow side.

Please note that while molten solder may support a bottom-side component during reflow, fresh solder paste most certainly will not. The flux and thixatropic vehicles will lose viscosity in solder preheating long before the solder reflows. Unless they are glued or otherwise fixtured, upside-down components must have already been reflowed in order to remain on the board during soldering.

Type 1 Characterized

Type 1 SMT excels in circuit miniaturization, simplicity of process flow, reliability, and use of high-pin-count devices. On the negative side of the equation, Type 1 usage limits component selection to surface-mountable varieties or special SM lead-formed THTs. And, Type 1 manufacturing requires the mastering of solder-paste application and reflow soldering techniques. The benefits and limitations of Type 1 technology are shown in more detail in Table 3-1.

Table 3-1 Benefits and Limitations of Type 1 SMT Assembly.

Features	Advantages	Limitations
SMT components only	Best miniaturization	Component availability
PLCCs/LCCs/µPaks	Simple automated process	Total new line required
TAB and Flip Chip	Adapts to batch of 1	Reflow and stencil printing learning curve
SO and SOL ICs	Single solder process	TCE match concerns (where leadless hermetic component packages are used)
Chip capacitors and resistors	High-reliability possible	
Connectors and miscellaneous	Potential high yield	
SMT lead-formed THT	Potential lower cost	
Very high density	Improved electrical performance	
High reliability	Noise immunity	
One or both sides populated	Speed	
	Allows uses of high pin-count packages	

3.2.2   Type 2 SMT Manufacturing

Type 2 Defined

Reviewing Type 2 SMT, we recall that it involves a mixture of SMCs and THTs on at least one side of the board. Type 2 assemblies may be populated on one or both sides. There are a multitude of ways to construct Type 2 boards, and each manufacturing method has its individual characteristics as to what component styles may be used in what combination(s). However, several generalizations can be made regarding Type 2 SMT. On at least one side, all styles of surface-mount components may be used, just as in Type 1 processing. And that side (the one with all styles of SMCs) will be reflow soldered. Several typical Type 2 process flowcharts are shown in Figure 3-21 and

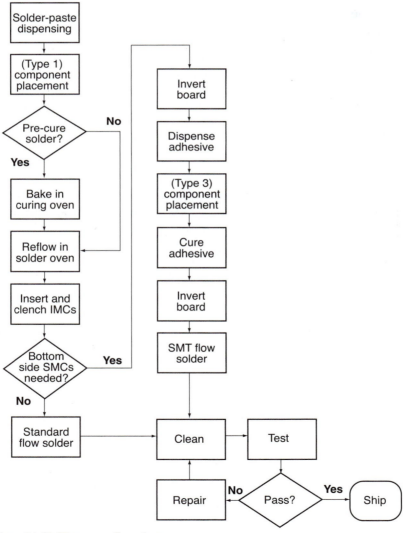

Figure 3-21   Type 2A SMT process flowchart.

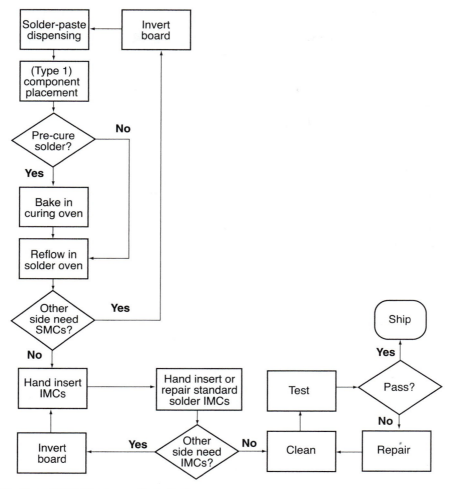

Figure 3-22   Type 2B SMT process flowchart.

Figure 3-22. As the IPC has done with its six types and six levels of SMT assembly, many variations of these charts could be developed from permutations of the SMT and THT process steps. The ones shown are offered as food for thought, not as a required blueprint.

Type 2 Characterized

Type 2 SMT allows impressive circuit miniaturization and the use of high pin-count devices, coupled with the flexibility of using insertion-mounted components. On the negative side of the Type 2 equation, the mixture of reflow and flow soldering may significantly complicate both assembly, test, and repair of circuits. And, like Type 1, Type 2 manufacturing requires the mastering of solder-paste application and reflow soldering technologies.

Type 2A combines the potential of using high pin-count chip carriers on the top side of a board with the manufacturing efficiencies of Type 3 SMT for mounting of passives and discrete transistors on the bottom side. Thus, Type 2A would be attractive where a circuit with a high R/C population count also used a few high pin-count (≥64 pin) ICs. The passives could be bottom-side mounted à la Type 3, and the high I/O actives reflowed topside in efficient SMC packaging.

Type 2B comes closer to Type 1 technology but allows through-hole mounting by hand or semiautomated means. Because of the labor-intensive approach used for THTs in Type 2B technology, it is best suited to SMT circuits where only a few through-hole parts will be needed.

The benefits and limitations of Type 2 technology are listed in more detail in Table 3-2.

Table 3-2   Benefits and Limitations of Type 2 SMT Assembly.

Features	Advantages	Limitations
Type 2 Generalities	*Type 2 Generalities*	*Type 2 Generalities*
Mix of THT and SMT components Some reflow solder All SMCs per Type 1	Component availability Component price shopping Good miniaturization High pin-count packages can be used	Reflow and stencil learning curve Multiple soldering processes Complicated test and repair
Type 2A Peculiarities	*Type 2A Peculiarities*	*Type 2A Peculiarities*
Type 1 SMT processing (top side only) Type 3 processing (bottom-side SMCs) Automated THT processing	Good density in circuit with high percentage of resistors and capacitors Relatively automated process Plus all of the general advantages	THT components on top side only Plus all of the general limitations
Type 2B Peculiarities	*Type 2B Peculiarities*	*Type 2B Peculiarities*
Type 1 SMT process (top or both sides) Manual THT processing (top or both sides)	High density in circuit with few THTs Allows THTs and SMCs on both sides Plus all of the general advantages	Labor-intensive THT process Requires operator finesse Plus all of the general limitations

3.2.3   Type 3 SMT Manufacturing

Type 3 Defined

To recap, we have said that Type 3 SMT is exclusively THT on the classical "top" component side and exclusively SMC on the bottom side. Only flow-solderable styles of surface-mount components are used in Type 3 processing.

Type 1 is clearly distinct from Type 3 in processing since it is fully reflow soldered. Types 2 and 3 may seem alike, as both involve mixed SMC and THT populations. However, they differ in two key areas. Type 2 mixes THTs and SMCs on the same side of the board. Type 3 separates them on different sides of the board, and Type 2 involves two different soldering technologies—reflow and flow soldering. Type 3 is exclusively one-pass flow soldered. A typical Type 3 process flowchart is shown in Figure 3-23.

Type 3 Characterized

Type 3 SMT allows some miniaturization, particularly for circuits having a high percentage of discrete components. Type 3 uses a simple automated process flow with only one soldering step. However, Type 3 limits SMT component selection to flow-solderable varieties, which rules

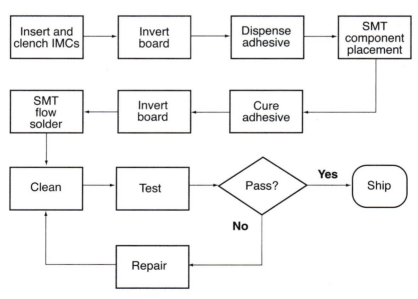

Figure 3-23   Typical Type 3 SMT process flowchart.

Table 3-3 Benefits and Limitations of Type 3 SMT Assembly.

Features	Advantages	Limitations
THTs on top side only	Simple one-solder process	Only flow-solderable SMCs allowed
SMCs on bottom side only	Automated process	Not compatible with high I/O SMC styles
Flow-solderable SMCs	Available components	
Chip capacitors and resistors	Good miniaturization of circuits with a high percentage of resistors and capacitors	Adhesives required
SOTs		Adhesive and SMC flow-solder learning curve
SOICs	Relatively simple test and repair procedure	

out the use of space-efficient chip carriers and high I/O packaging. Also, Type 3 manufacturing requires the mastering of adhesive application and curing with its inherent reliability concerns. The benefits and limitations of Type 3 technology are given in more detail in Table 3-3.

3.3 Process Steps and Design

In the preceding manufacturing discussion, we looked at the SMT manufacturing process in very broad terms, we covered differences between the THT and SMC manufacturing technologies, and developed a basic understanding of the process flows for various styles of SMT assemblies. All of this will serve as a background for the meat of this chapter that follows. How do manufacturing requirements impact design decisions? We will start the discussion by analyzing this question process by process.

3.3.1 Solder Application

Solder-Paste Screen and Stencil Printing

Screening and stencil printing are two widely used techniques for applying solder paste to assemblies for reflow soldering. Screening and stencil printing have been used for many years for solder application in the manufacturing of hybrid circuits, although most facilities are currently using stencil printing due its much higher and more consistent quality. For those reasons, screen printing will be mentioned only in passing. It is important to note that "screen" and "stencil" are used interchangeably in much of the SMT literature. Even some manufacturers of stencil printing machines use the term "screen" printer for their machines.

Silk-screen printing is a very old art, so named for the fact that the printing screens were originally made of silk stretched tautly on a frame. Areas to be printed were left as open mesh silk, whereas the areas that were to be free of ink on the printed image were filled in on the screen so that ink would not flow through them. Printing was accomplished by forcing ink through the screen's open areas by wiping the ink across the screen with a squeegee.

Today, most industrial screens are made of a more durable stainless-steel mesh rather than silk, and images are formed on the screen photographically using an emulsion coating that washes away in the print areas after the photo exposure. The basic process has not changed dramatically since its inception.

Screen printing of solder requires a planar circuit surface. Even a very small lead protruding from an opposite-side THT would destroy a printing screen. Therefore, screening is generally done before through-hole insertion. Also, there is a lower limit on the size feature that can be printed with screened solder paste. For solder-paste application, a comfortable rule of thumb is to print nothing smaller than 0.635 mm × 1.27 mm (0.025" × 0.050") with a silk screen, which makes the process unusable for most SMT IC packages available today.

Solder-Paste Stenciling

Solder-paste stencils are made for squeegee-applied paste and require a planar printing surface. Stencils are suitable for printing fine features below the typical limits of screens. Solder paste dots well under 0.254 mm (0.010 inch) square can be stencil printed reliably.

Stencil manufacturing is typically done by specialized shops. Stencils for solder paste deposition are available in brass and stainless steel, are durable, easy to clean, and require little on-line maintenance—mostly periodic wiping.

While originally stencils were manufactured by a chemical etching process using photographic images, (similar to the PCB etching process), stencil manufacturing technology has progressed to include laser cutting and additive processes. They may be the same thickness across the stencil, or they may be stepped to provide both high-volume and low-volume printing ability with the same stencil. It is not the intent of this book to provide details on either the manufacturing of stencils or the proper operation of a stencil printing machine. The designers and users should recognize these variables that can affect the quality of the print process, and these are why this book makes no attempt to dictate stencil designs nor the processes to be used:

- PCB warpage
- Solder paste viscosity
- Print thickness, typically varying by pitch of the components
- Squeegee issues: downforce, hardness, material, wear of the squeegee, print speed
- Stencil tension
- Shape of the apertures and aspect ratio (ratio of length to width of aperture)
- Formation technology of the stencil: chemical etched, laser cut, additive/electroformed
- Polishing or microetching of the stencil aperture walls or nickel plating to maximize wall smoothness and paste release
- Tapering of aperture walls (apertures slightly larger on board side of stencil) to maximize release of paste

As in other aspects of the solder paste deposition process, the solder paste manufacturer, the solder paste machine manufacturer, and the stencil manufacturer can all offer suggestions on controlling the above variable in ways to maximize the quality of the printing process.

Stencil Design

Stencil design for solder paste deposition (as opposed to adhesive deposition) relates primarily to the thickness of the stencil, the aperture sizes, and the finish of the stencil, as well as other variables mentioned in the preceding section.

Stencil thickness controls the volume held by each aperture. It also has some affect on the ability of the paste to be released from the aperture during the printing process. This ability is quantified as the relation between the forces holding the paste in the aperture and the attraction force holding the paste onto the pad. If the pad force is greater than the aperture force, the paste is released onto the pad. If the aperture force is greater than the pad attraction force, then some or all of the paste remains in the aperture. The relationship of these forces (worst case) is based on the ratio of the stencil thickness and the minimum aperture width and varies with the overall roughness and shape of the aperture walls. Generally, the ratio should not fall below the following:

- Chemically etched stencil: at least 1:1.5 ratio of stencil thickness to minimum width
- Laser-cut stencil: at least 1:1.2 ratio of stencil thickness to minimum width
- Additive/electroformed   at least 1:1.1 ratio of stencil thickness to minimum width
  stencil:

Based on these guidelines, for example, an 8-mil thick chem-etch stencil should not have an opening less than $8 \times 1.5 = 12$ mils wide. Other stencil types can be determined in a similar way.

General aperture opening guidelines closely follow pad sizes, with openings for fine-pitch parts slightly narrower than the pad, as shown by the examples in Table 3-4. As mentioned previously, the solder paste manufacturer and the stencil machine manufacturer can supply guidelines specific to their materials and machines.

Where apertures are narrower than the pad width, the apertures must be centered on the pads. If apertures are shorter than the length of the pad, they should be aligned with the outside of the pad to reduce the incidence of the formation of under-component reflow solder balls. An example of a shifted, narrow aperture is shown in Figure 3-24, as well as examples of other alternate aperture shapes.

The guidelines shown in Table 3-4 can be modified by the total volume desired versus the stencil thickness to be used and the length of the aperture.

Table 3-4   Stencil Aperture Examples.

Lead Pitch	Pad Width	Aperture Width	Lead Pitch	Pad Width	Aperture Width
1.27 mm	0.63 mm	0.63 mm	50 mil	25 mil	25 mil
1.0 mm	0.5 mm	0.5 mm	40 mil	20 mil	20 mil
0.65 mm	0.32 mm	0.32 mm	25 mil	13 mil	12 mil
0.5 mm	0.25 mm	0.2 mm	20 mil	10 mil	8 mil
0.4 mm	0.2 mm	0.15 mm	16 mil	8 mil	6 mil

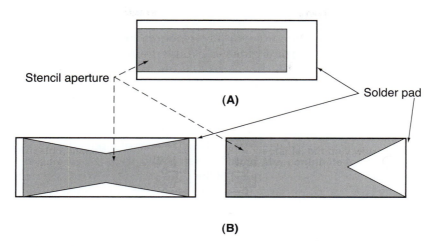

Figure 3-24 (A) Example of a narrow, shifted solder-paste stencil aperture. (B) Example of other aperture shapes.

Solder-Paste Dispensing

Solder paste may also be applied to the lands by dispensing. Dispensers generally use a timed pulse of pneumatic pressure to extrude paste from a syringe, or draw paste from syringes and dispense it using a positive displacement pump. Dispensing is inherently a sequential process, usually being done one solder site at a time. Hand dispense guns are often used for prototype and short-run assembly.

Since dispensing heads can be computer controlled on X–Y motions, dispensing allows the programmable selection of solder-application patterns. While it is inherently a slow sequential process and screening is a fast batch process, the flexibility afforded by programmable dispensers is attracting many manufacturers to this method of solder application. The relatively slow sequential process is often no slower than sequential placement, and thus may not form a bottleneck in a production line.

The primary advantages of the time-pressure dispense method include:

- Lower cost of dispensing equipment
- Can dispense on boards that already have mounted components
- Can be used to apply paste to individual components during a rework operation
- Can dispense large volumes in a small location, an advantage in the PIP process
- For low-volume, high-mix facilities, a change in board requires only a download of the appropriate software pattern—no hardware changes are necessary

The disadvantage is generally speed of deposition and potential for paste separation. Solder paste is 85 to 95 percent metal powder by weight. Under the heavy pressure necessary to extrude solder paste through a small dispensing nozzle, the more fluid elements of the paste (flux, vehicles, and thixatropic agents) may tend to separate out and be dispensed in disproportionate volume. When this condition occurs, the increase in metals content in the paste soon clogs the nozzle. Positive displacement dispensing is said to overcome this problem by applying dispensing stress to only a very small volume of paste, which is dispensed in a single shot.

Process-Driven Design Rules for Solder-Paste Application

Only certain pastes are suitable for a given application process. Generally, the viscosity and metals content of the pastes are varied to suit the application methods. Very viscous pastes with a metals content of 88 to 91 percent by weight stencil well. Pastes of moderate viscosity and metals content of 83 to 87 percent by weight are better for screening, and pastes of lower viscosity and metals content are indicated for dispensing. Solder paste and equipment vendors can furnish exact guidance in the selection of pastes appropriate for a particular process and application.

In reflow soldering, the amount of solder forming the solder joint is determined by the solder print area, the print thickness, and the paste metals content. The standard device land patterns suggested herein are tailored (in the case of 50-mil pitch SMT) for screen or stencil printing of 7 to 10 mils of wet solder paste and will provide the proper amount of solder to the joint under such print conditions. For dispensed solder, the dispensing volume should normally be equal to the volume in a 0.18 to 0.25 mm (7- to 10-mil) deposit exactly covering the land area. For fine-pitch components, it may be necessary to reduce the solder-print thickness to 4 to 7 mils. In order to provide sufficient solder to standard part leads on a fine-pitch assembly, the print pattern for the standard pitch parts may be expanded by up to 25 percent. The resulting overprint will not cause bridging but will withdraw into the solder bond during reflow driven by the cohesive and wetting forces of the solder.

Solder volume is sometimes varied from the above amounts for certain design and manufacturing objectives. By reducing solder volume, we can reduce tombstoning. Within limits, lowering the solder volume also increased the joint ductility. Conversely, adding solder will, within limits, increase the joint strength. However, the goal of assembly is not to produce the strongest possible joint, but to produce a sound ductile joint capable of absorbing stresses. Studies have shown that excessive solder contributes to component cracking and termination failures.

Pin-In-Paste Deposition

The pin-in-paste (PIP) process (also known as Alternate Assembly Reflow Technology [AART], Through-Hole Reflow [THR], and a variety of other acronyms) refers to the process by which solder paste is deposited at the location of a THT component through-hole and the THT component leads/pins placed through the paste into the hole. The rationale for this is that wave soldering of THT components is becoming an unacceptable process step for facilities that are primarily SMT. After all components are placed, the entire assembly is reflow soldered. This process eliminates the need for a wave soldering machine in the line, but it must be designed for. The general assembly process would then be:

print paste on "bottom" side > place SMT components > reflow > invert PCB > print paste for SMT and THT/PIP > place SMT components > insert THT components through paste > reflow

Lockheed Martin Naval Electronics and Surveillance Systems—Surface Systems (NE&SS—SS) redesigned their manufacturing process to use PIP (they call it "Pin-in-Paste Dual Technology") in 1998. Their tests showed that the process could be used on single-sided PCBs up to 0.120" thick and double-sided boards up to 0.100" thick. They found that the assemblies meet the requirements of J-STD-001 using a convection reflow oven with a nitrogen environment and that they reduced their cycle time compared to the reflow plus wave process formerly used by 40 percent. Lockheed Martin does use multiple paste stencils using multiple print steps to achieve the proper volume at the PIP sites. Others have used stepped stencils for the same purpose to achieve a thicker paste print at some sites and thinner prints at non-PIP sites.

Intel engineers presented information on the use of PIP specific to the use of non-silicon THT components such as connectors and battery holders.[16] They used 6-mil stencils to print the required volume of paste on both sides of the board, and the results showed solder DPM below 500 DPM for a 20-week production run.

The first issue with PIP is determining the correct volume of paste to be deposited. First the designer and the manufacturing engineer must recognize that solder paste manufactured for stencil deposition will shrink to approximately 50 percent of its paste volume during the reflow process, while syringe-deposited paste will shrink to approximately 40 percent of its paste volume. In general, the designer must determine the following dimensions to calculate the final amount of *metal* necessary for the soldered lead:

- Thickness of the PCB, T_{BOARD}
- Cross-sectional area of the lead, πR_{LEAD}^2, and V_{LEAD}, which is the volume of the lead encompassed by the through-hole in the board, $V_{LEAD} = \pi R_{LEAD}^2 * T_{BOARD}$
- Cross-sectional area of the through-hole, πR_{HOLE}^2
- Amount of required solder fill of the hole
- If solder fill requirement is 100 percent, the amount of additional solder fillet desired to make certain that solder fill occurs under all process conditions

An example of a through-hole lead for which a fillet is desired is shown in Figure 3-25.

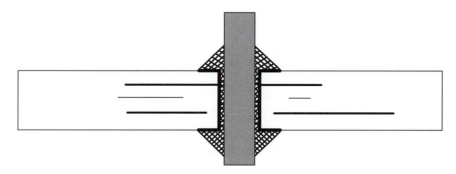

Figure 3-25 Example of PIP reflow result with idealized fillets and 100% hole fill.

For example, the basic calculations for the amount of final metal necessary for 75 percent hole fill are:

$$V_{FILL} = (V_{HOLE} - V_{LEAD}) * \text{hole fill fraction} = ((\pi_{HOLE}^2 - \pi R_{LEAD}^2) * T_{BOARD}) * 0.75$$

V_{PASTE} would then be ($V_{FILL} * 2$) for stencils and ($V_{FILL} * 2.5$) for syringe. Any additional volume for the fillet can be approximated by adding 25% to the V_{FILL} calculation. The designer must remember to multiply the desired metal volume by 2 for stencil-deposited paste and by 2.5 (1/0.4) for syringe-deposited paste. As a further example:

Round lead, diameter of 0.8 mm (0.032")
Round hole, diameter of 1.0 mm (0.040")
Board thickness 1.6 mm (0.062")
$V_{LEAD} = \pi * 0.4$ mm^2 * 1.6 mm = 0.804 mm^3 (49.86 E-6 in^3)
$V_{HOLE} = \pi * 0.5$ mm^2 * 1.6 mm = 1.257 mm^3 (77.91 E-6 in^3)

100% hole fill desired, plus 25% for fillet:

$V_{FILL} = (V_{HOLE} - V_{LEAD}) * 1.25 = 0.566$ mm^3 (35.06 E-6 in^3)
V_{PASTE} (for stencil paste) = $V_{FILL} \times 2 = 1.132$ mm^3 (70.12 E-6 in^3)
V_{PASTE} (for syringe paste) = $V_{FILL} \times 2.5 = 1.415$ mm^3 (87.65 E-6 in^3)

Note that plated hole diameter for the PIP process should be no more than 0.25 mm (0.010") larger than lead diameter to prevent voids in the hole fill. For leads with square cross sections, the plated hole should be no more than 0.005" (0.1 mm) larger than corner-to-corner distance.

An example of >100 percent hole fill is shown in Figure 3-26. This can be accomplished either by overprinting with a stencil or with the use of a syringe dispenser. One technique of stencil application for overfill requires a dual pass with the squeegee to force more paste into the corresponding stencil hole and thereby into the through hole. This has the downside of possibly causing paste smearing on fine-pitch pads during the second pass. Other stencil techniques use a stepped stencil with a thicker stencil in the PIP locations allowing the squeegee to force more paste into the hole, or use a pressurized squeegee.

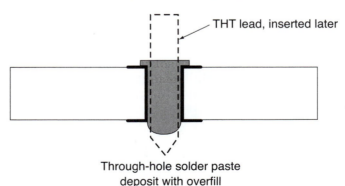

Through-hole solder paste
deposit with overfill

Figure 3-26   Example of >100% hole file for the PIP paste printing process.

Stencil design for printing paste in the PIP process demands consideration of a number of factors:

- Total volume needed for each THT site
- Volume of paste printed into the hole versus volume printed into the surface
- Pitch of holes to be printed
- Type of squeegee to be used

The type of squeegee means whether the stencil machine uses a "passive" squeegee that is pulled over the stencil or the system uses a pressurized paste-filled squeegee system. This latter type of system, e.g. DEK ProFlow, MPM Rheometric pump and the Fuji GP-641 printer, can force paste into the through holes far better than a passive squeegee can, reducing the amount of paste that must be printed on the surface of the PCB, thereby reducing the stencil opening required. Figure 3-26 shows a hole fill that would require very little paste deposit on the surface of the PCB.

Setup of any printer for the PIP process is best done working with the printer manufacturer's representative to set the most appropriate printer parameters for initial prints. Print parameters, such as print speed and pressure, are then adjusted as necessary based on the quality of the resultant solder joints. Some users of standard stencil machines find that hole fill is increased by a double pass of the squeegee. However, this may result in smearing of any fine-pitch print patterns on the board.

The volume of the stencil opening must take into account the amount of paste that is expected to be printed into the through hole. Only the amount in excess of that printed into the hole will need to be printed onto the surface. Generally, the shape of the stencil aperture should be the same as the shape of the pad on the board if the desired volume can be deposited. If the calculated volume cannot be deposited, e.g., through a round aperture for a round pad, then the aperture can be made rectangular and deposit the volume required. If the stencil opening is round, then the volume printed by the stencil itself will be:

$V = \pi R^2 T$, where R = the radius of the opening and T = the stencil thickness.

If the stencil aperture is rectangular, then the volume printed by the stencil will be:

$V = L * W * T$, where L = the aperture length, W = the aperture width, and T = thickness.

PIP users with non-pressurized printer heads will find that to obtain the total volume needed may mean that if printing through a rectangular aperture for a round pad, some of the paste is printed on the solder mask. It will then wick back to the metal when reflowed. The PCB fabricator needs to understand that this is the intent in certain areas of the mask. As discussed, users of pressurized printer systems find they can accomplish 100 percent+ hole fill, and that surface printing of paste may not be required. The capillary-action wicking of molten solder back into the hole is optimized if hole diameters are no more than 0.25 mm (0.010") larger than the pin diameter/cross section. Too large a hole can cause difficulty not only with printing since the paste may not adhere well to the walls of the hole, but also wick back may be less successful since there will be less capillary action if the clearances are too large.

Indium has Stencil Coach software available at http://www.indium.com that is valuable in calculating PIP stencil parameters. DEK has its own stencil calculator program at http://www.dek.com.

Jensen and Lasky[17] suggest that the volume of paste deposited for the PIP process will leave behind larger quantities of flux residue than normal SMT volumes, and that those larger volumes will be more difficult to clean. For this reason, they suggest that no-clean flux is the most appropriate paste chemistry for the PIP process.

Test runs with the chosen stencil design should result in reflowed solder joints that meet the joint inspection criteria of ANSI J-STD-001 and IPC CM-610C:

- No void areas
- No surface imperfections
- 95 percent minimum solder fillet
- Good wetting
- Good pad coverage
- Minimal test pad contamination by flux residue

It must be verified that the bodies of THT components are designed to withstand reflow temperatures, and since the THT component leads push a great deal of solder paste out of the bottom of the hole, they must be trimmed to extend no more than 1 to 1.5 mm past the bottom of the board when fully inserted to maximize the ability of the molten solder to wick back into the hole. Large THT components (such as connectors) must have standoffs so the body does not rest directly on the pads and smear the deposited paste, and so that the convective reflow heat can flow under the body to melt the paste. In profiling the board, temperatures under the center of large THT components should be monitored. If the SMT parts are moisture sensitive, they may need to be baked like SMT components prior to reflow.

3.3.2   Adhesive Application

For flow soldering, SMCs must be held in place by an adhesive until they are soldered. The adhesive is applied to the component sites prior to SMC placement and is cured immediately after placement. Manufacturing speed is dependent on the adhesive being quickly cured.

SMT flow soldering subjects the SM components to one or more passes through molten solder while hanging from the bottom of the board. Therefore, it is important that the cured adhesive have sufficient holding power for the task at hand and retains that holding power at the elevated temperatures seen during the soldering process. For the sake of rework, it is equally important that the adhesive not form so strong a bond that components cannot be readily removed.

Adhesives are generally left in place after the soldering process. Thus, long-term reliability depends upon the adhesive being fully cured (having no surface tackiness that could pick up contaminants). Adhesives must also be free from voids which may entrap solder during the flow operation and introduce a reliability concern.

Adhesives are usually applied by screen or stencil printing, dispensing, or transfer (offset) printing.

Adhesive Screening and Stenciling

Adhesives are screened and stenciled in the same manner described earlier for solder pastes. In order to be applied by these methods, adhesives must be thixatropic in nature, with an appropriate rheology.

As with solder printing, adhesive screening or stenciling must be done on a planar surface. Therefore, it is not an acceptable application method for bottom-side attachment on circuits where through-hole components are already inserted. For auto insertion, SMCs are usually added after the inserted components to allow maximum bottom-side clearance for cut-and-clench tooling and to protect the SMCs from the pounding of the auto-insertion process. Therefore, screen printing finds limited application in adhesive processing.

Adhesive Dispensing

Like solder-paste dispensing, time-pressure methods with syringes and positive displacement pumping are typical dispensing methods. Many X-Y placement systems designed for assembly of Type 3 discrete components have an upstream station running in slave format to the placement station where a single dot of adhesive is dispensed in parallel to each placement move. This arrangement takes advantage of the X-Y motion used in placement, giving the dispensing operation a free ride. However, unless equipped to implement nonslave motion, the slave station can only dispense in the exact pattern that the placement head follows. As we will see later under Design Considerations, other dispensing patterns may be desirable or even mandatory.

Adhesive Transfer Printing

Transfer, or *offset*, printing provides a batch printing method comparable to screening in speed but suitable for the application of adhesive on nonplanar surfaces. Offset adhesive printers use a finger (or multiple fingers) to pick up a metered layer of adhesive from an adhesive tray and transfer it to the printing site(s). A doctor blade is passed across the tray to reestablish a smooth adhesive layer of correct thickness after each printing pickup.

Offset printing may involve one finger running programmably in an X–Y motion, or multiple fingers individually set in a carrier plate at proper dimensions for each printing location. Single-finger printers operate sequentially and are relatively slow but highly programmable. Hard-tooled multiple-finger printers can print a large number of sites simultaneously. While they are much more setup intensive than the programmable type, their high throughput makes them attractive for large production-lot offset printing.

Process-Driven Design Rules for Adhesive Application

Again, Rule 1 is to choose a material suitable for the given application process. Epoxy and equipment vendors can furnish exact guidance in selecting the adhesives appropriate for a particular process and application.

Another important process design consideration using adhesives is how much glue to apply. Proper adhesive volume is determined by two variables. First, the height of component standoff (if any) dictates how high a droplet you will need to ensure contact between the component body and the dome of glue deposited on the PW board. Second, the rheology of the adhesive and dispensing-volume tolerance of the application equipment determines the quantity that must be dispensed to assure formation of the required height in that dome of glue. Figure 3-27 illustrates how these factors are weighed in determining, by empirical methods, adhesive dispensing specifications for a given application.

Never deposit more adhesive than necessary to satisfy the above test since excessive adhesive greatly increases the likelihood of the adhesive fouling the solder lands and causing resultant solder defects. Where components have no standoff, the glue deposit should be kept as small as equipment and adhesion considerations allow. Adhesive contamination of the solder is most likely to occur under components with no standoff, as shown in Figure 3-28.

Designing for adhesive application also includes the determination of where to put the glue dot(s) for a given component. With relatively small components and thermal-set epoxies, this is straightforward. The dot goes under the center of the component. However, for UV-cured epoxies, centering the dot would block the light in the curing operation and result in disaster. Figure 3-29 gives dot location recommendations for thermoset and UV applications. Dot location data should form part of the CAD component library database.

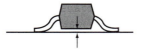   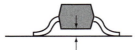

Minimum standoff, do not deposit so much adhesive that excess could be squeezed onto land areas.

Maximum standoff, enough adhesive must be deposited to ensure contact between component and glue dome.

(A) Standoff variance of component.

Maximum dot size. Should not foul lands under a component with minimum standoff.

Minimum dot size. Must contact and adequately attach a component with maximum standoff.

(B) Deposit volume variance of applicator.

Figure 3-27   Determining the proper glue-deposit volume.

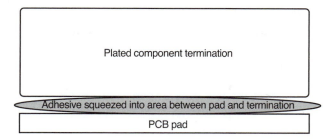

Figure 3-28   Solder defect caused by adhesive contamination.

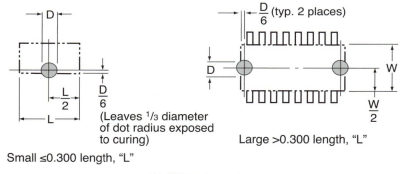

(A) UV-curing rules.

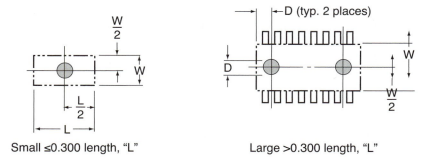

(B) Thermoset rules.

Figure 3-29   Adhesive dot-location recommendations.

3.3.3   Placement of Components

There is probably enough attention focused on robotics today, so we will start our discussion of placement by noting that components can be placed by hand. In fact, at times, components should be placed by hand in an SMT shop. Even the simplest SMT automated placement equipment requires considerable setup attention before addressing a new task. The engineering prototyping of one or two circuits and the repair operations are often much more efficiently done by hand, but this can only happen when someone has developed the necessary finesse. Note: Select SMT assembly technicians who are gifted with the patience of Job and with hands as steady as Gibraltar. If this is done, you will soon reach a pleasant surprise. The day will come when, for purposes of easy hand assembly and prototyping, you will prefer SMC to THT assembly.

Placement Equipment Defined

To assemble an SMT board, the proper components must be selected, oriented, and correctly located on the assembly. Placement equipment may be broadly defined as machines that assist in the assembly of SMT boards by helping select and/or orient and/or place components.

Placement Equipment Categories

Beyond the simple manual placement of SMCs, there is a wide range of equipment for SMT assembly. Equipment can be categorized by its mode of operation, degree of flexibility, machine intelligence, and level of automation.

Let's start by looking at the *mode of operation* categories. At the low end in cost and throughput are the semi-automated sequential systems, so named for their step-by-step assembly of one component at a time. At the low-volume end of sequential equipment are X-Y machines built around pen plotter motions. Then there are the moderate-volume sequential machines, many of which incorporate the latest in robotics technology (such as vision and tactile feedback) to improve their placement reliability. In addition, there are the high-volume sequential machines, often having multiple heads on a turret. Some of these machines are arranged so that several heads are performing separate functions simultaneously. There are very high-volume machines that use batch techniques, placing a large number of components in a single machine cycle. Mass placement machines are also called simultaneous units. Finally, there are the hybrid machines that combine the various features of both sequential and simultaneous placement.

On the low end of the *flexibility* ladder we find equipment that does very few tasks and concentrates on doing them very efficiently. At the top are the true robot, which can exchange end effectors allowing them to address an extremely wide range of assembly tasks. In general, increased flexibility comes at the expense of throughput and reliability of operation for any one task. However, robotics technology is making great strides toward a marriage of these three benefits of automation.

Categorizing placement equipment by *intelligence* may sound a bit silly to anyone who has spent a few minutes watching a robot assemble SMT boards. If robotic assembly machines get just one step out of sequence, hundreds of components are placed where their next door neighbor belonged. One of the authors was a witness when a robot forgot to index in the X and Y axes and started building a skyscraper of components at one side. Many times we have seen a machine drop a component in transit but go on to stuff its empty vacuum nozzle down in gooey solder paste and we haven't spent much time as a robotics observer.

One might justifiably ask if the phrase "machine intelligence" is a contradiction in terms, just like "jumbo shrimp" or "political integrity." The answer is "No," but only when we agree on the definition of intelligence in this sense. Intelligence here refers to a machine's ability to sense and react to its own operation and to the environment around it.

At the low end of the IQ scale, equipment has neither internal nor external feedback. The machine that built the component skyscraper described earlier was in this category. If it had been provided with internal feedback to verify whether or not the X–Y motion was occurring properly, it would have sensed the motor failure and called for help. Internal monitoring means sensing that the required actions within the system are occurring. Good robotic design practice demands that no machine action be inferred. All actions should be confirmed by feedback loops (monitored by encoders, etc.).

On the next step up the IQ scale is equipment with internal but no external feedback. The machine that dropped a component and still put its nose down into solder paste was not properly equipped to sense component presence on the pickup head. In fact, this particular machine did have a sensing mechanism, but the sensor did not always work. The sensing of external assembly conditions is an area where advances in robotics are rapidly improving the performance of automated machinery. Coupled with computer hardware and software advances (expert systems and artificial intelligence), sensing technology is dramatically extending the reach of machine intelligence today. Table 3-5 details some of the external sensing technology available in placement equipment/systems as of this writing.

Table 3-5 Types of Exerntal Sensing in Placement Equipment.

Sensor Category	Placement Task(s) by This Type Sensor
Delta Pressure	Component presence on vacuum pickup tip
Tactile Feedback	Component presence/size on pickup tip Placement pressure when placing a component
Component Testing	Verification of proper component-feeder loading Verification of polar comonent orientation Limited parametric testing of components (R/L/C)
Machine Vision	PWB orientation and referencing Skipping rejected sections of multi-board panels Solder paste or adhesive-dispensing inspection Placement site location Component form inspection and orientation Completed placement inspection
Bar-Code Readers	PWB recognition Component carrier recognition Setup versus program verification

Finally, as with all manufacturing equipment, we can separate placement machinery by *level of automation*. At the bottom of this staircase are the semi-automated workstations which simply serve to guide a manual operator through placement moves. A step above these, but still in the broad category of islands of automation, are the various standalone systems for populating boards. We say "standalone" to indicate that these systems provide no means of automatically loading workpieces from upstream processing stations, nor do they have means of passing completed workpieces to downstream machines. Next come the pass-through systems with automated board handling. And, at the top of the heap in degree of automation, we have the self-configuring robotic systems. These are systems which, to a marked degree, are able to automate the setup process. They may scan the bar code on an incoming bare board and then call up the proper assembly sequence and interchangeable tooling needed to build the given part number. These machines may be pass-through types or may be structured for robotic loading and unloading.

Process-Driven Design Rules for Placement

There are numerous areas where placement equipment requirements should influence design. We will discuss the more generic of these. Wisdom dictates that you quiz your placement system vendor or assembly house for additional equipment-specific design rules.

The clearances required between adjacent components vary dependent on placement equipment accuracy and, where required, tweezer clearances. Using the variables as illustrated in Figure 3-30, the SMT team should establish minimum clearance requirements for each component part number. The latest version of IPC CM-770 documents clearance requirements in detail, but typical values for these clearances are:

- Between chip components and chips and ICs: 1 mm
- Between SMT ICs: 1.25 mm
- Between components of different stencil print thickness when using a stepped stencil: 2.5 mm
- Between DIPs and SMT components: 1 mm

This information should form part of the CAD component library database. As adjacent components are located on the board, these clearance requirements should rule how closely they are arrayed.

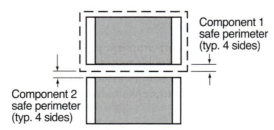

Component 1
safe perimeter
(typ. 4 sides)

Component 2
safe perimeter
(typ. 4 sides)

The Safe Perimeter extends an equal distance on all sides
from the nominal dimensions of the component (see text).

Figure 3-30   Clearances for collision-free automatic placement.

As illustrated in Figure 3-30, the minimum clearance requirements dictate the *safe perimeter* for each component. The safe perimeter extends an equal distance on all sides from the nominal dimensions of the component. To this end, the safe perimeter equals some of the factors, A + B + C + D, where:

A = The tolerance over the nominal component width or length, whichever is greater.

B = The component placement accuracy. (Note: Many placement machines specify this as ±0.001" or ±0.002". Actually their specifications seem to apply to the accuracy of their X-Y motion at the lead screw, not to the total error budget, which includes centering-jaw inaccuracies, Z-axis roll, pitch, and yaw inaccuracies, and theta inaccuracies. Few equipment suppliers can or will discuss this matter in detail. Actual observation and testing may be in order and may show that the total error budget at 3 sigma is closer to ±0.008" or ±0.010".)

C = The centering or jaw-carrying allowance, if these will be in contact with the component as it touches the board. (Some equipment breaks the contact of their centering jaws above the board and extends a vacuum nozzle only to place the component on the board. If this is the case, dimension C can generally be ignored.)

D = The tooling hole or optical registration error. (This item applies only to clearance of components loaded in a previous tooling setup.)

Placement machines with automated board handling use a multitude of PWB indexing schemes. Each system imposes its unique dimensional standards on assemblies. Areas of design and layout concern for automated board handling include, but are not necessarily limited to, those given in Table 3-6. Since the first publication of this book in 1989, one great advance has occurred on this front. The Surface-Mount Equipment Manufacturer's Association (SMEMA) has issued a standard design to allow integration of machines with automated board-handling features. Before buying machinery for a pass-through line, the reader should obtain the most recent version of this standard.

Table 3-6   Design Concerns for Automated PWB Assembly Handling.

Area	Concern(s)
Edge Clearance	Size of component-free area on PWB edges for conveyor handling
Registraton Feature	Tooling hole and/or fiducial mark size, shape, and locations
Top/Bottom Clearance	Highest allowable projection from the top and bottom surfaces of board
Flatness	Maximum allowable board warpage, including any sag produced by component weight. Avoid installing heavy parts until after placement.

Both the variety of component types and the quantity of different component part numbers are limited by the placement machine. It is advisable to keep the time from beginning an SMT assembly to the final soldering and cleaning to a minimum. Time limits vary dependent on the solder pastes or adhesives used, but any restriction speaks against multiple passes through the placement system. Where designs require component varieties and counts beyond the equipment capabilities, consider breaking the assembly into several boards or mother/daughter cards.

Vision systems may pose a number of restrictions on design. These include clear areas around features to be oriented or inspected and the required contrast between feature and background (which impacts materials selection).

The programming of placement equipment can be a tedious and time-consuming task, and errors can be introduced during manual programming. Therefore, engineering should provide as much help as possible to the manufacturing personnel responsible for programming. Depending on the placement system used, this might range from simple assembly details to post-processed CAD data suitable for direct downloading to the manufacturing line.

Simple assembly details should include a drawing showing the component center distance from board reference (tooling features) and the placement programming data from the CAD component library data base. The component library might include the clearance perimeter, placement pressure, pickup tooling, centering-jaw sequence and pressure, adhesive dot location and volume, and the component shipping-carrier information.

Post-processing for CAD downloading might include a conversion to placement machine language. Also, post-processing may involve the addition of details that are necessary to placement but are absent in the CAD data base, such as those listed for the component library, feeder location assignment, optimum assembly sequence determination, data on component rotation from feeders, and the conversion of component center distances from the abstract 0-0 point of the CAD system to the concrete 0-0 of the placement machine. Some placement equipment software includes the artificial intelligence necessary to simplify program development. Optimization routines might take care of some of the preceding steps, and various CAD systems differ in the level of detail that can be extracted for program generation. If CAD downloading is part of your manufacturing plan, start early and obtain the resources to develop the necessary post-processor. Your CAD system and placement machine vendors may offer some help, as may third-party CAM software vendors like Graphicode. But remember, it is likely that most software will require some tweaking by your manufacturing engineers.

3.3.4   Curing of the Attachment Media

As we established earlier, solder paste is sometimes cured, and adhesives are almost always cured before the reflow soldering operation. Solder-paste curing is done by a prolonged bake in a low-temperature oven. Adhesive curing generally involves baking, ultraviolet-light exposure, or a combination of the two.

Solder-Paste Baking

Curing of solder paste is common prior to vapor-phase and laser soldering. Without curing, the rapid temperature rises of these processes can cause an explosive outgassing of the volatiles in uncured paste. For small production requirements, the assemblies may be baked in a batch oven. For higher volume manufacturing, belt-type IR or convection furnaces are typically used. Where bake times are extended, temperatures are particularly high or oxidation-sensitive materials are used, a nitrogen or forming gas atmosphere may be specified. In general, however, ambient atmosphere works fine and keeps the curing costs to a minimum.

Adhesive Curing

Thermoset adhesives are cured in the same manner as that described for solder-paste curing. UV-curable adhesives are cured under ultraviolet light. Small belt and batch UV machines are available for curing printing inks and can be readily adapted for low production requirements. For higher production, belt machines combining UV and IR are typically used since heat greatly accelerates the polymerization of UV epoxies. Indeed, some UV adhesives demand elevated curing temperatures.

Design Considerations for Curing

In laying out boards that will require thermal exposure for curing, it is good practice to distribute the heat-sink masses evenly across the assembly. Large variations in thermal mass in an assembly produce hot and cold spots during a baking operation. Where uneven heat-sink masses cannot be avoided, the assemblies may be processed by lowering the temperature and increasing the soak time. Longer soak times allow temperature equalization across the assembly.

Assemblies that are to be UV cured should be designed so that ultraviolet light can easily reach its required target. Avoid any features with a parasol effect.

3.3.5   Soldering of SMCs

SMT Soldering Categorized

The first-order division of SMT soldering processes is between flow and reflow. Within each of these categories are various methods for accomplishing the soldering process. SMT flow soldering may be done on double-wave equipment, agitated wave machines, hollow waves, and specialized drag-soldering units. Reflow soldering is commonly done using conductive belt, convection, infrared, laser, and—rarely now—vapor-phase heating sources.

SMT Soldering Reliability Concerns

Many SMT soldering processes expose components to higher temperatures and steeper thermal gradients than wave soldering places on THTs. Therefore, it is reasonable to be concerned about component survival and reliability after processing.

Well-controlled SMT soldering processes using SnPb solder pose no threat to appropriate types of SMCs. Thus, as a designer, your job need not include any worry about soldering process-induced failures. Your job is to select the components appropriate for the soldering methods your assemblies will see. But again, changes in the soldering processes to accommodate lead-free solder need to be defined and quantified.

Soldering Process Impact on Design

First and foremost in design for solderability is the management of thermal mass. For flow soldering, this includes avoiding any large thermal vias which can connect a flow-type solder joint to a large heat sink. Such connections drain heat from the joint area and can produce poor wetting and open or marginal joints. Figure 3-31 shows how a thermal connection to a plane is reduced to avoid this condition with a PCB feature called a "thermal break."

We mentioned thermal path control in relation to flow-soldered boards because vias within SMC lands are permitted in flow-solder design rules. However, the same comment applies to reflow soldering, where special construction rules are followed to allow vias in lands or wherever lands have a direct thermal path to a nearby heat sink.

For reflow soldering, thermal-mass management also covers the board layout necessary to produce a relatively balanced thermal mass across the entire board surface. Having a large heat sink in one area of a board causes uneven heating in any mass reflow process. The solder lands near the heat sink do not reflow as quickly as those more distant from it. We know that tombstoning is aggravated when one end of a device reflows before the other. So, at best, large variations in thermal mass distribution may cause components to pop wheelies. At worst, the low thermal mass areas may exceed the safe time/temperature limits of assembly materials before the heavy thermal mass can be brought up to reflow temperatures.

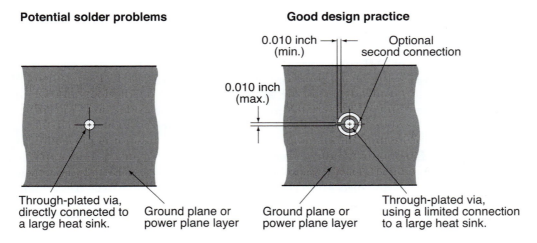

Figure 3-31   Reduce any thermal paths to large heat sinks.

If possible, leave high thermal mass objects off the assembly until after reflow soldering. If you must put a large heat-sinking device on an assembly prior to reflow, locate it near the edge of the card in a low population area and do not place potential tombstones near it. In all reflow processes, the card edges heat more rapidly than the center (assuming equal thermal mass distribution) because they have less thermal coupling and more exposed area for heat absorption.

3.3.6   Cleaning

Cleaning Equipment Categorized

Cleaners can be divided along two distinct lines. From a strictly process viewpoint, we can classify cleaners by the solvent used. Aqueous cleaners use water, usually with a saponifier (a wetting agent) as the cleaning solvent. Solvent cleaners use an organic-based solvent. Combination systems use an organic-based solvent followed by a water wash. From a materials handling view, we can separate cleaners into batch and in-line styles. Both batch and in-line systems are available in aqueous or solvent types, so our division produces a cleaner classification matrix.

Cleaning is basically a time/process-dependent equation. In other words, an aggressive cleaning process will accomplish a given task quickly, whereas some other process may yield equally clean assemblies if we are willing to wait a few years. Remember, just a little water and blowing sand was able to clean away millions of tons of solid rock and expose what we now recognize as the Grand Canyon. However, electronics manufacturing operations must usually reckon time in minutes or seconds, not geologic periods.

Batch Cleaners

Batch cleaners are generally devices for dipping assemblies into a cleaning solvent. The solvent may be in liquid or vapor form. In addition, sprays, heating, or agitation of liquid may be used to enhance cleaning. Excepting ultrasonic immersion cleaners, few batch-style machines clean rapidly and repeatedly enough to meet the demands of SMT. With the military concern over the safety of ultrasonic cleaning, most SMT cleaners are in-line-style machines.

In-Line Cleaners

In-line cleaners employ belt conveyors to transport the assemblies through a series of cleaning stages. These stages may include vapor baths, hot sprays, heated fluid immersions, ultrasonically agitated immersions, or combinations of these, followed by a drying stage at the end. Specialized machines, with aggressive cleaning processes such as ultrasonic immersion or very high-pressure sprays, have been developed to meet the challenges of cleaning surface-mount assemblies.

Aqueous and Semi-Aqueous Cleaners

Batch-style water-based cleaners operate like dishwashers. They may be far more sophisticated than the home appliances in their process methods and controls, but their basic operation is the same. In-line aqueous systems follow the description of in-line cleaners given above.

Arguments in favor of aqueous cleaning include the low cost of solvents and environmental or safety concerns which are less than those associated with some organic solvents. The negative arguments include contaminated waste-water treatment; the concern that water, even with a saponifier, may not penetrate under the minimal standoff of many SMCs; and the fact that special flux systems must be used for water cleaning. Certain water-removable flux residues are of more concern than rosin flux if not thoroughly removed in cleaning. Thus, homework is required in solder-paste selection.

Semi-aqueous systems bring a similar set of concerns to those listed above for the aqueous cleaner. In addition, some of the chemistries used are highly flammable and have a low flash point. Thus, automated fire control (again, without the use of ozone-depleting chemicals) may be a requirement.

Solvent Cleaners
Solvent cleaners use organic fluids that are particularly suited for removing organic soils, such as rosin flux. Because some polar contaminants may be entrapped in organic soil, stabilized organic/polar azeotropes are often used as solvents to remove both types of soils.

Batch solvent cleaners are typically either vapor degreasers or ultrasonic baths. In-line machines operate as described in our earlier discussion of in-line equipment. Most SMT solvent cleaners are of the in-line style because batch types generally require unacceptably long cleaning cycles to produce desirable results.

Combination Cleaners
Combination cleaners are in-line machines that use a series of solvent stages followed by a final aqueous stage to address the problem of removing both organic and polar soils.

3.3.7   Testing of Finished Assemblies

Testing is a topic that deserves a book all its own, and therefore cannot be fully addressed in a few paragraphs herein. Because of space limitations, we will present an overview of testing methods in this section. Then, in Chapter 4 through Chapter 7 we will talk in more detail about design for testability (DFT). For further discussions of test engineering there are numerous articles and books available. We suggest you start with reference information by Crouch,[18] Pyn,[19] Lawlor-Wright,[20] Parker,[21] and Gallagher,[22] and also refer to the additional references in Chapter 4 through Chapter 7.

Test Methods Categorized
Broad categories of testing can be drawn around the type of defect or parameters being inspected. We test for electrical defects and function using manufacturing defects analyzers, in-circuit testers, functional testers, burn-in systems, and mock-ups. Assembly integrity and design-rule adherence are checked using various manual and automated vision approaches. Solder-joint quality is inspected using certain vision systems and using laser thermometric analyzers. Assemblies are checked for harmful contaminants in cleanliness testing. All of the above, plus

basic engineering, may be tested using *destructive physical analysis (DPA)*. We will now briefly cover each of the types of testing we have identified.

Manufacturing Defects Analyzers

Manufacturing defects analysis (MDA) is a fast test for shorts, opens, and reversed-polarity components. Boards are probed using "bed of nails" fixtures. MDA does not confirm assembly or component operation, but it gives a rapid feedback on some areas of the manufacturing process. MDA testing is usually coupled with other tests to fully qualify the assemblies.

In-Circuit Testers

Ideally, in-circuit testers will fully parametrically test each and every circuit element by isolating each element and testing it as an individual unit. To reach this ideal, every circuit element's I/O must be able to be fully probed in a "bed of nails" fixture, and every circuit element must be completely in-circuit testable. SMT works against both these requirements. Dense circuit populations (the goal of most SMT design programs), close I/O pin spacings, and bottom-side devices can make the probing of all I/Os—or even all circuit nodes—difficult to impossible. In addition, the increasingly complex high-pin-count devices facilitated by SMT often cannot be fully tested in-circuit. Imagine testing for the entire truth table of a 32-bit microprocessor and all of its support chips.

The benefit of in-circuit testing, as opposed to MDA testing, is the finding of component as well as manufacturing defects. Thus, in-circuit testing may provide a guide to repair without costly troubleshooting by technicians. Of course, this benefit is limited to designs that allow in-circuit analysis of a major portion of a circuit's elements.

Functional Testers

Functional testing looks at circuit operation *in toto*, and/or in definable chunks. A chunk of a circuit may include one or more elements. Functional testers access the circuit using assembly I/O such as edge connectors or cabling and by the probing of circuit nodes on a bed of nails. Full access to circuit nodes is essential so a functional test can isolate a fault and furnish detailed troubleshooting guidance. Where probe access must be limited, the design team should devote attention to active partitioning and built-in tests in order to reduce the costs of troubleshooting on the bench.

Functional testing may be paired with other testing so that the functional test looks only for areas of circuit operation not verifiable in other tests. Manufacturing defects, component failures, and troubleshooting guidance are derived primarily from outside the functional test.

Assembly and System Burn-In Testers

Burn-in testing involves the monitored operation of the assembly or system at elevated temperatures for a defined period of time. Some companies call a simple on-line operation of a system without elevated temperature *burn-in*. More sophisticated burn-in tests can spot the time of failure and can provide significant detail regarding the failure mode when out-of-specification operation occurs. Like other board tests, access to circuit nodes determines the level of detail that can be obtained in burn-in testing.

As with burn-in testing of individual components, assembly and system burn-in are aimed at catching those functioning components that will fail (under power) early in the circuit's life. Such failures are called *infant mortalities*. Components that do not succumb to infant mortality tend to survive to a ripe old age or, in technical terminology, they will have a normal lifetime and will fail during their wearout phase. Therefore, eliminating weak sisters by operating under temperature stress will greatly enhance circuit reliability. Full-board or -system burn-in will also catch heat-related failures that might occur in the field when the system is under power at elevated ambient temperatures.

Mock-Up Tests

Mock-up testing may be as simple as powering a system up to see if it performs its specified functions. Or, mock-up testing may involve plugging the system into a black box to speed up testing and/or capture more information than would be obtained by just operating the system. Mock-up testing provides a low capital investment approach to determining if an assembly is a live one. However, it is best applied to testing systems with only a few operating modes and a limited number of response patterns. Complex systems, with many permutations of responses for various stimuli, would take an enormous amount of mock-up test time. Mock-up testing may also fall short in providing troubleshooting guidance.

Visual Testing

Vision is used in every facet of electronics for quality verification. Visual testing runs from simply "eyeball" inspection of a board to passing assemblies through sophisticated automated vision or x-ray systems. Simple operator inspection, with or without magnification, is used to detect missing or grossly misaligned components. Microscope inspection may be directed at solder quality and fine alignment. Automated vision systems using visible light are high production tools which, with carefully developed programs, will rapidly detect missing and misaligned components and some solder defects. X-ray systems may be used to spot hidden defects, such as voids in solder, delaminations in PWB material, cracked components, and broken internal wire bonds in IC packages.

Laser Solder-Inspection Systems

Laser thermometry uses infrared laser radiation directed at an individual solder joint to rapidly heat the joint. The inspection system then turns the laser off and monitors the radiated heat from the joint. Good solder joints provide a solid thermal path into the board, so the built-up temperature quickly dissipates. Defective joints are flagged because they retain the heat longer. Like x-rays, laser thermometry finds both visible and hidden solder flaws.

Cleanliness Testing

Conductometric testing has been the mainstay of PWB cleanliness testing in through-hole technology. A mounting body of evidence suggests that new methods, as discussed earlier in Section 3.1.5 of this chapter, will be useful in the cleanliness testing of SMT assemblies.

Nondestructive and DPA Testing

Finally, the physical analysis laboratory provides a powerful testing tool. *Destructive physical analysis (DPA)* is particularly useful in evaluating early SMT designs and detecting the causes of failures. DPA includes microsectioning, scanning electron microscope (SEM), and chemical analysis. Non-destructive techniques include hot-spot detection, x-ray radiography, particle impact noise detection (PIND) of packaged circuits, and high-powered microscopic observation.

By definition, DPA is not a production test, but it may be used to sample a small percentage of boards from a production line. Such sampling can yield early warnings of impending process problems.

3.3.8   Rework and Repair

Unless assemblies are very low cost, you will probably want to salvage those that fail test. The good news is that SMT boards can be reworked. In fact, you may find a 68-pin chip carrier easier to remove than a 64-pin DIP with clenched leads. The bad news is there are new and difficult challenges to meet in SMT repair.

Let's begin our repair discussion by looking at some of the challenges SMT presents. Obviously, one is the localization of repair reflow on densely populated boards to avoid damage to areas adjacent to the repair. A subset of this is that removal of SMT ICs is done by reflowing all leads simultaneously. While this can be accomplished quickly using a hot-air repair station, the PWB will remain heated much longer than it would if you were to desolder a DIP one lead at a time. Prolonged high-temperature exposure can cause delamination or measling in PWBs and can cause metal migration between board plating and solders. Another SMT repair problem is the visual identification of components that are small, may have no marking, and are easily confused with one another.

Now, we will look at typical repair processes and see how design can enhance their effectiveness.

Heated Tweezers and Special Repair Irons

These simple repair tools are soldering irons with tips that are shaped to cause the simultaneous reflow of all terminations of a certain size SMC. The tweezer variety can actually close on a component to grip it during the repair process. The specialized irons have no gripper but are shaped to heat all leads of the target device simultaneously. The cohesive bonds of the solder and wetting forces of solder on the iron cause the reflowed component to come away from the board when the iron is lifted.

Both tweezers and tong-tipped irons are usually used to remove and replace small components with leads on two sides. To remove a faulty device, the repair technician places the heated tongs of the iron on the solder joints of the component, waits for the solder to reflow, and then lifts the part with the iron. After removing a component, the joints should be cleaned of all solder, and fresh material is then used for replacement. Otherwise, solder contamination, particularly from the dissolution of traces from the board, may produce a marginal joint or worse. Replacement of components, using fresh solder and flux, is done with the same iron for reflow soldering.

Convective Reflow Stations

Convective reflow repair stations use either heated gas or heated air that is directed in a focused stream from above and/or from below the repair areas to melt the solder for rework. Solder joints of faulty components are heated. The operator removes the part with tweezers or a vacuum tool. Alternatively, some repair stations provide a vacuum head, that is used to remove components. After solder removal and cleanup, new solder and flux are added, and a new device is reflowed to the board using the same convective heating procedure.

Design for Specific Repair Procedures

First, the self-evident: repair operations are always easier on single-side populated assemblies and on boards with a little elbow room between the components. Adequate clearance between components is critical for rework when using tweezers or shaped irons. Plenty of clearance is also the key when shielding neighboring parts in a convective reflow operation. It is possible to convectively rework double-sided assemblies. Top-side heating only should be used. However, conductive heat transfer through the board will probably bring the bottom of the rework area over the reflow temperature. Solder surface tension will hold all standard SMCs on an upside-down board, so components are not likely to fall off unless they are upset externally while the solder is molten. The problem is that metal migration and solder contamination will occur, both during component removal and replacement. Thus, the part directly under a convectively reworked area may look fine but may have seriously weakened solder connections.

3.4 Degree of SMT Automation

To define degree of automation, let's imagine a degree of automation scale. At the low end of the scale is a total hand-assembly operation. On the high end is a machine that has a raw-material hopper at its input and finished, bagged, and boxed assemblies coming from its output. This topic has enormous impact on the manufacturing engineers of the team, particularly when we approach the high end of the scale. However, we will limit our discussion to an overview of design for automation in this text.

3.4.1 Prototype Assembly by Hand

The tiny, often unmarked components, dense layouts, and fine lead pitches of SMT can all spell headaches for assembly technicians who are learning to build surface-mount assemblies by hand. But, manual assembly frees the designer from many of the process design constraints listed earlier. Let's look at the design constraints that pertain to hand assembly.

Some designs would be severe burdens on automated lines because they would require backtracking for multiple passes through certain process steps. Such designs may be no particular problem for manual assembly. Humans are much more adept than machines at jumping back and forth between tasks.

Manual assembly requires sufficient component clearance so unsteady human hands can make placements without upsetting adjacent parts. Tweezer clearance must be considered. And, you should provide enough room so that technicians can clearly see how to orient miniature device leads on PWB lands and can visually inspect the finished work.

Manual assembly eliminates the concerns of automated registration and the handling of boards in placement equipment. Tall components, heavy components, irregularly shaped components all pose less concern in manual handling than in automated assembly. However, the process-dependent rules set forth in Section 3.3 for solder-paste application, adhesive application, curing of adhesives, soldering, and cleaning apply to all degrees of automation.

3.4.2 Islands of Automation

A word of explanation is in order before beginning our discussion of this topic. Islands of automation, as used here, means standalone machinery for the various process steps. We refer to equipment that is not able to or is not used to automatically handling and indexing assemblies between various process steps. Factory automation often uses standalone islands, called *work cells*, to do specific process steps. Robotic carts or smart conveyors may index work between these "islands." However, since we will consider these arrangements separately under factory automation, we will restrict our view here to the narrow definition given above.

When all workstations are standalone units, it is entirely possible that each will use a different approach to orient and handle the workpieces. Board clearances, indexing, and orienting features should be designed to accommodate all the pieces of process equipment that will be required for the assembly.

Islands of automation can accommodate multiple passes through a given process step. However, minimizing work in process (WIP), avoiding multiple setups, and the rapid completion of assemblies will all speak against pin-ball machine process flows. If possible, design the assemblies to flow in an orderly fashion one time through the manufacturing line just as if the line was connected by conveyors.

3.4.3 Line Automation

Line automation describes process machines (and steps) connected by fixed conveyors. In line automation, workpieces must visit each process step in the line whether any work is needed at that step or not.

Line automation dictates close attention to design to facilitate board handling. Free areas are usually required on the board or panel edges. Tooling holes and/or fiducial targets may be needed for alignment. Thus, design rules should be developed around the handling specifications of the line.

The design rules should also take into account the minimums and maximums for each process step during a pass through the line. In other words, if the placement system only allows feeders for 100 unique component part numbers, limit the assemblies to that number. It generally does not make economic or quality sense to make multiple trips across a pass-through line. Multiple passes requiring setup variations would be an economic disaster.

Manufacturing efficiency of line automation requires considerable manufacturing-engineering input from the earliest stages of system design. This comment is especially true when high-volume manufacturing is expected, which is often the reason for selecting a pass-through manufacturing approach. High-volume pass-through manufacturing adds emphasis to all the design-for-manufacturability comments discussed above, and in Chapter 4 (Section 6), and Chapter 6. Line automation doubles the importance of the manufacturing group's input to the SMT team.

3.4.4   Factory Automation

Factory automation is used here to mean flexible automation that requires little or no operator intervention to build a wide variety of PWB assemblies. Factory automation integrates the computer and machine intelligence with machine vision and sophisticated sensing of the manufacturing environment. The ultimate goal of this integration is to build a self-configuring flexible factory capable of throughputs that rival line automation. Such a flexible factory would be said to offer "economy of scope," whereas the pass-through line boasts "economy of scale."

The "design for manufacturing" comments made earlier regarding line automation are only amplified by factory automation, and the complexity of software and process control required for successful factory automation dictates the starting of projects far in advance of delivery dates and the allocation of substantial resources for prototyping the process and developing the line. We would not recommend factory-automated SMT be undertaken by any organization not thoroughly familiar with surface-mount design and manufacturing.

3.5   Manufacturing Volume of SMT

As we mentioned earlier, manufacturing volume has a significant impact on design directions. Next, we will look at various production levels and see how the SMT team adapts processes to suit production capacity.

3.5.1   Low-Volume Production

We will arbitrarily define low volume as the placement of less than 25,000 components per average month. By this definition, a placement system or technician placing about 175 components per hour could handle the required output in a single shift. Such a rate is reasonable for manual, semi-automated, or pen-plotter style placement. Note: Since placement is usually the pacing process in an SMT line, it makes a convenient benchmark against which we can gauge the manufacturing volumes.

Low-volume manufacturing shifts some of the SMT team's attention from design for producibility to parts procurement and quality issues. At low volumes, manual techniques can cover for many manufacturability sins. However, low-volume manufacturing robs an organization of clout with component suppliers and distributors. This can spell trouble for parts availability and quality, and low volumes make statistical process control somewhere between impractical to impossible. The SMT team should develop a strategy to deal with the unique challenges that low volume may pose.

Where low volumes are complicated by a high mix (a requirement to manufacture many different assembly part numbers), a simple primarily manual approach should be the target. Where possible, avoid excessive setup and programming burdens. For instance, bench troubleshooting will probably cost far less than the development of sophisticated tests to provide detailed fault analysis.

3.5.2   Medium-Volume Production

We will say that medium volume extends from 25,000 components to around 300,000 component placements per month. And, 300,000 placements per month equates to about 2150 parts per hour in a 20-day single-shift operation. There is a wide range of flexible placement equipment that is capable of supporting rates of 25,000 to 300,000 placements per month. In reference to placement rates, we will offer a word to the wise. There is a large troupe of manufacturing personnel who have been disappointed by equipment that would not deliver even 300K parts per month when the specification sheets promised three times that! What we are talking about here is day-in and day-out production on real-world assemblies, not an idealized laboratory test. And it is our experience that with one flexible placement machine, 300,000 parts per one-shift month is a high number in the real world.

Our comments on "design for process considerations" made earlier are aimed at just such a manufacturing volume. The only additional comment necessary is that a high manufacturing mix amplifies demands on the SMT team to develop solid design rules for the organization so that each successive design is manufacturable on the existing line.

3.5.3   High-Volume Production

For our discussion, high-volume manufacturing will begin at over 300,000 component placements per month. We chose this volume because it corresponds to just over 2,000 parts per hour in single-shift mode, and 2,000 parts per hour is a significant breaking point for flexible-placement machinery. Only a select few of the legions of flexible P&P machines available can sustain rates very far above 2,000 parts per hour over a long haul if all the downtime and setup times are considered.

Lines that are set up for significantly higher rates will often have several placement machines. In high-volume production lines, one of the machines may be a "chip shooter" (a very high-rate machine optimized for a narrow range of tasks, such as the placement of passives from 8-mm and 12-mm tape only).

All the comments in this chapter and in the rest of this book regarding design-for-manufacturability, test, and repair will apply tenfold when high volumes are planned. Special emphasis should be placed on standardized panel sizes and handling features so that any board may pass through the line without causing tooling changes. Standardized inventory will also pay off, particularly if the part number count can be kept low enough for stocking on the line. Multiple-pass manufacturing should not even be considered. If necessary, avoid it by routing problem boards to special work cells. A better approach would be to simplify the assembly by using standardized modules on daughter cards.

The SMT team should be up to speed on surface mounting at the prototype and moderate-volume levels before applying that knowledge to high-volume manufacturing of SMT. Likewise, a solid knowledge of manufacturing automation is a precious asset for any group launching into high-volume automated SMT assembly.

In closing, remember that all the rules and caveats given above are for guidance. After looking them over, if you feel you must violate one or more, that is all right. Many others have successfully done so. This book was not written to stifle engineering creativity. Instead, we hope the discussion of manufacturing rules will serve to guide those creative efforts.

3.6   Review

After reading and understanding this chapter, you should be able to answer these questions and note the reference location for the information in the chapter.

1. List the steps in the development and manufacturing of a PCB using SMT components.
2. Define "fiducial" and its correct shape.
3. What is the proper circuit board layer in which to place fiducials?
4. Explain the difference between a global fiducial and a local fiducial.
5. What is the most common composition of lead-bearing solder and solder paste?
6. What are the most commonly proposed compositions of lead-free solder? (see also Chapter 1)
7. Define "wetting."
8. Describe how wetting and surface tension are involved in the formation of a solder joint.
9. Define "intermetallic layer."
10. Describe how the intermetallic layer is formed.
11. Briefly describe the three types of commonly used fluxes and the solvents necessary to clean them off the printed circuit assembly if cleaning is necessary.
12. List two of the several 'unknowns' about the proposed lead-free solders.
13. Sketch the RDRP and RR profiles with time and temperature axes to scale.
14. Explain the purported advantages to the RR profile. Do you agree with them?
15. Explain issues with thermocouples that affect the quality of thermal profiling measurements.
16. Describe how a chip capacitor forms a "tombstone" during reflow soldering.
17. What is the purpose of each wave in a dual-wave soldering machine?
18. What is the purpose of a saponifier in a cleaning system?
19. Explain the problems caused by the surface tension of water during the cleaning process.
20. List three test issues that must be considered during the design phase.
21. For the following part and board characteristics, calculate the amount of solder paste that should be deposited by a stencil using the pin-in-paste (PIP) process:
    - A $1/4$-watt resistor will be placed in a through hole. What is the lead diameter?
    - The board is 0.062" thick
    - Visible solder fillets are required in the final solder joint
    - Make and state any assumptions you feel are necessary for the calculations

22. Describe four of the sensors that may exist on an SMT placement machine and the application of each sensor type.
23. For safe placement perimeter, determine the size and tolerances for a National Semiconductor Corp. QFP package and calculate the safe placement perimeter for this device. Make and state any assumptions and/or additional data you determine is necessary for this calculation.

3.7   Standards

Standards that apply to topics in this chapter include:

- IPC/EIA J-STD-001, "Requirements for Soldered Electrical and Electronic Assemblies," the industry-consensus standard for soldering, including military units since MIL-STD-2000 has been cancelled.
- IPC/EEA J-STD-002A, "Solderability Tests for Component Leads, Terminations, Lugs, Terminals and Wires," covering solderability of component leads, terminals, wire, and hardware.
- ANSI/J-STD-003, "Solderability Test for Printed Boards," covering dip, float, and wetting balance solderability tests for board surfaces.
- ANSI/J-STE-004, "Requirements for Soldering Fluxes," covering classifying and testing of rosin, resin, organic, and inorganic fluxes.
- ANSI/J-STD-005, "Requirements for Soldering Pastes," covering characterization and testing of solder pastes without regard to their performance in the manufacturing arena.
- IPC-A-600, "Acceptability of Printed Boards," covering acceptance of bare PC boards.
- IPC-A-610C, "Acceptability of Electronic Assemblies," covering acceptance of populated PC assemblies and including photos and illustrations of target, acceptable, and defect conditions in soldered assemblies.
- IPC-SM-840E, "Qualification and Performance of Permanent Solder Mask."

3.8   References

1. *Intel Packaging Handbook.* Available at http://www.intel.com/packtech.
2. Prasad, R. P., *Surface-Mount Technology Principles and Practice.* Chapman & Hall Thompson, New York, 1997.
3. Chip inductor "Notice (Soldering and Mounting)." Available at http://www.muRata.com.
4. *Intel Packaging Handbook.* Available at http://www.intel.com/packtech.
5. Leicht, H. W., "New Look at Vapor Phase." *SMT Surface-Mount Technology Magazine,* v. 9 #10, October ,1995, pp. 27–31.
6. Willis, B., "Vapour phase is back for lead-free soldering." Global SMT and Packaging, v. 3 #7, October, 2003, pp. 4–5.
7. Lee, N-C., *Reflow Soldeirng Processe and Troubleshooting.* Newnes, Boston, 2002.
8. Dahle, B., Lasky, R., "Optimizing Reflow," *SMT Magazine,* January 2004, pp. 40–42.
9. Suraski, D., "Reflow Profiling—Time Above Liquidus." Available at http://www.aimtech.com.

10. Shin, C.K., Baik, Y. J., Huh, J. Y., "Effects of Microstructural Evolution and Intermetallic Layer Growth on Shear Strength of Ball-Grid-Array Sn-Cu Solder Joints." *Journal of Electronic Materials*, v. 30 #10, 2001.

11. Rouband, P., Ng, G., Henshall, G., Bulwith, R., Herber, R., Prasad, S., Carson, F., Kamath, S., Garcia, A., "Impact of Intermetallic Growth on the Mechanical Strength of Lead-Free BGA Assemblies." Proceedings of IPC/APEX 2001.

12. Sunter, S., "Cost/Benefits Analysis of the 1149.1 Mixed-Signal Test Bus." IEE Proceedings of Ciruits, Devices, and Systems, v. 143 #6, 1996, pp. 393–98.

13. Lechner, A., Richardson, A., Hermes, B., Ohletz, M., "A Design for Testability Study on a High-Performance Automatic Gain Control Circuit." Proceedings of the IEEE VLSI Test Symposium, Monterey, CA, 1998.

14. Fujiwara, H., *Logic Testing and Design for Testability*. MIT Press, Cambridge, 1985.

15. Spigarelli, Donald J., "Design and Process Considerations for Soldering Surface-Mounted Components," *IEEE Eurocon 1982 Proceedings,* p. 365.

16. Aspandiar, R., Litkie, M., Arrigotti, G., "Pin-in-Paste Solder Process Development." Proceedings of the 1999 SMTAI. Available at http://www.smta.org/files/SMTAI1999-AspandiarRaiyo.pdf.

17. Jensen, T., Lasky, R.C., "Electronics Assembly: Practical Tips for Implementing the Pin-in-Paste Process." *Assembly Magazine,* Feb. 1, 2003.

18. Crouch, A. L., *Design for Test with Digital IC's.* Prentice Hall, Upper Saddle River, NJ, 2000.

19. Pynn, C. T., *Strategies for Electronic Test.* McGraw-Hill, New York, 1986.

20. Lawlor-Wright, T, Gallagher, C., "Development of a Printed Circuit Board Design for In-Circuit Test Advisory System." *Computers in Industry,* v. 33 #2–3, September, 1997, pp. 253–59.

21. Parker, M., Smith, J., "PCB Testing Goes Socketless." *Evaluation Engineering,* v. 43 #7, July, 2004, pp. 84–87.

22. Gallagher, C., Lawlor-Wright, T., "Review of PCB Design for In-Circuit Testability Guidelines and Systems." *Journal of Electronic Manufacturing,* v. 5 #3, September, 1995, pp. 175–81.

System Design Considerations for SMT

Objectives

After reading and understanding this chapter, the reader should be able to:

- understand the simultaneous engineering (concurrent engineering) approach to design
- understand the form, fit, and function of a product
- understand basic cost and capital investment issues
- understand the basics of the overall manufacturing environment

Introduction

This book is aimed at a technical subject and is intended for use by a technical audience. However, it is important to never lose sight of the people side of bringing the technical wonders of SMT to light. We'll cover some of the human facets of successful design project management in this chapter.

SMT is a complex multi-variable design problem. As such, good SMT designs require good systems engineering. In this chapter, we will develop a plan for an SMT systems-engineering effort. We will begin development of this plan by ranking the benefits that actual SMT users say they achieve from surface-mount technology. Our plan will follow an engineering approach in order to extract the best of what SMT has to offer.

Throughout 1987, Hollomon's company conducted a study of the electronics business in the upper Midwest. Even though much of the enabling technologies have changed, the basics are still similar enough that the following material from the data we collected may prove useful in guiding system design teams who are entering SMT for the first time. In our survey, we asked actual users of SMT to rate ten potential SMT benefits on a scale of 1 to 10, showing which had the greatest importance in their application. They were instructed that 10 equals the most important and 1 is the least important. Figure 4-1 details the averages for each benefit from among the 104 sites that responded.[1] Remember, these are ratings from actual SMT users.

In Figure 4-1, we see some of the benefits others have achieved using SMT. As your team begins systems engineering on a new project, remember these ratings and fight compromises in design that would fall short of the full promise that SMT offers. With our benchmark thus set, let's develop an organized attack plan for the systems engineering of SMT products.

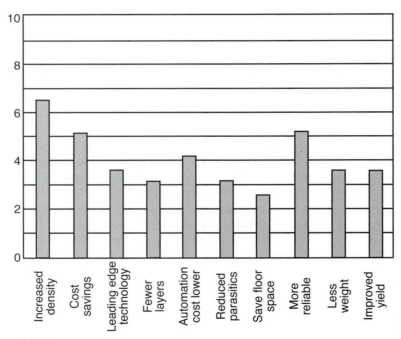

NOTES:
1. SMT benefits are rated on a scale of 1 to 10, with 10 being highest.
2. Results shown are the averages of responses received from 104 sites currently using SMT.

Figure 4-1   How users rated SMT benefits in importance.

One side note for the designer and/or design team that is new to SMT: While there is currently little written to bring true newbies up to speed on SMT, *Electronics Design News (EDN)* published a multi-part "Hand-on SMT Projects" in 1987. It may be worth finding this excellent series if you and your team are truly new to SMT. We refer to it several times in this chapter.

4.1   The Simultaneous Engineering Approach

"Simultaneous Engineering," as it is called at General Motors, is one name for the team approach to product development. Another is "Concurrent Engineering," with its subsets Design for Manufacturability (DFM) and Design for Testability (DFT). Concurrent Engineering can be succinctly described as:

> concurrent engineering requires that all aspects of design, manufacturability, testability and repairability be considered during the design phase of a product

Team engineering has come into vogue in the United States, as a result of cross-cultural fertilization from Japan. We have been deeply impressed by Japan's industrial success, taking its economy from post-war ruins to world-class performance in just a few decades. The U.S. auto industry, in particular, has studied and benefited from the Japanese model of automotive development and manufacturing. Chrysler, Ford, and GM all have team engineering programs. Let's begin our discussion of team engineering by looking at its application by the Ford Motor Company, where it is called "The Team Concept."

As originally designed and built, the Ford Taurus was a runaway success with both critics and consumers. Introduced in 1985, it was the first U.S. car built by team engineering. Other U.S. car companies soon followed. At General Motors it was called simultaneous engineering and involved every department that will ultimately bear responsibility for the product: Design, Engineering, Manufacturing, Marketing, and Finance, as well as outside suppliers.

How does this contrast with business as usual? In the sequential mode, the design department would develop plans for a new model. The plans would then go to Engineering where the available components, such as transmissions and engines, would be fitted into the new enclosure dreamed up by design. Often, friction would arise when existing hardware would not fit. The result would be a compromise that might leave each department somewhat frustrated. Next, the design would go to Finance where the cost of each component, engineering choice, and design direction would be called into question. Finance would usually find something objectionable, with the result of more compromises.

When Design, Engineering, and Finance finally came to an agreement on the new car, plans would go to the manufacturing division. Here, assembly line managers might say, "We can't build it," or "We can only build it if you give us half a billion dollars for new tooling." New and painful compromises would be the likely result. Finally, and only after all these parties came to a truce and decided to build the new model, outside suppliers were called in to provide new specified parts. Because of the compromised design, suppliers might be muscled into bidding so low that either their profits, their quality, or both, were seriously eroded.

The car that finally emerged from such a cumbersome design-by-negotiation process generally was too late to market, too poor in quality, and too far from what marketing needed to suit customer demands. Engineering lead time was up to 6 years, far too long for Marketing to accurately predict customer needs and wants in today's fast-changing world.

Recognizing that something had to be done to regain competitiveness, Ford's then-chairman Don Petersen hired Dr. W. Edwards Deming in the early 1980s. One of Dr. Deming's most quotable slogans is, "Quality is a management decision." He means quality comes from top management commitment, not from a forgotten few inspectors at the end of the line. In other words, quality demands that all departments in a manufacturing organization work as a team, with quality as a major goal. Out of this quality approach, simultaneous engineering has emerged as a powerful tool to improve engineering and manufacturing efficiency. Ford's original goal was to bring their normal engineering lead time down from 54 to 45 months and less. The resulting savings will be enormous, but the standard to measure against is even more impressive. Mazda boasts a 24- to 36-month cycle using simultaneous engineering.

Team engineering is here today and will undoubtedly become the accepted way of engineering any complex product. It is already the accepted way in all of the successful SMT operations that our consulting work has uncovered. From the preface of this book forward, we have spoken of the SMT team. At this point, let's look at what this team is. How does team engineering work? What proven history does team engineering carry in the electronics industry? Below, we will answer those questions. Let's start by looking at some SMT team applications from industry today.

4.1.1   Applications of the Team Approach

Delphi Delco Electronics Corporation

Part of the team approach to designing and manufacturing a product is to be certain that all members of the team understand and buy into the processes being selected. Dephi Electronics (née Delco Electronics Corp.) purchased their first chip placement machine for SMT in 1978 but did not enter into SMT production in volume until 1983. A team of twelve to fifteen players from diverse disciplines, such as Design Engineering, Manufacturing, Purchasing, Production, and Maintenance studied SMT for 5 years before turning on volume production. Delco took this time to study surface-mount technology even though they already had the benefit of some SMT experience from years of hybrid manufacturing.

According to Daniel K. Ward, Assistant Superintendent of Manufacturing Engineering at Delco's Kokomo production facility,

> "Purchasing and parts engineering groups obtained samples and did preliminary testing on component reliability. We also researched the best way to design our boards for SMT. We put both of these together in our products."[2]

Delco's facility at Kokomo, Indiana, was at one time the highest-volume SMT factory in the world. At that time, Delco was using over 20 million surface-mount components each day.

Hewlett-Packard Corporation

Continuing the concept that all players must understand and be involved in the process, Steve Hinch, Hewlett-Packard's Corporate SMT Program Manager, has noted,

> "It's important to get Purchasing involved as soon as you know you're going to do surface mount. We've found a wide quality difference among suppliers. It's clear that the selection of surface-mount component suppliers must be a partnership effort between the design engineers and Purchasing."[3]

4.1.2   The Need for the Team—A Complexity-Driven Problem

Once upon a time, a single skilled craftsman could turn out a beautiful and well-crafted product like a pair of boots. If you needed more boots, you employed more craftsmen. But, as manufacturing volumes grew and product complexity multiplied, the place of the generalist craftsman gave way to the technical expert, one who is highly skilled in a single facet of the creation of a product. Even a small SMT line like the one shown in Figure 4-2—with a solder-paste dispenser, a placement machine, and a reflow oven—needs skilled individuals to both design for manufacturing as well as actually run and troubleshoot the line itself.

This "Age of the Specialist," coupled with an exponential increase in complexity of manufactured goods, has made teamwork an essential element in product development and

Figure 4-2   A small SMT line.

production. This fact is particularly true in SMT. Design oversights often cannot be taped together on the production floor. If testability, repairability, manufacturability, etc., are not designed into an SMT board, the results may run from poor (higher manufacturing costs, lower yields, and more field failures) to abysmal (product not manufacturable within required costs, multi-million dollar write-offs, and a declaration of bankruptcy).

What happens when a company adopts surface mounting? Which of the corporate operations are likely to be affected? SMT will impact Management, Finance, Purchasing, Receiving Inspection, Materials Handling/Storage, basic circuit-schematic design, circuit-board layout, PWB manufacturing, assembly methods and tooling, test engineering and equipment, repair and rework methods, mechanical enclosures and board interconnect layouts, and warranty repair strategies to name just a few areas. Even the dust from the janitor's broom may be an issue, and SMT will go beyond the corporation in question to impact its suppliers as well.

Therefore, the single most important factor in successfully entering the SMT field is a smoothly operating team approach. To set up an SMT implementation team, do all of the following:

1. Establish top management involvement, understanding, and support.
2. Form a design team with all the functional areas involved, including the key vendors.
3. Give the team the time and resources to do the job.
4. Set and fully communicate realistic goals.

4.1.3   How Simultaneous Engineering Works

There is no absolute answer to the question, "How should team engineering work?" The New York Yankees differ from the Minnesota Twins. Each team is shaped by the culture of its town, the personalities of its players, and the attitudes of its managers, but both have demonstrated they can be winning teams. In the same way, engineering teams are molded by the culture, members, and management of their companies. However, there are a few absolutes to guide team formation, and there are some starting points we can suggest for your consideration of the optional points. We will cover the firm rules below.

Absolutes

Successful team engineering breaks down departmental battle lines. Team members no longer wear just one hat. They are not just designers, or manufacturing managers, or controllers. They cannot care only about the interests of one department. They put on a second hat of "team member." The team's assigned task is the responsibility of each and every member. The team either wins or it loses, and Most Valuable Player awards do not often go to losing teams. So the first absolute is that team engineering works by enforcing the Golden Rule,

> "Therefore all things whatsoever ye would that men should do to you,
> do ye even so to them."[4]

(For you younger folks, that's "Do unto others as you would have them do unto you.")

When trade-offs must be made, each player must consider the good of all the departments on the team rather than trying to protect the sacred turf. The focus must be on the overall goal.

Rule 1—Whatever the form your company chooses for simultaneous engineering, it should be designed to preserve the Golden Rule effect.

Rule 2—Do not leave any involved department off the team. If territorial squabbles interfered with design compromises before team formation, imagine the problem level when six departments gang up on one. Further, imagine that the one department feels both picked on and left out of the important work. Leaving out important members will absolutely not save money. It will waste money, and may result in the left-out department looking for monkey wrenches to throw into the system just for spite. Remember, the whole reason for simultaneous engineering is to boost efficiency.

Rule 3—Give the SMT program every chance to succeed by providing the team with all necessary resources. However, be judicious as it is a poor engineer who cannot figure out ways to use all the available space and spend outlandish amounts of money!

Rule 4—As noted in Deming's earlier quote regarding quality, management support and involvement is also critical in an SMT startup. A substantial level of man hours and corporate resources will be required for success. Commitment of such resources will involve top management. A middle manager trying to start SMT as a skunk-works project will have to cut too many corners and will certainly fall short of achieving SMT's full promise. The unbridled organizational backing needed to tackle surface-mounting technology only comes when top management is informed and excited about SMT.

Aside from providing the necessary resources, the other reason that top management support is vital is for communication. Since SMT impacts all functional areas of a company, the SMT program will run smoothly only with the support of all departments. And such support is not likely to come out of one excited engineer talking up SMT to Finance, Tool Engineering, Test, etc.

Rule 5—Management must assess what surface mount can offer the organization, set realistic goals for the adoption of SMT, and then communicate the goals and benefits of achieving them to the full population.

Options

All right, we have got a team. Every functional area is represented on the team. We have even invited key suppliers to give their input to our first design effort. Now, how does the team proceed to develop those first few critical designs? In Section 4.2 through Section 4.10, we will outline a systematic approach to steer the initial design decisions. Mr. W. C. Clark, a consummate engineer and a cousin of General Mark Clark, once said, "All engineering involves painting yourself into a corner. You just want to be sure it's the one with the door." Systems engineering is a navigational tool for finding the corner with the door.

In systems engineering, we will create a specification and drawing package that will guide each future step in the product's manufacture. This documentation package will grow as we move through the following listed steps. Each successive stage will build upon the past, and each will be guided by the previous steps. It is entirely possible that an advanced step like cost analysis or manufacturability consideration may point to a fatal flaw in the early planning. This can send the whole project back to square one. A really fatal flaw could even force management to abandon the project. Of course, we hope to avoid this by thinking each early move through, similar to championship chess. But even chess masters can be beaten. It is far better that System Engineering should highlight a serious problem than the problem be found on the manufacturing floor. So move ahead, endeavoring to win on the first roll, but do not feel that having to make a second pass is a failure—it is money saved.

4.2   System Performance

Step 1 in systems engineering is the development of a basic statement of system performance. This begins at the low level of simply describing just what the new product will do. We then define, in increasing layers of detail, how the system must perform to accomplish the function(s) required.

SMT is an exciting development to marketers in electronics because it offers product development possibilities that can solve problems and improve peoples' lives (to a good marketer, people and customers have the same meaning). In areas such as system performance, cost, and form/fit/function, the marketing side of the SMT design team should have a great influence on product development.

4.2.1   Product Performance to the Customer

We in high-tech industries have a tendency to lose sight of the customer's needs. Perhaps this comes from our instinctive engineer's love for the technology involved in building our products. We must remember that most customers do not care what circuit technology we used to build the computer, calculator, or modem that we sell them. They care what it will do, how long it will continue to do it, how easy it is to use, and how much it costs.

Engineering-intensive industries are not alone in the love of their product. Even low-tech businesses have occasionally forgotten what they are really trying to deliver to their customers. Perhaps a look at several real-life examples of this will serve to bring this vital facet of systems engineering into better focus.

Once upon a time, the Hollywood studios thought they were in the movie business, and, they loved the movie business. After all, the movies were an important part of American culture. Why, the Pentagon even hired them to make morale-boosting movies during the great war.

When television was invented, the movie moguls figured it had absolutely nothing to do with Hollywood because Hollywood was in the movie business. By the time that video recording was perfected for home use and satellite TV was developed, the Hollywood studios finally realized that these technologies had something to do with them and their customers. There were alternative types of entertainment that might attract customer dollars from the movie theaters. The studios actively fought to suppress these "bad" technologies. They fought other entertainment and hung on to being in the MOVIE business until they were almost a bankrupt industry. Then, the brighter studio heads figured out that they were in the entertainment business. They began to seek ways in which their expertise could be applied to delivering what their customers wanted, regardless of the medium. The result is today's healthy studios—churning out TV movies, satellite programming, video tapes, and yes, even box-office successes at the movie theaters.

In like manner, railroads once upon a time were the way that freight was moved over the land. The railroad companies were in the railroad business. Their executives had models of steam engines (like the huge, single-expansion, articulated Union Pacific 4-8-8-4) displayed on their desks. When the first few packages were moved by truck, the railroads had no interest in this. The suggestion that sensible men would ship freight in those newfangled flying machines brought guffaws on the railroad's mahogany row on more than one occasion. The railroads had to fade into near-economic ruin before they realized that their customers did not care about freight trains, they wanted freight transportation.

Nobody in America knew more than the rail industry about transporting large quantities of freight. But the railroads had to come to understand their strengths in relation to solving customer problems before they could find creative ways to solve their own marketing dilemma. When they realized that they were in the transportation business, they began to acquire well-positioned complementary businesses. They developed containerized cargo, interfacing the rails with sea, truck, and air transportation. And, they revived their industry.

Interesting though they be, what does the recovery of the railroad and movie industries have to do with SMT? Well, both effected a recovery by creatively applying their available skills and expertise to defining what their customers wanted and then delivering to that definition. Successful electronics companies succeed because they do the same, and SMT is a powerful tool that you can use to deliver what customers want and need from you.

For instance, computer users—the authors included—want more storage space for large database applications. We want the storage-unit size to be small for use in desktop applications, and we want to easily interface the storage device to whatever computer we are using. On a historical note, Control Data Corporation's 8-inch disk drive was the first 8-inch (20.32-cm) drive introduced into the market having greater than a 1-Gigabyte capacity. The 1.236-Gigabyte drive used SMT to pack all its control electronics into a "half a rack" space—unbelievable in today's technologies.

In more current products, silicon memory is the driving force in most PC-based and hand-held products today. The $20 128MB USB drive shown in Figure 4-3, and the ability of the Apple

Figure 4-3   A 128 MB USB drive.

iPod to hold 40 to 60 GB of data are examples of size being one of the most important attributes for today's customer base. SMT is the only technology that will support this type of design.

4.2.2   High-Speed Circuitry

Where performance means speed, the reduced lead capacitance and inductance of small SMCs are a driving force in their selection for many products. Smaller packages with shorter internal lead paths give faster rise times. Shorter traces on down-sized boards mirror these speed gains. But SMT performance benefits are not without a cost. Existing designs that ran perfectly in THT may develop timing problems when switched to surface-mount technology since SMT parts frequently included updated performance technologies with faster rise times.

Testing or modeling of SMT performance is a requisite for the timing of sensitive designs. Manufacturer's specification sheets typically make no distinction between the performance of a device in DIP or SMC, although their spice models may. And, to further complicate timing-test issues, the breadboarding of SMCs, using wire-wrap and sockets, may add enough parasitics to invalidate the test.[5] SMT requires a development of design confidence because designers must breadboard on the PWB. And, the cut-and-weld method of fixing a PWB error becomes tricky with 6-mil lines and 6-mil spaces.

Simulation on a CAE workstation is recommended where available. Virtually all parts now available in SMT have the appropriate models available to allow modeling.

4.3   Form, Fit, and Function

Step 2 in systems engineering is the form, fit, and function decision. While we usually write these three items in this order, we will actually consider function first, fit second, and form third. The topics flow much more naturally from one another when taken in this reverse order.

4.3.1   Function

Function is the most basic description of what we plan to develop in our system engineering effort. What will the new product be? What will it do? How can the product be designed to suit the customer's desires? Function is the stage where we formalize our preceding performance statement into a specification integrated with the beginnings of how we will shape the actual product hardware.

In considering function, we must decide, at a macro level, how each thing the product does will be implemented. For example, let's say the task is to create a timer for the remote control of household appliances. Will the customer want it to be an electronic or a mechanical timer? Once the electronic functions are identified, the team can move on to define how each will be implemented on a micro level. If the timer is to be fully electronic, would the customer's interests best be served by a monolithic approach, by a hybrid approach, an SMT board, etc.?

The output of the function stage of the systems engineering process is a detailed written description of all the required functions for the product and a plan for the basic implementation of these fnctions.

4.3.2   Fit

Fit is the way in which a product interfaces with the immediate outside world. The appliance timer we envisioned above might need to be connected between a 115-volt wall outlet and items like coffee pots or floor lamps. It might be situated on a kitchen counter or under an end table in a living room. The user would need to have some feedback as to its settings and operational status. These considerations would come under "fit."

Under fit, we would also consider how it will connect with other elements in its intended environment. Depending on task, this process may be driven internally or externally.

For a self-contained and self-sustaining product, it may be that no compelling fit constraints exist. The job is then a relatively simple inwardly driven one. We work outward from the circuit, determining what fit best suits its intended function. If the product's relationship to external elements is important, we work inward to fit our product to its expected environment. For example, suppose we are developing an SMT card to replace a through-hole board in a 19" rack-mounted VME-bus computer. The new board must have edge connectors and cable connectors in exactly the same locations as its predecessor, and it must mount from the same hardware.

The information we derive on circuit fit is added to the function description produced in Section 4.3.1.

Fit comes before form (size and shape) because this order of progression allows important external factors, which influence form, to be considered in advance of choosing the form. For instance, a circuit that must fit in a half-height space for a disk drive coexisting with mechanical drive hardware could be physically very different from a circuit of an identical function that is to fit in a 19" rack mount.

4.3.3   Form

Form refers to product size, shape, and weight. Form is, in many instances, responsive to fit. Our example of a household appliance timer would have a very different form factor from a timer for OEM heavy-equipment manufacturers.

As with fit, form may be externally dictated by the required fit and function or, in the absence of compelling forces, form may be internally determined by circuit requirements, cost considerations, etc. Externally dictated miniature form factors are often the genesis of surface-mount programs.

Consider our preceding illustrations. The half-height disk controller would have a rigid fit and form factor imposed upon it. Most small-computer disk-drive manufacturers are turning to SMT and VLSI to crunch needed functions into the tiny envelope available. The household timer would have a form and fit largely determined by pleasing outside package appearance, low cost, and minimal size. Miniaturization and cost reductions would be probable drivers of SMT for such an application, and the 19" rack-mounted VME card would have some latitude in form but its fit would be dictated by the rack-mount hardware and the electrical interfaces required. Surface mounting would probably not be a size factor, but would be a cost, reliability, or function-driven choice in such an application.

Documentation of the form determined for our product is added to the preliminary specification. Our form, fit, and function documentation will be the nucleus for the entire engineering project. At each successive stage, it will guide investigation. And, each additional step toward final product plans may amend the original form, fit, and function write-up.

System Packaging
The basic form, fit, and function description can be further refined now by going to a deeper level of detail on product design. Selection of the form, fit, and function will guide the system packaging design effort. How will each board be shaped, interconnected, and housed? This information forms the next concentric circle of data in our design specification.

System Interface
With system packaging understood, we move on to specify how the system will interface with the outside world. This area covers system I/O and user interfaces, such as switches, indicators, keyboards, etc. This forms the next growth ring on the project data tree.

Material Selection
Next, the materials, such as PW boards and solder, are specified. By this point, the team must have a reasonably solid idea of how the board will be built because certain materials will not be compatible with certain processes.

Component Selection
Right along with the board materials comes component selection. The two are so interrelated that they should be done together.

As SMT technology matures and grows in usage, component sourcing difficulties shrink exponentially. However, finding needed components for conversions occupied a considerable quantity of my time and energy in 1987, and I will be pleasantly surprised if the problem evaporates in the next 5 years. Therefore, early and diligent team attention (including purchasing) to supply is an obvious step. Also, SMC styles impact manufacturing method selection and so deserve consideration at the very beginning of a project.

Are SMCs the Only Way?
The first important component decision for the team is, "Will this be an all SMT board?" If the answer is yes, the component selection process requires very early attention. If, however, one or two components must be through-hole mounted, there is much to be said for relaxing the "SMCs wherever possible" rule. Resistor networks, R/C networks, and multi-chip modules in SIP or ZIP packaging are very space efficient and are often readily available at competitive pricing. They deserve due consideration on any board that will require wave soldering. And conversely, they can be handled with the pin-in-paste (PIP) process described in Chapter 3.

Components, a Multiple Choice Question
In addition to team effort, Engineering can often break component availability impasses by selecting alternatives that are electrically suitable to replace specified parts. The team should investigate equivalent devices from related device families (i.e., replace a 74LS part with a 74HCT or 74F part) and superset compatible parts (such as using a 2M byte memory chip in lieu of two 1M byte units).

Multi-Package Land Patterns

Another strategy that may ease component supply crunches is the incorporation of multiple land layouts on a PW board. Figure 4-4 shows several variations of multi-package pads integrated on a board.

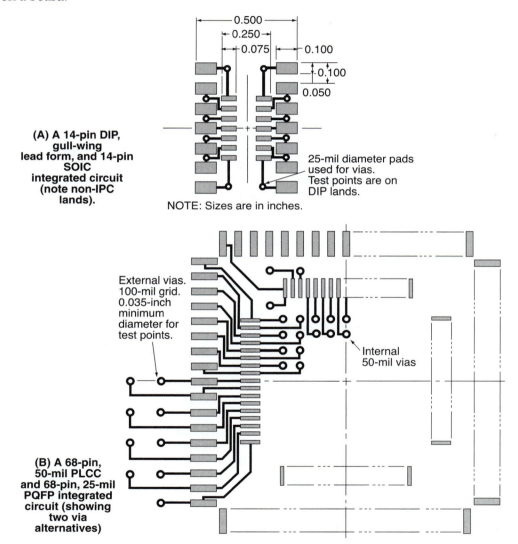

(A) A 14-pin DIP, gull-wing lead form, and 14-pin SOIC integrated circuit (note non-IPC lands).

25-mil diameter pads used for vias. Test points are on DIP lands.

NOTE: Sizes are in inches.

External vias. 100-mil grid. 0.035-inch minimum diameter for test points.

Internal 50-mil vias

(B) A 68-pin, 50-mil PLCC and 68-pin, 25-mil PQFP integrated circuit (showing two via alternatives)

Figure 4-4   Multipackage lands ease SMC availability concerns. *NOTE: These two examples are simply offered as food for thought. The same strategy has been used in through-hole technology for years. For instance, say we need a 1000 mF Radial-Lead Aluminum-Electrolytic Capacitor rated at 10 Volts. It has a lead spacing of 5 mm (0.197"). However, experience has shown that, at times, a 16 volt part, with a lead spacing of 7.5 mm (0>295") is either cheaper or more readily available. In such cases, space permitting, putting redundant patterns on the board is good design practice.*

4.4   SMT Manufacturing Style

In the product specification phase discussed earlier, the team will probably have developed a good idea of the SMT style that best fits the new creation. At this point, we will need to formalize thoughts on manufacturing style. The output of this step (Step 3) is a manufacturing flowchart showing each process step in order. As the flowchart develops, the team should review all available resources in light of the processes required. Planning must include both the flowchart and the manufacturing plan to execute the flowchart.

4.5   Manufacturability

Armed with our preliminary outline of a product and the process flow to build it, it is time to turn to manufacturability (Step 4). At this step, we will look for ways to make our product easier and less costly to manufacture. Strategies for yield improvement, cost reduction, work-in-process reduction, etc., will be part of this investigation.

For discussion, we have considered manufacturability, SMT manufacturing style, and manufacturing cost separately. In practice, however, they are taken together as a product manufacturing review. The topics are so closely related that they must be considered as a unit.

In Chapter 5, Chapter 6, and Chapter 7, we will discuss ways that design can improve SMT manufacturability. The team will use such information at this point to hone the manufacturing plan from the rough flowchart discussed in Section 4.4 to a tuned plan for efficient production. A review of design manufacturability is covered in Chapter 8, Section 8.2. Data on manufacturability and process details should be added to the process flow documentation.

4.6   Manufacturing Cost

In this phase of system engineering, the team should perform a careful cost analysis of the new product. This phase starts with a critical look at each piece part and value-added step. How can value be amplified? How can cost be reduced? After value engineering comes the tallying of predicted costs. The value analysis and cost estimate data are added to the product documentation package.

Both manufacturability (discussed earlier) and cost analysis at this stage produce their share of project U-turns. The team should not look at a reversal as a failure. Rather, it is a success of system engineering in preventing a costly failure in production. Quite possibly, the lessons learned in coming to the U-turn will steer a second effort to a stunning success.

4.7   Capital Investment Evaluation

Product plans that prove manufacturable and cost effective must now be reviewed as a prospective investment for the corporation. Each corporate finance department will have a standard yardstick it applies to this evaluation process. There are usually a number of investment alternatives competing for corporate resources in any fiscal period. The purpose of capital investment evaluation is to determine which of all the opportunities before the corporation deserve the investment of available funds.

The output of this stage of system engineering is the cornerstone of the business plan for the new product.

4.8   System Reliability

From both our own projects and our market research in analyzing other companies' SMT successes, we are convinced that surface-mount technology can actually be a vehicle to improved quality and reliability. It is equally clear that SMT, improperly applied, can wreak havoc with circuit reliability. Chapter 7 will be devoted to quality issues. Next, let's look briefly at quality-related differences between SMT and through-hole products, and then turn our attention to the goals of the system reliability step of system engineering.

4.8.1   How SMT Reliability Differs from THT

SMT is generally selected, among other reasons, for its density advantages. We have already mentioned that the significantly lower mass of SMT components results in generally fewer problems in environments with shock and vibration components, but SMT miniaturization can be a mixed blessing. While it allows the packing of more circuit function into a given space, this also means the packing of more power dissipation into that space, which can have a devastating impact on system reliability.

Through years of familiarity, many designers have developed a sixth sense of thermal management with DIP packages. Be aware that SMT and modern high-speed designs throw seat-of-the-pants thermal engineering into the ringer. You should allow time in the design process for thermal modeling. This will be discussed further in Chapter 5, but cooling options include everything from heat sinks[6] to liquid cooling systems.[7] Tried-and-true TTL designs might benefit from some experimentation with advanced CMOS when SMT changeover time comes around. Thermometric photography of a breadboard may be in order.

Breadboarding is not as simple with SMCs as it was with through-hole proto boards, and adapter boards may need to be designed or purchased, along with SMT sockets. High-speed designs should never be prototyped on protoboards. A prototype high-speed PCB must be designed, fabricated, populated and tested.

4.8.2   The Output of System Reliability Analysis

In the system reliability study, the team should calculate predicted reliability and B-10 life, lay plans for maximizing reliability, and develop a detailed test strategy to ensure that the reliability predictions become reality on the production floor. Test planning should not focus on final test, but on process control. The general wisdom that a poorly designed and manufactured product cannot be tested into excellence applies firmly to surface-mount devices.

4.9 Manufacturing, Storage, and Operating Environment

To build maximum reliability into electronic assemblies using SMT or any other technology, the design team must consider the full range of environmental conditions that assemblies and component parts will see during manufacturing, warehousing, and operation.

In this section, we will provide an overview of the issues to consider in design for environmental stress. The team should review environmental issues and document the intended manufacturing, storage, and operational environment as part of the product data package.

4.9.1 The Manufacturing Environment

ESD Control

In recent years, ESD has been identified by destructive physical analysis (DPA) laboratories as the cause of a large percentage of all electronic equipment failures.[8] Management often questions the worth of ESD programs when asked to invest in static protection. A great deal of ESD-related data and standards are available in the ESD section of NASA's Electronic Parts and Packaging Program (NEPP).[9] Table 4-1 provides statistical data gathered by Sperry Flight Systems, which can be used to calculate the likely cash benefits of ESD control.

Typical operating costs are between $50.00 and $100.00 in finding and replacing one failed component in the factory. Field failure costs vary somewhere between 10 to 100 times higher than factory-failure costs. The exact multiplier is dependent on the repair strategy used. The net result of all this is that in most manufacturing situations, ESD control will produce a significant reduction in failures. Intel, National Semiconductor, Texas Instruments, and other semiconductor manufacturers also feel ESD and its companion electrical overstress (EOS) are significant contributors to failure, to the point that they include a discussion of ESD and EOS issues in their literature, e.g., *Intel's Packaging Handbook*.[10]

Manufacturing Processes

Aside from external stresses, such as ESD, assemblies must survive the sometimes brutal environment of the manufacturing floor. In Chapter 3, we talked about the vibration and shock of component insertion, the heat of SMT soldering, and the action of aggressive cleaners. The team should review the process flow and ensure that all materials and components in the assembly are going to survive the stresses of the manufacturing environment. Any unusual care required in the manufacturing process should be clearly identified and should become part of manufacturing process instructions.

Table 4-1 Potential Savings from ESD Control.

Quality Problem	Percentage Attributed to ESD
Component failures	5% to 25% caused by ESD
DOA compnents	50% ESD caused
Infant mortality	>50% caused by ESD

4.9.2   The Storage Environment

Operating environments attract attention. Some products, like spacecraft electronics, face incredibly severe and wildly varying environments. Others, like office automation, live in relatively sedate and desirable habitats. However, designers considering thermal cycling must reckon with storage conditions as well as operating conditions. In the warehouse—or even in the office on weekends—the office automation product could see temperatures well below zero or above 100°F. The team should plan for the product to emerge breathing from the storage environment. The acceptable storage environment parameters should be added to the documentation library for the product, and this should serve as a guide in writing storage instructions, both for distribution and end users.

4.9.3   The Operating Environment

Finally, the team must consider the operating environment for the new product. What impact will temperature extremes, RFI, EMI, radiation, contaminants, moisture, shock, and vibration have on circuit operation and reliability? We list this item last only because it is the first and often the only factor we instinctively consider when the word environment is mentioned. The operating environment is far from last in importance, however. And scant data on SMC performance under environmental stress may make this phase a time-consuming part of the total systems engineering job. Testing will almost certainly be in order if the environment is at all harsh.

The operating environment specifications are added to the data package, and they are the key to generating safe limit literature for users and for setting testing parameters for QC.

4.10   Design Feedback

By design feedback we mean the formalized approach planned for analyzing and correcting the anomalies uncovered after production begins. A well-conceived design feedback system provides a wealth of data to improve the teams' skills, the design guideline library, corporate efficiency, and corporate profits. Design feedback is the cornerstone of a "solution-oriented problem-solving approach." Solution-oriented problem solving is a powerful new tool being directed at manufacturing efficiency improvement today.

4.10.1   Suggestions for Feedback Implementation

We now offer a few suggestions toward setting up the design feedback system. Feel free to customize these and innovate.

1. All employees involved with the product in any form should be encouraged to contribute suggestions, and they should be solicited for input about things they see going wrong. Nobody knows more about a product than the people who handle it day in and day out. When suggestions or comments are received, they should receive immediate attention before the team, and a report of action should be given to the originator. The absence of a response, taken as evidence that management doesn't care, will quickly crush any worker's constructive participation.

2. Develop a mechanism to unmask process control difficulties that are responsible for recurring line, test, and field failures. Add to that mechanism the apparatus to do something about the root problem. Communicate these efforts to the workers so they know reliability and efficiency are important to management.

3. Have a scheduled way to contact a random sampling of customers for their input about product performance. Each of us has had experience with some gizmo that would need a miracle to perform its specified function. Do you remember thinking that the guys who made that thing must never have used their own product or they would know better than try to sell you one? Remember how long it was before you bought anything else from the responsible (really, irresponsible) company?

The preceding are tried and true methods of improving manufacturing efficiency. These and other methods have been described in depth in works by such quality gurus as Deming, Juran, Crosby, et al. They will work in any manufacturing operation, but they are essential for complex problems like surface-mount technology. The important point here is to review all design-significant feedback that is set before the team and see that the SMT design database is a constantly growing and improving library.

4.10.2 Likely Areas for Concern

In our earlier study of the upper Midwest, respondents rated the following areas (shown in Table 4-2) as particularly troublesome in their SMT programs. We believe the answers are still valid, albeit with a decrease in percentage "component availablity" as an issue. Table 4-2

Table 4-2 Users Rate Their Unmet Needs in SMT*.

SMT Issue	Percentage Identifying Unmet Needs	
	Unmet in 1987	*Expected 1990*
Component availability	23%	21%
Standards/design guidelines	20%	7%
Automation in-house	13%	9%
Competitive manufacturing cost	11%	9%
Material inspection/control	5%	9%
Prototyping ease	5%	2%
All SMT needs unmet, leaded ceramic CC available	4%	0%
Design resources for SMT	4%	5%
Ability to rework/repair	2%	7%
Better PWB materials, local assembly houses available methods for fine-pitch parts	2%	5%
Faster deliveries, lack of expertise, P&P equipment better	2%	2%
Better market anticipation	2%	0%
Better design service, better cooling of fast ICs, ceramic board fabrication, flexible equipment	0%	2%

*Hollomon, James K., *Advanced Manufacturing Technology in the Upper Midwest—A Research Report,* Anatrek, Norfolk, VA, 1988.

represents answers from over 100 companies currently involved in SMT. Forewarned is forearmed. Your system approach should detect and prevent these "gotchas."

In closing, we should warn that, like good THT designs, SMT design is not impossibly complicated, but it is definitely not business as usual for those familiar with simple through-hole designs. Achieving design excellence in modern SMT and high-speed designs can be compared to walking a tightrope over the Grand Canyon. So long as you make no false steps, you reach great heights and enjoy an utterly breathtaking view. However, mistakes can be very costly indeed. *The safety net is the team*, and the dedicated attention to detail in engineering. Often, in the rush to move a product swiftly from inception to market, engineering is short-changed. This is always unwise, and doubly so when we are dealing with a complex integration of electronics and manufacturing technology such as SMT. Engineering typically represents well under 10 percent of the total investment required to deliver a product throughout its life-time, yet well over 90 percent of the decisions that determine the product's cost are cast in concrete during the engineering phase. Over 90 percent of the lifetime cost of a product is generally in cost of materials, manufacturing, distribution, and service. Yet these areas are usually able to contribute only minor reductions in product cost. The same equation holds true when we look at product quality and reliability. Engineering is in a highly leveraged position to contribute to product excellence. In order to improve product quality and lower costs, management attention must be focused on excellence in research and development (R&D).

That is not to say that Sales, Marketing, Manufacturing, and all the other departments have an insignificant role. Certainly, unless they do their jobs perfectly, cost and quality may suffer indelibly. However, their function is to actualize the potential cost savings and quality inherent in a design. Only engineering and R&D are in a key position to generate that potential.

4.11   Review

After reading and understanding this chapter, you should be able to answer these questions and note the reference location for the information in the chapter.

1. List the top three benefits to using SMT as rated by users of the technology.
2. Briefly describe the simultaneous or concurrent engineering approach.
3. Briefly describe whether management needs to be involved in a drive for quality.
4. What are the four rules of successful team engineering?
5. You are a member of team charged with designing a new DVD read/write drive for a laptop computer. Describe form, fit, and function for this product.
6. "Project U-turns" refers to changes in design as a result of the consideration of, e.g., manufacturability concerns. Are these U-turns good or bad? Why?
7. Briefly compare THT and SMT reliability issues.
8. Is there any cost associated with ESD failures?
9. Section 4.10.1 notes three ways employee feedback can improve a product. For the DVD drive that was considered in question #5, give hypothetical examples of how employee feedback might improve the product.

4.12 References

1. Hollomon, James K., *Advanced Manufacturing Technology in the Upper Midwest—A Research Report*, Anatrek, Norfolk, VA, 1988.

2. Duensing, S., "An Inside Look at Delco's Surface Mount Process, Part 1," *Surface Mount Technology*, October 1987, pg. 16.

3. Russel, John F., "Surface Mount Technology: Standards Finally Fall into Place," *Electronics Purchasing*, September 1987, pg. 52.

4. *Holy Bible*, King James Edition, Matthew 7:12.

5. Leibson, Steven H., "EDN's Hands-On SMT Project—Part 5: Automated Testing of SMT PC Boards," *EDN*, July 23, 1987, pg. 76.

6. Azar, K., "Cooling Technology Options, part 1." *Electronics Cooling*, v. 9 #3, August, 2003, pp. 20–25.

7. Azar, K., "Cooling Technology Options, part 2." *Electronics Cooling*, v. 9 #4, November, 2003, pp. 30–36.

8. ESD Failures. Available at http://nepp.nasa.gov/index_nasa.cfm/984/.

9. ESD Information. Available at http://nepp.nasa.gov/index_nasa.cfm/975/

10. *Intel Packaging Handbook*. Available at http://www.intel.com/design/packtech/packbook.htm, 2000.

11. Hollomon, James K., *Advanced Manufacturing Technology in the Upper Midwest—A Research Report*, Anatrek, Norfolk, VA, 1988.

Printed-Wiring Layout Using SMCs

Objectives

This is the chapter that pulls together a great deal of the information from earlier chapters and presents the concepts of printed circuit board (PCB) design and layout. After reading and understanding this chapter, the reader will understand the importance of considering the following during the design of a PCB:

- component orientation and clearances
- selection of appropriate trace widths and inter-trace spacings
- the footprint associated with each component
- inter-layer conductive interconnects, whether they are intended for the insertion of component leads (through hole) or intended solely to create a conductive path between layers (via)
- thermal issues that affect both components and substrates
- circuits using high frequencies and/or fast rise times

Introduction

SMT is one of the major design tools that is being applied to the shrinking of electronic products. How do we integrate tiny SMCs with chunky transformers and other insertion components? How do we orient parts, and how close can we pack them? What do good lands look like? In this chapter, we will answer questions relating to PWB layout using SMCs.

We have finally completed the preliminaries, and now we are ready for the main event—SMT PWB design. Since this book is intended for working engineers, the textural material on design is contained in this chapter where it fits naturally as one reads from cover to cover. Example drawings of the components and land geometries are collected in Appendix A. Note: The years since publication of the first edition have proven the folly of listing all components. While many of the mainstay packages are listed in Appendix A, no attempt has been made in this edition to provide a complete listing. New packages evolve far too rapidly for such an effort to succeed. Instead, we will caution the reader to consult component suppliers and standards bodies for the most current offerings. We will concentrate here on the pervasive packages and on how to develop land patterns for the newly arrived.

We will also touch on some of the newer technologies. However, as data rates work up to 10 Gbits/second, many design issues are being redefined as this edition is being written. Parallel bus structures are being replaced by serial asynchronous architectures (third generation I/O, also known as "3GIO") making issues like bit rate and jitter as important as crosstalk and overshoot are in more traditional designs. We will not attempt to lay down rules for the newest technologies, but give the reader solid fundamentals to build on.

More Lead-Free Issues

As you consider the topics in this chapter, consider also these issues that will impact PCB designs as lead-free solder is implemented:

- The growth of "tin whiskers" from pure tin coatings will become a much greater problem.
- Higher temperatures required by some lead-free solders may require substrates with higher glass transition temperatures than FR-4.
- The appearance of solder joint fillets will be different with lead-free solder.

5.0 PCB Basics

PCB designs should minimize trace lengths and control signal integrity issues. Good designs should also make manufacturing, testing, troubleshooting, and repair easier than they are without good design. Design methodologies like microvias, high-density interconnects (HDI), embedded passive components, and high pin-count ICs have entered mainstream design. This section will introduce the basics of PCBs so that later sections can talk about more complex issues.

5.0.1 Basic PCB Fabrication

For those truly new to PCB design, let's start with the anatomy of a PCB by briefly reviewing PCB construction and fabrication. Most standard designs start with a two-layer board: a rigid sheet of flame-retardant epoxy-impregnated fiberglass material with copper sheets glued to both sides, frequently known as "copper clad." This can alternately be considered as two conductors separated by a dielectric insulator—the definition of a capacitor. The most common substrate in commercial and industrial designs is FR-4 substrate in which the FR designates a "flame retardant" material, with the -4 indicating material with a nominal dielectric constant of 4. Multi-layer boards may use the same substrate material, with a layup of outer layers, inner layers, and ground and power planes. In this case, each layer would be thinner than those used for two-sided boards. The copper on each layer is chemically etched to create the traces that are designed as the circuit paths that make connections between component terminations. This is called a "subtractive" process in which the basic copper cladding is removed/subtracted chemically to leave only the desired traces and termination pads. We use the term 'terminations' since, as you saw in the discussion on components in Chapter 2, not all SMT parts have leads.

PCB Fabrication Basics

Let's look at a multi-layer example shown in Figure 5-1. Simplified PCB processing steps for a two-sided FR-4 layer would be:

- Copper foil bonded to FR material, one side for a single-sided board or layer, both sides for a double-sided board or layer.
- A photoresist material is deposited onto the copper layers. This 'resist' layer will be photo imaged in the next step and will protect the copper that will become the traces and pads of the circuit pattern.

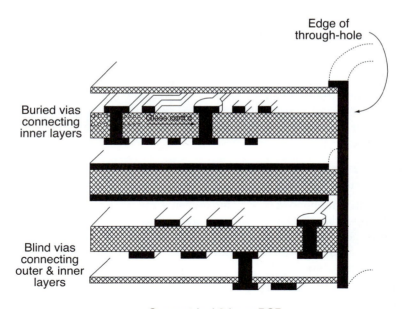

Symmetrical 8-layer PCB.
Layer stack-up: 3 signal layers, 2 planes, 3 signal layers

Figure 5-1   Example of a multi-layer PCB.

- Circuit/trace photomask or photoplot of the trace+pads pattern is placed on each layer of copper foil. Both layers of a two-sided substrate must be imaged before chemical etching is performed.
- The substrate/copper sandwich is placed in a vat of acid etchant where all copper not protected by the 'resist' is etched away. The acid is then washed off the board.
- The remaining resist is removed from the board, leaving traces and pads.
- Holes are drilled and deburred.
- Very thin copper is chemically deposited in the drilled holes—electroless plating.
- Resist is again placed on the board, imaged with a photo tool and hardened/developed by exposure to UV light. Resist is then removed from areas where we want thicker copper plating.
- Thicker copper is electroplated in the holes to form continuous electrical connections in either through holes, designed for THT component leads, or vias that have no component leads and provide electrical continuity between various layers of the board.
- Tin/lead plating is deposited on copper in the holes and on surface pads designed for component terminations whether THT through leads or SMT surface terminations. This tin plating serves as an etch resist for later plating and protects the copper from oxidation.
- Resist is then stripped leaving solder and exposed copper. Copper coated with solder will survive the next etching process.
- Using the solder as an etch resist, all exposed copper is now etched away, leaving the circuit traces and pads.
- Solder is stripped off the copper and the board is cleaned. If SMOBC is to be used, it is now applied. If other masks are to be used, the SnPb will be reflowed.
- Hot air leveling (HAL) is typically used to cover the solder.

For multi-layer boards, this general process is repeated multiple times to generate all the required layers. The material used to laminate the layers together is a "b-stage" epoxy-fiberglass laminate, called prepreg. Prepreg has the epoxy only partially cured. When all layers have been etched, the resulting layers of copper/glass and prepreg are placed into a curing press and oven and are cured together at moderate temperatures and pressures, resulting in a multi-layer PCB. This finishes the curing of the prepreg.

Substrate Materials

Other substrate materials include:

- FR-2: fire-retardant phenolic/paper, inexpensive and used for low-cost consumer products
- Polyimide/fiberglass: has a higher glass transition temperature and is harder than FR-4
- Flex circuits: polyimide or polyester, used in applications where weight, space, and flexibility are advantageous. Some flex circuits are used in applications like cellular phones and laptop computers

The choice of substrate materials depends on a number of concerns and is made primarily on design and performance decisions. The relative cost of the materials is important as are manufacturing concerns, but with the twenty-first century demands for maximum performance with minimum weight and bulk, the ability of the material to perform is frequently the overriding factor. More information on materials is discussed in Section 5.6.

PCB Surface Finishes

SnPb plating on the copper traces and pads of PCBs has been the standard for years. Not that other finishes have not been available, but SnPb is inexpensive, solders well, and the processes involved are well understood. For this discussion, PCB surface finish means the finish applied to the solderable areas of the substrate. As Barbetta notes, none of the existing finishes fits the category "universal surface finish."[1] The surface finish selected for a given PCB design depends on a number of issues:

- Compatibility with the solder to be used
- Compatibility with the component termination materials
- Compatibility with the solder process
- Compatibility with any wire bonding to be performed on the board
- Conductive surfaces for any contacts and switches in the PC assembly

Industry currently uses a number of finishes, as shown in Table 5-1.

Table 5-1   Comparison of PCB Surface Finishes.

Surface Finish	Advantages to Finish	Drawbacks of Finish
HASL with SnPb	Industry standard, process well understood Good bond strength Thermal cycles not a problem Easy rework	Inconsistent thickness Inconsistent flatness Coplanarity issues Bridging with fine-pitch devices Cracks with high aspect ratios
Reflowed SnPb	Inexpensive process, well understood Good solderability and bond strength	Coplanarity issues Reflow causes uneven hole deposits and hole size reduction
Immersion Sn	Planar surface Inexpensive	Limited rework cycles Process normally requires nitrogen atmosphere
Immersion Ag	Planar surface Easy process, eliminates Ni from the process Inexpensive Long shelf life if tarnish is controlled	Tarnish must be controlled Not good for use with NiAu pins
NiAu Plating	Can be wire bonded	Holes do not get plated exposing Cu Costly Au on board edges can corrupt solder joints
ENIG	Planar surface, even thickness Allows multiple thermal cycles Solders well and easily Long shelf life	Expensive High speed limits Black pad issues
OSP	Coplanarity excellent No effect on hole size Easy process, but not drop-in	Limited thermal cycles Limited shelf life Exposed Cu after assembly may have reliability issues Requires careful in-process handling

Plating can be deposited on bare substrate (e.g., FR-4) and in via holes by the electroless chemical process as well as on base metals and on top of electroless plating by electrolytic plating. Electroless chemical plating is used to create conductive surfaces on top of nonconductive surfaces. An example is after holes are drilled in the substrate and plating must be provided as the first step in creating a conductive through-hole or via. Chemical plating is performed by the PCB fabricator and is not in the control of the designer, except that the quality of the final plating may be specified by the designer or team and that quality depends on the quality of the initial chemical plating.

When SnPb plating is specified as the final finish, it will be deposited electrolytically on copper as an initial layer, then may be overlaid by a thicker electro plate, or may be covered by a dipping in molten SnPb and finished by Hot Air Surface (or Solder) Leveling (HASL) which will be discussed later. Numerous issues affect the quality of electroplate, such as uneven current density in the plating process and porosity. Again, the main concern of the design team is to specify the quality necessary, such as the coplanarity level of the pads, and require the PCB fabricator to supply boards to the appropriate specification. Porosity of the deposited solder plate is controlled by reflowing, or "fusing," the boards if necessary. Fusing also creates a strong intermetallic layer for mechanical strength. One issue with SnPb plating of entire boards is that the plating is then covered with solder mask. During the reflow process, the plating on the traces allows molten solder from the paste deposit on pads to migrate under the solder mask, creating the risk of starved solder joints. This will be discussed further in Section 5.9, "Solder Masks."

The HASL process involves dipping the board in molten SnPb, then passing the board over a hot air blower that levels and removes some of the dipped solder and any icicles that may remain. The main disadvantage to HASL is the unevenness of the resulting deposition of solder. Both too-thick and too-thin solder coatings have been reported after the HASL process, and it creates (in cross section) a dome-shaped pad. This is particularly a problem with components that require low coplanarity, like BGAs with their typical requirement for no more than 0.1 mm (0.004") coplanarity across the ball contact pads.

Organic solderability preservatives, or OSPs, are an alternative to tin-plating pads and have become commonly used in the PCB fabrication process due to their advantages. OSPs are applied to solderable pads for SMT components in place of the HASL finish and are an anti-oxidation finish. This creates a very flat finish with coplanarity as good as the copper itself, which is not necessarily equal to 0 mm! The OSP will disintegrate in the presence of activated flux at soldering temperatures. OSPs can be applied in a thin coating or a thick coating, ranging from 0.1μm to 0.5 μm

However, OSPs also degrade both with time as well as with each pass through a soldering process. Therefore, board handling becomes very important with OSP-coated boards. The solder paste supplier must be notified of the use of an OSP surface finish in order to provide a no-clean flux with enough activity to remove the OSP. The manufacturing process must assure that the paste/flux then covers the entire pad or there may be nonwetting or dewetting of a portion of the solderable surface. OSP finishes may also not completely cover the inside of vias, allowing oxidation of the copper plating and thereby creating problems with hole filling during wave soldering.

Electroless nickel immersion gold (ENIG) is a lead-free surface finish that is gaining in popularity. In 2004 IPC conducted a survey of rigid board fabricators in North America in which they found that 41 percent used HASL as their preferred solderable surface finish, 17 percent used ENIG, and 10 percent used OSPs. More recently, Cookson, looking at a broader section of the market, predicted that HASL usage would drop from 59 percent to 21 percent by 2007, OSP usage would increase from 19 percent to 35 percent, and ENIG would rise from 3 percent to 20 percent by 2007.

Obviously the surface finish market is changing. In large part this is being driven by the march to lead-free soldering. As has been noted earlier, we cannot cover this march in this book since it is occurring as the book is being written, but there are many conflicting reports and predictions yet to be sorted out. As an example, in 2002 Sohn reported on an iNEMI investigation of lead-free soldering, available at http://www.nemi.org/cms/projects/ese/lf_assembly.html. In "Are Lead-Free Solder Joints Reliable?" the investigation concluded that the conversion to lead-free could be accomplished without degrading solder joint reliability. In 2004 Borgeson and Henderson of Universal Instrument Corp. published a white paper, available at http://www.uic.com/whitepapers, that states the conversion to lead-free soldering brings with it "significant risks of solder joint fragility with all the commonly used solder joint pad surface finishes." While this may all be sorted out by the time the reader has this book in his/her hands, the authors would be amazed if that was the case.

With regard to surface finishes with lead-free, there is at least one report[2] that cites voids with lead-free finishes may take on different characteristics than voids with SnPb solder. This report indicates that voids on pads with OSP finishes are much larger with lead-free, and that voids with tin, silver, and HASL-finished pads are smaller and acceptable. At this writing, this is an initial report, and more study will be needed.

5.0.2   Design Steps for PCBs

While all PCBs are different and have different requirements, there is a general flow to the process of PCB design that works for most projects:

1. System specifications: define overall functional specifications
2. System block diagram
3. Partition system into PCBs if multi-board system
4. System design, including necessary interconnects both within the system and for external connections (field connections)
5. Schematic designs
6. Functional verification
7. Physical design, including partitioning of PCB(s) into analog, digital, RF, power supply, etc.
8. Build component libraries
9. Simulate design
10. Design for testability issues
11. Final functional verification, typically with a prototype

12. Design for manufacturability issues, including tooling holes, fiducials, and mounting holes
13. Place critical routes, including the critical signal path and I/O routing
14. Finish routing
15. Signal integrity and timing verification
16. EMI/RFI verification

Each of these issues requires consideration by one or more members of the design team. Most of them require consideration by several of the skill sets and/or departments represented on the design team. At some point in the process, most designs will be built up as a prototype to test circuit function before committing to a PCB although a prototype PCB may be part of the prototype itself, especially with high-speed designs. One must be careful to allow as much time as possible for the construction and evaluation of the prototype.[3] And if your circuit is expected to work under all possible component variations, the design team must consider that in the overall design, worst-case circuit design includes the effects of component tolerances.[4]

PCB Design Software

The computer-aided engineering (CAE) and computer-aided design (CAD) tools available to the design team are many. These tools may include:

- Schematic capture, for design of the circuit schematic and generation of the corresponding netlist.
- Simulation, using mathematical models of circuit components to allow for simulation of the circuit without the need for physical hardware.
- Synthesizers, which allow digital circuit designers to specify required logic functions in the circuit. The synthesizer will then extract those functions from a function library and connect them together as specified by the designer.
- Emulators, which contain collections of programmable logic elements that can be configured to form a complete operating system that operates much faster than a software simulation. An entire complex IC can be emulated prior to committing its design to silicon.
- Component placement tools, which allow manual or automatic placement of components on a PCB layout following orientation and spacing rules.
- Routing tools, that make the interconnections on a PCB which will be ultimately etched in copper on subtractive boards. The router starts with a "rat's nest," a complete set of interconnections based on the netlist of the circuit, then allows the users to manually route copper for each interconnect net or automatically routes each net based on orientation and spacing rules. Routes may be done with or without a grid using ground and power planes, etc.
- Design rule check tools, which complement the placement and routing tools to verify that none of the applicable rules have been violated.
- Post-processing tools, that generate Gerber or GenCam files for use by fabricators and assemblers and verify DFM rules.
- CAM tools, which can convert post-processing files into files useable by paste dispensing and placement systems. They may also be able to generate enclosure designs for the board.

There are a number of electronic computer-aided design (ECAD) software packages available that combine a variety of the above-mentioned tools. They range from relatively straightforward PCB layout packages to packages that will start from schematic capture, allow simulation, aid in both component placement and trace routing, and aid in the mechanical design of the board and its enclosure. Which package you should use is outside the authors' knowledge. Consider the tasks your design team must perform to complete the design successfully, talk to users of each package, and read the periodic evaluations of packages in *Printed Circuit Board Design and Manufacturing* magazine available through http://www.pcdandm.com. Software packages commonly considered by designers and their web sites include:

- Allegro, http://www.cadence.com, search for "allegro"
- AutoCAD—there are PCB design add-ons available for AutoCAD
- Board Station, http://www.mentor.com, search for "board station"
- Cadence, http://www.cadence.com
- Circuitmaker 2000, http://www.circuitmaker.com
- Eagle, http://www.cadsoftusa.com, free Lite version available
- EdWin, http://www.visionics.a.se/, schematic capture and layout
- ExpressPCB, http://www.expresspcb.com, provides free PCB design software if they make the board
- Orcad, http://www.orcad.com
- McCAD, http://www.McCAD.com, free Lite version available
- Mentor Graphics, http://www.mentor.com
- PADS, http://www.pads.com
- Pad2Pad, http://www.pad2pad.com, provides free PCB design software if they make the board
- Pantheon, http://www.intercept.com
- P-CAD, http://www.acceltech.com
- PCB Express, http://www.pcbexpress.com
- Protel, http://www.protel.com
- Pulsonix, http://www.pulsonix.com

If the final design is being performed out-of-house, then the team must know what package will be used by the designer and define their project within the confines of that software.

5.1   Component Orientation

SMT component orientation is determined by both constant and variable rules. Let's first cover the constants, those rules that apply regardless of the application or assembly style. Then we will discuss rules that are applied only to certain fabrication techniques or situations.

5.1.1   Steadfast Component Orientation Rules

As a rule of SMT or any PWB layout, the components should be oriented to keep the copper coverage nearly uniform across the surfaces of the board. Uneven distributions of copper can create problems in the electroplating of boards and can contribute to warpage of completed PWBs.

The most basic rules of good component orientation apply across the board, regardless of the application or manufacturing style. In fact, these rules apply to through hole as well as SMT assemblies. Thus, they apply to both SMCs and THTs on mixed-technology boards. While the following rules certainly can be violated, they should stand—except when there is a compelling reason to depart from them.

Orthogonality

Wherever possible, all components should face north/south or east/west. In other words, the sides or longitudinal axes of the components should be either parallel or perpendicular to one another. If the circuit board is generally rectangular or square, component axes should be parallel to the board edges.

Applications may force a modification of this rule to allow some components to be placed with axes every 45 degrees. Where even 45 degrees orientation is too constraining, the ruling factor becomes the placement equipment. If it is capable of resolving very small steps of component rotation, parts may be in any axial position. However, some placement machines can only rotate components in 5-degree or 1-degree steps; thus, they limit the axial orientation to only those points of the compass that the machine can visit.

Polarity and Pin-1 Alignment

Consistent component orientation is a boon to hand assembly, repair, and troubleshooting operations in the factory and in the field. The best practice, as shown in Figure 5-2, is to align all number 1 pins, all SOT single leads, and all cathodes the same. Actually, we should extend this rule to cover any component with a clear physical or electrical characteristic.

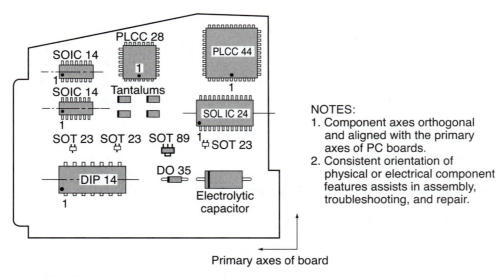

NOTES:
1. Component axes orthogonal and aligned with the primary axes of PC boards.
2. Consistent orientation of physical or electrical component features assists in assembly, troubleshooting, and repair.

Primary axes of board

Figure 5-2   Electrical and physical alignment of components.

Population Distribution

For SMT assemblies, good urban development planning is a world apart from the work of your city planning commission. SMT designs should avoid open areas like city parks. Instead of downtown congestion balanced by suburban open streets and grassy areas, we shoot for urban sprawl. The objective is to have a balanced distribution of heat-sink mass across all areas of the board. While this is critical in reflow soldering applications (see Chapter 3 for a further discussion of heat-sink distribution), it is good practice for any design to promote solderability and reflow repair techniques.

Accessibility Clearances

Good design practices squeeze components no tighter together than necessary. One millimeter of extra space between each component can be of great benefit to visual inspection and repair/rework operations. More details on "design for inspection" procedures are provided in Chapter 7.

5.1.2   Situation-Dependent Component Orientation Rules

Many of the rules governing the best component orientation for SMT are invoked only by certain applications or assembly processes. We will present these situation-dependent rules and discuss the circumstances that mandate their practice.

Probe-Ability/DFT

Probe-ability is a coined word describing the layout of PWBs so as to place probe points on easily accessible grids for test purposes. It is, therefore, part of the Design for Testability (DFT) process. The 100-mil (2.54-mm) grid of through-hole assembly leads to a predetermined design for probing. The board is laid out such that all component leads are centered on an imaginary 100-mil grid on the board assembly. Thus, we ensure easy test access using a "bed of nails" fixture, with probe pins on 100-mil centers.

Probe-ability is a process-determined design rule. Many simple boards, and even some complex memory boards, do not require probe-ability. They are either tested by a "hotshot" mock-up or are fully bus-addressable for I/O-connector test access. Probe-ability design rules are invoked by the need to probe for testing those circuit nodes not routed to external I/O.

When probe-ability is required for test, SMT poses new challenges for PWB layout. We can see why by comparing surface-mount probing with through-hole mounting. Placing THT components on a 0.100"-(2.54-mm) grid is simple enough because standard THT components are designed with leads on 100-mil centers or multiples thereof. SMCs, which are on 50-, 40-, 25-, 20-, and 10-mil centers as well as 1.27-mm, 0.8-mm. 0.65-mm, 0.5-mm centers or variables in between, make probe-ability design a more interesting task. We may bring traces out from fine-pitch SMC lands to coarse-pitch 100-mil grid vias or test points. These test points may be designed as addressable from the non-component side. Thus, an SMT board may be designed for a 100-mil grid bed of nails. This will probably halve the bare and loaded board testing cost compared to the same number of test points on 1.27-mm (0.050") centers. But the real estate required by these probe points, associated traces, and single-side assembly rules conflicts with the miniaturization goals of surface mounting. Fine-pitch bed of nails fixtures are available.

With these, probe points are located on more space-efficient grids, generally of 50-mil but increasingly, at 0.635-mm (0.025") centers. The tradeoff in specifying fine-pitch probes is the cost per probe point and the fixture durability. Probes are available down to pitches of 10 mils or below, but costs rise inversely and exponentially with decreasing center distance.

Another factor differentiating SMT and THT probe-ability is that we do not probe on SMC leads as we can on THTs. This is because probe pressure may force an open lead to contact a solder land and falsely pass a defective solder connection. Also, probe-to-lead misalignment can damage fragile SMT leads, crack ceramic chip components, and damage fine-pitch probes. Since we do not probe SMC leads, we have a far larger degree of freedom in locating components for SMT probe-ability than we enjoyed with insertion-mount devices. In theory, the surface-mount component could be anywhere on the board as long as we routed traces from its lands to test points on our probe grid. Smaller chip components could be mounted between gridded test points, as shown in Figure 5-3. However, before using this technique, be certain to determine the specified mechanical accuracy of the probe pins, also known as pointing accuracy, in your bed of nails test set. Components cannot be placed closer to the test points than allowable by the pointing accuracy + PCB artwork deviations. Probe contact with the ceramic body of a component is a guarantee of damage! Pointing accuracy is defined as the measure of ± radial movement of the probe plunger tip from the CAD centerline of the probe mounting hole to the actual contact point on the board.

However, the packing of components can be overdone, and there is little to be gained in locating components as close as paving stones only to give up the real estate saved by running traces out to test pads that must be on a grid. We would compromise visual inspection and repair operations while gaining little or nothing in miniaturization. Therefore, where probe-ability is an issue, it is common practice to place components around the test grid and thus minimize the track length from component lands to test points.

For instance, 1206 passives are workhorse R&C components of SMT industry in the United States. They are nominally 0.120" long by 0.060" wide (3.2 × 1.6 mm) and can be placed on 2.54 mm (0.100") centers, whereas the larger 0.126" by 0.098" (3.2 mm × 2.5 mm) 1210 devices must be placed on 3.81-mm (0.150") centers to allow a 1.27-mm (0.050") test grid, and 5.08 mm (0.200") to fit a 2.54-mm (0.100") grid. Driven by stringent real estate demands, the smaller 0805 and 0504

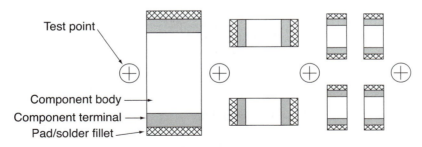

Scale representation of 1206, 0603, and 0402 components
between test points on 0.100" centers.

Figure 5-3   Chip components in between gridded test points.

parts are gaining some market share. However, test grids still dictate a placement on 2.54-mm (0.100") centers. Therefore, gridded designs may profit little from the assembly effort expended in using ultra-small parts except where a tight grid spacing, well below 50 mil, is used.

Assembly Clearance

The intercomponent clearance is application dependent, being influenced by the method of and the equipment used in board assembly. When assembling boards by hand, components could conceivably be laid side by side with only enough clearance to accommodate the maximum dimensional tolerances of each part. (In practice, more clearance is desirable to simplify inspection and rework.) Automated assembly introduces some additional clearance requirements. The placement equipment may be very accurate, but all machines fall short of absolute perfection. Chapter 3 covers the calculation of required clearances needed to ensure collision-free automated placement.

Assembly Flow Through the Solder Furnace

Solder defects may be minimized by proper component orientation in relation to workpiece flow through the soldering operation. Just what constitutes "proper component orientation" depends on the soldering method in question.

The chief gremlin in reflow soldering, assuming the proper solder-paste deposits are present, is tombstoning. This defect is produced by a combination of the wetting and surface-tension forces in molten solder. It occurs when forces at one end of a component are not reasonably balanced by forces at the other end. The imbalance may be produced by land-geometry variations, component-termination variations, or time-of-reflow variations. Whatever the source, when the solder-force imbalance is great enough, tombstoning is the predictable result. This was introduced in Chapter 3, and Figure 5-4 is a reminder of the unequal forces that may be at work on a chip component undergoing reflow.

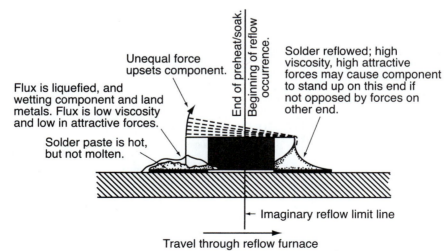

Figure 5-4   Diagram of uneven forces on chip terminations during reflow. These forces can lead to a tombstone defect if the "upset" force prevails.

We will consider imbalances from land geometries or component terminations in separate sections. Here, we are concerned with solder liquidus time variations as introduced in Chapter 3. This phenomenon may occur when an assembly migrates from the homogenous environment of batch soldering to the moving reflow band of the linear furnace. We can imagine a reflow limit line exactly where assemblies reach reflow/liquidus temperature as they travel through the oven. In reality, this limit line is a moving target pushed back and forth in the furnace by heat-sink loading and furnace temperature stability. We will take up its movement in a moment, but for now, let us think of the reflow limit line as stationary.

As an assembly moves through the reflow limit line, solder melts (reflows) precisely as it passes through the line. Some components on our imaginary board assembly are oriented so that a line through their two terminations is parallel to the reflow limit line across the furnace. Others are situated with their termination-intersecting line perpendicular to the reflow limit line. It follows that components so oriented that their terminations hit the line simultaneously will not experience the unequal forces. However, parts oriented so that one termination contacts the line first will be subject to the unequal solder forces we have identified as contributing to tombstoning. Figure 5-5 shows the preferred component orientation needed to minimize tombstoning and component misalignment defects for standard components in linear reflow soldering.

Occasionally, an application may demand that parts be reflowed without regard to the preceding preferred orientation. If tombstoning plagues such a reflow operation, you may

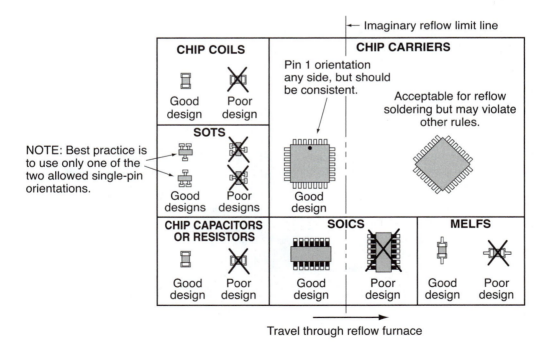

Figure 5-5 Diagram of preferred component orientation for linear reflow soldering.

resort to batch reflow to avoid the moving reflow limit-line effect, or you may glue or fixture the offending components to prevent their movement during reflow.

Lest the preceding rules seem too simple, remember that we discounted the effects of unequal thermal mass distribution for our discussion of component orientation. Inequalities in thermal mass distribution cause the reflow limit line to move within the reflow furnace and to waver and bend across the width of the board, much like isotherms on a weather map.

Therefore, thermal mass rule number 1 is to orient components to distribute the thermal mass as evenly as possible. This rule is of particular importance for linear reflow soldering, but gross departures from it will also produce solder problems in batch systems. At the extreme, local-ized heat-sink masses can make a board impossible to reflow solder without causing overheating and consequent damage to low thermal-mass areas. Where one component has a large thermal mass, leave some vacant area around it to minimize the heat-sink congregation in its neighbor-hood. In particular, components that are likely to pop wheelies should be kept at a distance from large heat sinks. Where they must be located in close proximity to the heat sinks, experimentation is suggested in order to determine the ideal orientation for the part which is prone to tombston-ing. Figure 5-6 suggests thermal mass dictated departures from the rules of Figure 5-5.

Detailed isotherm mapping of boards may be done with multi-channel recorders and heavily thermocouple-monitored samples. Such maps serve to guide any special orientation. In the absence of equipment necessary to produce an isotherm chart of the assembly, the "cut and try" approach will have to do.

Flow soldering invokes a completely different set of component orientation rules. Here, one of the driving considerations is in not creating eddy currents that can cause solder skips by blocking the solder from some terminations as the assembly moves through the molten solder. The other primary concern is to position the parts to avoid bridging. Since one defect involves too little solder deposited while the other is due to too much solder, the solution to one prob-lem may work against the other. The rules presented below have been empirically determined as the best compromise for maximum flow-soldering yields.

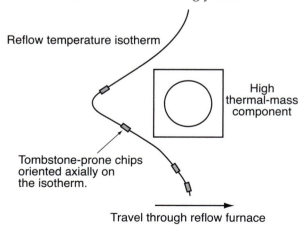

Figure 5-6   Special orientation may lessen the solder defects caused by thermal mass peaks.

Component orientation for solder-skip (or open) avoidance is pictured in Figure 5-7. The preferred anti-open rules should be rigidly enforced if solder opens are the prevalent solder defect in your flow-soldering operation.

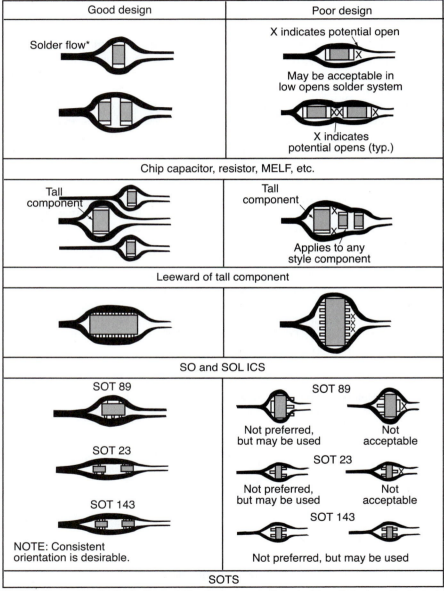

*Solder flow arrows indicate flow relative to component; typical for all devices illustrated.

Figure 5-7   Component orientation for minimizing flow-soldering opens.

Today's SMC wave solder machines may obviate the need for some of these rules. However, the production environment, state of repair of equipment, operator training, and assembly vendor selection throughout the entire life of a product are much less under the control of the circuit designer than is the layout board. Once these layout rules are well understood, their cost is minimal, and need be paid only once per design. The expense of violating them can be substantial and will be exacted many times per day over the life of the manufacturing cycle.

The rules for solder-bridging (or short) reduction are illustrated in Figure 5-8. Bridging is often a pernicious problem in the flow soldering of SMCs. Therefore, anti-short rules should be followed for all designs.

Certain flow-soldering systems are particularly prone to bridging problems. With these, bridging may be reduced on SO and SOL ICs by deliberate violation of the rule illustrated in Figure 5-8. If orienting the SOIC with its long axis perpendicular to the direction of flow will reduce bridging defects enough to offset the increase in opens, do it. Experimentation will tell. You can easily try it by rotating a sample board or section of board 90 degrees and passing it through the flow-solder system. If the experiment indicates a need, reorient only the SO and SOL ICs on the board. Solder thieves may be necessary and will be discussed in Section 6.2.3.

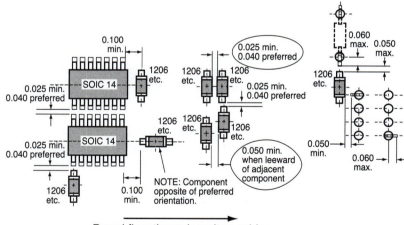

NOTES:
1. 4-sided I/O devices (PLCCs and LCCs) are not recommended for flow soldering.
2. These recommendations are for flow-soldering lands (refer to Appendix A, Section 1.3).
3. Dimensions are in inches.

Figure 5-8 Component orientation for minimizing flow-soldering shorts.

Figure 5-9   Wave solder differences at leading and trailing edges of chip components.

As Figure 5-9 shows, chip components that pass through the wave solder process can show the differences between wave effects at leading and trailing edges as they passed through the waves from left to right. Clearance of pads following the chips may need to be increased if bridging becomes evident.

Thermal Management
Thermal management impacts the component orientation for high-dissipation circuits. Where significant heat will be generated in the circuit operation, spread the high-dissipation devices evenly across the circuit board or as otherwise dictated by cooling strategies. Thermal transfer through the substrate will be improved by leaving as much copper as practical on each layer of the board. Thermal vias are discussed later in this chapter.

Double-Sided Assembly
Placing components on both sides of SMT boards doubles the real-estate savings, but placing components on both sides may more than double the headaches for design and manufacturing. The best component-orientation rule is to place the first side for low density if possible and with moderate density if needed. Then, place the second-side components to follow the rules given previously. For two-pass reflow soldering, locate all high-mass components on the side that will be up in the second pass through the furnace.

Additional concerns are the need to decide whether test will require access to one or both sides, then place components clear of all test points. Be particularly mindful of heat-sink mass distribution. Double-sided assemblies double your opportunities to create soldering nightmares with lumped-up thermal masses. Likewise, in designing to manage high-dissipation problems, double-sided assembly complicates the task. Clearly, we would not intentionally locate high-dissipation parts one under the other. But, it is easy to do if you are only concentrating on the population of the side on which you are currently working.

All the challenges of double-sided assembly design, test, and rework are indeed formidable, but the real-estate rewards are great enough that it would be pointless to argue for a single-side-only blanket rule. Do not go double-side if you can help it. Further, if you must use both sides, allow extra time to do the needed design homework.

Partitioning

Partitioning refers to the division of circuitry into blocks. Electrical and/or mechanical considerations may influence the component orientation for the sake of circuit partitioning.

Mechanically, partitioning may be done to facilitate assembly processing. I/O-intensive sections of components might be located near the edge connectors. Also, sensitive components may be partitioned into a protected area to shield them from hazards in the environment. High-dissipation parts may be located to suit heat-removal strategies.

Electrically, circuits are commonly partitioned:

- to separate analog and digital sections.
- to separate high-frequency areas from medium- and low-frequency areas.
- to divide the audio and digital sections.
- to position test points for functional tuning.
- for the management of power dissipation.
- to control noise and crosstalk. For instance, high-speed switching areas are often separated from low-level, high-gain amplifiers because the fast switching noise can be isolated in the process.
- to modularize the assembly to promote manufacturability.
- to separate functional blocks for testing.
- to reduce trace parasitics or cluster components of particular impedances on multiple-impedance boards.
- to place matching/tracking parts on one card in a system.

Each of the electrical partitions (analog, digital, RF, etc.) should be laid out separately, then combined after the mechanical items (such as fiducials, mounting and tooling holes, required connector locations, and heat sinks) have been accounted for.

Parts Placement

For analog circuits, particularly those dealing with low-level signals, identify the most critical signal flow path, keep inputs and outputs separated, and place the involved components first. Generally, the lower the signal level, the closer together the parts—including any I/O connectors—should be placed. Parts should be placed to avoid any parallel paths in the routing, then place other parts onto the critical path components.

Digital circuits should generally be most careful of clock generation and routing. Remember, it is not the frequency of the clock that is the main contribution to noise, it is the rise time t_r. The shorter t_r is, the higher the fastest component frequency is, and the more likely crosstalk and EMI/RFI is. Clocks must be routed carefully. If built-in self-test/boundary scan to IEEE 1149.1 standards are used on digital/micro circuits, routing for the test access port (TAP) must be considered in component placement and orientation.

In-depth discussions of RF circuits would require volumes. High-frequency design will be discussed further in Section 5.10.

5.2   Clearances Around Components and Test Pads

Several issues must be considered in deciding the clearance around components and test points. We will discuss these next and will develop rules for standardized spacing. Such rules should be built directly into CAD systems wherever possible.

5.2.1   Component Clearances for Visual Inspection and Rework

Component spacing must be, at a minimum, adequate for an error-free assembly operation. Chapter 3, Section 3.3.3 (illustrated in Figure 3-30) presents a formula for calculating this minimum assembly clearance. But any reasonably accurate placement system will safely place parts far closer than desirable for visual inspection and repair. It is important to recognize that this can create inspection problems since, for many lines, visual inspection is the first inspection to occur whether done by humans or an automated optical inspection (AOI) system.

Figure 5-10 and Table 5-2 give the clearances recommended between typical components for inspection/repair. Where real estate permits, the minimum spacings should be established by the 45 degree rules. If required, the tighter 60 degree rules will allow a higher density at the price of complicating visual inspection and repair operations. To use Table 5-2, select, for two adjacent components, the larger clearance required by the viewing angle rule in force. The dimensions given in Table 5-2 may need to be modified by any rules that exist on your manufacturing floor. There needs to be a meeting of the minds between the design and inspection groups.

Note that there are plentiful examples of violations of these clearance rules in production today. Some of these are quite manufacturable, others are not. Since many successful designs ignore the clearance rules, they are only guides, not laws. Where very high-density boards are manufacturable despite close-order spacing, clearance violations have been carefully considered in design. If density demands that the visibility rules be ignored, consider this a warning flag for the team to spend extra effort on manufacturing, inspection, and repair concerns, beginning at the earliest possible phase of design and continuing through production startup.

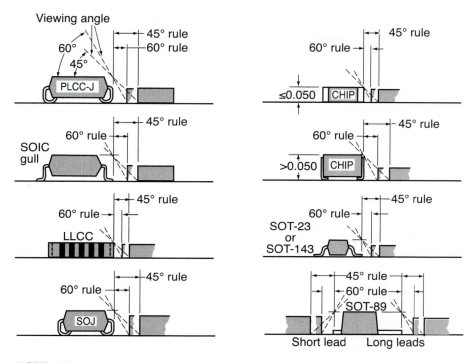

NOTE: All clearances are to be calculated from nominal JEDEC component dimensions and are rounded to the nearest 0.005 inch.

Figure 5-10   Component spacing rules for visual inspection and repair.

Table 5-2   Dimensions for Typical Component Spacing Rules.

Component	Clearance Recommendation*	
	45° high visibility	*60° red. visibility*
Chip component (≤0.05 inch Hi)	1.27/0.50	0.76/0.030
Chip component (>0.05 inch Hi)	Component height	0.6 × height
SOT-23/SOT-143	1.02/0.040	0.64/0.025
SOT-89 (both sides)	1.52/0.060	0.89/0.035
SOIC (gull-wing lead)	1.65/0.065	0.89/0.035
SOLIC (gull-wing)	2.54/0.100	1.40/0.055
SOJIC	3.56/0.140	2.03/0.080
PLCC (50-mil centers)	4.83/0.190	2.79/0.110
LLCC (50-mil typ. centers)	1.78/0.070	1.02/0.040

*All dimensions are mm/inch; inch dimensions control.

5.2.2 Spacing Minimums for Reflow Soldering

In Section 5.2 above, we presented spacing rules to minimize bridging in flow soldering. In reflow soldering, we also must observe minimum spacings to prevent bridging. Reflow bridging may occur on closely spaced lands for several reasons.

Solder Application Misregistration

Misregistration can leave solder paste printed between separate lands. If the lands are adequately spaced, such bridges will disappear during the reflow operation. However, lands that are spaced too closely may allow solder shorts to remain intact, producing a solder defect. This is particularly true under leadless components, such as chip carriers. Boards with leadless components having fine-pitch leads warrant special attention to solder print registration.

Solder Smearing

Solder smeared across several lands can act just like misregistered printing to form bridging defects. Solder smears are also a potential source of solder balls and should always be scrupulously avoided. This phenomenon, like misregistration bridging, is most prevalent under leadless components that have closely spaced terminations.

The Swimming Effect

The swimming effect refers to a tendency of SMCs to move during the reflow soldering operation. This occurs because most surface-mount components displace more than their own weight in molten solder and literally do swim or float. Components floating in solder move due to the influence of several interacting forces. On some components, like BGAs, this may be desirable, but frequently small components can swim to undesirable rotations.

As solder paste goes into a molten state during reflow, solder climbs up the sides of component terminations and spreads across exposed circuit-board metallizations drawn by wetting forces. Gravitational force acts in opposition to the upward movement of solder on device terminations. The result is the familiar meniscus formation. Finally, surface tension forces the liquid solder to seek its smallest possible surface area. The balance of these three forces determines where the swimming component goes.

To see how swimming might affect assembly, let's consider a simple 1206 chip placed on standard lands. Assuming all the metallizations are clean and easily wetted, the solder can achieve its smallest surface area when the component is symmetrically aligned within the land pattern. Thus, under ideal conditions, swimming works to correct the component alignment in reflow. But what would happen if only one side of the terminations were wettable? That one wettable area of the termination would move to the center of the lands. By the same token, uneven wettability of land metallizations destroys the centering operation of the swimming effect.

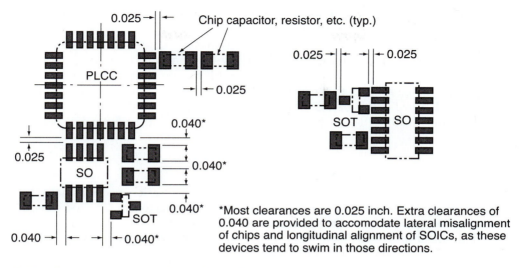

Figure 5-11   Clearances for bridge-free reflow soldering.

While it is true that reflow soldering generally aids component alignment through the swimming effect, we cannot rely on a perfectly even wettability of all device metallizations and lands. Therefore, we provide clearance so that components that misalign during reflow will not cross over to adjacent lands and produce shorts. Figure 5-11 shows minimum spacings needed to assure that swimming does not cause solder shorts.

5.2.3   Spacing Around Tall Components

Tall components usually create several concerns for component spacing. They block vision in a conical area around themselves, and they may interfere with placement head travel. To really simplify design and manufacturing, the rule with tall components (greater than 6.35 mm/0.25" high) is do not have them. Where this cannot be followed, Simple Rule 2 is to add them after all other things are said and done. If neither of these actions is practical, follow the guidelines below to ensure adequate clearance around tall components.

The Cone of Visibility

The size of the clearance core required around a tall component depends on the height of the part and the visibility level needed to accomodate inspection and rework. For very tall components, the rules of Section 5.2.1 would require an excessive clear area. For instance, a metal can capacitor that is only 1.27 cm ($^1/_2$") in diameter and 3.81 cm ($1^1/_2$") tall sits on only 4.84 sq. cm (0.196 sq. inch) of circuit board. Yet, using rigidly enforced 45-degree clearance rules, it would need a clear area 8.9 cm ($3^1/_2$") in diameter, or nearly fifty times its mounting area. Of course, this is ridiculous. We would relax the 45-degree visibility rule since an inspector could get a good view by a glance around that capacitor. But, significant visibility clearance would still be required. SMT boards must be inspected from the component side. This is quite a contrast from just flipping a through-hole board over and looking at the solder joints on its THT leads.

Clearance for Placement Apparatus

In Chapter 3, we considered component clearance to assure a collision-free automated assembly. Tall components complicate this picture. Most placement end effectors are shaped somewhat like a cone with its focal point on the component. Thus, there is a cone around tall components, determined by placement-head form factor, that must be free from obstruction during the placement operation. This clearance cone is graphically explained in Figure 5-12A.

Test Points Near Tall Components

Tall components also interfere mechanically with test probe fixtures. A tall component may require relieving the fixture, and SMTA Testability Guidelines note a 5.08 mm-wide (0.200") annular ring free from test points around the component in question.[5] Figure 5-12B depicts this clearance.

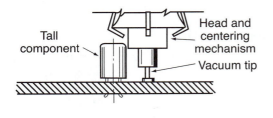

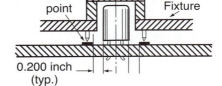

(A) Placement-head clearance.   **(B) Text fixture clearance.**

Figure 5-12   Clearances around tall components.

5.2.4   Spacing Around Test Points

Some clearance is required around each test pad, even when adjacent components are less than (or equal to) 6.35 mm ($^1/_4$ inch) tall. This concept was introduced in Section 5.1.2 where we defined the pointing accuracy of test probes. We must ensure that probes do not land on components for two reasons. First, the spring pressure in probes can cause an open connection to contact a land and falsely test as a good joint. Second, a glancing contact between a probe and component can damage either or both. And last, we do not want false failures caused by test probes landing on the wrong conductor. By allowing a 0.475 mm (0.018") clear annular ring around each test point, the SMTA Testability Guidelines again keep us in the clear.

5.2.5   Spacing Around Fine-Pitch Components

When FPT is mixed with 50-mil pitch on a single board, high-yield soldering often becomes a challenge. SMCs with a 50-mil pitch need plenty of solder (7 to 10 mils of wet paste is typical) to avoid defects such as insufficient solder and opens. FPT components require much thinner solder deposits (3 to 6 mils is common) in order to minimize bridging between their closely spaced leads. These disparate requirements may be dealt with by use of selective solder plating on the board, zippered printing, thin stencils, and overprint of the 50-mil pitch areas, or stepped stencils. The first three approaches do not impact board layout, but the use of stepped stencils may impose large increases in minimum space from the FPT lands to adjacent parts. Exact spacing requirements must be determined by process engineers and are a function of the step variation, the hardness of the squeegee, and printer dynamics.

5.3   Traces and Spaces

SMT lead spacings have forced us to shrink tracks and spaces to suit component dimensions. Tracks and spaces of 0.305 mm (0.012") are typical for THT PWBs. SMT, on the other hand, rarely uses such generous rules since there would not be room for a conductor between lands of a 1.27-mm (50-mil) pitch device on 50-mil centered vias. Remember also that any traces carrying higher voltages must meet the appropriate IPC standard, MIL standard, and/or UL standard.

One of the early decisions for the design team is the minimum space and trace dimensions to be used for a project. These must be determined in conjunction with the PCB fabricator. Below, we will discuss the issues influencing this decision and will present five rule sets for standard trace and space selections. These rules, too, are guidelines and not laws. Once understood, they are meant for guidance in selecting spacing and feature size rules.

The cost curve for fine geometries turns up at around 0.15 mm (6-mil) line-and-space rules. One PWB fabricator, commenting on yields in his shop, explained why. This shop was able to consistently achieve yields of between 85 percent and 90 percent on boards with 0.15 mm (6 mil) line/space, but the yields dropped to 60 percent for 0.1 mm (4 mil) line/space.[6]

A small percentage of contract PWB shops specialize in fine-line high-density work. They can produce 0.05 to 0.076 mm (2 to 3 mil) lines. In such shops, 0.127 mm (5 mil) lines, while not cheap, are not a problem. Use of 0.127 mm (5 mil) line and space rules allow designers to run two tracks between 1.27 mm center (50 mil) SO or PLCC lands. But, 0.127 mm (5 mil) rules will significantly reduce the number of board vendors available for a project. Figure 5-10, Figure 5-11, Figure 5-12, Figure 5-13, and Figure 5-14 show trace and space rules for varying densities. The team should pick from these or a hybrid of them the least dense rules that will satisfy the needs of the project at hand.

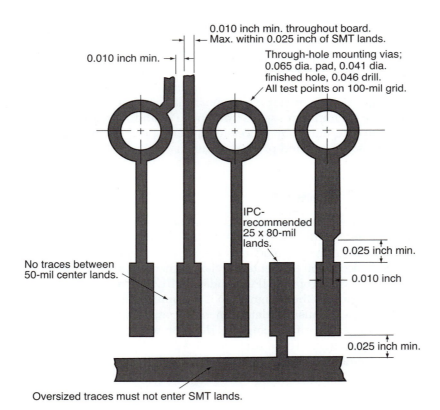

Figure 5-13   Trace and space rules—Level 1 density.

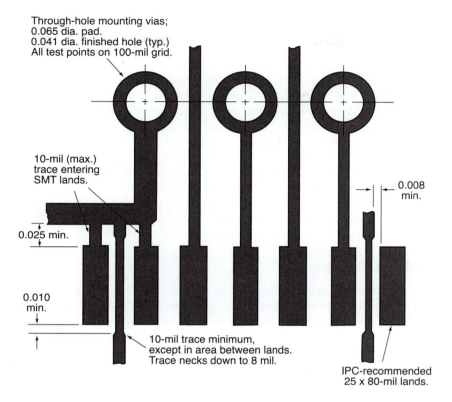

Through-hole mounting vias;
0.065 dia. pad.
0.041 dia. finished hole (typ.)
All test points on 100-mil grid.

10-mil (max.)
trace entering
SMT lands.

0.025 min.

0.010
min.

0.008
min.

10-mil trace minimum,
except in area between lands.
Trace necks down to 8 mil.

IPC-recommended
25 x 80-mil lands.

Figure 5-14   Trace and space rules—Level 2 density.

Fine-line board manufacturability can be improved in many ways. In the PWB etching process, some undercutting of copper occurs. The thicker the copper foil on the laminate, the more troublesome undercutting will be. Fine-line boards, particularly multi-layer designs, are generally specified as $\leq^1/_2$ oz. copper (0.018 mm or 0.0007" thick) instead of the 1 oz. or 2 oz. (0.036 mm / 0.0014" to 0.071 mm / 0.028") material commonly used in through-hole technology. By reducing undercutting, thin copper-foil laminates promote manufacturability of fine-line multi-layer boards.

Fine-line manufacturability is also enhanced by photoplotting lines slightly oversized and spaces slightly undersized of the final desired product. For instance, if the design calls for 0.20-mm (8-mil) lines and spaces, plotting 0.23-mm (9-mil) lines with 0.18-mm (7-mil) spaces will allow for line shrinkage during board fabrication and will produce a closer approximation of the desired result.

Some CAD systems produce excessive and unnecessary jogs in lines as part of their design optimization process. Where such features interfere with an even distribution of metal on a layer, the result will likely be uneven plating due to unequal current distribution. For fine-line designs, metal distribution should be kept even across the surface of each layer. For multi-layer designs, the "homogenous distribution" rule applies in the Z axis as well as the X and Y. Designs with one disproportionately heavy routing layer are more prone to warpage. Where heavy power and ground layers are required, they should be placed in symmetrical balance close to the center of the laminate.

By laser plotting the artwork, it is possible to avoid long runs of necked-down traces between via pads. Instead, full-size via pads may be shaved to provide clearance (where needed) for feed-throughs. Of course, minimum annular ring rules must still be followed.

Accurate phototools are a must for fine geometry work. Photo- or laser-plotted artwork and CAD-generated drill tapes should be used. Be sure that your photoplotter uses dimensionally stable film. Avoid exposing films to temperature and humidity extremes. Work requiring very high accuracy may be plotted on glass or special ultrastable film.

5.3.1   Level 1 Density

Level 1 density, per Figure 5-13, is recommended wherever it meets density requirements without forcing use of an undue number of layers. Level 1 rules make for very manufacturable PW boards and a multitude of suitable PWB vendors. Level 1 trace and space rules with Level 3 via rules are a good combination for pad cap layers of MLBs. Level 1 rules will only be suitable for routing layers on very low-density boards, however.

5.3.2   Level 2 Density

Level 2 rules, per Figure 5-14, are common for routing layers in SMT. These rules allow reasonable densities and yet are very manufacturable. Most quality PWB vendors can handle such dimensions in stride.

5.3.3   Level 3 Density

Level 3 rules, per Figure 5-15, are widely used for SMT routing layers. These rules allow reasonable densities and simple test probing. PWB shops can generally furnish Level 3 density, and there should be little or no cost premium in stepping up from Level 2 to Level 3.

5.3.4   Level 4 Density

Level 4 rules, per Figure 5-16, may come at a cost premium. However, they allow high densities and may not cost as much as the added layers that Level 3 rules would impose on a crowded design. Savings may be substantial where increased density translates into double-sided versus multi-layer, or one board instead of two.

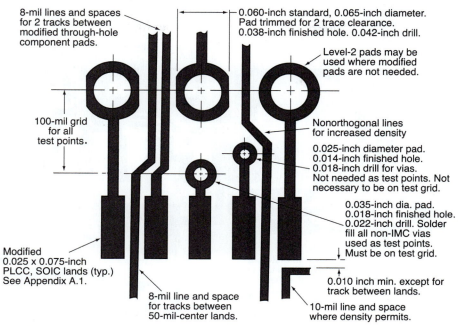

8-mil lines and spaces for 2 tracks between modified through-hole component pads.

0.060-inch standard, 0.065-inch diameter. Pad trimmed for 2 trace clearance. 0.038-inch finished hole. 0.042-inch drill.

Level-2 pads may be used where modified pads are not needed.

100-mil grid for all test points.

Nonorthogonal lines for increased density

0.025-inch diameter pad. 0.014-inch finished hole. 0.018-inch drill for vias. Not needed as test points. Not necessary to be on test grid.

0.035-inch dia. pad. 0.018-inch finished hole. 0.022-inch drill. Solder fill all non-IMC vias used as test points. Must be on test grid.

Modified 0.025 x 0.075-inch PLCC, SOIC lands (typ.) See Appendix A.1.

8-mil line and space for tracks between 50-mil-center lands.

0.010 inch min. except for track between lands.

10-mil line and space where density permits.

Figure 5-15   Trace and space rules—Level 3 density.

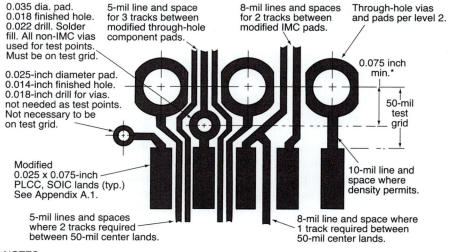

0.035 dia. pad. 0.018 finished hole. 0.022 drill. Solder fill. All non-IMC vias used for test points. Must be on test grid.

5-mil line and space for 3 tracks between modified through-hole component pads.

8-mil lines and spaces for 2 tracks between modified IMC pads.

Through-hole vias and pads per level 2.

0.075 inch min.*

0.025-inch diameter pad. 0.014-inch finished hole. 0.018-inch drill for vias. not needed as test points. Not necessary to be on test grid.

50-mil test grid

Modified 0.025 x 0.075-inch PLCC, SOIC lands (typ.) See Appendix A.1.

10-mil line and space where density permits.

5-mil lines and spaces where 2 tracks required between 50-mil center lands.

8-mil line and space where 1 track required between 50-mil center lands.

NOTES:
*   Assuming short clenched leads, auto insert and clench. This dimension may be reduced to 0.045 with solder-cut solder processing.
1. Reflow soldering only of high-density side.
2. SMOBC required for trace-bearing surface layers. Must be photo-imaged solder mask.

Figure 5-16   Trace and space rules—Level 4 density.

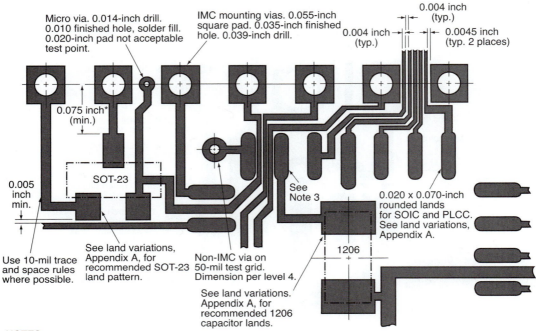

Figure 5-17   Trace and space rules—Level 5 density.

5.3.5   Level 5 Density

Level 5 rules, per Figure 5-17, may also carry a price premium. They should be invoked only when needed. As of this second edition, there are still a limited number of PWB fabricators capable of Level 5 manufacturing, but a growing number can handle even finer geometries. For standard PWB processes, the leading edge is around 0.025 mm (1 mil) line and space as of this writing. Finer geometries are made with thin films (see Chapter 9).

5.4   Component Lands

As we mentioned in the introduction to this chapter, examples of land pattern designs for various components, also known as footprints or land geometry patterns, are collected for reference in Appendix A. However, as you will see in Section 5.4.1, advances in components are happening so rapidly that the IPC has streamlined the manner of determining land patterns. Then, we will cover the generic rules for land geometries and the traces entering them.

5.4.1   Land Patterns

Standards that affect SMT land pattern designs are defined by two organizations, the Joint Electron Device Engineering Council (JEDEC) of the Electronic Industries Association (EIA), and IPC Association Connecting Electronics Industries. The IPC establishes standards for PCB designs, and their Surface-Mount Land Pattern Subcommittee of the Printed Board Design Committee has developed land patterns for devices defined by JEDEC.

In Appendix A, you will see examples of land patterns. However, as mentioned in the introduction to this section, several factors make it difficult (and in many cases unnecessary) to try and keep all possible land patterns on hand. The primary factor that makes it difficult is the continuing introduction of new components with new lead counts. The other is the continuing changeover (in the U.S.) from inch-dimensioned parts to the metric-dimensioned parts that are standard in the rest of the world.

In 2004, a joint effort between the IPC, the International Electrotechnical Commission (IEC), and various worldwide industry standard groups resulted in the creation of a new standard, IPC-7351 "Generic Requirements for Surface-Mount Design and Land Pattern Standard," for CAD land support. Developed also in conjunction with PCB Libraries were a set of land pattern calculators and viewers designed to interface with a variety of ECAD software tools and a freeware version of the IPC-7351 land pattern calculator. The calculator, along with a number of libraries, is available at PCB Libraries' website, http://www.pcblibraries.com. With this calculator available, continuing manual updates of land patterns may no longer be necessary.

Like its predecessor standard IPC-SM-782, IPC-7351 relies on proven mathematical algorithms that take into account fabrication, assembly and component tolerances in order to calculate precise land patterns. The standard improves upon the concepts developed for IPC 782 by establishing three land pattern geometries for each package type, describing the solder joint engineering goals for each package family, and providing the user with a naming convention that allows easy lookup of land patterns.

The three application-specific land pattern geometries are intended to support various levels of product complexity. These CAD library geometries will support the following variations for each device family:

    a. Least Environment Land Pattern for miniature devices where the land pattern has the least amount of solder pattern to achieve the highest component packing density.

    b. Nominal Environment Land Pattern for products with a moderate level of component density and providing a more robust solder attachment.

c. Most Environment Land Pattern for high component density applications typical of portable/hand-held products and products exposed to high shock or vibration. The solder pattern is the most robust and can be easily reworked if necessary.

It is important for the user to recognize that the primary purpose of a land pattern is not to provide a resting place for a component. It is to provide the proper copper area on the board to accomplish two tasks:[8]

- Provide for the creation of a strong and easily inspected solder joint
- Minimize the risk of solder bridging between joints

A minor detail is that, of course, these two criteria have conflicting dimensional requirements. Providing for a strong and easily inspected solder joint demands that we have a large copper area, and minimizing the risk of bridging demands as much space as possible between the copper features. Good footprint design must balance these two demands.

5.4.2   Traces Entering Lands

Where traces enter lands, special rules apply to the trace size and geometry. These rules cover influences on component swimming, control of thermal paths, and solder migration along traces.

Influences of Traces on Component Swimming

In Section 5.2.2, we discussed the swimming effect. We mentioned that one of the determining factors in the direction of swimming is the geometry of metallized area wetted by solder. Where a trace enters a land, therefore, it must either be solder masked to prevent its wetting, or it must be included in the swimming equation. Since screened-on masks must be set back from lands, traces entering lands from screened-on masks must be considered as unmasked traces. Figure 5-18 illustrates some rules for exposed traces entering lands. With the use of solder mask over bare copper (SMOBC) discussed in Section 5.9, these rules may not need to be followed, but if other mask techniques (or no mask) are used, they should still be followed. The use of standard mask or no mask on patterns B, C, and D may result in "see-saw," or component motion, during reflow.

For board designs using SMOBC, other angles of entry besides those shown in Figure 5-18 are acceptable, such as 45-degree entry. A caveat with 45-degree entry is that some board fabricators caution that this creates an angle of less than 90 degrees, which can hold acid from the etching process. While this is not the problem it used to be, it is still worthwhile checking with your fabricator before allow angles other than 90 degrees at trace/pad interconnects. Obviously, with round pads this is not an issue.

One pattern that should not be used is that of connecting two adjacent IC pads with a straight trace from pad to pad, in line with the component body line, as shown in Figure 5-18(E). This can be missed during inspection or can become confused with a solder bridge. It is better to bring the trace out from the pad, normal to the body line, than use two 90-degree angles to route it to the next pad.

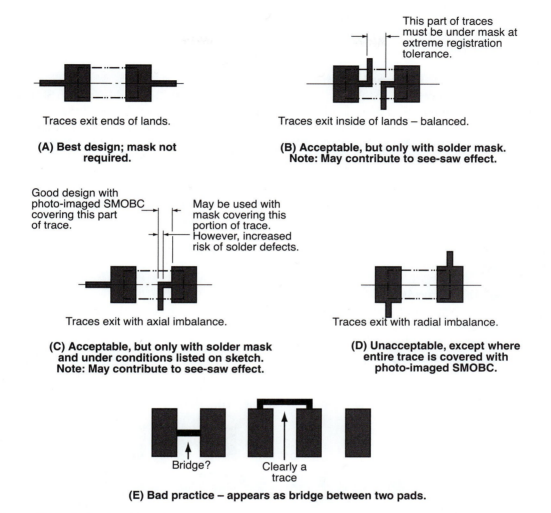

Traces exit ends of lands.

(A) Best design; mask not required.

This part of traces must be under mask at extreme registration tolerance.

Traces exit inside of lands – balanced.

(B) Acceptable, but only with solder mask. Note: May contribute to see-saw effect.

Good design with photo-imaged SMOBC covering this part of trace.

May be used with mask covering this portion of trace. However, increased risk of solder defects.

Traces exit with axial imbalance.

(C) Acceptable, but only with solder mask and under conditions listed on sketch. Note: May contribute to see-saw effect.

Traces exit with radial imbalance.

(D) Unacceptable, except where entire trace is covered with photo-imaged SMOBC.

Bridge?

Clearly a trace

(E) Bad practice – appears as bridge between two pads.

Figure 5-18   Rules for exposed traces entering lands.

A pattern with uneven pad areas is unacceptable in virtually all designs. These are sometimes used to allow a test point close to a component, but since there can be no mask over that type of pad, they are guaranteed to create unacceptable swimming. The preferred technique is to establish a round test pad on the trace a short distance from the component pad, with an intervening trace segment.

For ICs, traces may enter the lands from either end, but avoid those designs where a pre-ponderance of soldering forces are congregated on one axis. As long as the forces exerted by solder wetting remain equalized about the component axes, good alignment should result.

Control of Trace Thermal Paths

In reflow soldering, even small differences in thermal mass between the lands of a chip SMC will influence the time of reflow of those lands. If one land is late in reflowing, the result may be tomb-stoning or device misalignment. To keep reflow times uniform, avoid excessive thermal coupling with the lands. Do not connect traces more than 0.25 mm (0.010") wide to solder reflow lands. Where a wider trace must connect to a land (i.e., power and ground buses), reduce the trace to a 0.25 mm (0.010") width for at least 0.635 mm (0.025") before entry, as shown in Figure 5-19.

Other thermal issues include "How much current can this trace carry?" and "What will the temperature rise of a trace of a given size be when a specific current is passed through it?" Both of these are answered in IPC documents IPC-2221 and IPC-2222, and various websites have trace temperature calculators. Among these is the UltraCAD, Inc., website, http://www.ultracad.com. This site has a temperature calculator and a technical paper that deciphers the IPC trace/temperature curves.

Avoid Solder Starving

Tin/lead or tin plating on traces under a solder mask melts and forms a capillary during reflow soldering. Solder from SMC lands can migrate along such capillaries, leaving solder-starved joints and open circuits behind. Therefore, SMOBC is recommended for non-pad cap SMT boards. Specifying bare copper traces on pad cap designs also eliminates solder migration along traces and into adjacent vias. For multi-layer boards, the two additional layers required for pad caps often come at no premium over solder-mask costs for surface-routing layers, and pad cap designs may yield valuable economy on the manufacturing floor.

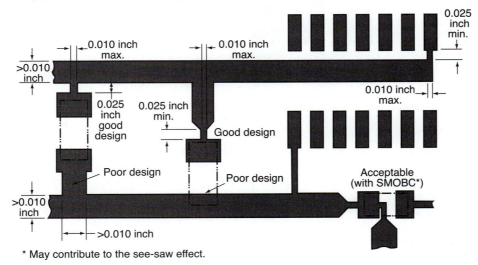

Figure 5-19   Reduce large traces before entry to reflow lands.

High Shear Stress Resistance

Thermal excursions, shock, or other environmental conditions imposing significant shearing stresses on solder joints will dictate the use of special rules for trace entry to lands. Abrupt 90-degree corners where traces enter lands can set up stress concentration points in solder joints. By rounding ends of the lands and adding a fillet radius to each trace at its entry point, stress risers are eliminated.

5.4.3   Land Specifications

Bare Copper Traces and Solder-Coated Lands

Bare copper trace boards, with tin/lead coating on the lands only, are produced by hot-air solder leveling over the solder mask, by selective tin/lead plating, or by selective tin/lead stripping. Since the hot-air solder leveling process requires a solder mask, it is not generally applied to pad cap boards. Solder-surface finish and height varies by process and vendor and should be critical in the process selection. Select only those processes which your PWB supplier can sufficiently control to guarantee you flat, solderable lands, and a smooth surface on the PWB. These include OSP and fused electro-tin plating.

Bare Copper Boards

Boards are also fabricated without any tin/lead plate on traces or land areas. Several coating processes are available to protect the exposed copper areas from oxidation, with SMOBC and OSPs being the most common. In some instances, where just-in-time delivery cycles can be tightly enforced, no coating is used. Finished boards are moved immediately from board manufacture, where copper is squeaky clean, to board assembly. If this is your assembly method of choice, be certain your solder paste supplier knows of the untinned pads.

Selective Plating Near Lands

Lands located in areas that will be dipped into selective gold-plating baths must be masked in order to prevent gold contamination. Gold forms gold/tin intermetallics during the soldering process. The intermetallic formed in this process is extremely brittle, destroying solder joint reliability. Wherever possible, locating lands above the selective plating areas, per Figure 5-20, saves masking.

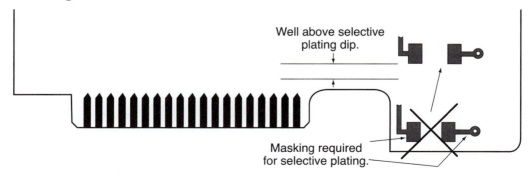

Figure 5-20   Locate lands above the selective plating areas.

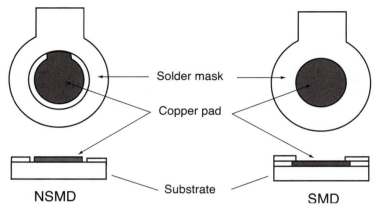

Figure 5-21 NSMD and SMD pads for BGA packages.

5.4.4 BGA Land and Routing

Two types of land patterns are used for BGA packages:

- Solder Mask Defined pads (SMD)
- Non-Solder Mask Defined pads (NSMD)

NSMD pads have a solder mask opening that is larger than the pad, whereas SMD pads have a solder mask opening that is smaller than the metal pad, as shown in Figure 5-21.

NSMD is preferred for BGA pads because copper artwork registration is more accurate than the accuracy and tolerance of the solder masking process. Moreover, SMD pad definition may introduce stress concentration points in the solder that may result in solder joint cracking under extreme fatigue conditions. NSMD pads require a 3-mil or greater clearance between the copper pad and solder mask to avoid overlap between the solder joint and solder mask due to mask registration tolerances. With regard to the size of the pad, a 1:1 ratio between the BGA package pad and the corresponding PCB pad is recommended. While a smaller pad, down to 0.8:1 can be used, a 1:1 pad is more reliable.

A typical PBGA has four or five rows of solder balls around the periphery of the package. The number of traces routed between the pads on the PCB is defined by the pad size and trace width and spacing capabilities of the PCB fabricator. In general, Table 5-3 shows appropriate routing guidelines.

Table 5-3 BGA Trace Routing.

	BGA pitch = 1.27 mm (50 mils)	BGA pitch = 1 mm (40 mils)
Trace width & space = 0.15 mm (6 mils)	1 trace between pads	Insufficient space
Trace width & space = 0.125 mm (5 mils)	1 trace between pads	1 trace between pads
Traced width & space = 0.1 mm (4 mils)	2 traces between pads	1 trace between pads

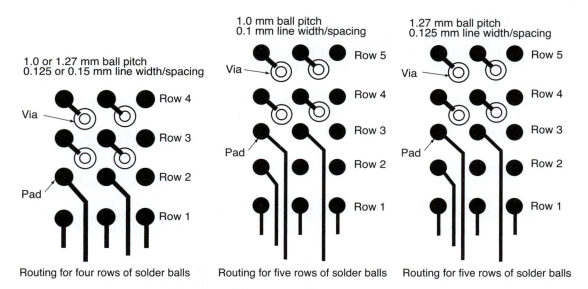

Figure 5-22A   Examples of BGA trace escapes. *(Courtesy of National Semiconductor Corp.)*

As BGA pitch is reduced down to that used in µBGA packages, the trace routing must all be done outside the BGA pad pattern. There is insufficient room between the pads; however, as will be discussed in Section 5.5.3, microvias do allow route escapes to take place within a BGA pad itself.

As with other complex packages, the component manufacturer will have recommendations on BGA routing and escape patterns. Additionally, IPC PCB documents have information on trace routing. Figure 5-22A shows examples of BGA trace escapes, and Figure 5-22B shows examples of BGA pad and via routings.

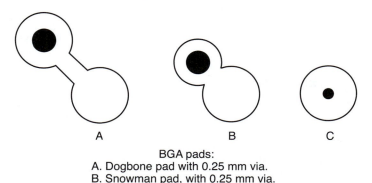

BGA pads:
A. Dogbone pad with 0.25 mm via.
B. Snowman pad, with 0.25 mm via.
C. Microvia in pad, with 0.10 mm via.

Figure 5-22B   Types of BGA pad + via routing.

As with most high-density package types, all manufacturers of BGAs include routing information in their data sheets and application notes on BGAs. As noted, the choice of any high interconnect ball count BGA package brings with it the need for certain advanced PCB technology.[9] The impact of the designer on the overall assembly process varies with the assembly issue in question:

Component coplanarity	limited designer impact	aided by largest interconnects possible
Paste printing	significant impact	use most appropriate land pattern
Placement/yield	significant impact	use largest interconnects possible
Inspection	limited impact	leave x-ray path as clear as possible
Test	limited impact	consider BIST

The pad designs shown in Figure 5-22B, whether dog bone, snowman, or via-in-pad, can affect PCB yield as well. The first two have minimal impact, but the latter requires close communication with the board fabricator. Note also that the via-in-pad option requires the use of microvias, and routing will go down one or possibly two layers where the larger vias used in dog bone and snowman pads can go through any required number of layers.

5.5   Vias; High-Density Interconnects (HDI)

In surface mounting, greater freedom of design is afforded by the fact that vias need not be directly associated with each component lead, but can be anywhere in a trace connecting that lead. In return for this freedom, the designer may have to worry about where to locate each and every via required for layer connection. In considering via specifications, we will deal with via size and via location.

5.5.1   Via Size

First, a via is a hole in the substrate intended only to establish electrical connectivity and is not intended to hold a component lead. Via size is much more flexible in SMT than in THT design where through holes are defined by the size of the lead they must accept. Real estate savings are a primary goal of most, and vias no longer need to be sized to accept a through-hole lead. Therefore, the question becomes, "How small can my vias be?" Limits to via size reduction come from two areas.

Relationship of Via Size to Via Cost

First, the cost of drilling via holes begins to increase in typical shops as drill diameters go below 0.635 mm (0.025"). The cost curve becomes non-linear in most shops at around 0.39 mm (0.015"), and thereafter, drilling costs may climb sharply with decreasing hole size. A 0.457-mm (0.018") drilled hole is a comfortable number for high-quality vendors, with 0.254 mm (0.010") being a typical lower limit for moderate-cost boards. Holes smaller than this are commonly created using lasers.

Pad Size for Vias

Second, tolerance buildups generally require that pads be a minimum of 0.89 mm (0.035") in diameter for test probing of bare and loaded boards. If a 0.46-mm (0.018") hole is drilled and a through-plating buildup of 0.05 mm (0.002") thickness is specified, the finished hole will be 0.36 mm (0.014") in diameter. Allowing an annular ring of 0.25 mm (10 mils) gives a pad diameter of ~0.86 mm (0.035"). Where such dimensioning does not interfere with necessary testing of inner layers before lamination, inner layer pads may have an annulus dimension of 0.13 mm (0.005") for a pad diameter of 0.64 mm (0.025"). Keyholes as shown in Figure 5-23 can be used when needed to prevent breakout of trace junctions. For any holes less than 0.5 mm (0.020") in diameter, a minimum plating buildup of 0.038 mm (0.0015") should be specified. The need for keyholes should be discussed with the PCB fabricator before the board design incorporates them.

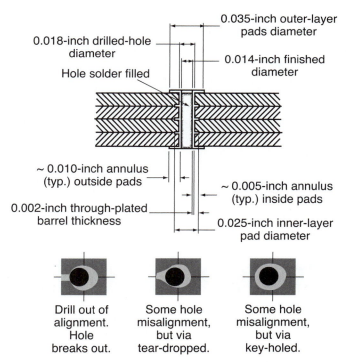

Figure 5-23   Via miniaturization for moderate cost boards. *Note: Tear drops, or the less attractive alternative of key holing, are used to reduce the risk of breakout of a misregistered drill where annular ring allowances must be small. Some CAD systems, numerous Gerber post-processors, and many PWB front-end CAM systems are able to add tear drops to key-holed via flashes.*

The filling of test-point vias prevents conical probe points from becoming lodged in holes during testing. Where vias are not filled, crown point probes are used. Filled vias improve the sealing on vacuum hold-down fixtures. However, solder-filled vias may fail due to on-board or environmental thermal cycling because the thermal coefficient of solder and that of copper are not well matched. Vias may be copper filled by plating shut small holes (0.33 mm/13 mil) or with solder by the solder fill and reflow method. Another alternative is tenting small vias with solder mask, discussed in Section 5.9.

Via Aspect Ratio

Another issue influencing hole-size specification for standard vias in thick boards is the *aspect ratio*. Aspect ratio refers to the board thickness divided by the via diameter. THT through holes did not have cracking problems because they were low aspect ratio (about 2.5:1), and the soldered lead in the hole added strength. High aspect ratio holes may be subject to cracking in through-plated via barrels because of the disparate TCEs of board and via barrel material. This is because the Z-axis TCE is much higher on fiberglass-reinforced boards than the X-Y axes' TCE since the glass fibers sit in the X-Y plane and constrain that expansion while there is no constraint in the Z axis except for the much higher TCE of the native epoxy.

Even low expansion boards may be unrestrained in the Z axis. In fact, polymer boards restrained in X-Y expansion by Kevlar™ fibers are actually worse in Z-axis TCE than unrestrained laminates of the same polymer. This is because Kevlar's apparent negative thermal coefficient of expansion results from the Kevlar fibers becoming shorter and fatter as they are heated. Since the fibers are oriented along the X and Y axes of the board, their contraction reduces the expansion of the board in the X and Y axes, but the fattening of the fibers adds to the substantial expansion of the polymer in the Z axis.

Unless special precautions are taken, the aspect ratio should not go above 6:1, and <3.5:1 is preferred for high reliability. All high-aspect hole plating should be specified as high-ductility copper. Where aspect ratios will go above 4:1, vias should be copper filled in the bare board. An alternative precaution against barrel cracking is the use of substrates with restrained Z-axis expansion approximating that of the copper barrel.

5.5.2   Via Location

As we mentioned earlier, SMT via locations are not predetermined by component lead penetrations as through holes are. The surface-mount via could potentially be located anywhere in the PWB. In practice, however, relationships to components, other features, test pads, and lands somewhat limit the choices. We tend to mitigate these limits by specifying filled vias. Designers should be very careful in locating standard vias in component land. An unfilled via located in a reflow solder land can cause defects in several ways:

1. Solder is pulled away from the joint and through the via by capillary attraction.
2. Uneven topography of the lands due to unfilled vias contributes to component swimming.

Figure 5-24 further illustrates via filling. For routing that requires in-land vias, a better choice is to use microvias, discussed in Section 5.5.3.

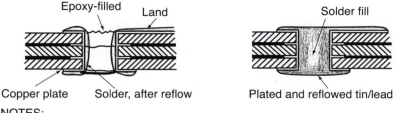

NOTES:
1. Vias filled by either of these methods available from some PWB vendors.
2. Open vias in reflow lands can cause solder defects.

Figure 5-24   Vias located in reflow solder lands.

Figure 5-25 details the relationships between via location and the location of lands, components, and features.

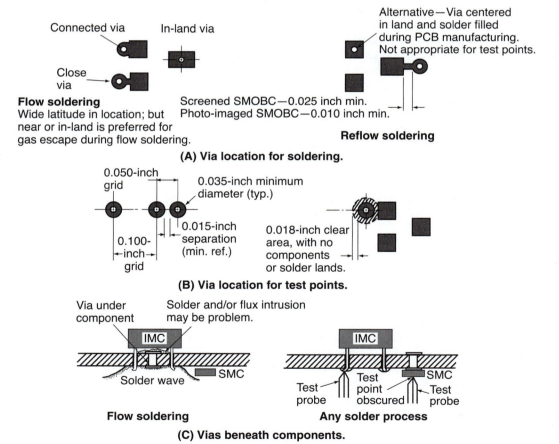

Figure 5-25   Relationships influencing via location.

Via Location in Relation to Lands

For reflow soldering, vias should not be located in or be intimately connected to solder lands unless special procedures are followed to prevent solder migration down the via. As Figure 5-23 and Figure 5-24 show, unfilled vias for reflow solder processes are normally isolated from the lands by at least a short trace.

This rule reverses for wave soldering. Open vias are encouraged near flow-solder lands because they provide an outlet through which any entrapped gasses can escape during soldering. Where vias cannot be close to the land, it is permissible to locate them directly in the land. However, since a component lead may block the gas escape in a land, the "connected" or "close" via is preferable to the "in land" design. Additionally, vias should not be located under components when the wave solder process is used since the gas escape can create accumulations under the components that may be difficult to clean.

Via Location for Test Pads

Unless custom probes are to be used, test pad vias must be located on the test fixture grid, as discussed previously. Use of a design for testability (DFT) strategy will optimize the location of test pads while maintaining signal integrity. Keeping track of which vias are so placed requires either CAD/CAT integration or a manual mark-up strategy of some sort. The wrong way to handle this challenge is to go back and locate the test points as an afterthought. Also, clearance must be provided between adjacent test points and the components so that no probes miss the intended site or hit the wrong target. For typical 2.54-mm (100-mil) probe cards, Uni-Fix recommends pad minimum centers of 2.1 mm (0.086"). For 1.27 mm (0.050") probe cards, Everett Charles Technologies recommends a 0.9-mm (35-mil) separation between test points, and a 0.457-mm (18-mil) annular ring of land-free area around each test point accomplishes this.

Remember also the costs involved with close spacing of test vias. Uni-Fix, a manufacturer of electronic test fixtures, notes the following relationships between fixture spacing and cost[10]:

- Standard test pads and probes on 0.100" centers, cost factor 1
- Test pads on 0.66" centers, cost factor 1.25, probe life and reliability equal to 0.100" center probes
- Test pads on 0.50" centers, cost factor 1.6, life approximately half that of standard probes

Vias Beneath Components

Before locating a via beneath an SMC, several questions need to be answered. Will the component block test probing of a needed test node? If so, locate the via elsewhere, place a separate test point somewhere else in the trace, or consider costly double-side probing. Will the board be wave soldered? If so, is there any possibility of solder or flux bubbling up through the hole and creating a short circuit under the device? If soldering defects could occur, either relocating the hole using a dry-film mask to tent the via or filling the vias with solder during board manufacturing will eliminate solder "bubble up" concerns. Vias beneath components and vias in component pads are discussed further in Section 5.5.3, "High-Density Interconnects."

Vias Near Gold Plating

Do not locate vias where they will be discolored by the selective plating of gold contact fingers. As shown in Figure 5-20, vias that are less than 2.54 mm (100 mils) from the fingers are affected. The defect is strictly cosmetic, so you are welcome to just grin and bear it if holes must be located in close proximity to the fingers. Or, the holes can tented with mask and thereby avoid discoloration.

5.5.3   High-Density Interconnects (HDI)

The use of IC packages such as ball grid arrays (BGAs), chip scale packages (CSPs), and chip-on-board (COB) require higher density of interconnects on the board than standard packaging technologies. To maximize the minimization opportunities provided by these packages, board designers need to make use of build-up technology boards, as opposed to subtractive technology, microvias, and high-performance materials—discussed in IPC-4104—when signal integrity is an issue. These technologies, along with smaller traces and spaces and thinner dielectrics/substrates, make up High-Density Interconnects (HDI).

Microvias are vias that:

- connect two adjacent layers
- have a 1:1 aspect ratio, that is, if the layer is 0.1 mm (0.004") thick, the via is 0.1 mm in diameter, although some HDI boards allow the microvia to go down two layers with a 2:1 aspect ratio.

Advantages to the use of HDI include:

- Use of microvias with the accompanying routing advantages and decrease in board area taken up by vias
- Ability to use vias in the component pads
- Size reductions due to smaller substrate size, weight, and volume
- Increased PCB density with closer component spacing
- Reduced parasitic L and C resulting in better circuit performance and signal integrity
- Reduced EMI and RFI

HDI then uses routing traces and spaces that are consistent with the size of the vias used. Microvias have the advantage that with a 1:1 aspect ratio they are unlikely to crack, and they are small enough that they can be placed within component pads to allow escape routing to an adjacent inner layer. We should note that while the 1:1 aspect ratio through one layer is the original definition of a microvia, there are some designers who are now extending them through two layers. It should also be apparent that a 0.1-mm via with its implied 0.1-mm thick layer means that the use of thinner dielectrics will be necessary.

The IPC has standards that define both HDI technology and its use:

- IPC 2226—"Sectional Design Standard for High-Density Interconnect (HDI) Boards"
- IPC/JPCA-2315—"Design Guide for High-Density Interconnects (HDI) and Microvia"
- IPC-4104—"Specification for High-Density Interconnect (HDI) and Microvia Materials"

- IPC 6016—"Qualification and Performance Specification for High-Density Interconnect (HDI) Layers or Boards"
- IPC 6801—"IPC/JPCA Terms and Definitions, Test Methods, and Design Examples for Build-Up/High-Density Interconnect (HDI) Printed Wiring Boards"

HDI is used to maximize board size reduction. IPC-2226, the HDI design standard, defines three types of HDI structures. From the simplest to the most complex HDI structures, Types I, II, and III go from the most straightforward construction to complex (and expensive) structures that include stacked microvias. Additionally, IPC-2226 and IPC-2315 have a design tutorial that focuses on the density tradeoffs that exist when a designer needs to understand the tradeoffs betweent he three types of HDI structures. In addition to designing the board, users must also select a capable fabricator. The more general standard IPC-9151, "Selection of Fabricators," specifies a capabilities benchmarking panel about which more information can be found on the 9151 website, http://www.pcbquality.com. This panel allows comparision of different fabricators' high-performance board capabilities. When designs call for the use of expensive substrates, such as low-loss laminates, the savings due to size reduction can be significant. In high-speed boards, Holden notes these major HDI drivers:[11]

- Lower costs due to fewer layers and smaller boards
- Reduction in layer count
- Miniaturization for portable products
- Improved electrical performance and signal integrity
- Reduced EMI and RFI
- Higher component density and component I/Os
- Integration of fine-pitch devices

One of the major advantages of HDI is allowing placement of vias in surface-mount pads, such as the 0.1-mm (0.004") via in a 0.6-mm (0.024") BGA pad shown in Figure 5-26. This allows

0.1 mm via in 0.6 mm BGA pad: dogbone or
snowman pad is not needed.

Figure 5-26   Microvia in BGA pad.

direct connection of pads to different signal layers without the need to take up routing and escape area on the surface layer.

HDI+microvia technology is still evolving, with some fourteen different technologies in use worldwide.[12] One of the major drivers to consider HDI technology is FPGAs. Components from Xilinx and Altera already include packages with over 1000 I/O. Along with the ability to route in-pad, blind microvias also have about 10 percent of the parasitic L and C of a standard through via. With regard to high-speed signals, through vias reflect about 10 percent of any signal under 2 GHz and become worse as the frequency increases. In Brooks' study of common 0.028" through vias in 50Ω traces on FR-4, each via in a trace caused a 20-picosecond reduction in rise time of a pulse, *whether the signal trace went through the via or just passed along the trace with vias in it*. The study also found that a via of this size presented a transient impedance discontinuity of about 6Ω.[13]

While blind microvias are in common use, whether as via-in-pad or as a via adjacent to a pad, buried microvias must be carefully considered. Capers reports a 20 to 60 percent cost premium on buried vias compared to blind vias.[14] Like standard vias, blind microvias must also be carefully considered. If an aspect ratio of 1:1 is to be maintained, then, e.g., a 0.010" blind drill hole should go no deeper than 0.010" into the board, and this only allows access to layer 2. If the board is to be an eight-layer board with power and ground planes in the middle, then no top or bottom traces will be able to access the planes. For deeper blind vias, a sequential build must be used where inner layers are drilled, imaged, plated, etched, and laminated. This sequence may be repeated several times depending on how the via connects must be made. This type of sequential work will add considerably to the board cost.

Last in this discussion on HDI, but certainly not last in designers' concerns, is the issue of test. The use of microvias and other facets of HDI make it too easy to design a board that is virtually untestable. The basic test tenet that each net should be accessible to a bed-of-nails ICT test set falls by the wayside in the face of HDI. Digital circuits can make use of IEEE 1149.1-compatible test designs (also known as boundary scan and JTAG), but this will not solve all the test issues. Test equipment manufacturers like Agilent and Teradyne are working to overcome the problems. As an example, Agilent has presented AwareTest, a technique that combines x-ray and ICT to reduce the required number of test pins by as much as 78 percent. According to Agilent, in AwareTest "X-ray inspection verfiies that all devices are present, that no device pins have open solder joints, and that there are no shorts on the PCB. Following this, the in-circuit test only needs to ensure that the component is: the correct type, oriented properly, and operational." This Agilent evaluation was presented by Kirschling at the February 27, 2001, EtroniX, and is availablee at http://www.agilent.com.

5.6 Substrates, Materials, and Very General Rules

Very early in the design process, the team should consider alternatives and specify substrate and material. Typical SMT substrate properties are shown in Table 5-4. Table 5-5 shows common applications of various substrate materials.

Table 5-4 Substrate Material Properties.***

Substrate Properties Substrate Materials	Glass Transition Temperature (°C)	TCE X & Y Axis (ppm/°C)	TCE Z Axis (ppm/°C)	Thermal Conductance (W/M°C @ 25°C)	Flexural Strength (K psi @ 25°C)	Dielectric Constant (@ 1 MHz and 25°C)	Volume Resistance (Ω-cm)	Surface Resistance (Ω)	Moisture Absorption (% of Weight)
Glass/Epoxy††	125	15	48	0.16	45–50	4.8	10^{12}	10^{13}	0.10
Glass/Polyimide	250	15	57.9	0.35	97	4.4	10^{13}	10^{12}	0.32
Kevlar/Epoxy	125	6.5	~50	0.16	40	4.1	10^{12}	10^{13}	0.10
Kevlar/Polyimide	225	5.0	~60	0.35	50	3.6	10^{14}	10^{12}	1.80
Alumina Ceramic	N/A	6.5	6.5	2.1	44	8	10^{14}	10^{14}	0.00
Beryllium Ceramic	N/A	8.4	8.4	14.1	50	6.9	10^{15}	10^{15}	0.00
Porcelain Coated Steel	N/A	10	13.3	0.06/—**	†	6.3–6.6	10^{11}	10^{13}	0.00
Porcelain Coated CCI**	N/A	7	†	0.06/57*	†	6.8	10^{11}	10^{13}	0.00
Polyimide/CCI Core	270	6.5	†	0.35/57*	†		10^{12}	10^{12}	0.35
Epoxy/Aluminum Core	125	15	†	0.16/203*			10^{11}	10^{13}	0.10
Epoxy/Graphite	125	7	~48	~0.16			10^{12}	10^{13}	0.10
Polyimide/Graphite	250	6.5	~50	1.5		6	10^{14}	10^{12}	0.35
Glass/Teflon	75	55				2.2	10^{14}	10^{14}	0.00
Glass/Triazine	220	12				4.5	10^{13}	10^{12}	0.015
Glass/Polysulfone	185	30			14	3.5	10^{15}	10^{13}	0.029
Quartz/Epoxy	125	6.5	48	~0.16		3.4	10^{12}	10^{13}	0.10
Quartz/Polyimide	270	9	50	~0.35	95	3.4	10^{13}	10^{12}	0.40

††Advances in modified epoxies are occurring very rapidly. Two areas are seen as primary objectives. One is the development of high-T_g materials to withstand the high temperatures in SMT assembly and rework. The other is the search for materials having a low dielectric constant suitablee for high-frequency boards but are inexpensive and similar to epoxy in all other properties.

*Surface coating/core material.

**Copper-clad Invar 20/60/20.

***Properties shown are average for comparison purposes only. For exact engineering calculations, consult the substrate supplier.

†Depends on core-to-surface ratio.

Table 5-5   Typical Applications of Various Substrate Materials.

Material	Potential Application
Kapton Film	Flexible circuits with flexural segments and/or rigidized populated areas. Circuits conforming to irregular areas, such as inside cameras and cell phones.
FR-4 and Other Epoxy/Glass	Standard commercial and industrial SMT circuits up to 3 GHz.
Alumina	Relatively small circuits requiring moderately high heat-sinking capacity or using leadless ceramic components. High-frequency circuits.
Beryllia	Relatively small circuits requiring very high heat-sinking capacity. High-frequency circuits.
Porcelainized Steel	High heat-sink capacity and the ability to form a structural element of system, such as in auto instrumentation panels.
Polyimide/Kevlar	Matching TCE of ceramic components by balancing the mix of Polyimide and Kevlar.
Copper/Invar/Copper or Copper/Molybdenum/Copper	Very high heat-sink capacity and rigidity. Core material for matching TCE or ceramic components.
Teflon	High-frequency boards. Difficult to work with.
Glass	LCD display and other see-through applications.
Polyimide/Glass	High glass transition (high operating or processing temperatures). Readily available, lightweight, and low cost.

5.6.1   Very General Rules

First, we would like to present some very general rules that apply to any substrate and virtually all designs. They come under the heading of Design for Manufacturability (DFM):

- Identifying marks: All boards should have unique markings in the copper layer so there can be no question about which board it is. This is particularly important when there are versions of the same board with only slight variations.
- Tolerances: The design team must ultimately determine the overall tolerances that affect the boards and can cause havoc, particularly with fine-line designs if ignored. These include layer-to-layer registration, hole size tolerances, film registration and expansion, and clearances (also known as antipads) around certain features.
- Board outlines: Should exist on all layers.
- Inner layer clearance: Copper on inner layers should stay at least 0.25 mm (0.010") from the outer edges of the board.
- Inner layer clearances: Allow at least 0.5 mm (0.020") clearance around all unconnected vias and holes on inner layer copper. Thermal pads may have closer clearances, but remember that the larger the clearances for any features, the higher the board yield will be.
- Conformal coating: Consider whether boards will need to be reworked or repaired. Do the proper equipment, materials, and personnel training exist to deal with coated boards?

- And finally: Any unusual clearances and any other unusual features under consideration during the design phase should be discussed with the board fabricator prior to finalization of the design. Understanding the capabilities of the fabricator may keep both production problems and costs down.

5.6.2   Substrate Properties and Applications

For usual applications, solder coatings on substrates should be approximately 63 percent tin and 37 percent lead (the tin/lead eutectic). Most wave solder and reflow operations are optimized for eutectic solder-coated boards. Where another solder formulation will be used for assembly, special matching solder coatings should be specified. When the environment dictates conformal coatings, the selection of coating materials should be guided by matching coating requirements and coating properties. Table 5-6 lists some coating-material properties.

5.6.3   Conformal Coating Materials

Table 5-6   Guide to Conformal Coating Material Selection.*

Characteristic	Acrylic	Urethane	Epoxy	Silicone	Paraxylylene
ELECTRICAL @ 23°C					
Dielectric stress (V/mil)	3500	3500	2200	2000	to 7000
Vol. resistance (Ω/cm) 50% RH	10^{15}	10^{14}	10^{14}	10^{15}	to 10^{17}
Dielectric constant 60 Hz	3.0–4.0	5.3–7.8	3.5–5.0	2.7–3.1	2.7–3.2
1 kHz	2.5–3.5	5.4–7.6	3.5–4.5	2.6–2.7	2.7–3.1
1 MHz	2.2–3.2	4.2–5.2	3.3–4.0	2.6–2.7	2.7–3.0
THERMAL					
Continuous heat resistivity	120°C	120°C	120°C	200°C	150°C
TCE in PPM/°C	50–90	100–200	40–80	220–290	35–70
Thermal conductivity in 10^{-4} Calories/(sec)(cm²)(°C)(cm)	4 to 5	4 to 5	4 to 5	3.5 to 8	3
CHEMICAL EFFECTS					
Of common solvents	Softens, dissolves	Generally resistant	Generally resistant	HydroCrb soften	None
Weak acids	None	Minimal	None	Minimal	None
Weak bases	None	Minimal	None	Minimal	None

*Conner, Margery S., "Technology Update—The Conformal Coating of Your PC Boards Will Enhance Their Environmental Resistance," *EDN*, June 11, 1987, pg. 89.

5.7   Handling Thermal Coefficient of Expansion Mismatches

Two major strategies have been applied to management of TCE mismatch. Either the substrate may be tailored to the TCE of components, or some form of compliant member may be introduced to absorb stresses between components and the substrate. We covered compliant leaded-component approaches in Chapter 2, Section 2.2.5. We will devote additional attention to substrates in Chapter 6, Section 6.4. The discussions from Chapter 2 and Chapter 6, together with the following material, give engineering guidelines for applications where TCE mismatches produce unrelieved solder-threatening stress levels.

Figure 5-27 details the thermal properties of common electronic construction materials.[15,16] This data is used to predict stress levels from a given environment.

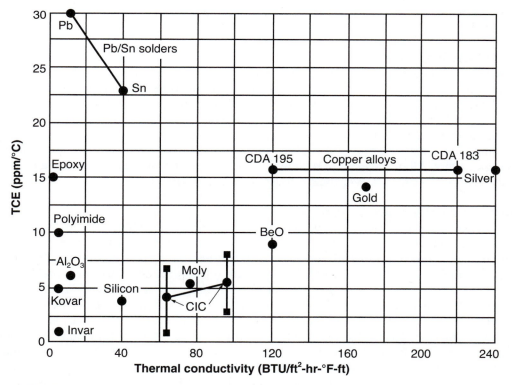

Figure 5-27   Thermal properties of electronics materials.

5.7.1 Restrained Expansion Boards

Matching Board and Component Materials

The most obvious way to tailor the substrate's TCE to that of the component is to use the same material for both. Since most CLLCCs are made of 96 percent alumina (polycrystalline Al_2O_3), which is also a common hybrid substrate material, this seems like an elegant approach. Indeed, alumina is used for SMT boards for just this application. Unfortunately, alumina has several drawbacks that prevent it from becoming the substrate of choice in mainstream SMT design. Ceramic boards are limited in size. Very few are produced in the 6-inch or 15-cm-square range, and almost none are produced in larger sizes. Alumina boards are relatively expensive. The material is about three times the cost per square inch of FR-4 epoxy/glass. Ceramic boards are very brittle, and this is a problem in environments where shock loads might break them. The ceramic boards are also costly to drill.

Where the above concerns dictate a nonceramic approach to TCE matching, common alternatives are the use of restraining layers or restraining reinforcement fibers to reduce the TCE of resin boards.

Metal-Core Boards

Low TCE restraining-layer materials, such as copper/Invar™/copper (CIC) and copper/molybdenum/copper used as power and ground layers, form functional units in the board and limit thermal excursions to match the components. These materials also add substantial heat-sink capacity, useful where cavity-up CLLCCs will dissipate heat into the substrate. Cost, weight, and limited board vendors are the negatives of this approach. However, as Table 1-2 of Chapter 1 indicates, the substantial weight savings of SMT easily overcome the weight penalty of restraining layers. All of the examples shown used metal-restraining layers, and the netweight savings ran from 2/1 to 5/1 over their through-hole counterparts.

Copper/Invar/Copper Cores

Copper/Invar/copper is generally supplied in 20 percent copper, 60 percent Invar, 20 percent copper laminated sheet stock. To be effective as a restraining layer, the CIC core (in the above percentage laminate) should be 25 percent of the thickness of the board. Note: Drill diameters through the metal cores must be kept above 0.5 mm (~0.020") to minimize drill breakage. The surface-layer copper is generally $^1/_2$ oz. (0.0007")[17] To comply with the MIL-STD-275 standard for rigid PWBs where boards must be tested to MIL-P-55110, a minimum dielectric thickness of 0.089 mm (0.0035") is required. General practice, even for commercial products, is to specify a 0.114 mm (0.0045") dielectric. However, by using MIL-STD-2118 for rigid/flexible PWBs, which will be tested to MIL-P-50884, flexible dielectric layers of 0.038 mm (0.0015") thickness may be used. For high-frequency applications, the resulting reduction in dielectric-layer thickness markedly lowers the impedance and propagation delays.[18] For a more complete discussion of high-frequency design, see Section 5.10 in this chapter.

Copper/Molybdenum/Copper Cores

Copper-clad molybdenum is supplied in 13 percent copper, 74 percent molybdenum, and 13 percent copper roll-bonded laminates. Standard thicknesses are 0.25 mm, 0.76 mm, and 1.52 mm (10, 30, and 60 mils). Strength for strength, copper-clad moly is lighter than copper-clad Invar.

Restraining Fibers

Filaments of Kevlar™, quartz, e-glass, or carbon may be used to control the X- and Y-axis expansion of epoxy or polyimide resin boards. Note that Kevlar filaments exhibit a negative coefficient of thermal expansion in the long axis. As mentioned earlier in our discussion of barrel cracking in vias, this is because the filaments shorten and grow fatter with increasing temperature. Thus, while Kevlar restrains the X- and Y-axis expansion, it adds to the Z-axis thermal expansion. For reliability concerns introduced by Z-axis expansion and cures for these concerns, see Chapter 7, Section 2, and Section 5.5 of this chapter.

5.7.2   Compliant Leads to Absorb Thermal-Expansion Stresses

The substrate matching method works well with moderate I/O devices. As chip carriers exceed 68 pins, however, compliance between component and board is necessary. Internal device heat can cause a great enough differential temperature between the board and device to overstress solder joints. You may have to turn to compliance between component and board to manage the thermal stress.

Compliant Lead Components

One method for providing compliance between the board and the component is the use of leaded components. Where leaded chip carriers are not available, post-lead attachment may provide the necessary compliant members. Post-leaded chip carriers are discussed in detail in Chapter 2, Section 2.2. Sockets may also be used to provide compliance between leadless chip carriers and a board.

Compliant Surface Layers

Compliance may also be built into the surface layer of the substrate. 3M has developed a copper-clad polyimide compliant layer which is laminated to the surface of a board using B-stage adhesive. The board is then completed using ordinary subtractive processing. The 0.127-mm (5-mil) polyimide provides compliance to absorb stresses generated by TCE mismatches between the ceramic components and organic PWB.

5.8   Thermal Management

On the positive side of the thermal equation, most surface-mount ICs are constructed on high thermal-conductivity copper-bearing lead frames instead of the less conductive Kovar and Alloy 42, which is commonly used in DIPs. For the same silicon, most SMCs exhibit a θ_{JA} lower than their THT counterparts, but they are surprisingly close.

However, surface-mount components have far less heat-sink mass than their through-hole counterparts. SMC lead frames are generally smaller, and the individual leads provide less of a thermal path than THTs. Also, the high densities inherent with surface mounting allow us to pack far more dissipation into a given square area. The net result is that surface-mount technology dictates a close attention to thermal management.

Next, we will look briefly at sources for thermal engineering data. Since this book is intended for working engineers, we have made no attempt to cover the basic engineering principles that apply to thermal calculation for either THT or SMC design. These principles are well explored in a number of engineering texts for through-hole design. Instead, we will cover thermal management strategies that are substantially unique to surface mounting. Hopefully most design teams will also recognize that in the design products for certain industries, such as automotive, it is particularly important to consider thermal issues.[19] Regardless of the industry, MIL-HDBK-217 shows that when operating at approximately 50 percent of their rated power, the failure rates of many types of electronic components can double when there is a 20°C rise in the temperature at the hottest portion of the package.

Designers should note that a number of computational fluid dynamics (CFD) programs are available that do an excellent job modeling thermal issues and the movement of heat in an electronics assembly. Brand names include Coolit, CFdesign, FloTherm, and Icepak. For more traditional calculations regarding hot spots and heatsinks on boards, Jouppi[20] presents an excellent introduction.

5.8.1   Interpreting Thermal Data From Component Specifications

Where data is available on θ_{JA}, θ_{JC}, θ_{CA}, and P_D, the task is a relatively simple one of interpolating data to application conditions in the same way that THT thermal design is handled. This discussion, many of the references, and most component data sheets assume the reader is familiar with this equation and schematic representation of the thermal-conduction process as shown in Figure 5-28:

$$T_J = T_A + (P_D \times (\theta_{JC} + \theta_{CS} + \theta_{SA}))$$

If assumption is incorrect, please see the references that relate to basic thermal issues.[21,22] And remember that in any thermal calculation, the ambient temperature T_A is the temperature immediately surrounding the part or board in question and is rarely—if ever—the room temperature of 25°C/77°F. Localized heating results in an ambient temperature for a hot component that is always higher than the room temperature or frequently even higher than the temperature elsewhere in the enclosure.[23]

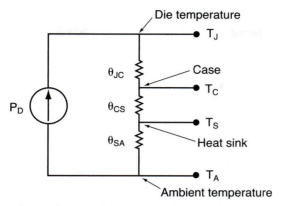

Figure 5-28   Equivalent electrical circuit for thermal conduction.

Where package data is unavailable for a given SMC, thermal design may take a bit more time. Data from another device may be used provided the other device is in the same style package, with the same lead frame and encapsulant materials, and same die size. At this time, most IC manufacturers provide thermal data for their components. In the event it is unavailable, thermal testing may be required. Examples of thermal test techniques and data are presented in the Intel Packaging Handbook.[24]

5.8.2   Heat-Sink Layers

Very early in the design process, the basic heat removal strategy must be selected. Will convective and radiant cooling alone remove heat from the system, or will forced cooling be required? Will heat from devices flow primarily into the board or into the surrounding air? These are the principle questions influencing strategy. Where high-dissipation circuits must be cooled without forced air, thermal layers in the board are often an answer.

For very high wattages per square inch, corrugated inner layers filled with a circulating coolant may be used. For less stringent cooling demands, a more common approach is the use of an outer or inner layer(s) of high thermal-conductivity material. Thermal layer materials, such as copper/Invar/copper or copper/molybdenum/copper, may serve several useful purposes as part of an SMT laminate. Do not underestimate the value of copper. SMT boards can be constructed using 1-ounce clad plated up to 2 ounces for outer, inner, or both layers. The high thermal conductivity and mass of copper make it a good place to dump heat. The low TCE of Invar and molybdenum make them useful in restraining the TCE of the polymer portion of the laminate. As discussed above, proper proportions of bimetal to polymer allow a close matching of the substrate and the ceramic chip-carrier TCEs, and the thermal layers double as ground/power planes, stiffen boards, and provide RFI shielding.

Single-side boards may also be sandwiched to a metal heat sink of a high thermal-conductivity material. Where this is done, symmetry of the full assembly is important. If only one board were bonded to a metal core, the dissimilar TCEs of the board and core would produce significant warpage during thermal cycling. Likewise, asymmetrical circuits on two sides of a metal core can cause warpage if the dissipation is concentrated in different locales on opposing sides.

The thermal properties of bimetal layers and metal cores are dependent on layers above the core and, also, the polymer thermal properties. Current data on specific core materials is available from thermal-material vendors and printed-wiring fabricator who specialize in metal-core boards.

5.8.3   Thermal Vias and Traces

In designs where heat will be removed primarily into the circuit board, particularly where metal-core layers are part of the thermal management strategy, thermal vias afford substantial improvements in θ_{JC} for closely mounted components. Figure 5-29 shows several thermal via designs and discusses their relative merits for heat removal.

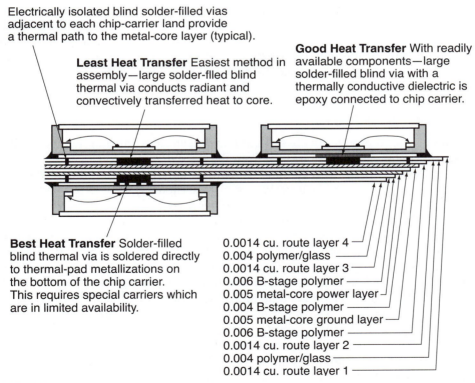

Electrically isolated blind solder-filled vias adjacent to each chip-carrier land provide a thermal path to the metal-core layer (typical).

Least Heat Transfer Easiest method in assembly—large solder-fIled blind thermal via conducts radiant and convectively transferred heat to core.

Good Heat Transfer With readily available components—large solder-filled blind via with a thermally conductive dielectric is epoxy connected to chip carrier.

Best Heat Transfer Solder-filled blind thermal via is soldered directly to thermal-pad metallizations on the bottom of the chip carrier. This requires special carriers which are in limited availability.

0.0014 cu. route layer 4
0.004 polymer/glass
0.0014 cu. route layer 3
0.006 B-stage polymer
0.005 metal-core power layer
0.004 B-stage polymer
0.005 metal-core ground layer
0.006 B-stage polymer
0.0014 cu. route layer 2
0.004 polymer/glass
0.0014 cu. route layer 1

Figure 5-29   Thermal vias carry heat to a metal core.

Additional heat may be carried to thermal layers by attaching unassigned or blank pins in order to direct vias to the ground plane/metal core. For this purpose, use blind vias and solder fill them during board manufacturing. Since thermal vias are generally solder filled for conductivity's sake, it is always good practice when using thermal vias to solder fill all the vias that are not required to be open for lead insertion. Again, solder-filled vias are not as prone to electrical failure due to through-plated-hole barrel cracking, and they provide better vacuum sealing in test fixtures.

The thermal vias under various BGA packages, shown in Chapter 3, perform these same functions. For both these BGA and other component designs, thermal vias are provided under the die and are typically connected to the package ground plane, thereby conducting the heat to the PCB ground plane.

Other thermal issues include "How much current can this trace carry?" and "What will the temperature rise of a trace of a given size be when a specific current is passed through it?" Both of these are answered in IPC documents, particularly IPC2221 and 2222 on PCB specifications. The temperature rise of a trace due to I^2R heating is not just a function of the current through it, but also of the local heat transfer properties. For instance, when a trace is internal to an FR-4 substrate, its current capacity is less since it cannot give off heat to the air like an outer-layer trace can. For additional thermal information, Flotherm has several white papers on its website, http://www.flotherm.com.

The current-carrying capacity[25] of vias can also be calculated. For component packages that require thermal vias, it is important to follow the manufacturer's recommendations or, in their absence, to calculate the number of vias required.

5.8.4   Component Package Styles

Chip Carriers

A wide variety of chip-carrier package types are available. Each has its own peculiar thermal characteristics. Designers must consider these characteristics for any package with significant heat dissipation.

Leadless chip carriers dissipate heat to the board far better than do leaded carriers. In selecting a leadless chip carrier, we must choose a package-style appropriate for both the intended thermal management strategy and the assembly process. Type A and Type D CLLCCs are designed for die-up/lid-down mounting. This places their heat-dissipating surface away from the board and makes them suitable for forced air cooling. Type A carriers are designed for socket mounting only, while Type Ds are meant for reflow soldering. Type B and Type C are designed for die-down/lid-up mounting, placing their dissipating surface close to the board for heat transfer to it. Type C is the mainstay of the CLLCC market today. Some Type C carriers are provided with electrically isolated metallizations arranged in a matrix across the bottom of the part. Soldering these to thermal vias or dummy lands boosts the package dissipation.

Leaded ceramic carriers may be made by post-lead attachment to Type B, Type C, and Type D CLLCCs. For dissipation primarily into the air in forced air cooling, Type D should be used.

For plastic-leaded chip carriers, lead-frame material is important to heat dissipation. High copper-content lead-frame materials, such as beryllium copper, are far more thermally conductive than Kovar or Alloy 42. Figure 5-21 shows the thermal comparisons for commonly used electronic materials. Leaded components and BGAs may also benefit from the use of heat spreaders and heat slugs. Examples of high-power BGAs, such as the National Semiconductor TE-BGA and E-BGA, were shown in Chapter 2.

Power IC Devices

Special SMC packaging is available to suit high-power applications. Numerous IC packages are available with integral heat spreaders that reduce thermal impedance to the substrate. In chapter 2 we discussed BGAs with heat spreaders. As another example, National rates its LM4668 in a TSSOP exposed pad package as high as 10 watts.[26] The exposed spreader or paddle on certain high power ICs is designed to be soldered directly to a copper pad on the printed board. The size of the pad and any necessary thermal vias must be designed by following the rules that the IC manufacturers provide on their data sheets and application notes. Additionally there may be specific soldering guidelines that must be followed in order for the rated thermal resistance to be achieved.

Power Transistors: SOT-89, TO-252 (DPAK), etc.

High-power SMC packages are also available for the smaller semiconductor devices, resistors, rectifiers, etc. The SOT-89 is a large die-cavity SMC package. Even larger dies and higher dissipations can be handled in the TO-252 package, Motorola's DPAK. Like all surface-mount devices, the actual power dissipation possible depends on the amount of copper on the board and any thermal vias.

High-wattage resistors are also available. Years ago, we worked on a project building 25-amp solid-state relays by surface mounting on a ceramic substrate. There are limits on the practicality of surface-mount technology for power handling, but if a high-power circuit would benefit from surface-mount conversion, a quick component availability review should be made through the distributors of power components.

5.8.5 Process Considerations

Some power components, such as the SOT-89, rely on a good solder bond between their heat-sink tab (also the die flag) and a land on the board for a large percentage of their heat dissipation. Since we cannot depend on flow soldering to provide a continuous solder board under the tab, safe design requires that we derate the SOT-89 to 110 percent of its free-air rating if it is wave soldered. In practice, this speaks against using flow soldering for boards with high-dissipation components such as the SOT-89s, TO-252s, and ICs with heat spreaders mounted on the board.

5.9 Solder Masks

SMT densities and fine-line spacings beg for environmental protection and insurance against dendritic growth. This may be provided by a properly designed pad-cap-only board, or by solder masking, or conformal coating. Under conformal coatings, solder masks are useful for controlling soldering results in non-pad-cap designs. Solder masks also lower the chances of the formation of unwanted solder balls by making it easier for the surface tension of any molten solder paste deposit to 'pull' the solder onto the metallic pads and terminations it is intended to adhere to.

5.9.1   Mask Technology Selection

Screened Masks

Screened masks are well understood, are available from most PWB suppliers, and are relatively low in cost. However, for dense SMT boards, misregistration or smearing of a screened mask can foul lands, leading to excessive solder defects. Also, a screened mask over tin/lead-coated traces tends to wander during reflow soldering. This phenomenon is evidenced by the mask taking on an orange peel appearance over the traces. When this mask rippling occurs, capillary traction may allow solder to migrate away from solder joints along the traces, producing solder defects. Excessive "orange peel" may upset the components and produce solder defects, or can actually part the mask, leaving unwanted exposure of the traces.

Liquid Photo-Imageable Masks

Liquid photo-imageable masks offer several unique advantages as a solder mask. Application speed is better than that of screened or dry-film masks. Coverage is excellent. There is no tendency to "tent," leaving openings around raised features, as shown in Figure 5-30. And, these masks provide the best of the inherent image accuracy of the photo-imaging process. The disadvantages include lengthy exposure times, costs, environmental concerns of some materials, and an inability to conveniently tent over large holes.

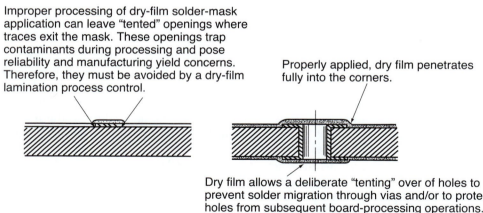

Improper processing of dry-film solder-mask application can leave "tented" openings where traces exit the mask. These openings trap contaminants during processing and pose reliability and manufacturing yield concerns. Therefore, they must be avoided by a dry-film lamination process control.

Properly applied, dry film penetrates fully into the corners.

Dry film allows a deliberate "tenting" over of holes to prevent solder migration through vias and/or to protect holes from subsequent board-processing operations. However, sealed tenting such as this entraps atmosphere that may expand and rupture the tent during soldering operations. Ruptured tents can entrap cleaning agents and contaminants which may lead to corrosive failure of hole plating.

Figure 5-30   The "tenting" definition illustrated.

Dry-Film Photo-Imageable Masks

Dry-film photo-imageable masks offer the photographic process resolution near that of liquid photo-imageable masking. Dry films also provide the ability to tent over holes and present no concern about unwanted penetration into holes. However, dry films are more costly than screened masks in many PWB shops. Also, dry films are generally too thick for use on SMT boards, where maintenance of a planar surface, including both masked areas and lands, is necessary in order to maintain acceptable soldering process yields. And, as shown in Figure 5-30, poorly applied dry films may leave openings for contaminants to enter around the traces. Good board manufacturing process control will strictly control tenting, however, and will leave no reason for lamination quality concerns when specifying dry film. With dry films, deliberate tenting may be used to mask tooling holes during through plating, thus preserving their drilled tolerance. Tenting also allows for protecting holes from selective plating processes.

Solder Mask Over Bare Copper (SMOBC)

SMOBC is a mask technology that eliminates the movement of solder under that mask, described above, since the mask is applied over untinned solder, and therefore there is no solder to move. Soldermask on boards that are tinned before the application of mask tend to wrinkle when wave soldered. This happens because the solder tinning on the pads melts at the same temperature as solder under the soldermask, melting it and wrinkling the soldermask in the process. SMOBC has no tinning under it, so there is no under-mask solder to melt.

During the reflow solder process, standard mask over tinning allows some of the molten paste to wick up under the mask following the tinning that is already there. SMOBC prevents this since it adheres to bare copper and there is no tinning to allow the paste to wick.

5.9.2   Mask Design Concerns

Screened Solder Masks

Screened solder masks may be applied over bare copper, thereby controlling the orange peel defect. But the inherent inaccuracy of screened masks limits their use for SMT. Where screened masks are selected, the mask should stop 0.25 mm (10 mils) from all solder lands to ensure misregistration of mask does not cause a fouling of the solder-land areas. The mask should also be omitted under passive components because bubbling and orange peeling of the mask during soldering can contribute to solder opens and defects.[27] These design rules are illustrated in Figure 5-31.

Some CAD systems generate mask artwork from SMT lands and do not allow the options of editing the setback from lands and placing the clear areas under passives. With such CAD systems, use a pad cap design or a photo-imageable mask instead of a screened variety.

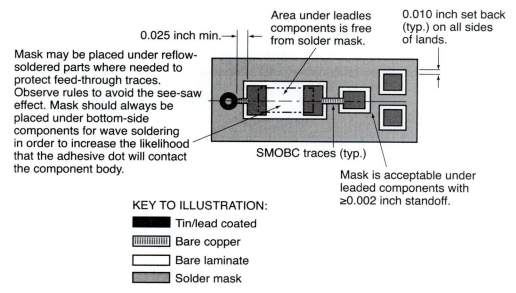

0.025 inch min. →

Area under leadles components is free from solder mask.

0.010 inch set back (typ.) on all sides of lands.

Mask may be placed under reflow-soldered parts where needed to protect feed-through traces. Observe rules to avoid the see-saw effect. Mask should always be placed under bottom-side components for wave soldering in order to increase the likelihood that the adhesive dot will contact the component body.

SMOBC traces (typ.)

Mask is acceptable under leaded components with ≥0.002 inch standoff.

KEY TO ILLUSTRATION:
Tin/lead coated
Bare copper
Bare laminate
Solder mask

Figure 5-31   Design rules for screened solder masks.

Liquid Photo-Imageable Masks

During the PWB manufacturing process, liquid photo-imageable (LPI) masks are registered in the same manner and use the same features as solder lands. Therefore, they may be produced in very close dimensional relation to the features that are important in SMT assembly. No set-back from lands is required for photo-imaged masks. The masks are generated from a reverse image of the lands or from solder-screen artwork.

There are some designers who use matte-finish LPI to reduce solder ball formation in no-clean wave solder assemblies. However, there can be problems with ionic contamination and the adhesion of conformal coats to the LPI.

Dry-Film Masks

Dry films are typically available in 0.025-mm to 0.102-mm (1 to 4-mil) thicknesses. Using a 0.76-mm or 1.02-mm (3- or 4-mil) film, vias up to 1.27 mm (50 mils) in finished diameter may easily be tented by the solder mask.[28]

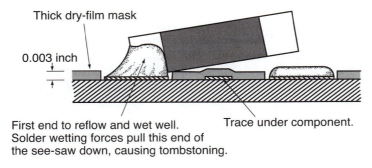

Thick dry-film mask

0.003 inch

First end to reflow and wet well.
Solder wetting forces pull this end of
the see-saw down, causing tombstoning.

Trace under component.

For reflow, best design procedure excludes
mask and traces from under small leadless
parts which are prone to tombstoning.

Figure 5-32   The see-saw effect.

One design concern raised by dry-film masks is the *see-saw effect* illustrated in Figure 5-32. The see-saw effect can be minimized by using thin dry films using $1/2$ ounce copper in lieu of 1 ounce, or, better yet, excluding the mask and traces under small components where they might contribute to the tombstoning problem. Thick dry films (>1 mil) generally aggravate the see-saw effect. Wet films in thin applications will reduce the see-saw effect.

Where none of these solutions are possible, tombstoning can be minimized by running two or more tracks under any problem component instead of just one track. Run a track as close as design rules will allow to each land, creating a balanced geometry instead of a fulcrum point. If you do not have two traces available to go under a problem component, substitute a short dummy trace. The dummy trace should extend just to the edges of the component. This set of features may be incorporated in the component library for all reflow-soldered chip components. A separate component library pattern is used for the part if it is placed on the bottom side for wave soldering.

Pad Caps in Lieu of Masks

For multi-layer boards, it is often possible to add two more layers containing all the solder lands connected to routing layers by closely coupled interconnect vias. Such designs are called pad cap boards, and do not need solder masks. In many cases, the two extra layers of a pad cap board will cost no more than the appropriate solder masks.

Failing to Plan Is Planning to Fail

You must determine what kind of solder mask you will achieve on their board or if you need any mask at all. Such important decisions should not be left to a board vendor who has no idea of the intended application or assembly processes for your product. For the high density and fine lines typical in SMT, solder-mask defects may account for more than one-quarter of all board rejects.

5.10 High-Frequency Design

As digital-circuit speeds increase and corresponding rise times decrease, circuit-board traces begin to act as transmission lines. Transmission lines affect the circuit operation just as components do. Therefore, great attention must be paid to the detail of trace design for high-speed circuits. And high-speed design continues to progress with the addition of complex devices such as field programmable gate arrays.[29]

It is beyond the scope of this book to offer an extensive treatise on high-frequency design. We will cover the general topics of particular concern for surface-mount technology and encourage engineers venturing into microwave frequencies for the first time to research the references listed in Section 5.11 at the end of this chapter. With regard to the team approach to design that we have mentioned throughout the book, it is particularly important in high-speed designs.[30] These design issues should be addressed during the design of a high-speed circuit and its PC board:

- PC board material
- Grounding strategy
- Interstage matching
- Input and output impedance matching
- Transient stability
- Thermal management
- Component packages to be used

In addition to the general rules presented in this chapter, designers need to verify all aspects of the design as they relate to the components being used. As a simple example, muRata says in their application notes for certain of their SMT inductors that "Large (PCB) lands reduce Q of the mounted chip. . . . A high Q value is achieved when the PCB electrode land pattern is designed so that it does not project beyond the chip coil electrode."[31] Hopefully this quote will remind readers of the importance of attention to component detail during PCB design.

5.10.1 Defining High Speed

As a first step in our treatise on SMT design for fast circuitry, let's define when signals must be considered high frequency. Designers sometimes speak of circuits with rise times below 4 nanoseconds (ns) as fast. But how much below? And should the cutting edge be at 4 nanoseconds, 2 nanoseconds, or some other number? The answer is, it depends. Basically, noise becomes a problem when t_r decreases to the point where parasitic inductances and capacitances on the board rear their ugly head and affect signal integrity. Again, somewhere around 2 ns.

Clock rates will yield even less specific insight to the high-speed boundary. Some circuits require transmission-line design at 50 MHz and others run at over 1 GHz, using ordinary interconnect design rules. Crosstalk is an issue since it intensifies by a factor of two every time the clock rate doubles.

These disclaimers are offered not to confuse the issue, but to demonstrate that a simplistic answer is not adequate. Lee Ritchey, president of Speeding Edge, Glen Ellen, CA, has offered a more useful rule of thumb for high-speed design-rule requirements. He explains that circuits should be considered ". . . high speed when the rise-time transition is completed before the signal has made a round trip along the conductor." When rise time is less than round-trip time, signal reflections and crosstalk are likely to degrade signal accuracy unless design rules are applied to control these factors.[32]

Ritchey applies the rule of thumb that signals travel on FR-4 or polyimide at between 5 and 6 inches per nanosecond. Thus, if a device with a 1-nanosecond rise time is at the end of a 3-inch signal trace, the 6-inch round trip may not occur before the device transition, and you are in the danger zone. A device with a 6-nanosecond rise at the end of an 18-inch line would similarly be a fast circuit.

Of course, in both cases, if we could reduce that signal trace to 2 inches, you would be back out of the circuit designer's twilight zone. Since SMT is such a champion at shrinking trace lengths on PWBs, surface mounting obviously has a great deal to offer designers of high-performance assemblies.

Other designers of high-speed circuits also offer rules of thumb, as well as more analytical guidelines. We do not have the space here to go into high-speed design in detail, so we will offer a sampler of these rules, along with their sources, just to give readers an idea of the wealth of information available about high-speed design. On his website, http://www.bogent.com and in his book *Signal Integrity Simplified*, Eric Bogatin offers 100 rules, which include:[33]

- Sheet resistance of 1-oz. copper is about 0.5 milliohms per square—a trace 0.010" wide and 1" long would have 100 squares, and therefore a series resistance of 50 milliohms = 0.05 ohms.
- The speed of a signal in air is about 12"/nsec, and in most laminates about 6"/nsec.
- The shortest risetime that can be propagated in FR-4 is 10 picosecond/inch × length. So a signal propagating down a 10" length of 50 ohm line on FR-4 must have a rise time of at least 100 psec.

Doug Brooks, in his writings at http://www.ultracad.com and in his book *Signal Integrity Issues and Printed Circuit Design*,[34] includes information on issues such as the following:

- Changes in trace impedance due to vias.
- Right-angle corners in traces do not impact signal integrity.
- Ground bounce.
- Calculators for temperature rise in traces due to currents, transmission line impedance, and crosstalk, among others.

Howard Johnson, in his book with Martin Graham, *High-Speed Design, A Handbook of Black Magic*,[35] and his website http://www.signalintegrity.com includes topics such as:

- Checklists that ask the questions an experienced designer would ask about a new system.
- Formulae for inductance, capacitance, resistance, risetime, and Q.
- Trade-offs between signal crosstalk, mechanical fabrication of tolerances, and trace routing density.

The above listing is, of course, only a teaser on the wealth of information available from these sources as well as many others, such as the aforementioned Lee Ritchey, with his book *Right the First Time, A Practical Handbook on High-Speed PCB and System Design.*[36]

From the preceding discussions, we can readily see that relatively slow (8 to 12 MHz) digital boards may have areas where the gate rise times qualify them as fast. Do not be one of the designers who, familiar with digital logic, is caught unaware by the technological advances of device edge speeds. Do not violate high-speed design considerations. The result may be a board that works in one slot of a backplane, but not in the next. Or, your board may work with one IC but fail when you trade it for what seems like another identical part number. Upgrading 74SL designs to 74HCT technology is not necessarily trivial—they are not the same in terms of rise time.

5.10.2   Transmission Line Design

Types of Transmission Line

When circuit miniaturization will not do the job, and Ritchey's rule indicates that circuit traces must be treated as transmission lines, several design alternatives are commonly used. Figure 5-33 illustrates four of these approaches.

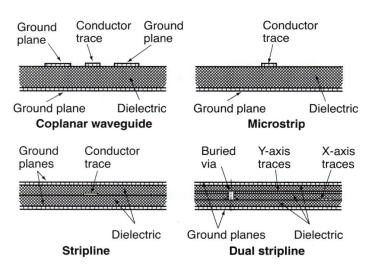

Figure 5-33   Classification of transmission lines illustrated.

5.10.3 Controlled Impedance

For fast devices such as ECL, terminated transmission lines must be designed to produce controlled impedances. Higher impedances limit current flow and, therefore, consume less power. But the impedance chosen must match the input impedance of the devices used on the board. Where multiple-device families are required, several layers or board areas with individual impedances may be necessary.

Impedance is influenced by the style of transmission line used, the dielectric constant of the board material, the dielectric separation of the line from its ground plane, and the width and thickness of the trace.

Effects of Transmission Line Style

For a given dielectric separation and trace dimension, microstrip transmission lines inherently exhibit less propagation delay and higher impedances than stripline or dual stripline. However, both stripline styles offer far better EMI shielding of fast switching lines. And, because of the shielding effects of the dual ground planes, stripline styles allow conductors to be placed closer together without crosstalk problems. Thus, stripline styles allow higher densities of circuit traces than does microstrip.

Dielectric Constant of the Board

Propagation speed along a transmission line is in inverse relationship to the dielectric constant (K) of the insulative material around the conductor trace.

Dielectric constant is a measure of a material's ability to store an electric charge when acting as a dielectric in a capacitor. K is the ratio of the charge that would be stored with air as the dielectric to that stored with the material in question as the dielectric. For this computation, air is assigned a K value of 1. Table 5-7 shows the K factors for some typical substrate materials.

It is important to note in Table 5-7 that the K value is not constant with frequency. Standard FR-4 epoxy/fiberglass board using the standard E-glass material has a K approximately equal to 4.4 at 1 MHz, but at 1 GHz K falls to approximately 3.9, a change in K of 11 percent. We say "approximately" since the actual K value depends on the specific epoxy used as well as the frequency of operation. It will also vary with temperature and humidity. Other resin systems may be more stable across the frequency range of operation. For example, E glass combined with allyated polyphenylene ether (APPE) resin results in a substrate with K = 3.7 at 1 MHz and

Table 5-7 Dielectric Constants of Typical Substrate Materials.

Material	K @ 1 MHz	Material	K @ 1 MHz
Alumina Ceramic	8.0	Polyimide/Kevlar™	3.6
Epoxy/Glass	4.4	Polyimide/Quartz	3.4
Epoxy/Kevlar™	4.1	PTFE/Glass	2.2
Polyimide/Glass	4.5	RO2800™	2.9

K = 3.4 at 1 GHz, a change of 8 percent. Other, more expensive materials will be more stable. While many materials can be used, Maxim Integrated Products notes for one of their 2.4 GHz power amplifiers, "The printed circuit board material should be either FR-4 or G-10. This type of material is a good choice for most low-cost wireless applications for frequencies up to 3 GHz."[37]

    We mentioned earlier that microstrip transmission lines exhibit smaller propagation delays than stripline types. This is because of the low dielectric constant of the air bounding one side of microstrip structures. Inner-layer microstrips are bounded on both sides by the substrate dielectric material. Thus, they do not display the speed of surface-layer traces.

Dielectric Separation from the Ground Plane

Dielectric thickness between a transmission line and its ground plane(s) has a direct influence on impedance. By decreasing the separation, we decrease both impedance and propagation delays. We can also reduce impedance and propagation delays by using a dielectric with a lower K factor. But, as the impedance formulas indicate, dielectric thickness has a more direct relation. Formulas for the exact relationships between impedance, propagation delays, separation, dielectric material, and transmission-line style are presented next, in the discussion of "transmission-line formulas."

Conductor Form Factors

Conductor width, and to a lesser degree, conductor thickness, determines the impedance of transmission lines. Formulas for calculating conductor impedance for a given size are presented next. Conductor geometry can be important in high-frequency design. Any separation from the ground plane will cause reflections, thereby creating unnecessary impedance, damaging system performance, and potentially degrading the reliability. One old saw that has finally been proven wrong is the concern for right-angle corners. When Brooks tested a variety of trace angles including 90 degrees, he found that no signal degradation could be detected within the limits of the instrumentation used down to pulse rise times of 17 picoseconds.[38] Figure 5-34 details some ground-plane treatments.

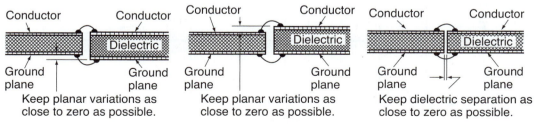

Figure 5-34   Ground-plane separation continuity.

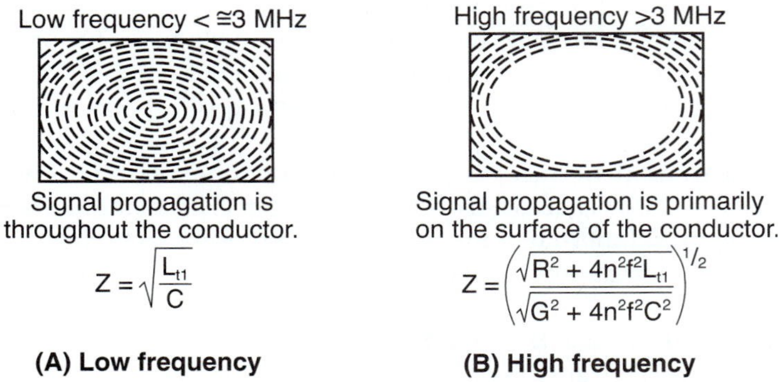

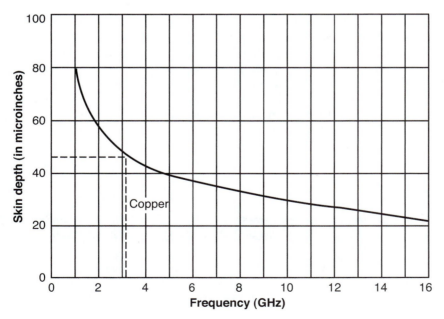

Figure 5-35   Skin effect in high-frequency signal propagation.

Transmission-Line Formulas

As frequencies move above 3 MHz, signal propagation along a conductor occurs increasingly at the surface. Figure 5-35 shows this phenomenon, called the skin effect,[39] and formulas to calculate it, while Figure 5-36 shows the relationship between frequency and skin depth.[40]

Figure 5-36   Skin depth vs. frequency.

Low-frequency signal propagation can be calculated using the equation*

$$Z = \sqrt{\frac{L_{t1}}{C}}$$

(Eq. 5-1)

where,
    Z is the characteristic impedance in ohms,
    L_{t1} is the inductance in henrys/unit length,
    C is the shunt capacitance in farads/unit length.
    while high-frequency signal propagation is calculated using*

$$Z = \left(\frac{\sqrt{R^2 + 4n^2f^2L2_{t1}}}{\sqrt{G^2 + 4n^2f^2C^2}} \right)^{1/2}$$

(Eq. 5-2)

where,
    Z is the characteristic impedance in ohms,
    $L2_{t1}$ is the inductance in henrys/unit length,
    C is the shunt capacitance in farads/unit length,
    R is the resistance in ohms/unit length,
    f is the frequency,
    G is the shunt conductance in ohms/unit length.

We can calculate impedances of microstrip and stripline transmission lines using the form factors shown in Figure 5-37 and the following formulas.

---

*Per MIL-HDBK-176.

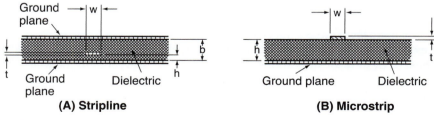

(A) Stripline   **(B) Microstrip**

Figure 5-37 Dimensional factors for transmission lines.

To calculate the propagation delays of microstrip and stripline, use the following formulas for transmission-line impedance. For microstrip, the characteristic impedance is calculated:

$$Z_0 = \frac{87}{\sqrt{K + 1.41}} \ \ln\left(\frac{5.98h}{0.8w + t}\right) \qquad \textbf{(Eq. 5-3)}$$

and

$$tpd = 1.07 \sqrt{0.475K + 0.67} ns/ft. \qquad \textbf{(Eq. 5-4)}$$

where,
Z_0 = the microstrip characteristic impedance,
K = the relative dielectric constant of the substrate material,
t_{pd} = the microstrip propagation delay.

The inductance per foot equals $L_o = Z_0^2 C_0$, where the capacitance, C_o, is taken from the graph in Figure 5-38.

For stripline propagation delays, the inductance per foot is $L_0 = Z_0^2 C_0$, where the capacitance, C_0, is taken from the graph in Figure 5-39, and the stripline characteristic impedance Z_o is:

$$Z_0 = \frac{60}{\sqrt{K_r}} \ \ln\left(\frac{4b}{0.67\pi w\left(0.8 + \dfrac{t}{w}\right)}\right) \qquad \textbf{(Eq. 5-5)}$$

and, the stripline propagation delay is:

$$t_{pd} = 1.017\sqrt{K_r} \ nS/ft. \qquad \textbf{(Eq. 5-6)}$$

High-Density Design

As we have established above, reducing the dielectric layer thickness reduces both the impedance and propagation delays of transmission lines. The impedance needed is determined by matching the input impedance with the logic family on the board. Therefore, thin dielectric layers may drop impedances low enough to allow reducing the line widths to gain the needed impedance, thus increasing circuit density.

Complex high-frequency design is a challenging chess game built around the above variables and a host of design considerations (discussed in the references in the following high-frequency bibliography). We have been restricted by space to only a cursory discussion of microwave design. Please refer to the following listed sources for more detail.

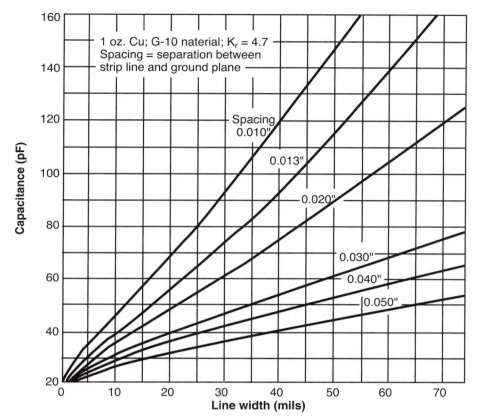

Figure 5-38   Capacitance of microstrip transmission lines. *(Courtesy Motorola, Inc.)*

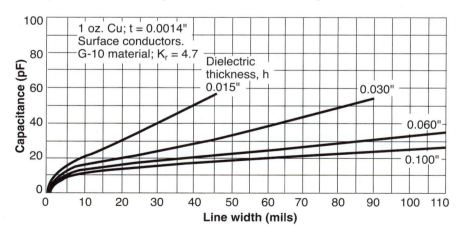

Figure 5-39   Capacitance of stripline transmission lines. *(Courtesy Motorola, Inc.)*

Vias in High-Speed Design

Vias affect signal integrity. A number of references document this[41,42] and note that signal integrity must take vias into account.

5.10.4　Decoupling and Capacitors; Noise and EMI

Power and ground planes are inherently noise-free, until some circuit creates noise and some other circuit allows that noise to invade a plane. It is important to recognize that since the planes start out clean, the purpose of decoupling is not to prevent noise on the planes from entering the circuitry, it is to prevent the planes from becoming contaminated with noise in the first place.

The noise, of course, comes from current transients due to high-frequency or high-speed switching circuits. Since $\Delta V = L\, dI/dt$, the faster the change in current, the higher the resultant voltages. As examples for typical slow and fast pair of logic families:

If $I = 10$ ma, $t_r = 12$ nsec, $L = 5$ nH, then $\Delta V = 4.2$ mv
If $I = 10$ ma, $t_r = 0.6$ nsec, $L = 5$ nH, then $\Delta V = 83.3$ mv

The difference in noise voltage generated by the faster rise-time circuit is obvious. Now consider the results if hundreds of gates are switching simultaneously. These inductive noise transients are typically dealt with by designed-in capacitance, whether plane capacitance or add-on decoupling capacitors.

The power and ground planes do have inherent capacitance. At slower rise times/lower frequencies where the wavelength of the generated signal is larger than either of the planes' X-Y dimensions, the capacitance assumption is correct. However, as with most capacitors, as the frequency increases (or rise time decreases), the characteristics begin to change.[43] The planes form a parallel plate transmission system, and the current that is typically carried by vias creates noise that will radiate away from the vias following the rules of a parallel-plate system—including reflections from the edges of the planes.

Decoupling capacitors always have some equivalent series resistance and inductance, although it is much less for SMT capacitors than it is for THT capacitors. Capacitor manufacturers have models for their components available, and these can be used to determine the high-frequency effects of the parasitic inductance and resistance of the capacitors. At higher frequencies, the inductance becomes predominant. For proper decoupling, it is not uncommon to use multiple decoupling capacitors. Brooks notes these conclusions to with regard to ESR of multiple caps and proper selection of bypass caps:[44]

1. As ESR goes down, the impedance troughs become deeper and the peaks become higher
2. The minimum impedance value of parallel caps is not necessarily ESR/n, where n is the number of identical parallel caps, it can be lower than that!
3. The impedance minimums are not necessarily at the self resonant points of the bypass caps.
4. For a given number of parallel capacitors, better results can be obtained from more capacitor values with moderate ESRs spread over a range rather than with a smaller set of capacitor values with very low ESRs.

Ground bounce is reduced by controlling the amount of inductance in the return paths. Thompson notes[45] that "as a practical matter, the decoupling capacitors serve to directly power digital devices, while the main power supply serves to recharge the capacitors." This means that if appropriate capacitance is not available, the following problems may occur:

- Digital ICs may slow down, causing timing errors.
- Current may not be available to support the output levels required, increasing noise and logic level errors.
- Voltage transients as a result of ground or power bounce may cause an increase in EMI.

And where do you put the bypass caps? It depends. Oh, on what you say? Well, there does not seem to be universal agreement on the issue. The general rules of thumb are:

- If the digital circuits you are designing have signals referenced to ground (TTL, ECL), they will be more affected by noise on the ground pin than they are by noise on the power pin. So, minimize the inductance of the leads at ground by keeping the bypass caps close to the ground pin.
- If the digital circuits are the type that switch between power rails (CMOS), they will be equally affected by noise on either ground or supply. So, place the caps equally between the ground and power pins.
- In general, when using planes, connect the bypass cap leads to the power and ground planes. Then connect the power and ground pins to the caps on the surface.
- For all applications, keep the traces connecting the bypass caps and IC pins as short and wide as possible since this minimizes inductance, which is the whole reason for the caps in the first place!

It seems there are as many articles on the use of bypass caps as there are designs using the caps. Read everything you can and make your best informed decision for your design.

Layout Rules

SMT follows the same generic rules for high-frequency board layout common in the through-hole world. However, we see so many instances of total disregard or plain ignorance of these rules that a brief discussion of them herein seems in order.

For EMI immunity, a good goal is to design to tolerate the H field generated by a $\Delta V / \Delta T$ of 10^{10} volts per second. On board relays, switching inductive loads can easily produce such events, as can any number of environmental events in close proximity to the circuit. Decoupling, shortening of traces, addition of protection diodes, filtration of input/output lines—all these strategies may be needed to achieve the 10^{10} immunity goal. Fortunately, SMT simplifies or at least reduces the real estate requirements needed for each of these approaches.

Ground integrity is vital to minimize EMI/RFI emissions from a high-frequency design and to ensure its proper operation. A very minimal impedance in a ground path will present absolutely no problem to a low-frequency circuit. To a high-frequency signal, that same path will appear to be an open circuit. Ground conductors must be as robust as circuit dimensions will permit. Ground routing must provide for the orderly flow of currents from the circuit to ground. The topology must be chosen to absolutely minimize voltage differentials from one point to another across the ground. Ground planes greatly simplify adherence to good topology rules and, when correctly placed, reduce H field coupling by several orders of magnitude. Bogatin gives these three rules for reducing the total inductance of the ground path:[46]

- Use as short a trace for the shared return paths as possible.
- Use as wide a conductor as possible for the return path. This may mean the use of a plane.
- Keep the signal path as close as possible to the return path. In packages and connectors this means keeping the returns interspursed among the signals rather than grouping returns and signals separately.

EMI/RFI emissions can easily be aggravated if allowed to pass from the board to the cabling and I/O connected to the circuit. The use of selective-pass filters or ferrite beads can help in this. Where size is at a premium, ferrites and filters are now available in surface-mount packaging. Decoupling is generally required for signal lines entering ICs. The closer the capacitor can be placed to the input it protects, the more efficacious it will be. SMT is great for decoupling for two reasons. Small SMT capacitors are easy to pack in adjacent to an IC. The fact that they can be placed on the secondary side of the board opposite the side where the ICs reside means that they can often be used with minimal impact to routing tracks approaching a busy IC. Still, in some cases, through-hole parts provide an easier avenue for trace routing. The designer should feel equally free to select either packaging format dependent on the needs at hand. Placement of decoupling is critical to success of EMI damping. Follow guidelines listed in the references at the end of this chapter. Note that when chaining SMT decoupling together, entry of the ground lines to the lands should be at the two sides of the land, not a single entry from the end with a T connection to the ground track.

Termination networks can be kept to a minimum of board real estate by selecting between THT or SMT networks or SMT discrete resistor chips. Ultra-miniature chips are sometimes used for terminators where dissipations will be low. For dense boards having large numbers of resistors with a close value range, buried resistor techniques may solve thorny miniaturization problems (see Figure 6-8).

Coupling of adjacent traces to fast-transition paths is a problem requiring all the vigilance the layout engineer can muster. The obvious rules of the road are just a starting point. Keep all runs on a given layer in essentially the same axis (N-S or E-W). Route adjacent layers in perpendicular axes. If two traces where crosstalk could be a problem must run parallel to each other, place a ground between them or route them on separate layers. If neither is possible, separate them. Where layer-to-layer crosstalk might be a problem, increase the separation of the offending traces, or place them on opposing sides of a plane. If trace width = W, trace separation for high-speed design should be 2W to 3W. Beyond these basics, it is a good idea to identify all the fast-transition paths on a composite check-plot, and make cross-talk possibilities the subject of frequent design reviews.

5.10.5   Overall Digital Design Rules

For high-speed digital design, these rules are widely followed because they work:

- Use multi-layer PCBs with ground and power planes. Device ground and power pins should be routed directly to the planes with vias.
- Decoupling capacitors should be used for every digital IC, placed following the rules noted in Section 5.10.4.
- Do not use capacitors directly connected to the output pins.
- Do not use sockets or wirewrap connections.
- Do not route control lines, such as set, reset, clock and chip select, through a device that drives data or address lines.
- Use board layout designs that minimize crosstalk and reflections.

5.11   Additional References on High-Frequency Design

Interested readers are directed to the following resources on high-frequency for additional information on high-frequency circuit design.

1. Bogatin, Eric (2002), *Signal Integrity—Simplified*. Prentice Hall, New York.
2. Brooks, Douglas (2003), *Signal Integrity Issues and Printed Circuit Design*. Prentice Hall, New York.
3. Johnson, H., Graham, M. (1993), *High-Speed Digital Design, A Handbook of Black Magic*. Prentice Hall, New York.
4. Ritchey, Lee (2003), *Right the First Time: A Practical Handbook on High-Speed PCB and System Design*, Volume 1. Speeding Edge, available at http://www.speedingedge.com.
5. *Motorola MECL System Design Handbook*, 1986, Motorola Semiconductor Products, Inc., P.O. Box 20912, Phoenix, AZ, telephone: (602) 944-6561.
6. Motchenbacher, C. D. and Connelly, J. A., *Low Noise Electronic System Design,* John Wiley & Sons, New York, NY, 1993.
7. Harper, Charles A., Ed in Chief, *Electronic Packaging and Interconnection Handbook,* McGraw-Hill, Inc., New York, NY, 1991.
8. Seraphim, Donald P., Lasky, Ronald, and Che-Yu, Li, *Principles of Electronic Packaging,* McGraw-Hill, Inc., 1989.
9. *FAST Applications Handbook*, 1987, National Semiconductor, Digitial, 333 Western Ave., South Portland, ME, telephone: (207) 775–8700.
10. *Applications and Performance of GigaBit 40 I/O Chip Carriers*, 1986, Gigabit Logic, Inc., 1908 Oak Terrace Lane, Newbury Park, CA 91320, telephone: (805) 499-0610.
11. *Guidelines for the Use of Digital GaAs Ics*, Gigabit Logic, Inc., 1908 Oak Terrace Lane, Newbury Park, CA 91320, 1987.
12. *Thermal Management of PicoLogic and NanoRAM GaAs Digital IC Families*, Gigabit Logic, Inc., 1908 Oak Terrace Lane, Newbury Park, CA 91320, 1988.

13. Ramu, S., and Whinnery, T. Van Duzer, *Fields and Waves in Communications Electronics*, 1984, John Wiley & Sons, 605 3rd Ave., New York, NY 10158, telephone (212) 850-6000.
14. Davidson, C. W., *Transmission Lines for Communication*, John Wiley & Sons, 605 3rd Ave., New York, NY 10158, 1978.

5.12   **Review**

After reading and understanding this chapter, you should be able to answer these questions and note the reference location for the information in the chapter.

1. Consider the steps of basic PCB fabrication. Which of them are impacted by a designer's decisions during the PCB design process?
2. B-stage material is also called "pre-preg." What is it? Define the state of the epoxy material in it.
3. What does the "FR" in FR-4 stand for? The "4"?
4. What is the purpose of electroless chemical plating?
5. Select four of the "Design Steps for PCBs." Define a hypothetical circuit board function(s) and describe the four steps you selected for a board with that function.
6. There are two plating processes used in PCB fabrication. Describe them.
7. Briefly describe the Hot Air Solder (or Surface) Leveling process.
8. A PCB design you are working on will incorporate a ball grid array (BGA). Give your manufacturing engineers two acceptable choices of surface finish that they can choose to use in the line.
9. What is the general rule for copper coverage on a PCB?
10. What is the standard test grid for a THT board that can also be used on a SMT board?
11. Why are test probes not allowed to contact SMT components?
12. Define "pointing accuracy."
13. What is the rule for the distribution of thermal mass across a PCB? Why?
14. Give an example of partitioning a PCB that includes analog, digital, and RF circuitry.
15. What is the typical first inspection to design for?
16. What is the IPC standard that defines land patterns and leads to a land pattern calculator?
17. When routing BGAs, which is the preferred land pattern, SMD or NSMD?
18. What does "escape" mean with regard to circuit traces and footprints?
19. Define "HDI," including the types of vias it uses.
20. Should test vias be filled? Why or why not?
21. What is via aspect ratio? What is the preferred ratio? Why?
22. What are the rules for vias located beneath components?
23. What are the rules for vias located in pads. What type of via is preferred for in-pad use?
24. List Holden's "major HDI drivers."
25. Does a via cause any change in a signal passing through it?
26. Define TCE, also known as CTE.

27. A semiconductor component is operating at an ambient temperature of 35°C with a power dissipation of 3W, a junction-to-case resistance of 15°C/W and a case-to-ambient resistance of 25°C/W. Calculate the junction temperature, and determine if the resulting junction temperature is typically acceptable.

28. For two otherwise identical traces, one on the outer surface of a PCB and one on an inner layer, which one has the greater current capacity? Why?

29. With regard to solder mask, what is "tenting"? For what purpose is it used? What type of solder mask is best when tenting is desired?

30. What is LPI?

31. What is SMOBC? What is its advantage?

32. What is one problem with the use of dry film masks?

33. What is a "pad cap"?

34. What is Ritchey's definition for a "high-speed" circuit?

35. The "skin effect" changes the current-carrying properties of a conductor. Is current depth greater at high frequencies or at low frequencies? How does this affect PCB design?

5.13   References

1. Barbetta, M., "The Search for the Universal Surface Finish." *Printed Circuit Design and Manufacture,* v. 21 #2, February 2004, pp. 34–38.

2. Bryant, K. (2004), "Investigating Voids." *Circuits Assembly,* v. 15 #6, June, 2004, pp. 18–20.

3. ———. (1994, Nov.). *Electronic Prototyping: Tips and Pitfalls.* Available: http://eeshop.unl.edu/proto.html.

4. Mancini, R., "Worst-case circuit design includes component tolerances." *EDN Magazine,* April 15, 2004, pp. 61–64.

5. "Testability Guidelines," *SMTA,* Edina, MN, 2002.

6. Patterson, Brian T., and Landolt, Roger H., "Fully Additive Plating with Permanent Additive Resist," *PC Fab,* June 1987, p. 54.

7. IPC-SM-782A, including amendments 1 and 2, "Surface Mount Design and Land Pattern Standard." *IPC,* Northbrook, IL, 1999.

8. Klein, W. P., "Solder Pad Recommendations for Surface-Mount Devices." Application report SBFA015A, available at http://www.ti.com, May, 2003.

9. Achong, C., Chen, A., "BGA Land Pattern and Assembly Issues." *Printed Circuit Design and Manufacture,* v. 21 #2, February 2004, pp. 24–28.

10. ———. "PCB Design for Testability Guidelines—SMT Products." Available at http://www.uni-fix.com/PDF/Design2.PDF.

11. Holden, H., "Does Your Design Need HDI?." *Printed Circuit Design and Manufacture,* v. 20 #10, October, 2003, pp. 20–25, 2003.

12. Holden, H., "Time for a Change: Innovation in Interconnects." Available at http://www.circuitree.com/CDA/ArticleInformation/features/BNP__Features__Item/0,2133,91510,00.html, 2004.

13. Brooks, D., "The Effects of Vias on PCB Traces." Available at http://www.ultracad.com, 1994.

14. Capers, C., "Cost-Effective Use of Microvias." *Printed Circuit Design and Manufacture,* v. 20 #3, March 2003, pp. 14–16.

15. Op. cit., ref. 1.

16. Op. cit., Chapter 1, Reference 12.

17. Knodle, John M., "Manufacturing PCBs with Invar Cores," *PC Fab*, June 1987, p. 58.

18. Cantwell, Dennis J., "Designing SMT PWBs for Manufacturability," *PC Fab*, June 1987, p. 69.

19. Myers, B. A., "Cooling Issues for Automotive Electronics." *Electronics Cooling*, v. 9 #3, August 2003, pp. 24–31.

20. Jouppi, M., "Hot Spots and Heatsinks." *Printed Circuit Design and Manufacture*, v. 21 #10, October 2004, pp. 28–31.

21. ———. (2000). "Appendix E: Understanding Integrated Circuit Package Power Capabilities." Available: http://www.national.com/.

22. Blackwell, G., "Heat Removal/Cooling in the Design of Packaging Systems," *The Electronic Packaging Handbook*, G. Blackwell, ed. Boca Raton, CRC/IEEE Press, 2000, pp. 11-16–11-26.

23. DuBois, K., "Thermal Management of Compact PCI Systems." *ECN Magazine*, February 2000.

24. ———. Packaging Handbook. Available at http://www.intel.com.

25. Brooks, D., "Current Carrying Capacity of Vias." *Printed Circuit Design Magazine*, pp. 25–30, January, 2003.

26. LM4668 Boomer 10W High-Efficiency Mono BTL Audio Power Amplifier. Available at http://www.national.com/ds/LM/LM4668.pdf

27. Leibson, Steven H., "EDN's Hands-On Project—Part 3: CAD and Surface Mount Technology," *EDN*, June 25, 1987, p. 212.

28. Goromdy, Emery J., "The Continuous Flow Dry Film Solder Mask Process," PC Fab, October 1987, p. 59.

29. Isaac, J. and Wiens, D. "The Future of PCB Design." Available at http://www.mentor.com/pcb/techpapers/index.cfm.

30. Burrell, D. (2004, Jan.). "Three-way talks make it happen." Available at http://www.industrienet.de/objekte/epp/heft/text14.htm

31. LQG/LQH series product data sheets. Available at http://www.muRata.com/catalog.

32. Markstein, Howard W., "Packaging for High Speed Logic," *Electronic Packaging and Production*, September 1987, p. 48.

33. Bogatin, E., *Signal Integrity Simplified.* Prentice Hall, New York, 2003.

34. Brooks, D., *Signal Integrity Issues and Printed Circuit Design.* Prentice-Hall, New York, 2003.

35. Johnson, H., Graham, M., *High-Speed Digital Design, A Handbook of Black Magic.* Prentice-Hall, New York, 1993.

36. Ritchey, L. W., *Right the First Time, A Practical Handbook on High Speed PCB and System Design.* Speeding Edge, Glen Ellen, CA, 2003.

37. "The MAX2242 Power Amplifier: Crucial Application Issues." Available at http://www.maxim-ic.com/appnotes.cfm/appnote_number/615.

38. Brooks, D., "90 Degree Corners—The Final Turn." Available at http://www.ultracad.com/articles.htm, 1998.

39. Blackwell, G., "Heat Removal/Cooling in the Design of Packaging Systems," *The Electronic Packaging Handbook*, G. Blackwell, ed. Boca Raton, CRC/IEEE Press, 2000, pp. 11-16–11-26.

40. ———. *Packaging Handbook.* Available at http://www.intel.com.

41. Riazi, A., "Via Modeling for High-Speed Simulations, Part 1." *Printed Circuit Design and Manufacture*, v. 20 #9, September, 2003, pp. 30–31.

42. Brooks, D., "The Effects of Vias on PCB Traces." Available at http://www.ultracad.com/articles.htm.

43. Fang, J., Zhao, J., "Low-Impedance Power Delivery Over Broad Frequencies." *Printed Circuit Design and Manufacture,* v. 20 #9, September, 2003, pp. 18–22.

44. Brooks, D., ESR and Bypass Cap Self-Resonant Behavior." Available at http://www.ultracad. com/esr.htm.

45. Thompson, J., "Decoupling Strategies for PCBs." *Printed Circuit Design and Manufacture,* v. 20 #10, October 2003, pp. 26–34.

46. Bogatin, E., "Bouncing out Ground Bounce." *Printed Circuit Design and Manufacture,* v. 20 #8, August 2003, pp. 22–27.

chapter 6

Assembly-Level Packaging and Interconnections

Objectives

This chapter concerns itself with system strategies, that is, with the overall assembly of the board beyond the layout of footprints and traces. After reading and understanding this chapter the reader should be able to:

- understand the continuing progression of IC size reduction and how this impacts PCBs
- understand various mixed-technology IC packaging methods
- understand board-level interconnects
- understand the use of imbedded components
- understand the use of different substrate materials

Introduction

In this chapter, we will turn our attention from the layout of the board to the layout of the system. Elegant packaging is no accident. Some SMT problems resolve much easier at the system packaging level than at the board level. For instance, repetitive circuit functions can be implemented on modular daughter cards. Thus, the designers and manufacturing people do not have to constantly reinvent. We will consider this and other assembly packaging issues in this chapter.

In Chapter 1, Chapter 2, and Chapter 3, we laid the foundation for SMT design decision-making. Chapter 4 and Chapter 5 developed on that foundation the framework for system and detail SMT design. Now we are ready to sheath that framework with tools for fine tuning our design. In this chapter, we will see how SMT cards are interconnected into working systems.

6.1 SMT Strategies to Reduce System Size

We have stressed team engineering as the key to surface-mount success. Here is how one SMT contractor helps clients apply teamwork in reducing package size. According to Gregory Horton, former president of the Chatsworth, California, aerospace and medical electronics contract house *SMTEK*, teamwork begins at the conceptual stage of a project. He recommends,

> "The team should develop the approximate component counts, the required board area for each function, power dissipations, I/O needs, etc. Then, they should partition the design into as few segments as possible to keep the board count down. This simplifies I/O, may improve electrical performance, and simplifies board-level interconnects and backplane design."

Horton states that this approach "boosts reliability and cuts cost. The limiting factor is the practical board-size limit." He also recommends the standardization of board and panel sizes where more than one module is required.[1]

The drive to save board real estate is certainly not newly arrived with SMT. THT packaging engineers have been in hot pursuit of miniaturization for many years, but some of the tools that SMT makes available for this quest are unique and bear discussion herein.

6.1.1 Common Ways to Gain Real Estate

Let's look now at some real estate-saving moves that work in many varied applications. Because these methods are broadly applied, they are worth consideration whenever a design might profit from a space or weight reduction.

Multiple Boards Into One Board

In changing an existing through-hole product to surface mount, it is often possible to crunch the functions of two or more THT boards onto one SMT card. Where this can be done without undue increases in test and repair costs, the savings are substantial. The raw board is cheaper because the single SMT assembly uses about half the PWB material of its multiple THT predecessors. Cabling and connectors are saved because there are no board-level interconnects required. And, the elimination of the board-level interconnects (a common failure point) improves system reliability. Figure 6-1(A) and Figure 6-1(B) show the use of SMT in a single-board system, with dense, mixed technology. This board would certainly have required 2 or 3 cards using THT technology.

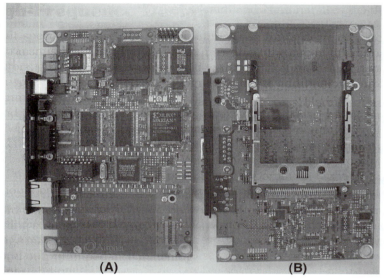

Figure 6-1   SMT allows a high-density design to be accomplished on one board. (A) Top view. (B) Bottom view.

SIP and DIP Modules

SIP and DIP modules, like those hybrid modules shown in Figure 6-2, serve miniaturization and economy needs in several ways. As we will see later, modules can satisfy standard circuit requirements or can simplify the conversion of just a part of a circuit to SMT.

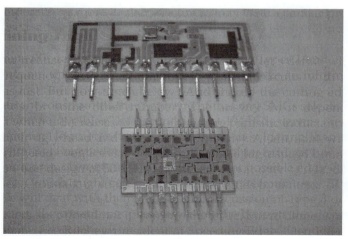

Figure 6-2   Hybrid SIP and DIP modules.

Where certain circuits, such as amplifier stages, comparators, etc., are used repetitively in designs, converting them to modules using SMT components can save considerable space and money. Instead of implementing the standard circuit with discrete THTs on each assembly part number where the function is needed, the SIP or DIP module may easily be integrated into existing and future through-hole boards, and the module may also be used as an inserted module on Type 2 and Type 3 SMT designs.

For designs that must remain partially through hole, SMT modules may simplify conversion of those areas that would most benefit from SMT. The SMT module can be built with cost-effective Type 1 technology. Its lead frame allows insertion into an otherwise through-hole board, and permits wave soldering the assembly on a standard THT production line.

Double-Sided Population

Double-sided SMT and mixed technology (Hall Type 2) boards are in common use today. Where space is not an issue, all-SMT boards may be designed with no components on the bottom side, as is true for the PCB shown in Figure 6-3(A). Products where density is truly an issue, such as in ultradense memory boards, just would not exist without the real estate benefits of SMT. Where two-sided boards are necessary, SMT makes them doable, like the network board in Figure 6-3(B).

Figure 6-3   (A) Single-sided SMT board.

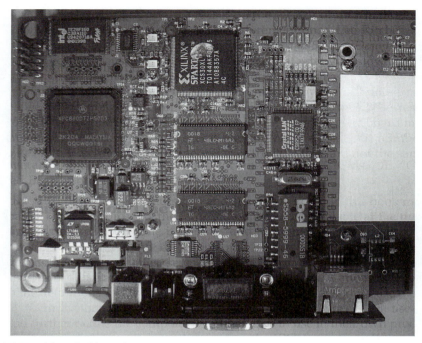

Figure 6-3   (B) Double-sided board.

6.1.2   Application-Determined Strategies

Here, we will look at real estate-savings moves that are peculiar to certain applications. These methods should be considered whenever a design of a type covered in the following discussion needs a space or weight reduction. Some of the lessons of these application-specific miniaturization strategies will benefit circuits other than pure memory arrays, or circuits with a large number of decoupling capacitors, or any airborne or spacecraft electronics. Thus, they bear study by all engineers interested in ways to shave inches and pounds from their products.

Memory Arrays

While no new memory ICs are available in THT technologies, several SMT component alternatives are available for memory boards. SMT components can take advantage of additional real estate savings by using SIMM™ or DIMM memory modules or other standard memory arrays. SIMM and DIMM memories are tiny daughter boards carrying one or more SO memory chips,

Figure 6-4 Typical 100 MHz DIMM memory module.

which form a complete memory array, as illustrated by the 100 MHz DIMM module shown in Figure 6-4. SIMM and DIMM memories greatly simplify the routing of memory-array boards. And, they offer socketability at reasonable cost and high density. Socketing individual memory ICs is expensive and wastes space.

6.1.3 Additional Technologies for Real Estate and Performance Improvement

Several real estate-miser methods are developing, which, while they hold significant promise, are not in the mainstream manufacturing practice as of this writing. Because they are not in widespread use, no industry-wide design standards have yet been adopted. They merit mention, however, as they are developing into standard approaches.

The speed with which technologies are advancing is best shown by the International Technology Roadmap for Semiconductors (ITRS), jointly developed by:

- European Semiconductor Industry Association
- Japan Electronics and Information Technology Industries Association
- Korean Semiconductor Industry Association
- Taiwan Semiconductor Industry Association
- Semiconductor Industry Association (U.S.)

The ITRS attempts to predict out as far as 15 years the progress of the driving technologies that affect the semiconductor industry and its customers.

As shown in Table 6-1, adapted from the ITRS Roadmap,[2] circuit performance and function requirements for high-speed circuits are driving the need for package performance improvements

Even where standardization is not forthcoming, these methods discussed in the following sections are available to designers and may be independently developed or acquired by licensing where needed. These methods are in addition to the overall performance increase and size decrease that has occurred as a result of the shift from bipolar transistor technologies to CMOS transistor technologies. IBM reported that in their flagship mainframe S/390 series, the redesign from bipolar to CMOS,[3] and the incorporation of other silicon-level strategies resulted in a reduction in volume and weight of over 90 percent![4]

Table 6-1   Package and Performance Requirements, Current and Projected.

	2003	2006	2009
IC Size (mm^2)			
Low cost, handheld, memory	100	100	100
High performance	310	310	310
Max. Power (W/mm^2)			
Low cost (watts)	2.5	3	3
High performance	0.48	0.58	0.64
Internal Voltage (V)			
Low cost	1.2	0.9	0.8
High performance	1.2	0.9	0.8
Max. Pin Count			
Low cost	480	600	800
High performance	2400	3800	4600
Bus Frequency (MHz)			
Low cost	100	100	100
High performance	400-2000	600-4000	800-7500
On-Chip Frequency (MHz)			
Low cost	500-3000	600-4000	800-5000
High performance	3000	5600	unknown, >6700

Silicon-on-Silicon

One emerging variation of SMT, silicon-on-silicon (SOS), promises ultrahigh densities rivaling those obtained by elusive methods such as wafer-scale integration. Silicon-on-silicon SMT is flip-chip and discrete-element placement on a hybrid circuit that is fabricated on a silicon substrate. By using a silicon substrate, many passive and active elements may be fabricated in situ on the circuit board. Where it is impractical to build up a particular component on the substrate, it is added as an SMT discrete.[5] Since this method is really a special variation of a hybrid circuit, design rules for SOS may be drawn from Chapter 9, SMT Hybrids. We mention SOS here because of the enormous miniaturization potential it offers.

The advantages of silicon-on-silicon are an extremely high-density, excellent heat dissipation, and, in very high volume, a low cost per functional module. Distinct from wafer-scale integration, SOS offers the advantage of discrete-device repairability. The limitations are sources for flip-chip components and technology, high front-end costs, and relatively long product development cycles.

System on a Chip (SOC)

SOC integrates all necessary functions of a sophisticated system (such as a video phone or a digital camera) into a single IC.[6] The chip functions may include digital, analog, mixed-signal, and radio-frequency functions. SOC design is complex silicon design and continues to evolve along with other IC technologies like silicon on insulator (SOI) which allows faster speeds with reduced power consumption. SOC incorporates ASIC and standard-cell design technologies. Since SOC is silicon-level design, we will not cover it further.

Multi-Chip Module (MCM)

MCMs, shown in Figure 6-5 are great space savers. We will deal with MCMs in more detail in Chapter 9, but the reader can note the obvious space savings that occur through the use of sev-

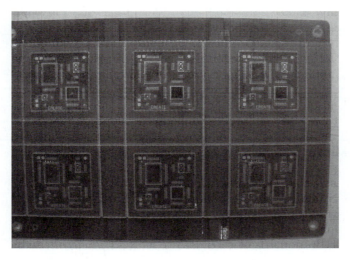

Figure 6-5   Array of six multi-chip modules.

eral IC die and chip parts in one BGA package. There are numerous works dealing with MCM designs, and while MCMs have not had the market impact initially expected, they are still a viable packaging alternative. We will deal with MCMs in more detail in Chapter 9. Suffice to say here that they provide the system engineer with a powerful set of miniaturization tools.

System in a Package (SIP)

SIP technology refers to a semiconductor package that incorporates multiple IC die as well as passive components that make up a complete electronic system in a single package. Unlike SOC, which requires the entire system to be designed on a single die, with the inherent time-to-market lag and accompanying non-recurring engineering (NRE) costs that go along with silicon design, SIP takes advantage of existing die. Similar to MCMs, the SIP allows the additional methodology of die stacking to maximize functionality in a given package volume.[7]

In high-speed digital and RF designs, SIP minimizes parasitics by adding passives inside the package immediately adjacent to the silicon dice. Intel has used SIP technology in certain Pentium designs by combining processor and cache when the process technology prevented doing this on one die.

The challenge in using SIP in designs comes in the design and assembly of the SIP module/package itself. This requires the ability to assemble and interconnect several die that are combined not only in the horizontal plane (side-by-side), but stacked on top of each other. This die stacking explains why SIP is known as 3-D packaging. Die stacking started out with a pyramid style of stacking smaller die on top of larger die, a technique that inherently keeps the peripheral wire-bonding contact pads on the periphery of each die clear at each level. Stacking has evolved that allows stacking of larger die on top of smaller die, as well as the technique developed by Tessera Inc., called "folded/stacked" technology. Along with these various stacking technologies come concerns about thermal and mechanical stresses that must be addressed.[8] Due to the nature of BGA packages, which already contain a routing substrate as part of the package, SIPs are usually packaged in area array packages. In addition to incorporation of multiple dies, a SIP would commonly include imbedded passive as well. These will be discussed further in "Embedded Components" below.

Ultra-Large-Scale Integration

While by no means unique to SMT, ULSI devices, often having very high pin counts, are often available only in SMCs or in a choice of SMC packaging or pin grid array (PGA). The PGA is often more expensive but it is easy to deal with for replacement. However, its internal design has not kept up with the more advanced SMCs. Thus, SMT, TAB, flip-chip technology, and MCMs play large roles in facilitating the move to ever-higher levels of silicon integration and the resultant enormous real estate savings.

6.1.4   Embedded Components

Resistive/conductive laminate layers, or resistive traces of polymer thick-film ink, may be built into inner layers of multi-layer boards. Capacitive and inductive components may also be formed in situ as described earlier and buried in a MLB. However, serious manufacturing, test, and repair-access problems must be addressed when using buried-circuit elements. Considerations include:

- What are the relative costs of embedded versus surface components?
- What is the availability of a board fabricator with embedded experience?
- Can speed and/or size constraints be met without embedding components?
- Are there a number of identical or similar components, such as pull-up resistors or decoupling capacitors?
- Will the added board complexity be offset by the surface layer area that will be freed up?
- Can a number of the embedded components, e.g., Rs, be fabricated with a single material?

If the results of the considerations lean to embedding components, then materials must be selected and parts must be created for the CAD libraries based on the component characteristics. Depending on whether planar or discrete resistors and capacitors are required, the board fabricator should be consulted to confirm the additional costs as the design progresses.[9]

Capacitors can be fabricated with distributed plane capacitance using power and ground planes or by multi-layer techniques, putting one plate on each layer and terminating the plates at alternating vias at the ends of the plates, per Figure 6-6. For calculation of capacitance, the standard capacitance formula is used:

$$C = \frac{((n-1) \times C_M \times K \times A)}{d}$$

where:

n = number of conductive layers (n = 2 for plane capacitance),

C_M = a measurement constant, equal to 0.2249 for A and d in inches, and 0.08854 for A and d in cm,

K = dielectric constant of the insulating material (See Chapter 5, Table 5-7),

A = area of the plates in either in² or cm²

d = thickness of the dielectric in either inches or cm (must be same unit as A).

Figure 6-6　In situ capacitors in MLBs. (Adapted from Harper, C. A., ed., *Handbook of Wiring, Cabling, and Interconnecting for Electronics.* McGraw-Hill, New York, 1972, pp. 8–18)

Spiral coil   Modified circles   Square coil

Figure 6-7 In situ inductor forms in PWBs.

Integral inductors are generally formed as spiral coils on the board's surface. While this same approach may be used on inner layers, use of the surface layer ensures test access and repairability. Figure 6-7 shows multiple turn loops for a spiral coil, modified circles, and a square coil.[10]

The inductance, L, of a single-turn inductor loop is calculated as follows:

$$L = 0.002 \, A(\log B - 2.451) \qquad \textbf{(Eq. 6-1)}$$

where,

$\quad A =$ The mean circumference of the loop,

$\quad B =$ $2A/(w + t)$ for rectangular conductors of width w and thickness t.

The *spiral-coil* inductance is:

$$L = 0.0215 \, a \, N^{5/3} \log \frac{8a}{c} \qquad \textbf{(Eq. 6-2)}$$

where,

$\quad a =$ The mean radius ($\frac{1}{2}(R_1 + R_2)$),

$\quad N =$ The number of turns,

$\quad c =$ $R_2 - R_1$ (the radial depth of winding).

(Note: To maximize Q, thus minimizing conductor length, use $c/a = 0.8$.)

Modified circles approximate the spiral-coil inductance and are easier to lay out.

Square spirals are easily laid out and maximize the inductance per square area of PWB. A square spiral of side A will provide 12 percent greater conductance (L) than a spiral coil of outside diameter A.

Polymer Thick-Film Components

Special conductive and resistive inks compatible with polymer substrates may be used to print resistors directly on circuit boards. Combinations of conductive ink layers covered by dielectric ink layers allow printing of low-value capacitors as well as resistors.[11] Conductive ink or standard circuit-board metallizations may be formed in spirals to make small inductances.[12] For design guidance in these techniques, see IPC D-859. Also, gather specific data on polymer ink behavior and properties from the material suppliers.

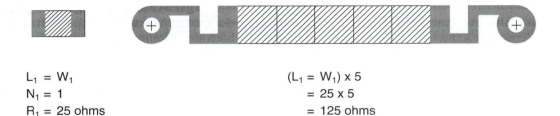

$L_1 = W_1$

$N_1 = 1$

$R_1 = 25$ ohms

$(L_1 = W_1) \times 5$

$= 25 \times 5$

$= 125$ ohms

Figure 6-8   Examples of buried resistor design using Ohmega-Ply. (Adapted from the Ohmega Tech., Inc. site, http://www.ohmega.com)

Buried resistors of 10 percent tolerance can be fabricated for lamination within a PWB. Typical applications are boards with large populations of low-tolerance resistors, such as pull-up and pull-downs, and termination networks. Buried resistors are made using special resistive-layer laminate or polymer thick-film printing techniques. These planar resistors are created using ECAD design rules on existing layers of the PCB.

The resistive laminate method involves electroplating a nickel/phosphorus-resistive layer onto a copper foil, then laminating the bimetal foil, resistive layer down, to a polymer sheet. PWB fabricators photoetch the exposed copper to leave only the conductive traces as required. The resistive layer is then etched to form individual resistors by leaving only the resistive connections desired between the copper conductors.

Alternatively, resistive sheets can be used. One of the more commonly used resistive material is Ohmega-Ply from Omega-Tech, Inc. Their website, http://www.ohmega.com, has complete design details for the design and application of buried resistors using their laminate. Examples of the layout of embedded sheet resistors are shown in Figure 6-8.

6.1.5   SMT Patches on Through-Hole Boards

Isolated SMCs

Occasionally, a solid THT design requires one or two components that are available only in SMT packaging. To the SMT purist, this smacks of heresy, but there are sockets that adapt SMCs to 2.54-mm-center (100-mil) THT mounting. They may be useful in downsizing a THT board by replacing a handful of chips with custom silicon. Often, high pin-count custom silicon is available only in SMC packaging. The adapters are not cheap and may be far more expensive than designing the board for the SMT part. However, if only wave soldering is available, sockets and/or adapters may be the only alternative.

For circuits where timing is not critical, sockets may provide a convenient breadboarding tool for SMT prototypes as well. Sockets on 2.54-mm (0.100-inch) centers adapt SMCs to 100-mil proto boards. However, the additional parasitics introduced make this approach useless in high-speed and timing-sensitive designs.

Figure 6-9   Javelin microprocessor module.

High-Performance Sections

Surface-mount performance is particularly attractive in designing high-speed circuitry. Where only a portion of a board will run at high speeds, a SIP or DIP SMT module allows taking advantage of surface-mount performance where it is needed, while sticking primarily to through-hole technology. An example which allows this is the Parallax PIC module shown in Figure 6-9. This module is an excellent example of a DIP assembly that allows a microprocessor and associated components to be breadboarded or plugged into a THT assembly.

SMT Daughter Boards

Some SMT modules with SIP and DIP lead frames were shown in Figure 6-2. These circuits were used as standardized functional modules in otherwise through-hole board manufacturing. The same approach may be used to solve other design challenges. Figure 6-10 shows a DIP module

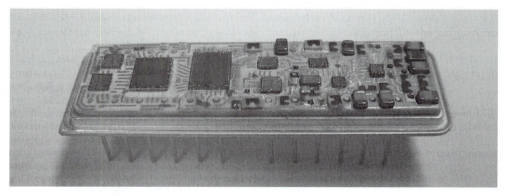

Figure 6-10   SMT/ DIP module.

built on a ceramic substrate. The ceramic substrate matches the TCE of the chip carrier, thereby eliminating thermal cycling concerns. The ceramic board also allows high-performance resistors to be printed directly on the circuit and laser trimmed to close tolerances.

6.2   Board-Level Interconnects

We covered packaging for miniaturization, reduction of board count, and convenience in dealing with manufacturing and design challenges above. Now, let's turn to the interconnection of SMT boards to one another and, also, to the outside world. We will look at those areas of board and system interconnects that are unique to SMT. We will not attempt to cover through-hole connectors and interconnect strategies, since these topics are well documented in other texts. However, we will start our discussion with a note to not rule out through-hole connectors because of surface mounting.

6.2.1   SMT and THT Connectors Compared

One of the first design decisions regarding interconnections is answering the THT versus SMC question. SMT and THT connectors differ in both their construction and in their required processing. We will explore these variations now. They are instructive in both the SMC/THT selection process and the design of SMT board interconnects.

Soldering Approach
Like all through-hole components, flow soldering THT connectors is no particular problem. While the connector solder tails pass through molten solder on the board's solder side, the delicate plastic portion of the connector is safely out of harm's way on the benign component side of the board. Not so in surface mount. SMT wave soldering involves passing components directly through the solder wave. Connector bodies cannot survive such treatment. Therefore, SMT connectors are usually designed for reflow attachment. Or, through-hole connectors are used with the pin-in-paste (PIP) process described in Chapter 3. A number of connector manufacturers offer connectors specifically tailored for the PIP process, as well as process-specific information. These include Phoenix Contact, Harting Connectors, Connecta, Vita, AMP, 3M, and others.

Reflow soldering subjects connectors to temperatures beyond the safe limits for many housing dielectric materials. Therefore, connector manufacturers have turned to high-temperature plastics. Be cautioned, however, that some high-temperature plastics are not as mechanically rugged as standard plastic housing materials. Also, connector material choices will have a direct bearing on the choice of various reflow-soldering techniques and temperature profiles. As you begin to understand some of the connector-related issues presented here and in other sources,[13] the connector suppler is the best source of specific information regarding these issues.

Mechanical Integrity

THT-connector solder tails are attached firmly to boards by solder fillets on both sides of through-plated holes. The solder joints of SMT connectors afford significantly less holding power. Estimates vary with the style of solder connection, but THT connectors typically provide 2.5 to 20 times more holding power per pin. Even so, SMT-connector solder joints have plenty of holding power for a few cycles of mating and unmating. However, where connectors will be cycled often, some additional hold-down is generally in order when surface mounting connectors. Hold-down methods include bolting or riveting, heat stake rivets, snap-in detent rivets, and press-fit bosses. All involve special assembly considerations and accurate hole alignment in the board for the mechanical attachment devices.

Zero- and low-insertion-force connectors are quite popular in SMT because they keep insertion forces well within the range surface-mount solder joints can accommodate.

Another solution is to use connectors that are provided with solder preforms shaped like donuts on their solder tails. This allows through-the-hole soldering to be accomplished in a reflow operation. An example of a THT reflow connector is shown in Figure 6-11.

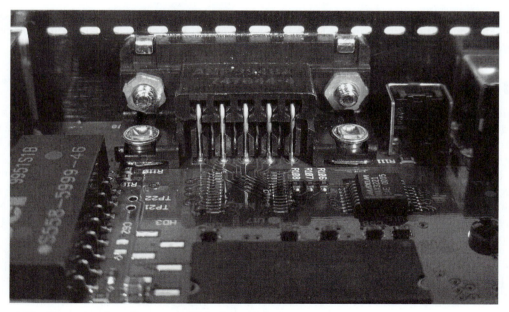

Figure 6-11   Example of a reflow-mountable through-hole connector.

6.2.2   SMT or THT Connectors?

As we mentioned above, a major board-level interconnect decision is whether or not to use surface-mount connectors. Decisionmaking factors include:

- Required mating connector, if any
- Connectors specified by the customer
- Connectors necessitated by the fit of the product
- Availability of connectors in SMT or THT
- Space available
- Soldering process to be used
- Assembly method
- Pitch of connector pins

These must be reconciled early in the design, as certain connector styles may rule out specific manufacturing styles and process selections which, in turn, will determine the design rules. There are a wide number and variety of SMT connectors available, and some provide very dense I/O. It is important to determine if the connectors can be placed with standard placement equipment or if they require dedicated robots or hand assembly. A few SMT connectors take more board real estate than their through-hole counterparts, and some carry a price premium. THT connectors may be a better choice on a board that will require flow soldering for other THTs. However, if selecting an SMT connector will allow an exclusive reflow soldering of the assembly, the benefit in process simplification will probably offset these negatives.

6.3   Board Flexure

Flexure of boards produces stresses on solder joints, land-to-board bonds, device terminations, board lamination bonds, device bodies, etc. In short, it stresses the entire assembly. Flexure may be produced by gravity, shock loads, uneven heating, uneven thermal expansion, board handling, and other mechanically imposed loads.

Board flexure is not exclusively an SMT issue. THT boards give concern, also, in severe environments or situations where they must be plugged in and unplugged often. But the relatively weak SMC solder joint, coupled with the lack of any mechanical attachment of SMCs, makes flexural integrity particularly important to SMT assembly reliability. In the following discussion, we will cover methods of improving the flexural resistance of SMT boards.

Flexure is an acute disease when it produces solder-joint stresses that exceed the safe limits of the joints on an assembly. The acute form is easily recognized; components pop off the board like tiddledywinks as the board bends. The more common chronic form of the disease slowly destroys joints by work hardening and crystallizing the solder in them. Eventually, such joints fail, just as a coat hanger fails and breaks when we flex it back and forth to make a tool for opening a locked car door.

Where calculated bending stress levels approach limits of concern for a given solder material and desired B-10 life, the following directions should be employed to ensure solder-joint integrity.

6.3.1   Board Stiffeners

Let's start with the simple and direct approach of stiffening the PW board. When a completed design proves to be too flimsy for its intended task, this may be the only avenue open—short of scrapping the design and starting over.

External to the Substrate

Stiffening hardware may be added directly to the board, or the card carrier may be stiffened to resist board flexure. If the flexure concern arises from card insertion and removal, the board-stiffener route should be selected as its protection travels with the assembly. Before reinforcement design may be undertaken, the electrical and circuit-board layout engineers must give the mechanical engineer an accurate description of the board material, its thickness, the weight distribution of components, and the handling stresses envisioned. Given this data, the modeling of external stiffeners becomes a straightforward mechanical engineering task.[14] The Machinerys Handbook lists stress formulas for stiffener members acting as simple beams, cantilevers, etc.

Within the Substrate

External stiffening is perhaps the easiest approach to beefing up boards for flexural stress. But what if there's no room for external ribs? We may still be able to save an existing design by adding a metal core or cores to replace the polymer ground/power layers, or to simply stiffen and restrain the board without any electrical function. Copper/moly/copper is particularly effective in strength-to-weight ratio, but any metal core will add considerable structural integrity and will also markedly improve the thermal characteristics.

There is a cost premium associated with metal-core boards, and PWB vendors are limited. However, the alternative of redesigning an assembly because of flexural concerns may be a larger negative than the cost and availability of the PWBs. Where this is the case, as with low manufacturing volume projects, metal cores may well satisfy the need for board stiffness.

6.3.2   Substrate Materials

When a flexural problem is identified prior to completion of design, more avenues are open to build structural integrity into an assembly. One method is the use of a high-rigidity substrate. Table 6-2 presents the mechanical data for some common substrate materials.

Table 6-2   Mechanical Properties of Substrate Materials.

	Mechanical Properties			
Material	Flexural Strength in PSI $\times 10^3$	Modulus Elasticity in PSI $\times 10^6$	Tensile Strength (ft-lb/in)	Izod Impact Notch Test (PSI $\times 10^6$)
Epoxy/Glass	G-10 45–50	2.5–3.0	5.5–7.5	2.5
Alumina	45–50	44	Brittle	25–30
Beryllia	39–49	50	Brittle	22
Porcelain/Steel		5–8P./25–30S.		
Glass 7059 Corning	1	9.8	Brittle	1
Quartz Fuse	7	10.5	Brittle	7
Sapphire	60	51	Brittle	60

In some instances, trading polymer-based devices for more rigid substrates may satisfy flexure concerns in existing designs as well. However, selection of relatively rigid substrate materials will often have a significant impact on design rules. A circuit designed for FR-4 may not translate directly to rigid substrate materials.

Ceramics

Most ceramic substrates are of high-purity alumina (polycrystalline Al_2O_3, with traces of metal glasses to tailor properties). Alumina is widely available, provides nearly fifteen times the thermal conductivity of epoxy/glass, and is very resistant to bending. Alumina substrates are the workhorse of the hybrid-circuit industry. Their refractory nature permits use of stable high-firing-temperature thick-film inks critical to many hybrid applications.

However, alumina is limited in SMT usage by several factors. Alumina generally restricts the substrate size to about 230 sq. cm. (36 sq. in.), and the per-square-inch costs begin to climb with substrates above 100 sq. cm. (16 sq. in.). Size for size, small alumina substrates are about three times the price of FR-4 epoxy/glass. Alumina is brittle, must be handled with reasonable care, and must be protected from high-impact forces while in service. And the dielectric constant of alumina is roughly twice that of epoxy/glass, making alumina unsuitable for some high-frequency applications.

Beryllia ceramics (polycrystalline BeO, with traces of metal glasses to tailor properties) share many of the positive properties of alumina. They are chosen over alumina for applications where their high thermal conductivity (nearly 100 times greater than epoxy/glass) is needed to manage thermal dissipation. The lower dielectric constant of beryllia, coupled with its thermal conductivity, makes it well suited for microwave circuits with high dissipations.

Beryllia substrates also share alumina's negatives. They are even more costly than alumina. And, unlike alumina, beryllia is highly toxic when inhaled in dust or fume form. Therefore, beryllia is used primarily where its unique properties are called for.

Ceramic-Coated Metal Core

Coated metal-core substrates offer a number of attractive features. They may be fabricated in any size and virtually any shape. Standard metalworking operations, such as bending, forming, and punching, may be done prior to coating. Thus, not only stiffening ribs, but circuit and housing mechanical features may be built directly into the substrate. In some cases, metal-core boards form both substrate and housing for an electronic system.

Metal-core boards are shock resistant, strong, moderate in cost, and provide a built-in ground plane. By selecting a metal core with a TCE close to that of ceramic chip carriers, thermal-expansion mismatches between substrate and component may be minimized.

Where in situ components will be added by hybrid thick-film processes, inks must be of the low-firing-temperature varieties. Tooling costs make metal-core boards unattractive for short-run and low-volume products.

Glass

Glass substrates are generally selected for applications such as displays, where their transparency and optical clarity are needed. Glass is highly available, can be produced in any size, is low in cost, and rigid. Glass can be produced with very smooth surface finishes, low camber, and minimal warpage, and in qualities useful in thick- or thin-film hybrid processing.

However, the low thermal conductivity of glass (alumina is thirty to fifty times more thermally conductive) rules out the use of glass for high-dissipation circuits. Like coated metal-core boards, glass requires low-firing-temperature inks for in situ fabrication of thick-film components. And, repeatable, reliable, thick-film processes are a challenge on glass.

Quartz

Quartz substrates are primarily made by heating silica to form fused silica quartz. The material is readily available and may be fabricated into boards of large sizes. The low K factor of quartz makes it of interest in high-frequency applications.

The limitations of quartz as a substrate restrict its use, however, even in the microwave area. The thermal conductivity of quartz, while higher than glass, is still just one-twentieth that of alumina and less than one-hundredth that of beryllia. Thus, quartz is limited to low-power-dissipation applications. Quartz is also highly brittle. Care must be used in its handling, and it must be protected from shock and impact while in service. Single crystal quartz, as opposed to fused quartz, is used in *standing acoustic wave (SAW)* devices, where its ordered crystalline structure justifies its high cost.

Sapphire

Sapphire (monocrystalline Al_2O_3) is produced in boules similar to silicon for electronic applications. Substrates sliced from a boule are limited in size by the boule diameter. An alternative process, *edge-defined film-fed growth (EFG™)* allows the fabrication of substrates of about 10 cm × 15 cm (4" × 6") as of this writing, and larger sizes are planned. Sapphire is used where its radiation hardness is needed and in hybrid microwave devices where alumina's camber and surface finish are inadequate for element fabrication. Applications also include hybrids, using silicon-on-sapphire devices. Sapphire's relatively low thermal conductivity (about one-tenth that of alumina), high K factor, and very high cost restricts its usage.

6.3.3   Lead Compliance

In many cases, with attention to device package selection, designers can accommodate anticipated board flexure with compliant leads on the components. This strategy is particularly viable where flexural excursions are small (not over 3.18 mm/0.125" per 25.4-cm/10" linear board surface, bidirectional), and components are not large (not larger than 44-pin chip carriers). Other methods, discussed herein, are called for where more severe bending will be encountered, or in moderate flexure applications where reflowed components are particularly large (high pin-count devices and long connectors, for example).

The design team must remember that non-leaded chip IC components have much less tolerance to bending. The manufacturers' specifications must be consulted in designs where either assembly or use will subject the substrate to bending forces.

6.3.4   Long Connectors

Long connectors, reaching across more than 38.1 mm (1.5") of the circuit board, are a particular challenge in high-flexure applications. Several manufacturers offer floating-lead designs. Floating leads supply extra compliance to absorb flexure and thermal-expansion stresses without placing undue strain on the connector solder joints.

6.4   Environmental Concerns and Packaging

SMT brings with it a heightened packaging concern about environmental stresses. We will cover the ways that SMT differs from conventional circuit approaches herein. We will not review any standard engineering practices that apply equally to surface mount and inserted mounting. This SMT bias does not, however, exempt surface mounting from basic electrical engineering practices regarding packaging for survival in the intended environment.

SMT calls for special environmental engineering attention to the problems of thermal cycling, high vibration, G forces and shock loads, connector cycling, frequent card insertion/withdrawal, radiation resistance, and operation in threatening ambient conditions.

6.4.1   Frequent Temperature Cycling

We discussed temperature cycling and engineering solutions to the stresses it produces in Chapter 5, Section 5.7. At this point, let's review why surface-mount technology differs from through-hole technology in thermal cycling, and develop a yardstick to determine when thermal cycling is likely to be a problem. Then, we will briefly review engineering solutions covered in detail in the previous chapter.

As most designers have noted, in addition to temperature cycling, one must also take into account the basic temperature rating of the ICs themselves. The ratings can be adjusted, or "uprated," in certain circumstances that allow operation at higher temperatures.[15]

Differences in THT and SMT Thermal-Cycling Resistance

The entrance of surface-mounting technology brings, along with its many benefits, concerns about long-term reliability in thermal cycling. These concerns arise from two significant differences between SMT and THT.

First, SMCs generally have no mechanical connection to the board outside of the solder joint. The SMT solder joint must furnish both mechanical rigidity and electrical contact with the circuit. THTs have an inherent mechanical attachment due to leads being inserted through holes. Clenched leads provide an even more robust connection for through-hole components.

Second, SMT differs in the compliance furnished by leads. The long DIP lead is quite compliant in X- and Y-axis motion relative to the board, although this comment is limited to thermally-induced X-Y motions since we have already noted one case where the longer DIP leads resulted in harmonic vibrations that severed the leads. In environments that do not have vibrations issues but that have severe thermal cycling issues, the shorter PLCC and SOIC leads generally provide less of a flexural element than DIP leads. In the most extreme cases—the leadless SMCs—there is no stress-relieving flexural element at all.

For these two reasons, designs that were quite reliable with through-hole components may have problems if we just switch the THTs for SMC parts without considering TCE engineering.

A Yardstick for Estimating Thermal-Cycling Concerns

This raises the question, "When should we become concerned about thermal cycling?" Several rules of thumb can help provide a yardstick by which to measure designs for potential TCE mismatch concerns.

We can start by drawing lines between various environments and labeling each by degree of temperature-cycling severity. Office products, consumer appliances, and some other electronics enjoy the luxury of a "moderate thermal environment" by the right of coexisting with temperature-sensitive humans. High-reliability industrial products are generally designed to survive 400 cycles from 0°C to +100°C. While this is less than brutal in temperature swings, it would not make for a cozy office, so we will call this a "rigorous thermal environment." Military products complying with MIL-STD-202, Method 107F Condition B and industrial products intended for harsh temperatures are generally designed to endure 400 cycles from –65°C to +125°C. We will call this a "harsh thermal environment." For long-term survival in truly punishing environments, SMT boards may be designed to survive 1000 cycles from –65°C to +125°C. We will call this an "extreme thermal environment." Using these environmental severity definitions, Table 6-3 shows several concern levels as determined by application variables in a given thermal-cycling severity. These data are not intended as a categorical statement that there will or will not be difficulties, but they should serve as a guide to flag engineering

Table 6-3   A Yardstick to Gauge TCE Concern.

Environment Severity Rating	Application Characteristics		
	Component	*Substrate*	*Concern Level*
Moderate	Leadless Ceramic, L* = ≤0.500 inch. SO and SOLIC, PLCC, <84 Pins. Leaded Carriers (all types).	Polymer	Very Low
	Leadless Carriers, L = ≤1.250 inch.	Match TCE	Very Low
	Leadless Carriers, L = ≥1.250 inch.	Any	Add leads to high-dissipation component. Match substrate TCE for low dissipation.
Rigorous	Leadless Ceramic & BGA, L* = ≤0.500 inch. SO and SOLIC, PLCC, ≤68 Pins. Leaded Carriers (L = ≤1.000 inch).	Polymer	Low
	Leadless Carriers, L = ≤1.000 inch.	Match TCE	Low
	Leadless Carriers, L = ≥1.000 inch.	Any	Add leads to high-dissipation component. Match substrate TCE for low dissipation.
Harsh	Leadless Ceramic & BGA, L* = ≤0.375 inch. SO and SOLIC, PLCC, ≤44 Pins. Leaded Carriers (L = ≤0.750 inch).	Polymer	Concern rises as component size increases.
	Leadless Carriers, L = ≤0.750 inch.	Match TCE	Add leads to high-dissipation component.
	Leadless Carriers, L = >0.750 inch.	Any	Adds leads to high-dissipation component. Match substrate TCE for low dissipation.
Extreme	Leadless Ceramic & BGA, L* = ≤0.250 inch. SO and SOLIC, PLCC, ≤28 Pins. Leaded Carriers (L = ≤0.500 inch).	Polymer	Concern rises as component size increases.
	Leadless Carriers, L = ≤0.500 inch.	Match TCE	Add leads to high-dissipation component.
	Leadless Carriers, L = ≥0.500 inch.	Any	Add leads to high-dissipation component. Match substrate TCE for low dissipation, i.e., alumina.

concern. As such, the statements of Table 6-3 tend toward the conservative. Also note that lead-less ceramic and BGA components are grouped. Concerns for these components will vary depending on the specific component in question, the size of the component, substrate material(s), and known thermal issues with the part. The IC manufacturer's data sheets and application notes must be consulted to make an informed decision.

Where thermal cycling, either from the environment or dissipation within a device, produces thermal expansion differentials between the substrate and component, stresses may damage the solder joints. Later, we will review the ways covered in Chapter 2 and Chapter 5 for bringing stresses back within safe limits. Exactly what is a safe limit is determined by the solder alloy used. Consult your solder vendor for data.

Compliant Leaded Components

Compliant leads on components flex to absorb thermal expansion-induced stresses, protecting the solder joints. For leadless components, it is possible to attach leads. There are services available for postlead attach, or it may be done internally. Where post leading is not desirable, sockets may add compliance. Special flexural columns are also available for solder attachment between the board and leadless component. Properly engineered, any of these methods can accommodate typically encountered TCE mismatch stress levels.

Matching Component and Substrate TCEs

Thermally generated solder stresses come from two sources. Mismatches in TCE of the substrate and component materials produce stresses when temperature changes from the environment or circuit dissipation act on the mismatched pair. Due to differential temperatures when under power, circuit dissipation also produces solder stress, even in situations where the component and board TCE match. While tailoring substrate TCE to that of components, will not eliminate internal temperature differential stresses; these forces are low enough to be of no concern for moderate-sized IC packages and typical dissipation levels. As long as very large packages and/or very high internal dissipations are not involved, matching or approximating the component and substrate TCEs will keep stresses within safe limits.

6.4.2   Vibration

SMCs are generally more vibration tolerant than equivalent THTs. At first blush, this is somewhat surprising. After all, surface mounting provides no mechanical attachment to the PWB except through the solder joints, and the SMC's leads are narrower and thinner than the DIP's. But, a 16-pin SOIC weighs only 130 milligrams, while a 16-pin DIP weighs 1200. The differential in mass translates to far less vibration-induced stress in the leads and solder joints

Table 6-4 Approximate Weights of Various SMT Packages.

Component	Weight (mg)	Component	Weight (mg)
SOIC-14	100	PQFP-44	500
SOIC-16	130	PQFP-100	1600
SOLIC-20	480	TQFP-176	1900
PLCC-28	900	BGA-256	2200
PLCC-44	1200	BGA-728	6200
CLLCC-20	470	μBGA-56 (CSP)	100
CLLCC-44	1750	μBGA-280	500

for the lighter surface-mount device. Table 6-4 shows weights of typical SMC packages for vibration and shock force analysis. Note that there are also significant mass differences among various SMT component packages.

Putting lighter weight to a practical test, one major off-road equipment manufacturer was astonished when a destructive vibration test appeared to shear the leads of a 40-pin DIP just above the solder fillets while an adjacent PLCC-44 remained unscathed. Mechanical modeling showed that the DIP leads were actually deforming into a waving "S" curve under vibration. The natural frequency of the heavy DIP was near resonant with the vibrations produced by a large reciprocating engine. The DIP's leads soon work hardened and broke. The low mass and short protected leads of the PLCC left it on board and rattling away unharmed. Figure 6-12

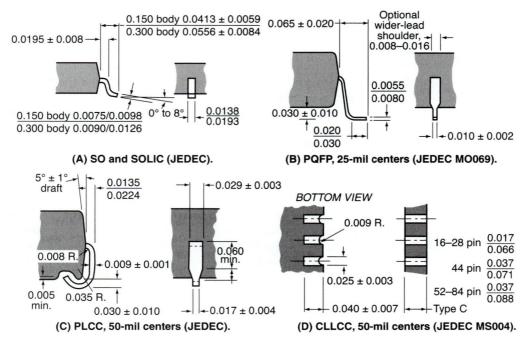

Figure 6-12 Standard SMC lead dimensions.

shows dimensions of standard SMC leads for use in stress analysis calculation. Note that the lead material may vary according to vendor. Consult component manufacturers for verification of the physical properties of the lead-frame material. With the dimensions of Figure 6-12, the component weight from Table 6-4, and data on solder material properties, calculation of solder-joint strength is allowed. See Manko[16] for formulas for various joint configurations and application factors affecting the calculations.

Natural Frequency of the Assembly

The natural frequency of an assembly may be tuned away from resonance or harmonics with environmental vibration. This is most often done by adding weight to lower the assembly's resonant point. The addition of a metal core can serve this purpose while simultaneously adding stiffness and improving the thermal properties of the assembly.

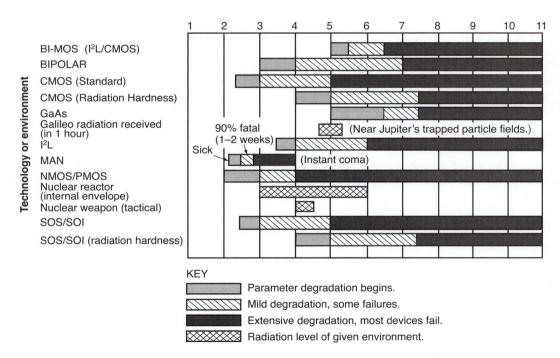

Figure 6-13 Radiation resistance of various technologies. (Adapted from *Rad-Hard/Hi-Rel CICD Data Book*, Harris Custom Integrated Circuit Div., Palm Bay, FL, 1987, pp. 1–8, and from Gauthier, M. K., and Dantas, A. R. V., "Radiation-Hard Analog-to-Digital Converters for Space and Strategic Applications," *JPL Publication 85-84*, NASA Jet Propulsion Laboratory, Pasadena, CA, 1985, p. 5-3.)

Mechanical Integrity Considerations

Since the solder joints of SMCs must supply the total connection to the PWB, soldering quality and repeatability are of particular concern in mechanically threatening environments. Special consideration should be given to process controls, nondestructive testing, and destructive sampling of the solder bonds.

6.4.3 High G Forces and Shock Loads

We have just discussed how their low mass and low profile make SMCs more vibration resistant than equivalent-type THTs. These same properties work to SMT's benefit in high G-force and shock-loading environments. And, the same engineering solutions work to ensure survival in high acceleration or impact loadings. Table 6-4 and Figure 6-13 provide data that can be used to calculate shock and G-force resistance.

6.4.4 Connector Cycling

In Section 6.2, we mentioned the frequent cycling of connectors as a major factor in connector selection. The repeated operation of connectors can easily break SMT solder connections, where the stronger through-hole bonds of inserted connectors would comfortably survive. SMT connectors are particularly at risk when they are large, have a high pin count, are without mechanical attachment, or are not of low- or zero-insertion-force design. See Section 6.2 and Chapter 2, Section 3 for details on engineering solutions to maintaining connector integrity in high cycling environments.

6.4.5 Frequent Card Handling

The frequent insertion and removal of SMT cards raises concerns about connector-bonding and handling-induced bending of the card. Threats to the SMT assembly increase as the component and card size increases, and also as the card stiffness decreases. See Section 6.3 for a full discussion of card flexure.

6.4.6 Radiation Resistance

Harsh radiation environments are encountered in space, in nuclear reactor design, and in radiation equipment. While we pray the feature is never needed, we design military electronics to survive such radioactivity as well.

It is well known and understood that radiation can destroy or impair the function of semiconductors. Devices that depend on junction isolation rather than on a dielectric or the silicon-on-insulator isolation of on-chip elements are particularly at risk.[17] So Step One in engineering for high-radiation environments is the selection of components with rad-hard silicon inside. Figure 6-13 shows the radiation resistance of various silicon semiconductors compared to carbon units.

While selection of rad-hard silicon is vital, it is not the totality of designing for radioactivity survival. Shielding, grounding, and decoupling must all be designed to accommodate the worst-case rad dosages that the system may see. The system should be designed with error correction. And, in cases where periodic high-intensity radiation is a possibility, the design should provide for a controlled shutdown of the system to protect sensitive areas.[18]

Radiation in the form of high-energy cosmic rays or electrons traveling in the Earth's magnetic field is typical of the space environment. Such radioactive particles can induce ionizing radiation in circuits, momentarily imparting enough energy to change the state of a bistable element. Unless special rad-hard technology is used, CMOS devices are particularly sensitive to this effect, known as a "single event upset error." Such events may result in a momentary or a permanent failure.[19]

Radioactive-particle bombardment can also displace the atomic structure of semiconductor and electronic-material latices. "Displacement damage," as this defect is called, can temporarily or permanently alter semiconductor performance.[20]

Large dosages of x-rays produce electromagnetic pulses (EMPs) in system transmission lines and interconnections. Unchecked, EMP can fuse semiconductor metallizations and can permanently damage or destroy device operation.[21]

Solar ultraviolet radiation is also capable of degrading electronic assemblies. An example is found in the electric power meter which resides (in a glass dome) on the back of your house. Power-meter circuit designers derate the solder tensile strength of the circuits by 50 percent for the long-term effects of ultraviolet radiation.

6.4.7   High Operating Temperatures

A high-temperature operating environment complicates thermal management. Elevated ambient temperatures, coupled with temperature rises from device dissipation, can also significantly weaken solder joints. Figure 6-14 shows the fall off in strength of 50/50 tin/lead solder compared to the higher-melting pure tin. Data on other alloys, including lead-free alloys, are available from solder paste manufacturers. Users must determine this data during the design phase if their product will be used or stored at elevated temperatures.

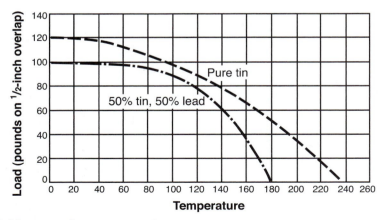

Figure 6-14   Solder strength versus operating temperature.

Tall components are of concern where they might impede the flow of cooling air around SMCs mounted close to the board. Locate high-dissipation components where they will receive a full flow of cooling air.

6.4.8 Humidity

High humidity raises concerns about surface conductance between the close leads of SMCs and dendritic growth, corrosion, and penetration of moisture into devices. Use of pads-only surface layers reduces the PWB concerns, but it does not eliminate them. And pads-only layers provide no protection above the board. For severe conditions, conformal coating is recommended. Table 5-6 in Chapter 5 lists the performance characteristics for various coating materials.

When designing for humid environments, particular care should be given to metals combined in the assembly. Metals with a significant difference in electromotive potential between themselves or the solder joining them are prone to corrosion. The electromotive potential of common electronic materials is shown in Table 6-5.

Table 6-5   Properties of Common Electronics Metals.

	Property			
Metal	Electromotive Potential (V)	Resistivity (@ 20°C μΩ/cm)	TCE PPM/°C	Tinsel Mod. (lb/in² × 10⁶)
Aluminum	+1.67	11.5	25.74	10.0
Antimony	−0.10	39.0*	11.34	11.3
Beryllium	+1.70	106.8*	12.42	4.6
Bismuth	−0.20	119.0*	13.32	4.6
Cadmium	+0.40	3.43*	29.88	3.0
Copper	−0.34	1.726§	16.38	16.0
Gold	−1.68	2.19*	14.40	12.0
Indium	+0.38	8.37*	32.40	11.0
Lead	+0.13	20.65	28.80	2.6
Nickel	+0.25	6.84	13.68	30.0
Palladium	−0.82	10.8	11.88	17.0
Platinum	−0.86	9.83*	7.74	21.0
Silver	−0.80	1.59	18.90	11.0
Stainless Steel	−0.09	74.0	16.56	29.0
Steel	−0.58	18.0	12.06	Dependent on Alloy
Tin	+0.14	11.5	23.40	6.0

* At 0°C.

§ At 23°C.

6.4.9   Reactive Atmospheres

Some materials present in atmospheric environments are particularly threatening to electronic assemblies. Threats may range from the muddy, salt-spray-filled air around shipboard and military gear, to the highly acid atmospheres in some chemical processing plants. Such atmospheres are particularly difficult to control when high humidity and high temperatures are also part of the potential environment. Atmospheric contaminants are dealt with by the choice of materials, conformal coatings, and protective hermetic housings.

6.5   Review

After reading and understanding this chapter, you should be able to answer these questions and note the reference location for the information in the chapter.

1. Why does Horton recommend partitioning the board into "as few" segments as possible?
2. What is the ITRS, and why is it important to PCB designers?
3. What are the ITRS projections for on-chip frequencies?
4. What is an MCM?
5. Briefly describe an example of SIP technology.
6. What types of components can be embedded in a PCB design?
7. Comparing SMT and THT connectors, which would you expect to be more vibration resistant?
8. What is one problem when considering the use of reflow soldering with THT connectors?
9. What sort of problems can board flexure cause?
10. Briefly describe the differences in THT and SMT thermal cycling problems.

6.6   References

1. Ball, Michael, "SMT: A Demanding Master." *EDN News,* July 16, 1987, p. 62.
2. "ITRS 2003 Edition." Available at http://public.itrs.net/Files/2003ITRS.
3. Isaac, R.D., "The Future of CMOS Technology." *IBM Journal of Research Development,* v. 44 #3, May 2000. Available at http://www.research.ibm.com/journal/rd/443/isaac.pdf.
4. "Z-series 900." Available at http://www.lugs.ch/lib/vortraege/Linux_auf_S390.pdf.
5. Ibid.
6. Doerre, G. W., Lackey, D. E., "The IBM ASIC/SoC Methodology." *IBM Research Journal,* v. 46 #6, 2002. Available at http://www.research.ibm.com.
7. Lerner, S., and Truzzi, C., "The Move Towards System-in-a-Package Solutions." *Semiconductor Fabtech,* 11th edition. Available at http://www.semiconductorfabtech.com/journals/edition.11.
8. "System in a Package (SIP)." Available at http://www.semiconfareast.com/sip.htm.
9. Snogren, R., "When to Embed." *Printed Circuit Design and Manufacture,* v. 21 #2, February 2004, pp. 40-42.

10. Harper, Charles A., Editor, *Handbook of Wiring, Cabling, and Interconnecting for Electronics,* McGraw-Hill Book Co., New York, NY, 1972, pp. 8–18.

11. D'Sousza, L., "Embedded Passive, RF Design." *Advanced Packaging,* November 2004, pp. 34-38.

12. DESIGN STANDARD FOR THICK FILM MULTILAYER HYBRID CIRCUITS IPC D-859. IPC, Evanston, IL, 1989.

13. Aspandiar, R., Litkie, M., Arigotti, G., "Pin in Paste Solder Process Development." Proceedings of the 1999 SMTAI Conference. Available at http://www.smta.org/files/SMTAI99-AspandiarRaigy.pdf.

14. Oberg, Erik, et al., Ed. Henry H. Ryffel, *Machinery's Handbook,* 27th Edition, Industrial Press, New York, NY, 2004.

15. Mishra, R., Keimasi, M., Das, G., "The Temperature Rating of Electronic Parts." *Electronics Cooling,* v. 10 #1, February 2004, pp. 20-26.

16. Manko, H. H., *Solders and Soldering,* McGraw-Hill Book Co., New York, NY, 1979.

17. *Rad-Hard/Hi-Rel CICD Data Book,* Harris Custom Integrated Circuit Div., Palm Bay, FL, 1987, pp. 1–8.

18. Pearson, Bob, "Designing Low-Power, Rad-Hardened Satellite Systems," *EDN,* August 21, 1986.

19. Andrews, J. L., Schroeder, J. E., Gingerich, B. L., Kolanski, W. A., Koga, R., and Diehl, S. E., "Single Event Upset Error Immune CMOS RAM," *IEEE Transactions on Nuclear Science,* Vol. NS-29, December 1982, pp. 2040–2043.

20. Srour, J., "Basic Mechanisms of Radiation Effects on Electronic Materials, Devices, and Integrated Circuits," *Proceedings of the IEEE Annual Conference on Nuclear and Space Radiation Effects,* 1982.

21. Ibid.

chapter 7

Quality Assurance in SMT

Objectives

This chapter concerns itself with quality and testing of printed circuit assemblies (PCAs). After reading and understanding this chapter the reader should be able to:

- understand the necessity for maintaining quality assembly techniques
- select incoming inspection techniques for components
- assure reliability of components and materials used for assembly
- assure quality and reliability of PCAs through appropriate testing techniques
- understand appropriate testing techniques

Introduction

Surface mounting is not just another way to slap parts into assemblies. This is true in all facets of SMT, but nowhere is it more glaringly obvious than in quality control and quality assurance (QC/QA). What have always been interesting slogans, "Design for Manufacturability" and "Design for Test," have become a way of thinking for companies trying to achieve good SMT manufacturing yields. As a result, phenomena that were seldom considered in THT manufacturing are of high importance in SMT manufacturing. For instance, solder-joint integrity and intermetallic growth are emphasized. Processes that are poorly engineered or controlled will serve to launch this chapter's topic, *Quality Assurance in SMT*.

One of the great prophets of *statistical quality control (SQC)*, W. Edward Deming, has noted that both Japanese and American companies implementing SQC have repeatedly proven its value as follows:

- a tenfold to hundredfold (or more) reduction in defects
- a tenfold improvement in manufacturing cycle time
- inventory is reduced to one half the previous levels
- manufacturing floor space is cut by one half[1]

It is not our intention to offer a course on SQC herein. We do want to note, however, how vital SQC is to surface-mount manufacturing. Many of the tricks that worked to cover sloppy process controls in through-hole assembly will result in enormous disasters on an SMT line. Successful SMT usage demands that you have your SMT processes under control.

We should also note that it is also not our intention to recommend a specific quality technique above all others. There are many quality techniques available and, from both our readings and our experience, many of them will work IF all areas of the company from top management to the secretarial staff believe in them and implement them. Failing to "believe" leads to failure of any quality technique.

In the following discussion, we will cover areas of specific interest to SMT quality and reliability. In an earlier chapter we mentioned that "quality is a management decision." The SMT team should be charged by management to consider techniques in "quality control for surface mount" as part of the assignment to integrate surface-mount technology into a manufacturing organization. Included in the topics should be Design for Experiments (DOE), statistical process control (SPC), Taguchi, and others. The team members most responsible for quality should decide the appropriate quality techniques to use, and make certain that all other members of the team understand the quality processes to be used, the justification for the processes chosen, and how the function of each member of the team is valuable to the overall quality of the product. One of the better books in the quality arena is *World Class Quality— Using Design of Experiments to Make it Happen* (2nd edition), by Keki Bhote and Adi Bhote.[2] The first edition of this book was written while Keki Bhote was head of quality for Motorola's Automotive and Industrial Electronics Division, so many of the examples involve discussion of electronics manufacturing technologies.

7.1   Component Reliability and Quality Assurance

How can we assure SMT component quality and reliability? First, we must understand and agree on what these terms mean. Such agreement does not come without effort. To illustrate, let's look at a perfect example where a confusion with terms caused trouble. A major corporation called in a high-power QC consultant to find why its quality program was not yielding the desired results. The consultant found when he asked various people "What is Quality?," he heard a different answer from each individual surveyed. One person thought it was conformance to specifications. Another saw it as defect-free manufacturing. Some believed it was long-term operation without a failure. No wonder the quality program was disappointing when the people implementing the program had not come to a basic agreement on the meaning of its most critical word.[3] As Deming and others have said, quality must start with management, and the first of management's responsibilities is to be certain that all players are using the same rules!

Being forewarned, let's begin by stating what this text means by quality and what is meant when we say reliability. For our purposes:

- A quality electronic product is one which, at the time of test immediately after manufacture, works per specifications when it is turned on.
- A reliable product is one that continues to work to specification during its intended lifetime.

Given these definitions, let's look now at how SMCs stand up in quality and reliability when compared to through-hole devices.

7.1.1   How SMCs Compare with DIP Quality

The drive to push component quality defect-density into the low PPM range was fueled, in part, by SMT developments such as tape-and-reel packaging. End users who had been content to buy 1100 ICs to have 1000 live ones could no longer afford such wasteful practices. Components in tape cannot be conveniently cycled through a receiving test. And the cost of correcting failures caused by 10 percent defective components would be enormous. Typically, costs of correcting defects escalate about tenfold, as shown in Table 7-1, with each successive step in the manufacturing process.

Table 7-1   Cost of Correcting Component Defect versus Detection Point.

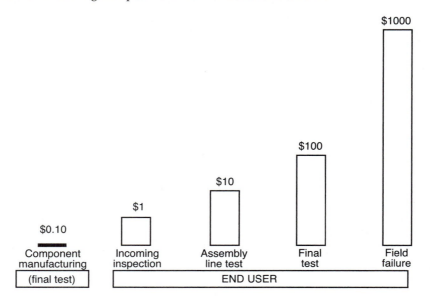

The answer, of course, is to work with suppliers to be certain that the required quality levels exist when components are shipped rather than waiting until failures show up in the manufacturing processes. In today's competitive markets, component suppliers today are willing to work with their customers to reach a specified quality goal.

7.1.2 How SMCs Compare with DIP Reliability

Since SMCs are so much smaller than DIPs, one of the first questions engineers unfamiliar with SMCs raise is *reliability*. Doesn't it make sense that a major reduction in plastic encapsulant material spells less protection for the semiconductor die? The answer, as simple logic would lead us to guess, is a qualified yes.

Table 7-2 shows the results of early testing conducted by Texas Instruments Incorporated on PLCC, SOIC, and DIP packages housing both bipolar and linear ICs. These tests, selected from MIL-STD-883 and industry standards, are of particular interest because they are directed at package integrity.

At first glance, the data in Table 7-2 are somewhat amazing. How could SMCs be more reliable than DIPs? In answer to the need for high-integrity small-outline packaging, semiconductor manufacturers introduced new encapsulant and new lead frame materials for SMCs. However, the most dramatic reliability improvements have been in the semiconductor die passivation (protective glass covering) itself.

Additionally, TCE issues have improved with SMT. Epoxy encapsulants expand with heating at about three times the rate of silicon. Lead frame materials in DIP packages are of materials like Kovar and alloy 42, where lead frame material in most SMT packages is copper alloy. The SMT lead frame material expands at nearly the rate of epoxy. TCE mismatches between materials in IC packages set up internal stresses in the package. Although the SMC and DIP TCE differentials are similar, the SMC has less internal stress because of its smaller absolute size.[4]

Table 7-2 Comparison of Package Reliability.†

| Test | Failure Rates | | | Units |
	SOIC	*PLCC*	*DIP*	
Life Test, 125°C	2	2	5	Fits*—60% UCL
85°C/85% R.H.	0.6	0.4	0.5	Percentage/1000 Hours
Autoclave	0.3	0.0	0.5	Percentage/240 Hours
T/C, −65/+150°C	0.03	0.2	0.5	Percentage/1000 Cycles

*Derated to 55°C, assuming eV activation energy.
†Hutchins, Dr. Charles, and Ganden, Howard, "Pitting SMT Against Standard Mounting," *Electronic Engineering Times*, October 19, 1987, p. T8.

7.1.3   Assurance of Component Quality

Receiving Inspection Electrical Test

In Section 7.1.1, we established that surface-mount components are available in qualities matching that of through-hole devices. We mentioned that tape reeling of SMCs was a factor forcing quality improvements. It is inefficient to inspect 100 percent devices supplied in tape and reel. For incoming tests, relatively robust packages, such as 1.27-mm (50-mil) PLCCs and SOICs, may be handled stick to stick like DIPs. If such handling is intended, Purchasing should be sure to obtain components in plastic tubes, not tape reels. This minimizes both the handling and purchase costs of components.

For more fragile components, there are special handlers for contacting devices in protective shipping carriers. While carriers are more expensive than tape reel and may afford little more in shipping protection, they allow test contacting of a fully protected device. The considerable risk of lead damage in the receiving-inspection handling of dereeled parts and the high cost of coplanarity loss may well outweigh the cost premium for carriers over tape.

Clearly, where practical, the most cost-effective strategy is to use no incoming verification. Simply purchase high-quality parts sealed in protective antistatic tape, and review the vendor's quality procedures, not the parts. This strategy is made all the more effective when vendor SPC data are available electronically. The handling involved in just passing 150 PPM components through a test handler may add the same number of failures as it weeds out.[5]

Subjective Solderability Testing

A quick and simple to perform solderability test is the dip/visual inspection. This test requires nothing more exotic than a solder pot filled with 63/37 solder (or equal), a pair of tweezers of nonsolderable material, and a stereoscopic microscope having between 10- and 20-power magnification. The SMCs are dipped as shown in Figure 7-1.

After a 10-second immersion in solder at 240°C, the parts are observed under magnification. Any visual dewetting is a concern. A greater than 5 percent nonwetted area of termination

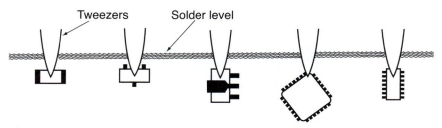

NOTE: Shaded areas are points for inspection.

Figure 7-1   A simple immersion wetting test, using various types of components.

is certainly unacceptable. Figure 7-2 compares three leads and shows the relative degrees of nonwetting.

Good wetting is shown in Figure 7-2A. The tinning is nearly defect free. There are no areas of congregated defects. It is 100 percent wetted. In contrast, Figure 7-2B shows a cause for concern. There is visible dewetting. This may be acceptable, as dewetting affects less than 5 percent of the termination area. Figure 7-2C illustrates a condition that is clearly unacceptable. Over 5 percent of the termination area is dewetted. Further information on wetting tests can be found in IPC J-STD-002, "Solderability Tests for Component Leads, Terminations, Lugs, Terminals and Wires," and IPC J-STD-003, "Solderability Tests for Printed Boards."

Solder surface tension and cohesive forces can mask considerable areas of poor wettability. Boards should always be inspected after the soldering process, especially hot-air solder leveling (HASL). For high yields, any dewetting on chip components or boards is a cause for concern.

Dewetting of the tips of SO and gull-wing carrier leads is not of great concern. Dewetting lead tips indicates the ends of the lead-frame material that were left untreated for solderability after lead-frame trimming. Other areas are of varying significance and should be considered in the context of Class 1 (general consumer), Class 2, or Class 3 (high reliability) products. Again, the pertinent IPC documents will define this. Figure 7-3 diagrams the areas of a gull-wing lead and indicates the level of concern for solderability of each distinct region.

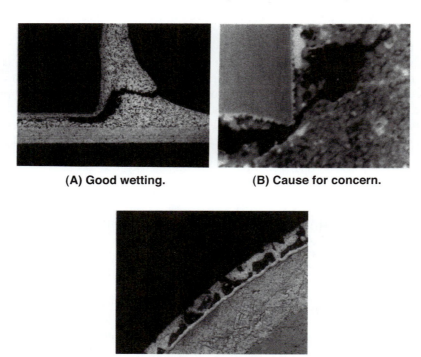

(A) Good wetting.   **(B) Cause for concern.**

(C) Clearly unacceptable.

Figure 7-2   Visual keys to the solderability immersion test. *(Courtesy Signetics Corp.)*

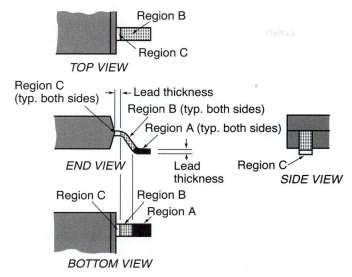

Figure 7-3   Wetting concern varies by regions—Gull-Wing Leads.

Region A, in the bottom and side views of Figure 7-3,[6] includes the side faces of the lead to within a lead thickness of the body and the bottom of the lead upward around the bend to a height equal to the lead thickness (the heel). This area is critical to solder-joint strength. It must be fully wetted with a smooth, shiny coating of solder. A small amount of scattered imperfections in the smooth surface may be accepted. Imperfections concentrated in one area are a cause for rejection.

Region B covers the upper sides of the lead, the bottom of the lead above the heel, and top of the lead—all to within the lead thickness of the body. Region B is not so critical to solder-joint strength as is Region A. However, Region B provides a large exposed and visible area for use in the observation of lead solderability. Any significant dewetting of this area may indicate a solderability problem and should be investigated. Dewetting may be produced by random contamination, fingerprints, overly oxidized lead finishes, or improperly finished leads. A physical and chemical laboratory analysis can determine the nature of soils causing solderability problems, and such information will often point to adjustments in the handling and processing procedures, which bear great fruit for process yield improvement.

Region C, as shown in Figure 7-3, includes the noncoated cut end of the lead and the area near the body where the lead may be contaminated with mold flash. Solder wetting is not required in these areas.

Objective Solderability Testing

MIL-STD-202, Method 208 and MIL-STD-883, Method 2022 are the commonly applied tests of solderability for DIPs. However, these tests are not appropriate for leaded and leadless SMCs. Martin Marietta has developed a method that defines procedures for proven wetting balance testing of SMDs.

Purely objective solderability testing, in lieu of the simple but more subjective immersion test, is needed where documented results must be available for after-the-fact inspection. The Martin Marietta test is a solderability test using a modified wetting balance (Meniscograph), a vibration-free environment, and precision tooling to measure solder-pot insertion and withdrawal force and solder-pot weight gain/loss.[7]

7.1.4   Assurance of Component Reliability

Earlier, we considered the problem of assuring component quality. To build SMT products that work the first time, we must have components that work the first time, or we face enormous troubleshooting and rework costs. A closely related but more challenging problem is the assurance that the components will continue to work properly for the intended life of the product. Fortunately, the tried and true THT QA technologies can be adapted to surface mount. Next, we will investigate the adaptations and look at strategies for SMT component-reliability assurance.

Burn-In Testing

An electrical test of components operating in elevated ambient conditions, *burn-in testing*, will weed out most of the infant mortalities in incoming components. Burn-in sockets are available for all major SMC package families. Testing may be performed using standard burn-in equipment with SMC sockets. Dependent on the burn-in board and the placement equipment available, burn-in board loading may be automated using assembly pick-and-place systems.

Destructive Physical Analysis

Destructive physical analysis (DPA) is a powerful tool for quality and reliability assurance. A sampling of components can be sectioned. Photomicrographs, SEM pictures, spectrographic material analysis, etc., can show potential problems and point to their source long before systems start failing in the field.

Electrostatic Discharge (ESD)

In Chapter 4, Section 9, we covered the need for ESD protection and the potential savings that a well planned static protection program may provide. But which components really need protection? Table 7-3 lists the electrical over-stress sensitivity of typical families of devices. Most IC

Table 7-3   Semiconductor Static Susceptibility.

Type of Semiconductor	Danger Level/Volt
CMOS	250/2000
Diode, Schottky	300/2500
ECL Hybrid (PWB level)	~500
EPROMs	100/300
GaAs FET	100/300
MOSFET	100/200
Op Amps	190/2500
SCR	680/1000
Transistors, Bipolar	380/7000
TTL Schottky, LS and ALS	100/200

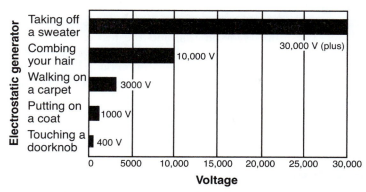

Figure 7-4   Static potentials developed by various ESD sources.

components are ESD stress susceptible, and therefore all should be treated as sensitive in order to develop proper handling as a way of life. It is false economy to save a few pennies on a resistor by buying it in unprotected tape. While the resistor may be quite static tolerant, the cover tape being dereeled from it may zap a $35.00 ASIC next door in the assembly operation.

Compare the data from Table 7-3 with the electrical potentials generated by typical sources of electrostatic charge in dry environments, as shown in Figure 7-4,[8] and you can see why quality experts raise so much static about ESD. Data are available from many journals, ESD protection suppliers, and consultants for static-control program help.

7.2   Printed Wiring Board Quality and Reliability

We have just covered the quality and reliability of electronic components. Ensuring the quality and reliability of printed wiring boards is equally important. Bare boards are normally tested for three factors. These are isolation (detecting the fault, shorts), continuity (detecting the fault, opens), and contamination. We will now discuss quality and reliability assurance for bare boards.

7.2.1   Printed Wiring Board Quality

Bare-board testing is generally required today for products of any significant complexity, and is so recommended for SMT boards. Bare-board fault distribution is, in order of occurrence, as follows:

1. Shorts (most often the result of slivers)
2. Opens (often from overetch)
3. Cuts and holes
4. Excessive leakage (features too close)
5. Contamination.[9]

A good testing program should be designed to intercept all these flaws.

Bed of Nails Testing

Bed of nails testing may be performed at low voltages to find shorts and full opens. The test will detect defects on all layers of multi-layer boards, and the test can readily be performed on boards with test-pad grids of 1.27 mm (50 mils) and greater. However, test costs are not insignificant for 2.54-mm (100-mil) grid boards, and costs are in inverse relation to grid size. For grids below 1.27 mm (50 mils), test costs are high, and there are a limited number of vendors capable of supplying test services.

High-stimulus voltages and high isolation-resistance resolutions allow bed of nails detection of all five of the faults listed above, but the cost is significantly greater than for a simple continuity check.

Automated Optical Inspection

Automated optical inspection (AOI) is useful in finding opens and some shorts. While it is not as certain as bed of nails probing for these purposes, AOI carries the added benefit of detecting design rule violations, such as traces too close to features or holes out-of-concentricity-specification with annular rings. For multi-layer boards, the PWB manufacturer may apply AOI to each layer before lamination and do a continuity check on a final bed of nails probe to find shorts and gross opens only. This can reduce bed of nails test costs while delivering very high-quality boards.

X-Ray Inspection

X-ray inspection is well adapted to verifying the internal alignment of layers in multi-layer boards. Such inspection forms a powerful tool when combined with AOI and bare-board continuity/isolation testing.

Visual Inspection

Visual inspection under low-power magnification can be used to detect PWB flaws, such as delamination, measling, etc. However, visual inspection alone is far from perfect at catching other flaws, such as opens and shorts. For SMT boards, visual testing of board coplanarity should be specified.

Cleanliness

The cleanliness of PWBs should be monitored throughout the manufacturing process. The final testing for contamination is generally a 100-megohm-resolution isolation-resistance test.[10]

Destructive Testing

The forcing of board failure by stress may detect latent defects before they become field problems. DPA also allows a board manufacturer to monitor the processes and ensure that they stay in control.

The IPC provides standard test patterns for single-sided, double-sided, and multi-layer boards. By testing standard test patterns as well as the actual product, test results may be compared to a large bank of statistical data, and variations between individual PWB suppliers may be determined.

7.2.2   Printed Wiring Board Reliability

Printed wiring board reliability is assured primarily by following proven design practices and using materials suited for the product's operating and storage environment. Short of monitoring a product over its lifetime, environmental testing is the most accurate method of proving whether a given design achieves the mark. Where environmental contaminants or dendritic growth under power are found to interfere with PWB reliability, conformal coatings can add the necessary protection to assure reliability.

7.3   Quality and Reliability of Materials and Supplies

Many materials enter into the electronic assembly reliability equation. Space will simply not permit a full discussion herein. Also, individual variations between commodities are clouded by marketing claims and are confused by constant product innovations. Therefore, we will stick to items of a generic and established nature.

7.3.1   Solder Mask Quality and Reliability

Solder mask reliability is specified in IPC-SM-840A for Class 1 (Consumer), Class 2 (Industrial and Computer), and Class 3 (High Reliability) boards. Table 7-4 compares Class 2 and Class 3 requirements.

As we have previously stated, photoimagable solder masks are preferred for all but the coarsest SMT densities. Both dry-film and wet-film photoimagables are available. Properly applied, there does not appear to be a reliability bias for either wet or dry. However, wet films are said to be suitable for finer geometries than are dry films. To be reliable, there must be no openings around conductors produced by dry-film tenting around the raised conductor topography. Wet films must achieve 100 percent coverage of the raised features specified to be masked.

Table 7-4   IPC-SM-840A Specifications, Class 2 and Class 3 Solder Masks.

Test	Class 2 Requirements	Class 3 Requirements
Adhesion Over Copper	≤5% Loss	No loss allowed
Over Laminate	≤5% Loss	No loss allowed
Over Tin/Lead, Tin	≤50% Loss	≤10% Loss
Hydrolytic Stability	No degradation	No degradation
Insulation Resistance	1×10^8 ohms minimum	5×10^8 ohms minimum
Humidity and Insulation Resistance	1×10^8 ohms minimum	5×10^8 ohms minimum
Electromigration	Not applicable	None allowed
Dielectric Strength	500-V DC peak/mil	500-V DC peak/mil
Flammability	94V-1 or better	94V-1 or better
Chemical Resistance	No degradation	No degradation
Thermal Shock	Not applicable	No blistering, measling, crazing, or delamination

7.3.2   Solder Quality and Reliability

The most commonly used solders in SMT have traditionally been tin/lead combinations near the tin/lead eutectic. However, as of this writing, most component manufacturers are providing pure tin coatings, and assemblers are beginning to move to lead-free solder to comply with the EU's directives on reduction of lead and waste elimination.

As we have mentioned before, the changes are in process as of this writing, and this prevents us from making specific recommendations.

Solder Bar Stock

Quality assurance for solder bar, or slab, stock for flow soldering can be approached several ways.

By a simple wetting test on clean metal, we can determine the wettability of newly procured solder. The wetting test involves reflow of a small chunk of the solder material on a test coupon of clean metal. Provided the wetting test is preformed with controlled time and temperature on a perfectly clean smooth coupon of specified base metal, no flux is used. Thus, the flux activity level does not need to be considered. We look for a full, even spread of solder on the coupon and observe the dihedral angle of the sample with the substrate after cooling. If total wetting occurred, the angle, as shown in Figure 7-5,[11] should approach 0 degrees. If

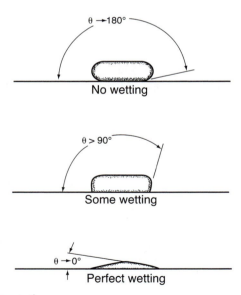

Figure 7-5   Wetting-test interpretation.

the angle is greater than 90 degrees, little wetting took place. An angle of 180 degrees (which could only be observed by sectioning the sample to see under the droplet) would indicate no wetting.

Interpreting the wetting test requires a foreknowledge of solder and coupon materials. With clean CDA copper finished to a reflective surface, the dihedral angle should be well under 90 degrees. A commonly set limit is no greater than 75 degrees, but experience in the given production environment should be the determining factor. A much lower limit, perhaps 15 degrees, might be called for in reflow soldering.

Materials analysis is used to detect potentially harmful contaminants in solder and also to flag incorrect alloys. Two particularly harmful contaminants in tin/lead solders are sulphur, which should be limited to no more than 0.001 percent, and phosphorus, which should not exceed 0.01 percent.[12]

The reliability assurance of solder materials is accomplished by a destructive analysis of actual solder joints. Such testing comes as a by-product of thermal cycling, vibration, shake/rattle/roll, and board-flexure tests which stress the whole assembly system.

Solder Pastes

In addition to the wetting test described above, solder paste quality assurance includes inspecting for flux activity, viscosity and rheology (critical to screening and dispensing), and shelf life.

The ceramic coupon test is a quick check for solder paste that bolsters the wetting test described above. A small dab of fresh paste is placed on a clean, smooth, unmetallized ceramic coupon and reflowed. Reflow should be at a correct time-temperature profile for the given paste. With good paste, virtually all the material should coalesce into one shiny, nearly round ball. Only samples passing the ceramic coupon test, as described in IPC/EIA J-STD-005 "Requirements for Soldering Pastes," should be used in SMT. Any significant solder balling of newly received paste is unacceptable.

Viscosity is tested automatically using a viscometer, a device resembling a cake mixer, which measures the resistance to stirring the paste. Viscosity must be monitored throughout the use of the paste since it is subject to change due to the evaporation of volatiles in the paste.

The rheology of the paste is a function of its viscosity, thixatropic nature, and solder particle size. Proper rheology is necessary for screen, stencil, or dispenser operation. The simplest of tests for determining screenability is to deposit paste samples in the assembly pattern using the assembly application method. Results should show uniform depths of deposit with clean edge definition of the patterns and full coverage of the print area. Unfortunately, this test, while simple, is highly subjective.

For more quantifiable testing, a uniform pattern may be generated. The pattern in Figure 7-6 serves to numerically specify stencil print quality. The *print quality number (PQN)* of a paste sample is defined by the average of the three smallest features reproduced with good quality.

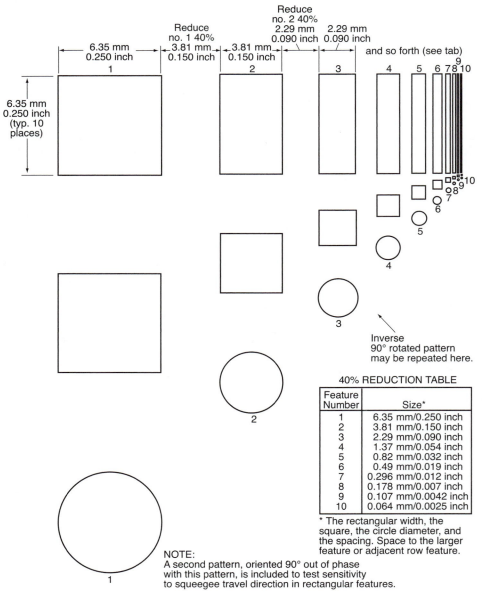

Feature Number	Size*
1	6.35 mm/0.250 inch
2	3.81 mm/0.150 inch
3	2.29 mm/0.090 inch
4	1.37 mm/0.054 inch
5	0.82 mm/0.032 inch
6	0.49 mm/0.019 inch
7	0.296 mm/0.012 inch
8	0.178 mm/0.007 inch
9	0.107 mm/0.0042 inch
10	0.064 mm/0.0025 inch

40% REDUCTION TABLE

* The rectangular width, the square, the circle diameter, and the spacing. Space to the larger feature or adjacent row feature.

Inverse 90° rotated pattern may be repeated here.

NOTE: A second pattern, oriented 90° out of phase with this pattern, is included to test sensitivity to squeegee travel direction in rectangular features.

Figure 7-6   Stencil-print test pattern.

By good quality we mean uniform depths of deposit equal to the stencil thickness, clean edge definition of pattern, no bleeding between patterns, and full coverage of the feature area. Inspection under 10X magnification will show where this is achieved. The results of this test may be related to the print quality required for a given assembly job, and a minimum number can be set for inspection to verify paste printability.

Paste shelf life is a function of material properties and solder particle size. Spherical particles of nearly uniform size minimize the surface area for a given metals content and a specified particle size. Elongated particles, or excessive particle size variation resulting in a large number of "fines," will both reduce screen printing quality and increase the surface area, which can increase the rate of paste oxidation. However, some pastes prove to print better with the inclusion of a few "long" and a moderate number of "fines" to increase the angle of repose of the paste particles. High-power microscopic, or SEM, inspection will allow particle inspection.

7.3.3   Adhesive Quality and Reliability

Adhesives are used for a wide range of electronics assembly applications. Most are covered sufficiently in texts on adhesive technology. However, the unique application of adhesives to Type 3 SMT does bear some discussion.

Single Part

With a single part adhesive system, our primary concerns are its dispensing qualities, its curing ability, and its properties once cured.

An adhesive's dispensing quality is influenced by material viscosity and rheology, and these properties are tested using the methods described in Section 7.3.2 for solder paste. Also, adhesives must be protected from whatever it is that initiates their curing process if they are to retain their dispensing properties over time.

How thoroughly a material polymerizes can be assessed by dispensing a measured drop and curing it in the production curing process. The test is most valid when conducted with worst-case production parameters. In other words, if a U-V curing epoxy must cure with only one third (nominal) of the dot exposed beyond the component, try curing it with one fifth dot exposure (see Chapter 3, Figure 3-17). After curing, component pull testing and sectioning will yield data on how thoroughly the material polymerized.

The strength of adhesives can be tested at room temperature with tensile and shear testers. We also must verify the high-temperature strength of the adhesive. The most critical function of the material will be to hold a component on board during a trip through molten solder. After it has succeeded in that, the adhesive could just as well go away. In fact, engineers and chemists have been working on the development of an adhesive system that would do just that in the cleaning process. The reason is that the adhesive residues after assembly present both reliability and rework concern. Reliability issues center around the adhesive outgassing, or remaining partially tacky, and trapping contaminants on the board. Rework and repairability demand that the adhesive not have such a tenacious grip that, at reflow temperature, it prohibits component removal or damages the board during component removal.

Two Part

In addition to all the issues presented above for single-part systems, Receiving must ensure that the right two parts are on hand for a two-part system. In an application with assembly line equipment, does the material catalyze and cure properly?

Thermoset

With thermally activated adhesives, quality control includes a determination that the specified time/temperature profile will fully cure the material. For reliability, storage and manufacturing environmental temperatures are an issue.

U-V Activated

In addition to all of the preceding, U-V activated adhesives must be protected from environmental U-V radiation during storage and use. The comments regarding environmental temperatures for thermoset adhesives apply also to U-V systems since their curing process is dramatically accelerated at elevated temperatures.

7.3.4 Socket Quality and Reliability

Where sockets are required for production boards, both leaded and LLCC sockets of proven reliability are available. Sockets are available that have passed MIL-STD-1344A, Method 2005.1 vibration testing and Method 1003.1 thermal cycling from -50°C to +125°C. Users should recognize that any use of sockets adds to reliability issues, since all sockets have known failure mechanisms such as oxide growth and spring termination issues.

7.4 Test and Process Control in Manufacturing

Since we are dealing primarily with design in this book, we cannot delve too deeply into process controls and other areas of assembly and manufacturing. It is useful, however, for each team member to have a basic idea of how the manufacturing processes are monitored and know what test methods are available to ensure the processes are in control.

7.4.1 Inspection of Materials at Receiving

Section 7.1, Section 7.2, and Section 7.3 lay out the rules for quality assurance for incoming components, boards, and materials. Specific tooling in receiving inspection equipment has been tailored to suit SMT demands, but the basic inspection methods used in THT manufacturing apply to surface-mount processes as well.

The difficulty of testing tape-reeled SMT components is perhaps the greatest difference actually impacting the philosophy of surface-mount receiving inspection. This, plus a drive for Just In Time (JIT) manufacturing and successes with supplier partnerships in quality, is combining to push component quality responsibility to the component manufacturer and out of the assembly front end.

7.4.2   Storage of Materials

In Chapter 6, Section 4, we covered environmental concerns and materials. We further discussed ESD concerns in Section 7.1.4. ESD sensitivity of SMCs has not been shown to be significantly different from that of THTs. However, several storage and handling differences with surface-mount devices are worth noting. These are covered below.

Coplanarity

DIPs were forgiving bugs in the hands of warehouse personnel. Their leads can be slightly bent during handling and still work without a glitch on the factory floor. Even if the leads were bent beyond allowable limits, the parts are readily salvageable with a lead straightener. SMCs advantages bring with them requirements for less careless handling. New habits may need to be learned by all who touch or move fragile surface-mount ICs. Maintaining coplanarity is a necessity, and proper handling will prevent issues that will affect production yields.

The basic way to maintain required lead coplanarity is to buy good parts and then teach all the people who might ever handle them (including engineers who rummage through the stock room) how fragile they are and how much to respect them. Particular notice should be paid to the handling of fine-pitch parts. Semiconductor manufacturers and component users alike employ scanning equipment to weed out nonplanar parts. Many placement machines have the ability to scan for acceptable coplanarity of ICs.

Lead fragility has pushed some designers to reject PROM burning in SMC sockets for fear of losing coplanarity and yields. On-board PROM programming, use of E²PROM, and other on-board techniques will solve this problem.

Solderability and Moisture

Solderability of component leads has always been a concern in electronics manufacturing, but DIPs with poor solderability often yielded to aggressive fluxes, high solder temperatures, and multiple passes through flow soldering. None of these "band-aid" procedures fit surface-mount products. Parts must be solderable or they cannot be used on a surface-mount line. This places a limit on how long SMCs should be stored and suggests storage in a cool low-humidity environment. It also implies that solderability testing be undertaken for the surface-mount process if it was not already in place for IMC.

Moisture absorption by plastic component bodies presents another storage concern. As discussed in Chapter 2, entrapped moisture can cause components to rupture in a soldering process. Low humidity or desiccant storage is a preventive. SMT parts, with their great need for coplanarity and solderability, are often shipped in desiccated bags with ESD protection and mechanical packaging to prevent lead damage. Opening the seal on the bag negates the effect of the desiccant in preserving lead solderability and preventing moisture absorption. The "open time" limitations on the sealed shipping bag must be observed and, in the case of Class 6 products, baking must be done on the floor just prior to assembly and reflow.

Kitting

Organizing the parts to build a dozen through-hole boards was a minor chore. Discrete components were available in bulk, and ICs came about twenty to a stick. If you didn't need one full stick of ICs, you could take the required number out and stuff their pins down into the conductive foam. Voilá, a kit. With the neat reel packaging common to surface mount, kitting can be more of a problem. Unused components in reels must be restocked, and the stockroom will want to know how many are still in the reels when they come back. Since some may have been wasted during the assembly operations, just deducting the "bill of materials" requirement is not necessarily going to answer that question. Equipment is available that counts and respools partial reels.

7.4.3   Solder Paste and Adhesive Application

The basic issues governing control and inspection of attachment media are the same whether the media is adhesive or solder paste. Therefore, we will cover both solder paste and adhesives together, issue by issue.

Material Concerns

Whatever the method of media application, the attachment material properties must be within limits or the application process will be out of control. QC and QA testing of attachment materials are described in Section 7.3.2 and Section 7.3.3.

Dispenser Control

All dispensing operations share one element in control. Dispensing must occur within a predictable distance from the board surface. If we were dispensing on a granite surface plate, this would be simple. It is a little more difficult on a warped and bouncy printed wiring board. We must either put the board into a known planar location or sense the true relation of the board to the dispensing mechanism.

Other areas of dispenser control are specific to dispenser style. Dispensers may generically be divided into two categories, the time/pressure type and the positive displacement type.

The time/pressure type is controlled by monitoring the pressure applied to the media and how long that pressure is applied. Some systems simply apply pressure to a syringe to dispense and then shut off the pressure to end the dispense cycle. Others use a brief reverse suction to reduce stringing. Still others time the dispense operation with a pinch valve near the nozzle. In any case, time/pressure dispensing is not a closed-loop system. The desired action, dispensing of a specified volume of material, is inferred from the fact that a given pressure for a given time dispensed that amount in previous trials with the same equipment and nozzle size. Obviously, any change in material viscosity or rheology or reduction of nozzle opening (clogging) will invalidate this assumption. Therefore, after-the-fact verification of time/pressure dispensing is wise. (See the discussion on "Inspecting Attachment Media Application" given at the end of this section for details on verification methods.) This is particularly true for the time/pressure dispensing of solder pastes because solder pastes are prone to a change of material properties under dispensing pressures. Pastes may (as noted in Chapter 3, Section 3.1) separate and become increasingly high in metals content. When this condition occurs, clogging of the dispensing nozzle will result.

The positive-displacement method of dispensing uses low pressures and large line sizes to feed media to a screw pump that dispenses metered volumes of material on demand. Since the screw is turned by a stepping motor, this method approaches closed-loop operation. We say "approaches" because it is still possible that a glitch might occur in the feed line to the piston pump. It is also remotely possible that the dispensed material might escape from a rupture in the nozzle or pump assembly and not really be going onto the board. Experience teaches that verification of media supply to the pump is generally adequate to verify high-yield dispensing. Where this is provided in dispensing equipment, after-the-fact verification of dispensing may be relegated to visual spot checks rather than an automated scanning of each operation.

Transfer Printing Control

Transfer printing involves the laying out of a metered depth of media on a smooth surface, and then picking up a dot of media on a transfer pin and offset printing it from the pin to a desired location. Accurate transfer printing is established by a complex balance of material rheology, affinity of the media for the metered layer surface, pin surface, and printing site surface, and the printing speed and pressure. Regarding printing pressure, remember our earlier comments about the nonplanar and unpredictable surface topography on which we are printing. The metered depth of media is usually established by a doctor blade being swept across the surface of the media before each print. As long as an adequate flood of proper rheology material remains before the blade, the results should not vary.

Screening and Stenciling Control

Screen and stencil printing of media are done on the same piece of equipment. Only the setup and pattern mask are varied. Thus, many of the process controls are identical. We will discuss the two topics together, noting where they differ.

Snap-off height refers to the distance between the relaxed mask and the material to be printed, as illustrated in Figure 7-7. For screen printing, this distance usually is between 0.25 mm (0.010 inch) and 0.635 mm (0.025 inch) and is specified by the printer manufacturer. It can be measured by a feeler. Because the stencil mask will not easily deflect, stencil printing is generally done with a zero snap-off, or "on contact." However, some shops mount stencils on open-center, highly tensioned polyester screens to allow a minimal snap-off.

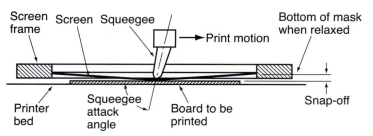

Figure 7-7 Snap-off height defined.

Squeegee material also has an impact on printing quality. The printer manufacturer can suggest durometers that work well for specific applications. Ranges from 30 to 90 durometer are common. Fortunately, the effect of durometer changes is minimal and does not usually need monitoring. The point is to make the proper choice in the beginning. The blade tip of the squeegee will wear during printing. The rate of wear is influenced by printing pressure, blade hardness, tip shape, and mask material. Worn blades should be periodically restored as a part of routine maintenance.

Squeegee attack angle refers to the angle of the squeegee to the screen, as shown in Figure 7-7. This angle is adjustable on most printers and should be set between 45 degrees and 90 degrees based upon the manufacturer's recommendations. Attack angle does not vary unless the adjustment is changed, so it is not normally monitored as part of the process control.

Squeegee speed across the print area is important to the print quality and may be monitored as part of the process control. Process control might also include verification of a full sweep of the squeegee and flood bar strokes across the mask. (The flood bar is a secondary squeegee that wipes paste across the screen between each print without applying printing pressures.)

Squeegee pressure against the screen is adjusted on the screener and should be periodically checked due to changes caused by blade wear.

Paste rheology has a major influence on printing quality. Environmental temperature and humidity variation must be controlled to keep rheology variations predictable.

The screen mesh size must be selected to suit the paste used and the minimum feature size to be screened. Clearly, the openings in the screen mesh must be larger than the solder particles in the paste. In fact, to avoid any chance of clogging, the screen openings should be large enough for at least three average-size paste particles (the 3X rule) to pass through simultaneously, as shown in Figure 7-8. Where more than one mesh size fits this rule, the finer wire mesh

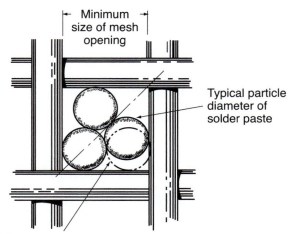

NOTE: For ease of estimating, two times the average ball diameter gives a very close approximation of the 3:1 rule.

Figure 7-8   Defining the 3:1 particle-to-mesh rule.

Table 7-5 Paste Particle and Screen Mesh Dimensions.

Paste Specifications		304 Stainless-Steel Screen Specifications				
Mesh	Ball Diameter**	Mesh	Mean Size	Open Area (%)	Mesh Thickness	Minimum Print†
100	150 micron (0.0059 inch)	80	269.2 micron (0.0106 inch)	70.6	106.7 micron (0.0042 inch)	76.2 micron (0.0030 inch)
200	75 micron (0.0029 inch)	80	223.5 micron (0.0088 inch)	49.6	203.2 micron (0.0080 inch)	101.6 micron (0.0040 inch)
325	45 micron (0.0018 inch)	105	165.1 micron (0.0065 inch)	46.9	177.8 micron (0.0070 inch)	83.8 micron (0.0033 inch)
400	38 micron (0.0015 inch)	150	104.1 micron (0.0041 inch)	37.2	157.5 micron (0.0062 inch)	58.4 micron (0.0023 inch)

* Screen mesh data courtesy IRI Division of BMC, Gardena, CA.
** Less than 1% balls greater, 90% this to next smaller size. (ANSI/IPC-SP-819, *Solder Paste for Surface-Mount Applications*, IPC, Lincolnwood, IL.)
† Print thickness with no emulsion extension past mesh.

will generally reproduce finer features, and the coarser wire screen will last longer. Table 7-5 shows the dimensions for typical paste particles and screen meshes, and pairs them to accommodate the 3X rule.

Screen mesh to frame orientation is important, particularly where small feature sizes must be printed. As Figure 7-9 illustrates, mesh wires are oriented on the same axis as the screen frame, and features may partially obscure fine-print openings. Mounting the screen mesh at 45 degrees to the frame avoids any chance of this happening.

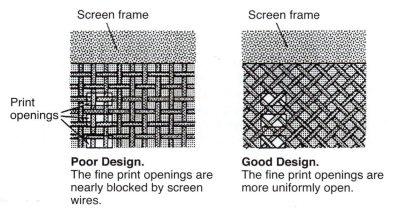

Poor Design.
The fine print openings are nearly blocked by screen wires.

Good Design.
The fine print openings are more uniformly open.

Figure 7-9 Screen mesh orientation.

Emulsion thickness is the main factor in determining print thickness for screened solder paste. The print thickness of paste is controlled for several reasons. First, too little paste will cause an increase in opens from reflow soldering. Second, too much paste increases risks of bridging and tombstoning defects. Also, overly stiff joints caused by too much solder are more prone to stress-related failures in thermal cycling, vibration, etc. For standard-pitch SMCs, wet-paste thickness is generally controlled between 0.15 mm (0.006") to 0.25 mm (0.010"). The lower number is preferred for all leadless assemblies, while the thicker paste deposit reduces opens when using leaded parts with coplanarity problems. For fine-pitch components, paste deposits generally range between 0.05 mm (0.002") to 0.15 mm (0.006").

Texas Instruments suggests the following formula to determine the emulsion thickness required for a desired wet-print thickness.[13]

$$Et = Pt - (Mt \cdot Ao)$$

where,

Et is the emulsion thickness (beyond mesh),
Pt is the desired print thickness (wet paste),
Mt is the mesh thickness ($>2 \times$ mesh filament diameter),
Ao is the proportion of open area in the screen.

Using this formula, if a print thickness of 0.20 mm (0.008") is desired and we are using a standard 80-mesh screen with 0.09398-mm (0.0037") diameter filaments, a mesh woven thickness of 0.2032 mm (0.0080"), and the screen has 49.6 percent open area, we achieve, in millimeters,

$$Et = 0.20 \text{ mm} - (0.2032 \times 0.496)$$
$$= 0.09931 \text{ mm (about 4 mils)}$$

To avoid having to calculate print thickness with the formula, we have included theoretical minimum wet-print thicknesses for the screen meshes listed in Table 7-5. By simply adding the emulsion extension thickness past the mesh to the figures given under "Minimum Print," you derive the actual print thickness for a given emulsion extension.

In practice, some emulsion extension below the bottom (print side) of the screen is needed by some screen manufacturers to smooth the rough surface of the weave. Typical rules of thumb call for 12 to 25 microns (0.0005" to 0.001") of emulsion past the bottom surface. However, Frank Greenway, a research scientist for Advance International of Chicago, reports good results down to 9 microns (0.0004"). And, IRI reports a process yielding "zero" emulsion extension.

The mesh characteristics, as defined in the preceding formula, plus the minimum required bottom-side emulsion extension, determine the minimum wet thickness that a given screen material will accurately serve to print. Within limits, an emulsion thickness past the screen-mesh outline adds wet-paste thickness at a 1:1 ratio, as the formula indicates.

Stencil thickness is a bit easier to deduce. If the desired wet-paste thickness is 200 microns (0.008"), use stock of that thickness to construct the stencil. Print thickness should follow a 1:1 relationship with stencil thickness.

Inspecting the Attachment-Media Application

Whether automated or semi-automated printers are applying attachment media, a visual inspection by an automated optical inspection (AOI) system or the printer operator is necessary for on-line verification. Since solder-related problems typically account for around 70 percent of in-process quality issues, inspection must be done at this point. AOI systems will use camera- or laser-based systems to check for print volume (and are typically set to do spot-checks. That is, not check every board but rather check features based on the expected failures.) If a certain area on the board is known to have printing problems, only that area is checked. If the main failure is plugging of the stencil, periodic boards will be checked.

With operator inspection, pad coverage can be verified, and wet-print height is checked periodically using a height gauge available from various paste manufacturers. Like the AOI systems, operator inspection may only be made on certain board areas or of the entire board periodically.

Using either type of inspection, print failures may include:

- Slumping of paste, an indication of poor paste rheology.
- Excess paste, shown by either slumping of the paste or excess height.
- Bridging between pads.
- Insufficient volume on a given pad, shown by too low a height or an unprinted area of the pad.
- Skipped pads, sometimes caused by excess pad height as a result of hot air solder leveling. It can be exacerbated by high rubber squeegee pressure.
- Misaligned print, where the paste deposits do not match the outline of the pads.

Once errors are determined, they must be corrected. All the possible printing errors and their corrections are beyond what we can discuss here. Printer manufacturers and paste suppliers can provide suggestions to minimize and correct these errors.

7.4.4   Component Placement and Insertion

Component placement and component insertion actions are monitored and inspected as stated in the following discussions.

Board Registration

Boards are registered for automated assembly operations using mechanical pins, flat surfaces against board edges, or a combination of mechanical and optical schemes. Pins usually register the boards by engaging tooling holes in the board's periphery. However, for cases such as ceramic thick-film boards, two reference edges abutting the artwork zero point for the board are oriented against stop pins. Typical registration methods are as shown in Figure 7-10.

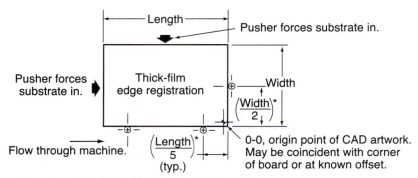

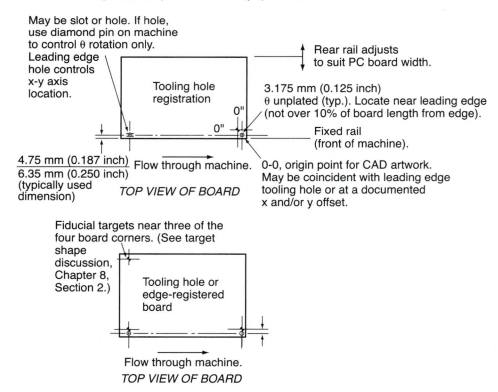

Figure 7-10 Typical mechanical and optical registration methods.

Feature Registration

In insertion machines, pin registration of the tooling holes has long been the standard. For surface-mount technology, however, tooling hole registration alone is only accurate enough for very coarse assembly tasks. Placing chip carriers (particularly high pin-count and fine-pitch parts) requires a more accurate location of the features on the board. Tooling holes and pins, or reference edges, may be used to establish a rough position of the board; then, AOI equipment is used to accurately locate optical targets on the board and interpolate component land orientations. Sophisticated placement systems can automatically adjust their placement routine to suit routine manufacturing tolerance variations detected by such AOI equipment.

For the extreme accuracy required to place high pin-count, fine-pitch parts, AOI is used to inspect both the component leads and the component lands, and then actively target one to the other. Such process controls can also be used to flag damaged leads and incorrect form factor components and lands.

A special application of board-registration methods is required to handle very small boards, flexible substrates, and irregular-shaped substrates. In these instances, one or more substrates are mounted on a tooling plate. Pin or edge registration is used to locate the tooling plate, and pin or edge registration is also used to position the substrate(s) on the tooling plate. AOI may still be used to orient optical targets after their rough position is established mechanically.

Component Inspection

Some insertion and placement equipment incorporates a limited component-testing capability, performing a quick test before the assembly operation. For DIPs, this test is usually a simple short or open test between the V_{SS} and V_{DD} pins of an IC. Both SMT and THT passives may be verified by RLC measurement. The purpose is to guard against backward ICs and diodes and to detect setup errors where an incorrect value discrete has been loaded on the machine.

Placement-Action Monitoring

Surface-mount placement equipment may monitor important parameters during operation to verify its own behavior, check the validity of the setup, and ensure certain types of conformance of assembly parts.

Sensors may detect the arrival and verify the proper orientation of a board. Bar-code readers may even verify that the correct board part number has arrived, or provide information on the assembly part number to automatically govern the machine assembly actions.

Sensors can also measure the Z-axis excursion during the placement stroke and the Z-axis placement pressure applied. Thus, overly warped boards, failures to contact the board, and collisions with unexpected obstructions can be flagged.

Smart placement heads can provide feedback from the centering jaws, verifying the presence and size of a component. Failure to pick up a component can initiate a recovery sequence. Repeated failures can trigger an empty feeder alarm, cause the machine to switch to an alternate feeder location, or initiate machine shutdown. Component size data can be compared to a lookup table to see if it matches expectations for the desired component.

Optical inspection may verify the features of a picked component and can compare these with the data stored in memory on a known good sample. Barcode scanners can verify the location of a feeder and the markings on tape reels during machine loading.

Placement Verification

After the assembly operation, AOI systems are available to verify placement operations. At the low end, these systems perform a simple check to see that all programmed sites are populated. More sophisticated equipment may measure component sizes and placement offsets, flag out-of-tolerance conditions in both areas, find extraneous components and debris on the assembly, and even provide some confidence that the right component is in each location. AOI systems may also provide valuable data to guide the repair of errors they detect.

7.4.5   Curing of Media

Thermal Processes

Solder paste curing (when needed for vapor phase processes) and thermoset-adhesive curing are generally done in a linear-belt IR furnace where both the belt speed and temperature are monitored. Temperature measurements are usually made by means of a thermocouple located in close proximity to the travel path of the workpieces. Note that this does not measure workpiece temperature. For instance, a small furnace could be set at 250°C, but boards passing through it at 100 meters per minute would come out cool to the touch. Boards going through at $1/10$ meter per minute might well be overheated. What we need to monitor is the time/temperature profile that the workpiece encounters.

To actually measure the time/temperature profile, thermocouples may be inserted in a range of points across the board, as will be discussed in the next section. Then the board is processed at a given set of parameters to characterize the furnace for the workpiece. This is not practical for an on-line measurement, however. Instead, process control consists of monitoring the furnace settings and the workpiece characteristics to see that no significant changes occur. Loosely controlled time/temperature profiles are sufficient for solder paste curing. Closer control may be necessary for certain adhesives. Curing parameters will be indicated by the material supplier.

UV Curing Processes

The curing of ultraviolet-activated adhesives is generally monitored by spot checking the curing operation results. Often, elevated temperatures are used to accelerate UV curing. Where temperature is part of the equation, it is controlled as described in the preceding paragraph.

7.4.6   Reflow and Flow Soldering

Reflow Soldering

Reflow oven processes are primarily monitored periodically to maintain process control and quality. Thermocouple-monitored boards are sent through the furnace to characterize the setup.

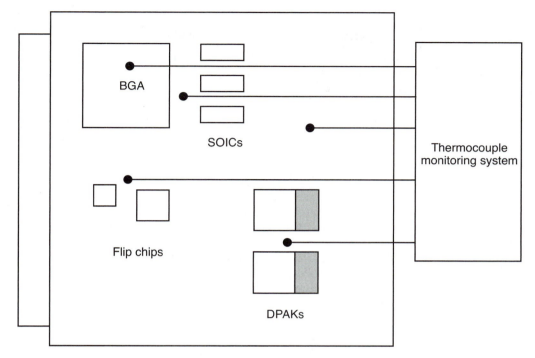

Figure 7-11   Examples of monitoring points with PWB thermocouples.

Special instrumented boards, like the one shown in Figure 7-11, with monitoring systems can ride through the furnace with the test board, simplifying the handling of thermocouple wires in linear ovens. Note that a variety of areas on the board are monitored, including:

- Under BGAs
- In lightly populated areas of the board
- In high thermal mass areas such as the DPAKs
- Around sensitive components such as the flip chips

After characterization, the temperature of each zone is automatically controlled by the oven's systems, and the belt speed is maintained to ensure the desired time/temperature profile.

Although advancing sensor technology is being focused on the task of on-line measurement of the actual workpiece temperature in the reflow furnace, most furnaces today still rely on close process control of the atmospheric temperature which leads the user to monitor actual board surface temperature.

Vapor-phase oven controls ensure that enough heat is introduced to maintain an adequate vapor blanket for the work load, and they monitor the time that the work is in the vapor blanket. For batch systems, the elevator descent and ascent are programmed, and its operation is verified to ensure proper timing. For linear systems, belt speed controls the time at temperature. Additional controls may monitor contaminants in the system and cycle the machine through its filtration process periodically. The control of water flow in the cooling coils may be used, particularly in batch systems, to restrict primary and secondary vapor-blanket height and also prevent vapor escape.

Convective reflow ovens have fewer issues in the control of the process. The major issue to be determined is whether to use inert atmospheres or to reflow in air. The solder paste manufacturer is the best source of initial guidance on the issue surrounding nitrogen reflow.

Flow Soldering

Flow soldering surface-mount devices usually involves passing boards across a high-velocity or agitated wave, or a jet of solder, to ensure application of solder to all the leads. In many cases, this operation is followed by a second pass across a wave having a smooth low-velocity laminar flow. The second wave remelts and removes the excess solder left by the high-velocity/high-pressure wave.

System controls include the liquid-fluxer temperature and operation, the preheat panel or emitter temperature, the solder temperature, the height and velocity of solder wave(s) or jets, board travel speed, and the airflow to the exhaust. Air knives are used to remove any solder bridges not cleared by the laminar wave. Air-knife air pressure, air temperature, flow rate, distance to work, and attack angle are all controlled to maximize this effect.

7.4.7   Cleaning

Batch Solvent Cleaning

Batch-style solvent cleaners, or vapor degreasers, are not widely used in SMT because of their slow cleaning action under low standoff SMCs.

The controls of degreasers include vapor-blanket condensers and spray-wand pressure. Some systems also provide a hot fluid bath with controlled temperature. Fluid baths may be ultrasonically agitated to dramatically improve tight-space cleaning speeds. Where ultrasonics are incorporated, both frequency and amplitude should be tightly monitored to ensure the safety of the ICs. Frequencies near an IC's resonance, or excessive amplitudes, can damage or destroy IC wire bonds within the devices.

Linear Solvent Cleaning

Linear solvent systems transport assemblies through either one or a series of vapor blankets, spray zones, and heated fluid baths. Like batch systems, immersion sections may be ultrasonically agitated to enhance cleaning under tightly spaced SMCs. Conveyor belt speed is controlled to produce the correct length of cleaning cycle.

Vapor-section controls include both vapor-blanket height condensers and vapor-generation heater temperature.

Spray controls monitor the pressure on both the top- and bottom-side spray heads. (Top-side pressures are set slightly higher than the bottom side to prevent blowing boards off the belt.) Spray nozzles set the type of spray that each generates. The nozzles may be individually angled in some systems to maximize their effectiveness. The fluid temperature to the nozzles is also controlled.

Batch Aqueous Cleaning

Batch aqueous cleaners use a combination of saponified water-spray/-immersion and deionized water rinses to clean assemblies. Controls monitor the wash and rinse water temperatures, the spray pressures, and the saponifier mix.

Linear Aqueous Cleaning

Linear aqueous cleaners use immersions and sprays of saponified water washes and deionized water rinses to clean. Controls include the wash, spray, and rinse temperatures, the spray pressures, the saponifier levels, and the conveyor belt speed.

7.5   Testing Finished SMT Assemblies

We have touched on "test" previously, both in our discussion of the manufacturing process in Chapter 3 and again in the intervening chapters on design. Developing testable designs should be one of the SMT team's top priorities, and so the topic has been seeded throughout our study of SMT design. Here, in our QC and QA chapter, testing will receive our full attention. Below, we have categorized tests used in surface-mount (and IMC) manufacturing, and we present data on each. Some of the following material is a review of Chapter 3, but the importance of SMT testability convinced us that we should cover the topic both under manufacturing and under QC/QA.

7.5.1   Functional Test

Functional testing looks at the circuit as a complete circuit. Functional testers access the circuit using assembly I/O, such as edge connectors or cabling, sometimes known as the field connectors. These are the connections that will be used by the board in normal operation in its intended product configuration. For example, at the final test station for a laptop computer mother board, the board may be plugged into and tested with:

- a display and an external monitor.
- a modem card and/or a wireless card.
- a hard drive.
- a CD drive.
- its intended software.

This type of testing maximizes the probability of finding any functional the board may have. If testing is to isolate a fault and furnish detailed troubleshooting guidance, then further testing may be needed if the board fails functional test. If probe access must be limited, active partitioning and built-in test may reduce the costs of troubleshooting on the bench.

Functional test is often coupled with other testing. Efficiency may come from using functional test only for those areas of circuit operation not verifiable in other tests. Less programming-intensive test methods may then be applied to detect manufacturing defects, weed out component failures, and provide troubleshooting guidance.

Figure 7-12 shows an SMT board with a double-sided bed of nails being used for test access. Full access to the device under test is a boon in both in-circuit and functional test. The design in Figure 7-12 minimizes wiring runs.

7.5.2 In-Circuit Tests

The ideal target of an in-circuit test is to individually isolate and parametrically test each and every circuit element. To do this, the I/O of every circuit element must be fully addressable for the probes in a bed of nails fixture. And, every circuit element must be completely in-circuit testable. The rule of thumb for this is that every circuit node must have a test point.

High-density designs often work against both these requirements. The close I/O pin spacing of SMCs place difficult demands on probe card design. One solution is to route the traces to test points on 2.54-mm (0.100") centers, but this wastes valuable real estate. High-density designs with small component clearances, tight track spacings, and hidden/buried vias further complicate test. And, bottom-side populations interfere with access to the necessary circuit nodes. Finally, SMT is often chosen because it facilitates the very high pin counts needed for VLSI custom ICs. But, these complex devices often cannot be fully tested in circuit.

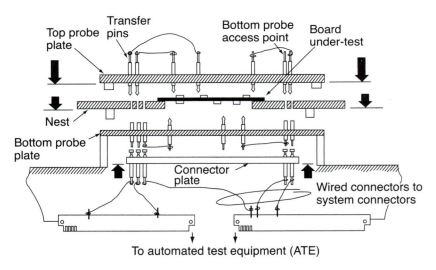

Figure 7-12   Test fixture provides two-sided access.

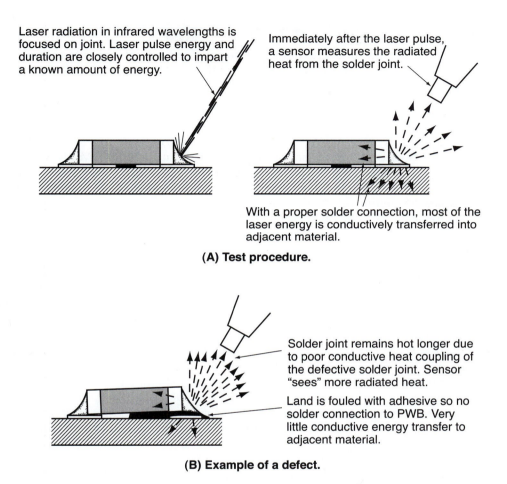

Laser radiation in infrared wavelengths is focused on joint. Laser pulse energy and duration are closely controlled to impart a known amount of energy.

Immediately after the laser pulse, a sensor measures the radiated heat from the solder joint.

With a proper solder connection, most of the laser energy is conductively transferred into adjacent material.

(A) Test procedure.

Solder joint remains hot longer due to poor conductive heat coupling of the defective solder joint. Sensor "sees" more radiated heat.

Land is fouled with adhesive so no solder connection to PWB. Very little conductive energy transfer to adjacent material.

(B) Example of a defect.

Figure 7-13   Laser thermometry used to inspect SMT solder joints.

The major benefit of in-circuit testing is in finding all of the component—as well as the manufacturing—defects. Where it can do this, in-circuit test provides guidance to the repair of defects without expensive troubleshooting by technicians. However, this benefit is limited to those designs allowing in-circuit analysis of a major portion of the circuit's elements. The value of the test data decreases as the number of elements that cannot be in-circuit tested increases.

7.5.3   The Functional/In-Circuit Combinational Test

Some testers can combine functional and in-circuit modes. The strategy here is to provide one-shot fault isolation via an in-circuit test, and then add in the advantages of functional test in the detecting of timing-critical faults, the testing complex VLSIs, etc.

7.5.4   Manufacturing Defects Analysis

Manufacturing defects analyzers (MDAs) provide a quick test for shorts, opens, and reversed polarity components. Boards under test are probed on a bed of nails fixture. MDAs give rapid feedback on solder opens and bridges, components loaded backwards, and missing components. But, MDAs do not confirm that the correct components were used or that those components are operational. Therefore, MDA testing is usually used to quickly flag certain production difficulties. It is often coupled with other testing to fully qualify the assemblies.

7.5.5   Burn-In Test

Burn-in testing has different meanings in different circles. Some companies call a simple on-line operation of a system, without elevated temperature, a burn-in. If it does not begin to smoke during that period and keeps performing its specified function(s), it passes burn-in. For our purposes here, burn-in means somewhat more than that. Burn-in "involves the monitored operation of the assembly or system at elevated temperatures for a defined period of time. Sophisticated burn-in tests can spot time of failure and provide significant detail regarding the failure mode when out-of-specification operation occurs." The exact level of detail that can be extracted from a burn-in test is, like the other test methods mentioned earlier, dependent on access to all the necessary circuit nodes.

By stressing circuits under operation, assembly and system burn-in tests catch infant mortalities. Infant mortalities are circuits that work immediately after manufacture but contain a defect that will cause failure early in the circuit's life. Once such infant mortalities are weeded from the circuit population, the remaining circuits predictably survive to a ripe old age. Therefore, eliminating infant mortalities by operation under temperature stress greatly enhances circuit reliability. Since some such failures occur only intermittently under stress, constant monitoring during burn-in is necessary for high-confidence-level testing. Full-board or system burn-in will also catch failures instigated by heat buildup. Without such testing, these failures would not occur until the system was in the field under power at elevated ambient temperatures.

7.5.6   Assembly Integrity Tests

Mockup Tests

Mockup testing provides a quick verification of circuit operation to some level of its specified requirements. Mockup tests may be a simple powering up of a system to see if it "comes on".

Mockup testing does not require a high capital investment, nor an extensive test programming development time. It provides a low-cost, fast-response method to find dead-on-arrival circuits.

Most gross manufacturing errors, such as a wrong component, a backward component, a missing component, and shorts and opens, will mean a dead-on-arrival circuit. Therefore, mockup testing provides a fast and low-cost way of monitoring the manufacturing process. However, mockup testing is best applied to testing systems with only a few operating modes and a limited number of response patterns. Complex systems, with many permutations of responses for various stimuli, would take an enormous amount of mockup test time to fully qualify. Desktop computer users can well imagine that their machine might power up without fault but fail to properly access a certain disk or respond to a certain command. Trying out everything the machine is specified to do would take days. Mockup testing also produces very little in troubleshooting guidance.

X-Ray Analysis

X-ray systems may be used to spot hidden defects, such as voids in solder, delaminations in PWB material, cracked components, BGA ball soldering problems, and broken internal wire bonds in IC packages. Both standard x-ray and laminography systems are available. The standard x-ray system is similar to that used to check for a broken arm—it shows an image completely through all layers of the material. The laminography system is much more expensive and slower but allows the operator to visualize different levels of the image, e.g., at the level of the pad or at the level where the ball attaches to the BGA body.

Laser Solder Inspection

Laser thermometry uses thermal signature analysis as an inspection method for solder joints. The system directs infrared laser radiation at an individual solder joint, rapidly heating the joint. The laser is then turned off, and the inspection system monitors radiated heat from the joint. Good solder joints provide a solid thermal path into the board, so the built-up temperature quickly dissipates. Defective joints are flagged because they retain heat longer. Like x-rays, laser thermometry finds both visible and hidden solder flaws. For instance, the defective solder joint of Figure 7-13B produced a dramatically different signature than the acceptable joint illustrated in Figure 7-13A.

7.5.7   Nondestructive Physical Analysis

Thermographic Analysis

Infrared photographs can display hot spots on operating circuits and can flag trouble before smoke signals make the trouble all too obvious. Where temperature limits are critical to a board's performance, online testing can be done by operators watching television monitors (displaying the IR results) or by automated inspection techniques.

Cleanliness Testing

Conductometric testing is the standard of THT PW board cleanliness testing. Cleanliness studies of SMT assemblies indicate that new methods, as discussed earlier in Chapter 3, Section 3.1.5, will be needed in the cleanliness testing of SMT assemblies.

Vision Testing

Just about every inspection line includes at least a simple eyeball inspection of the boards. A simple operator inspection, with or without magnification, is used to detect missing or grossly misaligned components. Then, microscope inspection may be directed at the solder joint quality and fine alignment. Sophisticated high-volume lines may employ AOI or x-ray systems for fast objective testing. Visible-light AOI equipment, with complex accept/reject algorithms, will rapidly detect missing and misaligned components and certain solder defects.

7.5.8   Destructive Physical Analysis

Solder Strength

Shear- and tensile-strength testers are available for the destructive testing of solder joints. A solder tensile-strength tester can pull a solder joint apart to record the stress levels required to rupture the joint.

Board Flexure

Flexure tests involve the repeated controlled bending of boards to determine the solder-joint and assembly resistance to recurring mechanical stress. The test may be conducted on home-built equipment, as diagrammed in Figure 7-14.

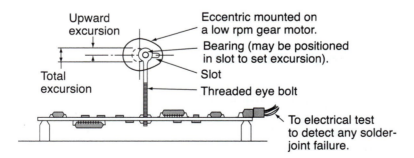

Figure 7-14   A board flexure test apparatus.

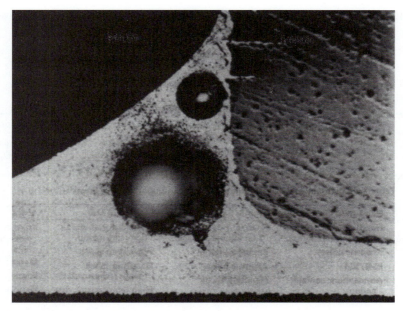

Figure 7-15   Gas entrapment void revealed by microsectioning.

Microsectioning

Sections taken through solder joints, boards, and components can provide valuable information on failure mechanisms. Figure 7-15 shows large voids produced by gas entrapment during the reflow process. The culprit here was probably inadequate paste curing. Undetected, such outwardly good-looking solder joints, might inexplicably fail in service.

Environmental Aging

Environmental aging may be produced by putting circuits in high-temperature, high-humidity environments, such as 85/85, by extended burn-in, by radiation, etc. Environmental aging tests are designed to accelerate the real-life environmental impact to circuits in order to verify, within a reasonable time, the survivability of designs for the specified life of the product.

MIL-STD-810 lists shipping, storage, and operating environments for a number of categories of military products. Environmental ruggedness requirements for military hardware are admittedly stringent and may constitute overkill for some commercial applications. However, the basic environmental concerns for military equipment engineering are very instructive in guiding any high-reliability design effort. These issues are listed in Table 7-6, Table 7-7, and Table 7-8.

Table 7-6   Environmental Hazard Summary of Shipping and Transportation Stress.*

Truck Transport	Rail Transport	Air Transport	Sea Transport
Road shock from bumps/potholes	Road shock from bumping	In-flight vibration (engine/turbine)	Wave-induced vibration (sinusoidal)
Road vibration (random vibration)	Rail vibration	Landing shock	Wave slam shock
Handling shock (drops/overturn)	Handling shock (drops/overturn)	Handling shock (drops/overturn)	Handling shock (drops/overturn)
High temperature, dry and humid	High temperature, dry and humid	Reduced pressure	High temperature, dry and humid
Low temperature	Low temperature		Low temperature
Rain/hail	Rain/hail		Rain/temporary immersion
Sand/dust	Sand/dust		Salt fog

*MIL-STD-810, *Military Standard Environmental Test Methods and Engineering Guidelines,* July 19, 1983, p. 7.

Table 7-7   Environmental Hazard Summary of Storage and Supply Logistics Stress.

Handling Logistics	Storage in Shelter	Storage in Open
Road shock from bumps/potholes	High temperature, dry and humid	High temperature, dry and humid
Road vibration (random)	Low temperature/freezing	Low temperature/freezing
Salt fog/solar radiation	Salt fog	Rain/hail
Handling shock (drops/overturn)	Fungus growth	Sand/dust
High temperature, dry/humid	Chemical attack	Salt fog
Reduced pressure		Solar radiation
Low temperature/freezing		Fungus growth
Rain/hail/sand/dust		Chemical attack

Table 7-8   Environmental Hazard Summary of Operating Conditions Stress.

Hand-held Equipment	Mobile Equipment	Shipboard Use	Airborne Use
Handling shock (drops/ overturn, slamming)	Road/off-road vibration (bumps/treads)	Wave-induced vibration (sinusoidal)	Runway vibration (aerodynamic/turbine)
Acoustic noise	Engine vibration	Wave slam shock	Maneuver buffeting
EMI/RFI	Acoustic noise	Handling shock (drops/ overturn)	Engine vibration
Atmospheric contamination	Handling shock	High temperature, dry and humid	Handling shock
High temperature, dry and humid	High temperature, dry and humid	Acoustic noise	Aerodynamic environment (high temperature/ dry/humid)
Low temperature	Low temperature	Low temperature	Acoustic noise
Rain/hail/dust	Rain/hail/dust	Rain/hail/dust	Low temperature/thermal shock
Solar radiation	Solar radiation/salt fog	Solar radiation/salt fog	Rain/hail/dust
TCE storage to use	Fungus/chemicals	Fungus/chemicals	Solar radiation/salt fog
			Fungus/chemicals
			Reduced pressure

Thermal Cycling

Early research identified the TCE mismatch problem between large leadless ceramic components and polymer substrates as a major SMT reliability concern. Thermal cycling of assemblies provides a method of testing solutions to this problem and has become de rigueur in the engineering development of high-reliability SMT products.

Some engineers have asserted that 1000 cycles from –65 to +150°C is unrealistic for most product environments and that, indeed, THT boards will often fail the same test. We have no hard evidence in hand to show that THT assemblies typically fail rigorous thermal cycling tests. However, even if they do, the issues of solder-joint integrity and strength are so much more important to SMT than THT that the argument seems vacuous. Thermal cycling is a valid tool to ensure correctness of engineering solutions to TCE challenges.

7.6   How to Build Testability Into Designs

Many of the design guidelines for the testability of SMT boards mirror those for through-hole technology. However, SMT often moves tried-and-true practices from desirable to vitally important. In the following discussion, we will cover those areas that uniquely relate to surface mounting, and we will recap the basic rules made doubly valid by the new demands of SMT.

7.6.1 Basic Rules

The basic rules include:

1. Provide extra gates to control and back-drive clock circuits.
2. Insert extra gates or jumpers in feedback loops and where needed to control critical circuit paths.
3. Tie unused inputs to pull-up or pull-down resistors so that individual devices may be isolated and back-driven by ATE.
4. Provide a simple means of initializing all registers, flip-flops, counters, and state machines in the circuit.
5. Build testability into microprocessor-based boards wherever possible.

Design and test engineers can now use built-in test regularly in digital designs. Built-in test (BIT), Built-in self test (BIST), and boundary scan are variations on design and test techniques that reduce the need for test fixtures in the test process and correspondingly reduce the need for test points in the board layout. Testing is typically faster and board layout more flexible, which together more than justifies the additional design efforts.

Components for these techniques are available from many manufacturers using the test access port and scan gates defined by the IEEE 1149.1/JTAG standard.

7.6.2 In-Circuit Test and MDA

Probe-ability design guidelines are important where bare boards will be tested for opens and shorts, or where in-circuit or manufacturing defects analysis testers will be used on finished assemblies. Because THT component leads are on 2.45-mm (100-mil) centers, through-hole assemblies naturally adhere to a 2.45-mm (100-mil) grid for test points. Also, inherent in THT technology, every single circuit node is readily accessible from the solder side of the board. SMT does not automatically impose nodal access and generous probe grids. Therefore, for in-circuit and manufacturing defects analysis testing of SMT boards, be sure that every circuit node connects to at least one accessible test point. If at all possible, this test point should be probe-able from the back side of the board.

THT probe points are generally at the lead penetration on the solder side of the board. In SMT, leads may not be probed, and top-side component leads are not available on the bottom side of the board. In order to avoid costly double-side probing for tests, leads are generally connected to on-grid vias leading to the board's bottom side. When this is done, some method must be used to identify those vias that will be used as test points and to record the node to which they are connected. Test engineers will need this listing and the X-Y offset from tooling holes in order to develop a test fixture.

Since a 2.45-mm (100-mil) grid is not enforced by component lead spacing, we are free to select a test grid for SMT boards. Be aware that 1.27-mm (50-mil) on-center probe cards may cost two times as much per test point as 2.45-mm (100-mil) cards. That assumes single-side probing; costs escalate for dual-side fixtures, and, 1.27-mm (50-mil) center probes have a service life of about one half to one third that of the 2.45-mm (100-mil) variety. If you must probe assemblies

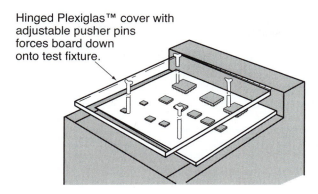

Hinged Plexiglas™ cover with
adjustable pusher pins
forces board down
onto test fixture.

Figure 7-16   Board test using a hold-down cover.

with test grids smaller than 1.27 mm (50 mil), specialized and exotic probe cards and micro manipulators may be required. Use a 2.45-mm (100-mil) grid for test vias wherever possible. Also, whenever possible, route all test points to the bottom side of the board, eliminating the need for double-side probing.

Note that device leads or terminations are not acceptable test probe points. The spring pressure of a test probe might cause an unsoldered or disturbed joint to make good contact for the duration of the test. Thus, a failed board may pass testing. Also, fragile spring-loaded test probes may be bent as they glance off poorly aligned SMCs.

Whenever vacuum fixturing is part of the board test strategy, all via holes should be solder or epoxy filled. See Chapter 5, Section 5 for further discussion of via filling. Figure 7-16 shows a mechanical top cover that is used by Hewlett-Packard's Manufacturing Test Division to replace the vacuum hold-down for an SMT board with unfilled vias. While this gets the job done, it adds cost and slows the production loading of the tester. It also excludes access to the board for functions such as mechanical cycling of switches during test.

7.6.3   Functional Testing

With double-side-populated boards where high densities prevail, it is often impossible to design for full probe-ability for in-circuit testing. In other cases, in-circuit testing may cost more than it saves. Circuits with a bus lend themselves to functional testing since devices may be individually addressed on the bus. Circuits with even a few devices that cannot be in-circuit tested are also candidates for a functional test.

The circuit design team should not consider any one form of testing as a quality-mandated issue. *Quality should always be designed in, not tested in.* In-circuit testers, functional testers, and manufacturing-defects analyzers are tools that may reduce the test and repair costs involved in producing a quality product. When they do not reduce the costs of producing a quality product,

Table 7-9   Advantages and Disadvantages of Various Test Methods.

Test Issue	Type of Test			
	Mockup	*MDA*	*Functional*	*In-Circuit*
Typical uses	Go/No go	Manufacturing defect detection	Performance verification	Manufacturing and component defect detection
Go/No go decision	Very fast	Slow	Fast	Slow
Fault sensitivity level	Manufacturing defect	Manufacturing defect	Manufacturing, component, and engineering defect	Manufacturing and component defect
Fault isolation level	System	Component	Node	Component
Multifault isolated	No	Yes	Maybe	Yes
Repair guidance	Poor	Good	Poor	Good
Programming costs	Low	Low to moderate	High	Moderately low
Timing sensitivity	Critical	Noncritical	Critical	Noncritical
Capital equipment cost	Low	Low	High	Medium

use some other QA strategy that will. Alternatives include a functional test and hot-shot mockup testing. Table 7-9 lists the general advantages and disadvantages of several test strategies.

In those instances where nearly 100 percent of the circuit components can be tested, in-circuit testing shines. In such cases, an in-circuit test provides a full defect analysis for repair. However, most complex VLSI components cannot be fully tested in-circuit, so boards with a large complement of VLSI devices are candidates for a functional test. The timing sensitivity of functional testers allows them to detect timing defects that in-circuit tests will not find.

Where functional test is the method of choice, circuit designers can minimize test and repair costs by providing for the probing of critical circuit nodes. Internal probing allows test engineers to monitor functional test results within the circuit as a signal propagates from device to device. Thus, with complete probe access, it is often possible to isolate specific faults in functional testing. Finding these defects would require considerable technician time if all access were from the edge connector.

Functional test is most commonly performed using the "field connectors" that the product will use in its final form. As an example, an automotive powertrain control computer would be plugged into a test set that would provide simulations of the inputs the computer would expect to see, such as engine temperature and mass air flow, and into outputs that would be expected from the computer, such as fuel injectors. The unit is then tested as though it were in its actual application.

While functional testing in this manner does tell whether the product will perform to specification in its final form, this type of testing gives very poor diagnostic information. In our automotive example, functional testing may report out that injector #3 is not opening for the proper duration, but may not be able to determine if the problem is in the computer software, a faulty MOSFET driver, or other portions of the supporting circuitry.

Table 7-10   Test Methods and Design for Testability.

Test Strategy	Pertinent Design Guidelines						
	Extra Gates	Tie-down	Initializing	On-Board Test	Probe-ability	Vacuum Sealing	Active Partition
Mock-up	N/A	CMOS	Maybe*	Yes	No	No	No
MDA	N/A	N/A	N/A	N/A	Yes	Yes	N/A
Functional	Yes	Yes	Yes*	Yes	Yes**	Yes**	Yes
In-circuit	Yes	Yes	Yes*	Maybe†	Yes	Yes	N/A

Our thanks to Ted Tracy of ATE Labs, Garden Grove, California, for his guidance in the construction of this table.

* Initializing circuits might be faster than forcing the initialization externally. In some cases, ICs can only be initialized by reset lines once they are installed in-circuit.

**To provide access to help partition the board diagnostically. Signals may be forced from the edge connectors, but probe monitoring helps track where the faults occurred.

† Use where on-board or chip testability solves the problem of the components being untestable via in-circuit test.

7.6.4   Test Strategy and Design

To simplify application of the preceding testability guidelines, we have condensed the rules into the outline provided in Table 7-10. Once you have selected a test method(s) for your product, this table will highlight pertinent design rules to simplify testing.

7.6.5   Boundary Scan

As mentioned in the introduction to this chapter, digital and microprocessor components using boundary scan to the IEEE 1149.1 (or JTAG) standard are readily available. The standard calls for:

- a test access port (TAP).
- a boundary cell at each logic input and output pin.
- a bypass register.
- a TAP controller.
- an IDCODE register.
- four or five extra pins on the devices for the I/O required to implement scan.

Texas Instruments provides an excellent introduction to scan in their publication "IEEE 1149.1 (JTAG) Testability Primer."[14] National Instruments also makes design for scan easy with their information at http://www. national.com/scan.

7.6.6   Design for Repairability

Perhaps the most important point in design for repair is to always be mindful that the average human (presumably the description of the repair person) cannot bend and attenuate their limbs like Gumby™. Discipline yourself for this by recalling the horror of changing a spark plug or oil filter on an option-loaded, big engine Detroit car.

Beyond allowing room for hands and tools, you can also simplify troubleshooting by providing a silkscreen legend identifying the components to the circuit diagram. Component markings help on larger components, but 0805s and SOTs do not provide room for much in descriptive nomenclature. Provided manufacturing will use tape-reeled components, it is worthwhile to specify markings on the resistors and capacitors. Resistors in particular must be assembled with the correct side up in order to be able to dissipate heat to their full wattage rating. If installed upside down, they will not. Also, locate all nomenclature so that it is readable when the board is on an extender or in the system. Even the best markings are difficult to read while looking into a dentist's mirror. Provide sufficient clearance around the components for test-clip attachment and for heat shielding from adjacent components during rework soldering.

7.7 Design for Reliability

All the design guidelines herein are aimed at production of reliable assemblies using surface-mount technology. The following suggestions go beyond the norm in that they call for cost or availability trade-offs. Consider the cost versus the reliability gain.

Leads were originally added to chip carriers to enhance solder joint reliability by absorbing thermal-expansion and board-bending stresses. As chip carriers became larger, such stresses became more extreme. At some point of increasing size, the PLCCs leads would not provide sufficient resilience to prevent expansion and bending forces from breaking solder joints. As Figure 7-17 shows, there is an inverse relationship between package size and solder joint reli-

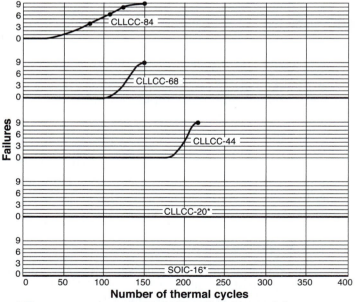

Figure 7-17 Relationship of package parameters to solder reliability.

ability in temperature cycling tests. The following presents a solder-stress formula that relates package size and other parameters to solder stress (see also text and Figure 9-9 in Chapter 9).

7.7.1 Power Cycling Produces Complex Stresses

The stresses produced by power on/off cycles are complex. TCE mismatches between ceramic components and polymer boards act like a bimetallic strip, producing a bending stress. Differential heating in components and boards tends to oppose this bending. Acting on substances of matching TCE, differential heating would produce a bimetallic bending effect in the opposite direction. Power dissipation from a mounted chip produces an out-of-plane displacement.

7.7.2 The Manson-Coffin Equation Modified

Expanding on the original work represented in the generalized Manson-Coffin equation, Norris and Landzberg of IBM modified the formula to define and model these complex bending stresses to account for the effects of maximum temperature and frequency of temperature cycling. Shah and Kelly added consideration of the dwell time at high temperature.

Based on his own work plus research from Hall and his associates, Englemaier of Bell Laboratories modified the original Manson-Coffin equation, producing the following model:

$$N_f = \frac{1}{2} \left(\frac{\Delta\tau}{2\varepsilon'_f} \right)^{1/C} \quad \text{(Eq. 7-1)}$$

where,

N_f = Mean cycles to failure,
ε'_f = Fatigue ductility coefficient (0.325),
C = Fatigue ductility exponent ($-0.442 (-6) \times 10^{-4} \overline{T}_s + 1.74 \times 10^{-2} \ln(1 + f)$),
$\Delta\tau$ = Shear strain range $\left(\dfrac{L}{\sqrt{2h}} \Delta(\alpha\Delta T)_{ss} \times 10^{-4} \text{ \{in percent\}} \right)$.

and, where

$\Delta(\alpha\Delta T)_{ss}$ = The in-plane steady-state thermal expansion mismatch
   $(\alpha_s(T_s - T_o) - \alpha_c(T_c - T_o))$,
α_c = TCE of LCC,
α_s = TCE of substrate,
T_c = Temperature of chip,
T_s = Temperature of substrate,
T_o = Temperature in °C (power off, steady state),
$\overline{T}_s$ = Mean cyclic solder joint temperature (°C),
f = Cyclic frequency ($1 \leq f \leq 1000$ cycles/day),
L = Chip carrier size,
h = Height of standoff.

Lead form also enters into the reliability equation. The lead form factor was covered in greater detail in Chapter 3.

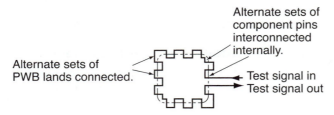

Figure 7-18   Daisy-chain continuity test interconnects.

7.7.3   Test Methods

Tests producing the results shown in Figure 7-18 were conducted by Texas Instruments. Nine units of each device type were constructed with internal daisy-chain interconnects, as defined in Figure 7-18. These devices were placed on PWBs with appropriate land patterns—also daisy-chained for continuity testing. Test points were included on the boards to facilitate the location of failures. A failure was defined as an essentially open circuit (whenever the monitored current through the daisy-chain path was interrupted).

The PWBs used were built up of one 0.508-mm (0.020") G-10 epoxy-glass wiring board laminated to two substrates of the same material and thickness. Assemblies were temperature cycled in a relatively benign 0–80 °C excursion for a period of 1 hour. Cycling was continued until a failure occurred. Upon a failure, the test was interrupted. The failure was diagnosed, recorded, and repaired with a jumper wire. Testing was then resumed, and the procedure repeated until all devices had failed.

7.7.4   Solder-Joint Reliability

The same relationship is seen in the comparison of solder joint reliability in board-bending stress, as shown in Table 7-11. For this test, three 254-mm (10.0") long boards of each design were bent 6.35 mm (0.25") in one direction only and then allowed to relax. This approximates a smooth radius of 101.6 cm (40.0"). The boards were bent and relaxed five times every 2 hours.

Table 7-11   Solder Joint Reliability in Board-Bending Stress.*

Assembly Type	Number of Cycles	Failures
18-pin PLCC	1000	None
	6500	None
68-pin PLCC	1000	15 opens
14-pin to 24-pin SOIC	1000	None
	5000	1 open

Recognizing that large SMCs place great demands on solder joints, we can enhance surface-mount reliability by using smaller packages. Where devices of greater than 44 pins are required in a design, use 0.635-mm-center (25-mil) PQFP, TapePak, or Micropack devices. A 132-pin PQFP with 0.635-mm (25-mil) centers is the same size as a 68-pin PLCC with 1.27-mm (50-mil) centers. The 0.508-mm (20-mil) TapePak would provide 168 leads in the same space.

7.7.5   ESD/EOS Resistance

ESD/EOS hardness follows the same rules whether the board has SMT, IMC, or a mix of the two. We do see numerous examples of products where these rules are ignored, however. Therefore, we will review the rudiments here.

Electrostatic discharge (ESD) is a common occurrence in any dry environment and can easily subject parts designed for 3 to 6 volts to momentary surges in the order of 15 to 30,000 volts. When a spark jumps from your hand to some exposed piece of metal on a circuit (a shaft of a control, a board-mounted bezel, etc.), it will follow the path of least resistance to ground. Predicting that path in the absence of deliberately designing it is nearly impossible since it will be affected by such imponderables as ionization paths in the air, contaminants or humidity on the surface of the board, spacings of parts, clenched leads, and so on. The only way to harden the design against damaging ESD is to eliminate entry points or to arrest the discharge and drain it through a low impedance path to ground before it enters any areas where it might harm circuit elements. To do this, zener diodes can be placed across lines that might sustain a hit. Guard-band traces can be routed around entry points and connected to ground. Spark gaps can be designed into the board to provide a path to ground around I/O. A clever design for a spark gap is shown in Figure 7-19.

Electronic overstress (EOS) can occur when a sudden overvoltage—or surge—enters through the AC mains or from the driven load. Line surges occur within household, office, and industrial settings, and may be produced by a number of sources. Disconnecting an inductive load causes an inductive kickback that can be quite substantial in voltage and rapid in rise.

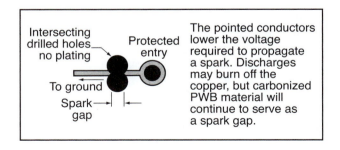

Figure 7-19   The pointed conductors lower the voltage required to propagate a spark. Discharges may burn off the copper, but carbonized PWB material will continue to serve as a gap.

Capacitive and resistive loads cause inrush surges in amperage when switched on. Lightning strikes can induce very large voltages on power lines. More details about both ESD and EOS sources and preventive techniques for the workplace can be found in the Intel Packaging Handbook, at http://www.developer.intel.com/design/packtech/packbook.htm, Chapter 6 "ESD/EOS."

The designer must evaluate where inrushes may enter the circuit and provide protection such as spark gaps, MOVs, sidactors, or other voltage-clamping devices. For sensitive circuits, it is common to use a combination of protective devices. Spark gaps may be used to short-out high voltages, typically over about 150 V, MOVs for mid voltages, and if necessary zener or standard diodes to protect against low voltages.

7.8   Review

After reading and understanding this chapter, you should be able to answer these questions and note the reference location for the information in the chapter.

1. What one "failure" leads to the failure of any quality technique?
2. Define *quality*.
3. Define *reliability*.
4. What is the relative cost of finding and correcting a component defect at incoming inspection versus finding and correcting it at final test?
5. Why are SMT IC packages more reliable than DIP packages?
6. With modern components, is protection against electrostatic discharge (ESD) still necessary?
7. Briefly describe four tests that may be used to determine printed wiring board (PWB) reliability.
8. Briefly describe four tests that may be used to determine solder paste quality.
9. How can factory personnel maintain the solderability of components?
10. How can factory personnel minimize moisture absorption by components?
11. Briefly describe four factors that affect the quality of solder paste stenciling (not screening).
12. Briefly describe four factors that affect the quality of component placement.
13. What are the typical tests performed by a manufacturing defects analyzer?
14. What is a bed-of-nails unit?
15. What are two ways to perform burn-in testing? Which way do you think is more likely to weed out potential failures?
16. What can x-ray tests detect?
17. Briefly explain how a laser thermometry system detects the quality of solder joints. Note two things that can prevent a valid measurement.
18. Briefly describe three ways to build in ICT and MDA testability.

7.9 References

1. Roome, Diana Reynolds, "Management Techniques & Processes—Cutting Through Complexity: How SQC Can Help?" *Printed Circuit Assembly,* July 1987, p. 16.

2. Bhote, A., Bhote, K., "World Class Quality—Design of Experiments," *AMACOM,* New York, 2000.

3. Richter, Ed, "Let's Talk Business," *Quality,* September 1986, p. 31.

4. Hutchins, Dr. Charles, and Ganden, Howard, "Pitting SMT Against Standard Mounting," *Electronic Engineering Times,* October 19, 1987, p. T8.

5. Hollomon, James K. Jr., "The Case for On-Line Test on Placement Systems," *SMART* 1985, New Orleans, LA.

6. Anderson, Dan C., "The Doorknob Syndrome," *Assembly Engineering,* September 1986, p. 14.

7. Lish, Earl F., and Weber, John O., "Solderability Testing of Leaded and Leadless SMDs by Means of a Modified Wetting Balance," *11th Annual Electronics Manufacturing Seminar Proceedings,* Naval Weapons Center, China Lake, CA, February 1987, p. 283.

8. Anderson, Dan C., "The Doorknob Syndrome," *Assembly Engineering,* September 1986, p. 14.

9. Hroundas, George, "PCB Test Strategies for Manufacturing Yield Improvement," *Proceedings of NEPCON West 1987,* Cahners Exposition Group, Des Plaines, IL, February 1987, p. 913.

10. Ibid.

11. Manko, Howard H., *Solders and Soldering,* McGraw-Hill Book Company, New York, NY, 1979, p. 44.

12. Neal, Thomas N., "Production Testing of Board Mounted PLCCs Made Easy," *Evaluation Engineering,* October 1987, p. 28.

13. Mullen, Jerry, *How to Use Surface-Mount Technology,* Texas Instruments Incorporated, Dallas, TX, 1984, pp. 2–8.

14. MIL-STD-810, *Military Standard Environmental Test Methods and Engineering Guidelines,* July 19, 1983, p. 7.

15. "IEEE1149.1 (JTAG) Testability Primer (Rev. C)." Available at http://www.ti.com, search for "testability primer."

chapter 8

SMT and Design for Manufacturability

Objectives

This chapter concerns itself with the design for manufacturability (DFM) of printed circuit assemblies (PCAs) using surface mount technology (SMT). After reading and understanding this chapter, the reader should be able to:

- understand how equipment availability impacts design decisions
- understand what equipment needs to be available for SMT manufacturing
- understand what inspection equipment needs to be available
- understand what break-offs are used for
- understand issues about the panelization of boards
- understand why coplanarity is important

Introduction

SMT made its U.S. debut in the 1960s and 1970s in the laboratories of the major electronics manufacturing firms. In a lab environment, the manufacturability and testability of designs were all but irrelevant, but the twenty-first century designs require DFM and design review, this chapter's topics. Any high-volume product with manufacturing or test difficulties could have a devastating financial impact on its maker; therefore, all such designs demand a careful design review. The software for design analysis, thermal modeling, and high-frequency performance analysis, etc., are major tools directed at design review.

Chapter 1 introduced SMT, and Chapter 2 covered the components. In Chapter 3, we discussed SMT manufacturing methods and the impact of manufacturing style on design. With that as a background, Chapter 4, Chapter 5, and Chapter 6 carried us through the design process. Then, Chapter 7 covered quality assurance. The next step might be called Design Review. This is not to suggest, however, that the design review should be the last step in the

R&D process. To be effective, design review must occur frequently from the first phases of conceptual design forward throughout the design effort. If design reviews are delayed, by the time problems are identified in the review process, the cost and schedule impact of corrections will likely be so great that the review will be pointless.

Design review really includes several issues such as cost analysis, a manufacturability study, resource review, the build or buy decisions, etc. This chapter will cover design review, but we will not recap all the rules we have laid out in the foregoing chapters. Further, we will not cover any application-specific SMT hybrid rules, which will be the subjects of Chapter 9. What we are presenting here are some additional considerations for design review, points that add to the rules established in our other chapters. So, if design fits all of the requirements established outside this chapter and passes the following suggested review points, it should be good, manufacturable SMT design.

We will cover the design directions used to promote manufacturing ease, cost effectiveness, testability, repair simplicity, etc., and we will touch on the nonmanufacturing issues which make a design either a *win* or a *sin*.

8.1 Design Constraints and Manufacturing Resources

Even the best design may be of little value if you cannot build it. And, if nobody can build it, it is of no value at all. Some manufacturing-imposed design constraints are easy to work around, and some others can be solved by subcontractors. Other constraints stand firm until advancing technology solves them. In the following discussion, we will relate manufacturing capability to the requirements imposed by various SMT designs. We will sort between limitations, challenges, and the advantages of various process steps.

Some of the tests and testing equipment have already been discussed as part of the quality issues in Chapter 7. However, we will briefly mention them again here to make a complete presentation on manufacturing issues.

8.1.1 Available Equipment

Available manufacturing equipment does not dictate what can and cannot be done in SMT, but it does determine what can successfully be done in-house without capital expenditure. And, creating the need for surprise capital expenditures can dictate the premature end of a career. Thus, manufacturing resources review deserves to be high on the priority list in analyzing a new design.

Receiving Inspection Equipment

Receiving inspection can generally be contracted to vendors for a price. And, under some conditions, with proper purchasing clout, it may be replaced by a *No Incoming Verification (NIV)* program. NIV substitutes source inspection and review of vendor quality data for receiving inspection. However, the operative word in effective NIV is usually "purchasing clout." Without this, projects must be guided in component selection by a knowledge of what the QC department can inspect, or inspection must be farmed out. Below, we will look at the special resources required for SMT inspection.

Component Test

SMC handlers are needed for anything more than the sample testing of incoming components. Component handlers are designed for receiving inspection use, and flexible handlers have interchangeable accessories so they can be tooled for the test sorting of various integrated circuits. Separate equipment will be needed to handle discrete components.

Solder Paste Test

Solder paste is inspected for particle size using high-power optical scanning electron microscopes (SEM). The paste must be checked for its dispensing or screening quality. Also, the paste is examined for contaminants. In Chapter 7, Section 7.3.2, we presented testing methods for these aspects of solder paste inspection. Viscosity is also measured at receiving and periodically throughout the paste's use. A viscometer is used for measuring the shearing stress while stirring a sample of solder paste.

Printed Wiring Board Test

The need to inspect printed wiring boards is certainly not unique to surface-mount technology. Bare-board probing is a common test aspect for all board types. Many SMT boards require fixtures with probe centers closer than 2.54 mm (0.100"). Tighter 1.27-mm (0.050") probe cards are available at an increased cost. Even finer pitches can be probed, but below 100 mils, the test point cost is an inverse function of center distance, as was discussed in Chapter 3. Board probing can be with a bed of nails fixture or a fixtureless bare-board tester, which uses a moving probe to solve access and gridless test problems.

Glass transition temperature, or *Tg* (i.e., the temperature at which a material begins to flow), is important to SMT. Fine features and small-aspect ratio holes are easily damaged when the board Tg is exceeded and the substrate expansion triples. Astute inspection departments have found that some laminate and prepreg suppliers have material Tg in tight control, and others do not. Also, Tg is a function of the level of the PWB supplier's curing of B-stage material.

For fine-line and tight-geometry boards, Tg may need to be inspected. *Thermomechanical analysis (TMA)* measures the dimensional effect of temperature changes on materials. *Thermogravimetric analysis (TGA)* measures the weight change due to outgassing, decomposition, or solvent removal under temperature changes. *Differential scanning calorimetry (DSC)* determines the cure state of prepreg resins by measuring the Tg of a sample. *Dynamic mechanical analysis (DMA)* measures the stiffness of prepregs and laminates as a function of heat. Complete PC-based inspection setups are available from several sources for benchtop TMA, TGA, DSC, and DMA testing. Services are also a source of thermal analysis tests for infrequent requirements.

After all the QC data in Chapter 7 and the above additions on QC/QA, one might ask, "Isn't this a bit of overkill?" We will answer with a military anecdote that aptly illustrates what usually happens when you leave things to chance. As a student in the Navy's torpedo maintenance school, Willard Paul, Jr., was impressed with the instructor's repeated discussions of the importance of built-in safety interlocks. At one point in the training, Mr. Paul asked the instructor if there was an interlock to prevent the opening of the inner door while the torpedo tube was filling with water. The instructor replied haughtily that, "A torpedoman *always* checks his level indicators *first*." He then promptly turned around and opened the breech of the training tube, dumping 3500 gallons of water in the classroom.[1] If anyone's surface-mount program ends up all wet, we would rather it be because of tasks that they choose to leave out, instead of something we left out of this text.

PROM Programming

Programming PLDs and PROMS in SMT packages is generally no problem. Technologies such as E²PROM and flash also allow for in-circuit programming with minimal issues.

One word of caution, however. Where conversion from DIP to SMT required a pin-count change (for instance, a 24-pin DIP PLD converted to a 28-pin PLCC), vendors are at liberty to choose how they will pin out the new device. They may exercise this liberty to achieve best wire-bond routing on their silicon rather than to standardize devices for you, the user. One example of this is Intersil's ADC+LCD driver ICL7106. The pinouts for the 40-pin DIP and the 44-pin metric QFP part are completely different. So several sockets, one for each pinout, may be required for a single package size. There are many different adapters for a variety of conversions.

8.1.2   Board Size

For automated assembly, there is some upper limit of board size that can be processed by the manufacturing line. This limit includes maximums of how wide, how long, and how thick a board may be. There will also be a lower limit, past which boards must be carried on tooling plates. If tooling plate handling can be avoided by making a board just 5 mm (0.197") bigger, the change will save considerable tooling and handling costs. Lower limits may also be avoided by processing several circuits in a panel. More information on panel handling is given in Section 8.2.1.

Upper board size limits cannot be exceeded without risk of serious grief in the manufacturing operation, so they are quite important in design review. The upper and lower limits are imposed by each required piece of process equipment. Even when all the equipment comes from one vendor, it is quite unusual for there to be much agreement on board size limits, machine to machine. Thus, the line limit is set by the least common denominator of maximum and minimum width, length, and thickness for the individual equipment in the manufacturing system.

Note that size of the boards may have a far greater impact on per-square-inch cost than one might expect. This is because raw materials for PWB manufacturing are typically supplied in standard sizes, such as 1 meter square, 36 inches square, or 36 by 48 inches. Let's look at an oversimplified example to illustrate this principle. If you designed a board as 19 by 25 inches, only one circuit could be fabricated out of a 36 by 48-inch sheet of laminate. Of course, the board shop could use the scrap from the laminate to make other, smaller boards, but who will pay for the material? If you guessed, "Both customers," give yourself 100 points. By simply reducing the board dimension to 19 by 24, you would sacrifice 4 percent of the board area but cut the cost of raw material in half. If you could squeeze the board down to 18 by 24 inches, you would give up around 9 percent of your 19 by 25-inch board's area but cut the raw material cost by 75 percent! When boards include small projecting areas, elimination of these, or panelization in nested orientation, may produce even more dramatic results. Now, as we said above, this is an oversimplification. In the real world of board shops, raw stock is generally cut into a more manageable size (also called a panel) for circuit manufacturing. If raw stock of 36 by 48 inches is used, panels may be 24 by 36 or 18 by 24 inches. Shops vary. Also, because of manufacturing process requirement in the board shop, some perimeter of the standard panel must be left clear of the imaged circuit. Therefore, to intelligently select a board size for maximum material usage, you must know a good deal about the practices of the vendors you will use for board manufacturing. Good designers will learn these facts. Purchasing departments do well to build a strong partnership with a limited number of quality suppliers rather than paper the world with RFQs searching for today's bargain-basement pricing each time a new requirement arises.

8.1.3   Board Shape

Irregularly changed boards require special handling methods, just like undersized boards. Odd shapes may be processed in multiples inside rectangular panels, or loaded on tooling plates. Alternatively, odd-shaped boards may be made in standard sizes with cutout or break off areas that will be dealt with at the end of the manufacturing line, as the board in Figure 8-1 shows. This board shows a square cross-hatched break off area and a long-edge break off (with the word "ARRAY" on it) that will result in the proper final board shape. The reader will realize that the purpose of the breakoffs is to result in a board that fits properly in the enclosure and around any large components also in that enclosure.

Figure 8-1   Irregularly shaped boards can be "standardized" in size.

8.1.4   Panelization

If panel handling with multiple boards per panel is part of the manufacturing plan for a product, depaneling methods must be available. For a very small number of boards, breakaway tabs may be cut with tin snips or heavy diagonal pliers, and the rough edges removed by grinding (while wearing appropriate breathing protection to avoid inhaling the dust). Alternatively, snap-off depaneling is also available for low volumes. For higher volumes, however, automated depaneling equipment should be available, or a depaneling contractor should be secured.

8.1.5   Clear Areas on Board

Each automated processing station in the line will impose certain requirements for areas of board clear of components. Such clearances may be required for guide rails, pushers, and locating pins for automated board handling and registration. Figure 8-2 shows typical clear area requirements. Individual equipments vary, so establish clear area requirements for your line based on the worst case for all stations in the line.

We will note that there are two common industry terms used to define areas. A "keep-out" area means that no components may be mounted in the defined area, and in some cases traces may also not be run through this area. A "keep-in" area means that all components must reside in that defined area of the board. Determine early on in the design phase if you have a need for either or both areas in your board design.

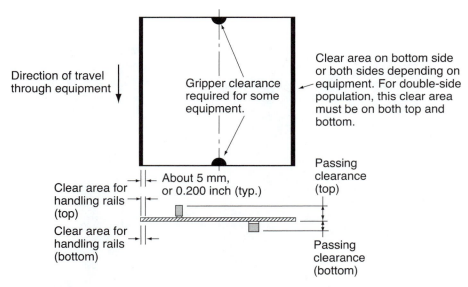

Figure 8-2   Typical clear-area requirements for automated handling.

8.1.6   Passing Clearances, Top and Bottom

Additional clearance considerations are imposed by the proximity of the undercarriage and overcarriage of a piece of equipment to the boards passing through it, as shown in Figure 8-2. Remember that the under and overcarriages include any moving parts mounted thereon that may pass across the board during processing. Also, for boards requiring components on both sides, remember that they must pass through the line in both orientations. Thus, top-side components in the first pass will need to comply with bottom-side clearance requirements in order to clear the second pass. Likewise, if guide rails require clear edges on the bottom, the inverted second pass will dictate that the same clearance is on the top.

8.1.7   Heavy Components

Rule 1 for heavy components is the same as Rule 1 for rework. "Don't have any." Unfortunately, both rules have, at times, proved difficult to follow. Where heavy components must be used, add them after the automated assembly operations if at all possible. If the chunky component absolutely must be on the board for the trip through the SMT line, locate it near the board's edge where it will have maximum support. Watch out for heavy parts on the board during the wave-soldering operation. They can cause the board to act like a submarine if it falls off the line!

Board carriers may be used to add stiffness. For panelized boards, if the panel is wide and/or has heavy components, stiffening webs help minimize warpage during soldering processes. As a last resort, "pop-up braces" may be installed on each machine to raise up after the board is registered and support the sagging assembly. Of course, this is a costly and time-consuming solution that is to be avoided whenever possible.

8.2   Design for Manufacturability

In the following paragraphs, we will discuss rules for reviewing and enhancing the manufacturability of SMT assemblies. Enforcing these rules while the assembly can still be changed with an eraser or a strike of the 'delete' key can save many a dollar and can prevent needless gun fights between manufacturing and engineering.

8.2.1   Standards

Great gains in manufacturing efficiency are available in the area of internal standards. The following are some suggestions regarding the areas to standardize for your SMT program and the ways to derive standards for them.

Databases

Maintain a common database throughout design and manufacturing. If necessary, use protocol conversion software to connect nonstandard equipment. Hewlett-Packard's Entry Systems Operation in Cupertino, California, has implemented a common database, and they can shed some light on how it helped in their case.

The schematic capture and data checker on the front end, as well as the component libraries and the post-processing file conversion for creating programs for production and test equipment, required the greatest effort, but the result is well worth the trouble. The standardized database permits programming of the entire operation from one post-processed CAD/CAE database. Over the life of a factory, programming costs for line equipment and test can quickly exceed the capital acquisition cost. Therefore, cost-reduction efforts focused on programming expenses can yield great savings.

Printed circuit board files have for years been sent to PCB fabricators and manufacturing as Gerber files (see Chapter 5). The IPC has been working on the newer GenCAM standard for files which allows easier file conversions. In 2002, the IPC released GenCAM v. 2.0 in the XML format, which allows users to describe a printed circuit board as well as its assembly and test fixtures in one file. For more information on GenCAM, see B revision of the IPC-2511 "GenCAM Standard" at http://www.gencam.ipc.org.

Components

Standardize on as few component part numbers as is humanly possible while allowing construction of the required circuits. Many designs electrically call for the same value resistor or capacitor with several different tolerances. If you specify several tolerances for the parts, each part must be a separate part number that must have a unique feed location on the manufacturing placement equipment. There are physical limits to the number of positions on an assembly machine. When these limits are exceeded, double-pass placement is required, and this not only drives setup costs through the roof, it can destroy first-pass yields as well. There are also hidden costs in purchasing, inspecting, warehousing, and the handling of five different part numbers where one would suffice. Limiting all designs to 1 percent tolerance resistors can cut the resistor part number count by over 50 percent. The substantial manufacturing cost savings are well worth a slight cost premium in components.[2]

Many ICs are available in several package styles. The same memory may be offered in SOIC, SOJ, PLCC, CLLCC, and TapePak. Consider the alternative packages, and standardize on one type for all ICs in a given pin count unless other factors absolutely force a deviation. In all cases, endeavor to have no more than one part number for a given IC.

SO packages are more space efficient in devices with less than 20 pins while 20- and 24-pin SOICs take slightly more real estate than their PLCC counterparts, but they provide a clearer routing access on board. From 28 pins up, the quad I/O devices have a clear size advantage. For parts requiring extra handling steps, such as PROMs or PLDs, SOICs may be a poor choice due to the fragility of their leads. To avoid lead damage during programming, a PLCC is a better package for such devices since it is easily socketable. However, many IC manufacturers are phasing out their PLCC packages due to both package height and perceived solder integrity issues.

ANSI/IPC CM-770C calls for coplanarity of ±0.05 mm (±0.002") for leaded IC packages. To IC manufacturers, this total variation of 0.1 mm (0.004") is in serious conflict with packaging cost-reduction efforts. In fact, at least one major U.S. semiconductor manufacturer has attacked the coplanarity problem by forming the J leads of PLCCs tightly over molded-plastic mandrels under the device body. The net effect is the production of a highly coplanar package with virtually no lead compliance. This manufacturer sells compliant-leaded devices to the few users aware of the need to ask for them and boosts their own yield by making coplanar noncompliant devices for everyone else. Compliant J leads do work to relieve damaging TCE-induced stresses to solder joints. The noncompliant devices produced by this one supplier may well be at the core of perceived problems with solder integrity of J leads.

While the component supplier may feel that the coplanarity specification is too tight, the end user suffering with excessive solder opens sees ±0.004" as unconscionably loose. For reasonable manufacturing yield, coplanarity variations should be no greater than ±0.025 mm (±0.001").

Component coplanarity specifications determine the assembly parameters required to achieve high solder yields. Figure 8-3 shows a model for determining the required wet-paste print thickness and the lead penetration into that wet paste for solder joint formation with a given coplanarity error. The three parameters—device coplanarity, wet-print thickness, and lead penetration—can be monitored. So, given the data from the model shown in Figure 8-3,[3] we have the tool for specifying and controlling a high-yield soldering process.

The required lead penetration can be calculated using the following two equations:

$$\text{Penetration Percentage (\%P)} = 100 \times \frac{D}{T} \qquad \textbf{(Eq. 8-1)}$$

and

$$\text{Penetration Percentage (\%P)} = 100 \times \frac{\text{Maximum Coplanarity Error (Ce}_{max})}{T} \qquad \textbf{(Eq. 8-2)}$$

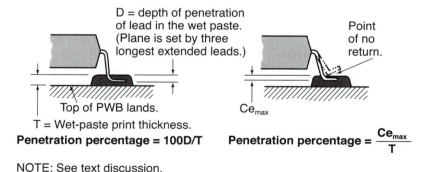

D = depth of penetration of lead in the wet paste. (Plane is set by three longest extended leads.)

Point of no return.

Top of PWB lands.   Ce_max

T = Wet-paste print thickness.

Penetration percentage = 100D/T   **Penetration percentage = $\frac{Ce_{max}}{T}$**

NOTE: See text discussion.

Figure 8-3 The determining processes needed to prevent solder opens.

Therefore, in order to apply sufficient placement pressure to yield the proper penetration and for a wet-print thickness of 0.2 mm (0.008") with standard parts having a coplanarity specification of 0.1 mm (0.004"), the device placement system should be set to supply sufficient placement pressure to yield a percent penetration as follows (Equation 8-1):

$$\%P = 100 \ \frac{D}{T} \ = 100 \times \ \frac{0.004}{0.008} \ = 100 \times 0.5 = 50\%$$

Note that this is a simplified model that ignores the effects of lead form and that assumes good solderability and good wetting forces of the materials. Also, the joints formed by deformed leads which barely touch the surface of the paste will probably not be as strong as joints formed with leads that are not deformed. If joint strength is important in a given application, testing should be used to establish the maximum coplanarity variation that will produce acceptable joints.

Uniform Device Polarity and Orientation

Chip resistors should be oriented with their resistive element side up. In this position, as noted in Chapter 2, their resistive element can give off heat to the atmosphere. This mounting position is also more stable because the resistive element side is rougher than the alumina substrate side. Figure 2-11 in Chapter 2 shows the construction details that explain the variation in flatness. Also, component markings are generally on the resistive element side. Thus, orienting the device with the element upwards leaves the markings visible after mounting. Taped resistors are supplied with the element side up as the tape is dereeled. Therefore, resistive element upward orientation comes automatically when using taped components on automated placement engineering. However, bulk resistors must either be optically oriented in special feeder equipment, oriented by hand placement, or randomly placed. When random placement is required, a silkscreen legend, visible after component placement, helps with the troubleshooting, field service, and repair of boards.

ESD Protection

If your shop has a true ESD plan in place and static charges are controlled to less than 50 volts, you can design freely with all classes of components. Unfortunately, too many manufacturing organizations still speak in terms of being able to "get by with" limited ESD protection. This is only true where a company wishes to "get by with" poor yields, shoddy merchandise, dissatisfied customers, a falling market share, and an eventual demise. Chapter 4, Section 4.9 and Chapter 7, Section 7.1.4 present useful data in determining static control strategies. And as mentioned in chapter 7, further information about both ESD and EOS sources and preventive techniques for the workplace can be found in the Intel Packaging Handbook, at http://www.developer.intel.com/design/packtech/packbook.htm, Chapter 6 "ESD/EOS." An engineering review should include a consideration of the ESD protection levels of the devices in the assembly.

PWB Specifications

All bottom-side chips should use wave soldering lands, as shown in Figure A-48 and A-49 of Appendix A. One of the first design decisions for a multi-board system is how the board should be panelized. The most common approach is to place each board in its own panel. If the design pushes board tolerance limits or has many layers, board yield may be low. Each panel may have a number of X-outs (rejected individual circuits). In such a case, separate panels is the best strategy. However, where board yields should be high, it may make sense to place one or several sets of boards for the entire system in a panel. By doing this, so long as your testers have sufficient pin capacity, it is often possible to complete both in-circuit and functional tests of the board in the flat. Thus, reworking failures from functional tests is much less difficult. It is sometimes possible to use axial inserters to "staple" interconnect wiring between circuits in a multi-board panel. After depaneling, the boards are formed to fit the final packaging. This trick provides a very low cost interconnection that is inherently more reliable than a header and also facilitates the in-the-flat system test we mentioned above.

Giving a manufacturing engineer a fistful of board sizes in small-lot manufacturing is about as likely to win his friendship as giving him a one-way ticket to Siberia. On many pieces of process equipment, each unique board size requires custom tooling. There is also the cost for setup time needed to adjust for board variations. And, with each adjustment, there is the risk of missing the correct mark. When these costs are submitted to top management, the design team may be the ones traveling to the Siberian salt mines. To avoid undesirable relocations while still retaining the right to design the board sizes that an application demands, panelize the designs for batch processing. Most shops can standardize on one form-factor type of panel which will accommodate at least one of their largest boards. Smaller boards are then built in multiples on the panel. The outer edges of the panel may be standardized with free areas for handling conveyors, tooling holes, fiducial marks, and panel identification markings or bar codes.

In panelizing designs, always put each circuit within the panel on 2.54-mm (0.100") incremental steps so that probe points are on such centers, board to board, within the panel. This factor is often overlooked and causes unnecessary grief when making bare-board test fixtures.

A panel will typically have a perimeter of circuit-free material, or selvage. This vacant ground need not be entirely wasted. By clever forethought, the selvage can serve several functions. Most obviously, it provides an unobstructed area for handling on sliding rails, nesting in magazines, moving by grippers, fixturing by tooling holes, and orienting by fiducial marks. But the uses need not end here. If copper thickness is an issue, an extended serpentine pattern in the selvage can be used to measure the copper by doing a quick resistance measurement on each manufacturing lot. Test coupons can be included to allow testing of throughplating in holes, peel resistance of copper, and cleanliness of the board surface. Coupons are defined in IPC-D-330 and MIL-STD-275. Another valuable addition to include in selvage areas is dummy lands. These are lands for mounting SMT parts. As a spot check of soldering adhesions, you can periodically hand-place components on these sites, run the board through the soldering process, then use a shear and/or tensile test to determine how well your soldering process is working. We have even used the break-away tabs to carry tracks from ground planes into the selvage and back through breakaways of adjacent circuits where we were panelizing a multi-board system in a single panel. This simplified testing of the entire system with high-integrity grounds. One word of caution, though. The level of consternation this generates with PCB front-end personnel tells us how uncommon it is. If you do not want to receive phone calls, place clear manufacturing notes on the mechanical layer. This will not stop people from calling. After all, who reads the notes first? But it will put you in a position of power with those who call.

Individual circuits within the panel may be partially defined by routing edges, but leave a few bridges of material (called breakaway tabs). Each board shop has individual requirements for panelization, and their suggestions should be weighed in the selection of actual panel dimensions.

Generally, four tabs per board are used. Tabs are typically only on two sides and should not be on sides that must go into card guides. For small boards, two tabs may suffice. Where two tabs are used, boards are less prone to warp out of the plane of the panel if they are located near opposing corners rather than directly across from one another. Round or irregular-shaped boards may use three tabs.

Board stiffness is of concern when using breakaways, particularly where wave soldering will be required. For large panels needing stiffness, the 12.7-mm (0.5") separator strips improve transverse rigidity. Where stiffeners are not required, individual boards should be separated only by a single route line and tabs.

The perimeter, or frame, serves to give the panel stiffness, provides free space for assembly handling, tooling holes, optical targets, and bar-code labels. Circuit board manufacturers also require a frame. If your panel frame meets their minimum requirements, they can save board material by using your frame. Discuss frame and tooling-hole size and copper-free areas around the optical features with your PWB vendor early to take advantage of such savings.

The removal of tabs is often approached as an afterthought. Involving manufacturing engineering early in the process can save headaches. A 90-pound assembly operator with a pair of diagonal pliers small enough to fit into the routed slot is a poor but frequently applied alternative to forethought on this matter. If no automation is available, a clever solution is to use carbide-blade nibblers. Unfortunately, we do not know of any available from the shelf. You will probably have to buy steel-bladed nibblers and have a toolmaker replace the blades. This effort produces a tool that self-locates in the slots and removes tabs with a minimum of effort.

Alternatively, individual boards may be depanelized after assembly by specialized depanelizing shears or by other cutting methods such as lasers or high-pressure water jets. All boards on a panel should be oriented identically. Panelization within the CAD system and photoplotting of the panelized phototools will minimize any individual feature registration errors.

Additional common methods for panel separation include scribe or notch and break, perforation (a string of closely spaced unplated drill holes around each circuit), and punchout-and-reinsert. Scribe or punch techniques are particularly suited to punchable materials such as CEM-1 or CEM-3. Vendors who supply appropriate laminates can give guidelines for depaneling schemes fitting their shop practices.

Whether or not boards are panelized, the tooling holes should be standardized per manufacturing equipment requirements and PWB fabrication considerations. The tooling holes should always be left unplated to produce closely toleranced diameters. For the sake of accuracy, tooling holes should always be drilled during the first drill operation prior to through-plating of the holes. To leave the holes unplated, they are tented with the plating resist. Holes larger than 3.1785 mm (0.125") cannot be reliably tented and, therefore, should not be used.

Tooling Holes and Fiducial Marks

Clear areas are generally required around tooling features, such as locating holes and fiducial marks. The amount of clearance required is determined by the equipment used in your line. Equipment vendors can give you these specification for incorporation in your "house standards." Fiducials should be designed to meet SMEMA Standard 3.1, "Fiducial Mark Standard."

Warpage

Board warpage presents problems to the automated handling equipment, from the front to the back ends of the line. Warpage is a particularly troublesome problem for wave soldering, where extreme cases may cause voids or submarine diving. Any warpage in wave soldering will affect the time that the solder side of the board spends in the wave. Warped boards also may cause problems for adhesive or solder-paste dispensing operations and for placement equipment. Within board cost constraints, warpage should be kept—by specification—to an absolute minimum.

8.2.2 Reflow Soldering Yield Enhancement by Design

Even Heat-Sink Mass Distribution

Keep the heat-sink capacity of the vias nearly even. Where a via makes connection to an internal power or ground layer, create thermal breaks to prevent excessive heat sinking through the via if that via must be soldered or is near a solder land. The recommended thermal path break to an internal layer is shown in Figure 8-4. This strategy may be built into CAD design rules so that it is always followed unless there are reasons to override the rule.

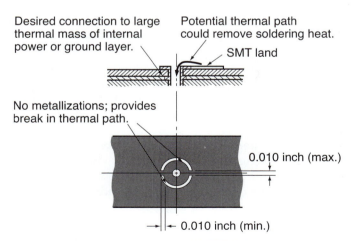

Figure 8-4   Breaking the thermal path through a via. *NOTE: Wagon wheels are commonly produced with four spokes where greater electrical conductivity is required.*

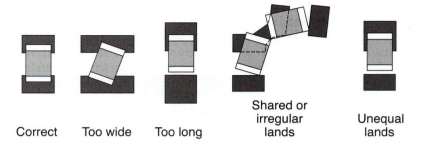

Correct   Too wide   Too long   Shared or irregular lands   Unequal lands

Figure 8-5 Reflow-soldering land geometry-related defects.

Land Geometries

Certain manufacturing defects relate specifically to land geometries. Lands that are too large or too small will produce a high number of reflow soldering defects, as illustrated in Figure 8-5. Use these illustrations to help the team understand the results they will see in prototyping.

Large lands bring more solder to the joint, creating an overly stiff structure. Tin/lead solder has a TCE of 23 to 30 ppm/°C. Ceramic components, such as chip capacitors and resistors, have a TCE of about 8.3 ppm/°C. With the common practice of using a nickel barrier to open up the soldering process window regarding leaching (see the discussion on chip capacitor terminations in Chapter 2), we introduce a very hard and brittle layer between two objects moving at very different rates. When the solder joint is sufficiently small, the inherent ductility of the tin/lead solder accommodates this mismatch fairly well. However, if too much solder is applied, expect to see cracking of components, the delamination of end terminations, etc.

When using nickel in a solder termination, be aware of the potential for the formation of tin/nickel intermetallics. There are two intermetallics of concern. Stable Ni_3Sn_4 forms very slowly at room temperature, but its formation speed increases significantly with temperature. Ni_3Sn_4 is a concern where its formation during soldering or rework compromises the solder joint integrity. A metastable intermetallic, $NiSn_3$, forms at room temperature. Its formation accelerates with increasing temperature up to about 145°C and then decreases with any continuing temperature increases. The metastable intermetallic may be a concern for device solderability where devices are stored for long periods.

The stable Ni_3Sn_4 tin/nickel intermetallic is extremely brittle, and, under thermal cycling stress, this can contribute to cracking along its boundary with adjacent structures. If, when microsectioning a failed part with a nickel barrier you see a thin intermetallic at the nickel boundary and a tin depletion in the grain structure of the adjacent tin/lead, the Ni_3Sn_4 intermetallic should be a suspect. For more information on the various intermetallic formations, see Reference 4.[4]

Tombstoning

We have mentioned drawbridging, or popping wheelies. We covered design directions to minimize this defect, but what do you do when your prototype comes out of the reflow oven covered with chips that are standing up like skyscrapers (the "Manhattan Effect")? As the formulae in Figure 8-6 show,[5] component motion is influenced by the aspect ratio of the part and the size of the pad. The formulae provide a mathematical model for predicting and controlling tombstoning.

Three forces interact to determine chip component motion during reflow soldering. Gravity pulls the chip downward and tends to oppose tombstoning. Surface attraction forces of solder under the chip also pull it downwards. Opposing these, the surface attractive forces of the solder fillet (to the left of the chip in the sketch) create a turning moment, acting to lift the chip about its cantilever point.

To predict the motion of the chips without cumbersome calculations, the following model does not take into account minimal effects of fillet surface tension to the right of the fulcrum point. The model also assumes that the fillet meniscus is a straight line throughout.

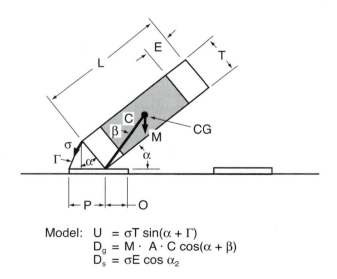

Model: $U = \sigma T \sin(\alpha + \Gamma)$
$D_g = M \cdot A \cdot C \cos(\alpha + \beta)$
$D_s = \sigma E \cos \alpha_2$

Figure 8-6   A model to predict and control tombstoning.

The model formulae are:

$$U = \sigma T \sin (\alpha + \Gamma) \qquad \textbf{(Eq. 8-3)}$$

$$D_g = M \cdot A \cdot C \cos (\alpha + \beta) \qquad \textbf{(Eq. 8-4)}$$

$$D_s = \sigma E \cos_2^\alpha \qquad \textbf{(Eq. 8-5)}$$

where,

A   = Acceleration due to gravity,
C   = Distance of CG to cantilever,
CG = Center of gravity of part,
D_g = Downward moment from gravity,
D_s = Downward moment of surface tension under the chip,
E   = Length of bottom termination,
L   = Length of component part,
M   = Mass of part per millimeter of width,
O   = Overlap of component over land,
P   = Projection of land past part,
T   = Thickness of component,
U   = Upward moment from fillet surface tension,
α = Angle of bottom of component to substrate,
β = Angle of bottom to dissector of CG; or, arctan (T/L),
Γ = Angle of fillet and chip end; arctan $\dfrac{(P - T \sin \alpha)}{(T \cos \alpha)}$

Tombstoning occurs when the upward exceeds the downward moment, or $U > D_g + D_s$. From the formulae, we see that a chip of large T/L ratio will have a greater tendency to tombstone. Also, chips with a very short E-termination extension are more prone to the defect. With chips of typical aspect ratios and termination dimensions, the balance between upward and downward moments is tenuous. In such cases, minor influences from vibration, condensing soldering fluids, evaporating gases, etc. can cause a defect.

Reflow: Soldering Component and Trace Orientation

We covered this topic in detail in Chapter 5, Section 5.1, but it is important enough that it bears mention in any discussion of design review. Are the components oriented and spaced for good reflow yields? Have correct trace size and orientation rules been followed? Are the reflow lands sized properly for the process?

Dummy traces, such as shown in Appendix A, Figure A-46, are a necessity where parts must have track routed underneath. Of course, if two tracks are to be routed under the part, the tracks can simply take the place of the dummy tracks.

Materials Specifications

Specify solder pastes based on the manufacturing process to be used to dispense solder on your boards. Table 8-1 shows typical solder paste parameters for the three primary paste application methods. Additional information may be obtained from paste and application equipment suppliers.

Table 8-1   Guide to Solder Paste Specifications.

Paste Property	Application Method		
	Screen	*Stencil*	*Dispenser*
Mesh size	−270 +500	−200 +325	−270 +500
Viscosity KCps	600 900	700 1100	300 600
Percent metal weight	86 89	89 92	85 90

8.2.3   Wave-Soldering Yield Enhancement by Design

Gas Escape Holes

Gas bubbles can form around components as they pass through molten solder in the wave-soldering process. Bubbles can come from cavitation in the wave, outgassing of flux volatiles, or outgassing of heated materials. Whatever their source, if there is nowhere for the bubbles to go, you will think they came from the devil. They will cause solder skips in the wave-soldering operation. The result is that a painstaking and expensive solder touch-up will be required.

By providing escape holes for any entrapped gases, solder yields may be dramatically improved. Some recommended locations for gas escape holes are shown in Figure 8-7. *Do not* place breather holes under a top-side component where flux or solder from the soldering operation might bubble up and hide, creating both problems with cleaning as well as a place for contaminants to congregate over time.

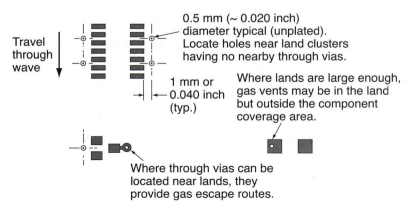

Figure 8-7   Gas escape holes for wave-soldering entrapped gases.

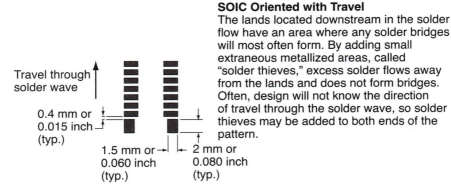

SOIC Oriented with Travel
The lands located downstream in the solder flow have an area where any solder bridges will most often form. By adding small extraneous metallized areas, called "solder thieves," excess solder flows away from the lands and does not form bridges. Often, design will not know the direction of travel through the solder wave, so solder thieves may be added to both ends of the pattern.

Travel through solder wave

0.4 mm or 0.015 inch (typ.)

1.5 mm or 0.060 inch (typ.)

2 mm or 0.080 inch (typ.)

Figure 8-8   Solder thieves soak up excess solder.

Solder Thieves

In the preceding paragraph, we dealt with a design trick to reduce solder skips in wave-soldered joints. Curiously, right next door to this problem we may find solder bridging because of too much solder on the same assembly. During the wave-soldering process, it can be seen that commonly the last two pads on, for example, an SO pattern, will have a solder bridge. Since this seems to be an artifact of the wave process and one for which there does not seem to be a robust way of preventing them, we must design to minimize the problem. *Solder thieves* give us a design trick that can be used to allow the bridges to form between the last SO pad, a pad that is not electrically connected—the thief. Solder thieves are small extraneous metalized areas that cause the excess solder to flow away from the lands so it does not form bridges on electrically active pads. They work in the same way that plating thieves act to distribute the current densities during board plating. Figure 8-8 shows some recommendations for solder thieves for flow soldering.

Wave-Soldering Component and Trace Orientation

If component orientation, spacing, and land geometry deserved mention in our previous reflow soldering discussion, they deserve double attention here. Proper orientation is a critical issue to flow-soldering yields. In your design review, check that all previously developed rules have been rigorously enforced.

8.2.4   Cleaning Enhancement by Design

The design team should be aware that, minute by minute after the boards leave the soldering oven, board cleaning becomes more difficult. Think of SMT assemblies like dirty dishes. Right after dinner, a quick dip in a little hot soapy water will leave your good china sparkling and squeaky clean. Leave those same dishes on the counter to dry for three days, and then washing them becomes a monstrous chore.

Ideally, the product should flow into the cleaner while it is still warm to the touch from soldering. Flux residues polymerize over time after soldering. Therefore, avoid designs that must be routed to other processes between the soldering and cleaning.

Component standoff has a dramatic impact on how quickly an assembly can be cleaned to a specified level. Where alternatives are available, the package with the greater standoff will enhance the cleaning processes. A side benefit is that a greater standoff will usually spell more compliance to absorb TCE and bending stresses.

8.2.5 Design for Inspection

Whether a product will be inspected by operators during hand assembly or by sophisticated vision robots, automated test equipment (ATE), etc., an adherence to basic inspectability design rules will simplify the job and improve the results.

Visual Inspection

For visual inspection, the clearance between devices must be sufficient to allow a good view of solder fillets on the device leads. Repair operations, as well, demand a clearance, both for safe reflow and for visibility. Figure 8-9 shows clearance design rules for ensuring visibility during inspection and repair. Table 5-2 in Chapter 5, Section 5.2.1 lists the clearances determined with these 60-degree/45-degree rules for typical components. The rules may be used to calculate clearances for any parts not included in the table.

Note that solder joints on J-lead PLCCs may be given a quick visual inspection easier than can gull-wing SOICs. Too bad PLCCs are being phased out! Where there is insufficient or no solder on a J lead, the immediately discernible point of light reflected by a good fillet is replaced

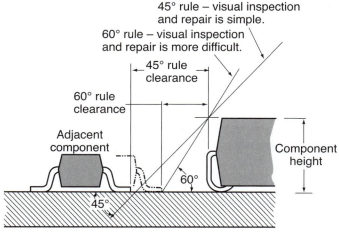

Figure 8-9 Design rules for inspection visibility.

by an obvious black cavern. With the gull-wing lead form, it takes close observation to determine if the reflected light is coming from the lead itself or from a solder fillet. And, for Class 3 products, the heel must be inspected. Therefore, where real estate permits, it is good practice to always use the 45-degree rules or better for gull-wing leads, even on boards using the 60-degree rules for other components.

Automated Vision Inspection Systems

The automated optical inspection (AOI) systems used in manufacturing can provide valuable process-control data for SPC programs and an early warning of drift in a process. Early knowledge of process drift lets manufacturing recognize and cure machine wear, misadjustment, etc. before it produces defects at final test.[6] Several design rules may be imposed by vision systems. We will consider them by category, based on the type of vision inspection in question (Table 8-2).

PWB registration systems use vision to accurately register the surface features of SMT workpieces for a process operation. In THT assembly, the boards were normally registered by pins engaging tooling holes in the board. This strategy works well for through-hole devices because the holes we are trying to register are in close registration to the tooling holes having been fabricated in the same close-tolerance drilling operation. However, surface features, such as SMT lands, are not fabricated in the same operation as the tooling holes. Thus, there may be considerable misregistration between the lands and the tooling holes. Board registration vision automates the orientation of the board's surface features by finding targets (called fiducials) fabricated in the same high-accuracy photographic processes used to construct the lands.

For good results, fiducials should be placed close to three of the four corners of the board. For panelized boards, targets may be on three corners of the panel if individual circuits are closely registered with the panel. However, any error from step-and-repeat operation in panel-

Table 8-2   Typical Applications of Vision Inspection in SMT.

| Applications | Design Considerations | |
	Clearance Rules by Application	*Requirement for Fiducials*
PW board registration	Clearance required by vision system, features to fiducials	Size, shape, location, and contrast as specified by vision equipment
Dispensing inspection	Generally, vision inspection can accept standard design-rule clearances	Same as PWB registration, where vision inspection is used to register workpiece
Vision-guided auto placement	Systems aligning to fiducials by component need extra space	Fiducials may be required near component for targeting
Post-placement inspection	Generally, vision inspection can accept standard design-rule clearances	Same as PWB registration, where vision inspection is used to register workpiece
Solder joint inspection	Generally, vision inspection can accept standard design-rule clearances	Same as PWB registration, where vision inspection is used to register workpiece

ization may be canceled by placing fiducials on each individual circuit. The three-fiducial approach allows for the determining of separate offset and size corrections for the X, Y, and 0 axes, and for orthogonal errors (perpendicularity errors in the X-Y grid system). Fiducial registration systems assume that all these errors are linear across the board. For 1.27-mm (50-mil) center components on ordinary sized boards, we can mathematically demonstrate that this is a safe assumption. For fine-pitch component handling, systems allowing localized feature recognition and component lead-to-pad correction are a necessity.[7]

Whether fiducials are being registered by the vision equipment or actual SMT lands are used as targets, sufficient clear space (free from circuit features) should be provided around the vision target to preclude the vision system from becoming mixed up. Note that the clear area requirement must often be applied to inner as well as outer layers. IR wavelengths are quite distinct from visible light in their propensity to penetrate polymer materials and reflect off inner layers of metallization. Registering from the wrong feature will probably cause undesirable quirks in the assembly process. Consult your vision system vendor for required clearances.

Contrast between the target and board material is important in vision system accuracy. Those of us who do not deal in vision equipment tend to think of the video camera as an eye and the computer that decides the image as a human brain. Actual experience with vision system capabilities has given us a true appreciation for the wonder of God's creation. Whereas human eyes can distinguish minute density and hue variation, machines routinely see a green PWB mask, a white silkscreen legend, and silvery metallizations as "one continuous field." Human eyes easily adjust to the color temperatures of incandescent or fluorescent light or sunlight. Machines are miraculously confused by minor changes in light levels or color temperature. The best motto is: "Provide all the contrast possible between target and surrounding board." Be sure that the contrast is evaluated by comparing the reflectiveness of the board surface (mask, etc.) and target material at the wavelength of illuminating light in the vision system. Materials that look very different to us may look very similar to an infrared camera, and vice versa. This is the main reason that most AOI systems carry their own lighting with each camera.

Target shape is determined by the vision system used. Better systems allow a wide latitude of feature shapes. Where the choice is up to you, consider some variation of the shape shown in Figure 8-10 since it provides visual verification of PWB etch control as well as a good optical tar-

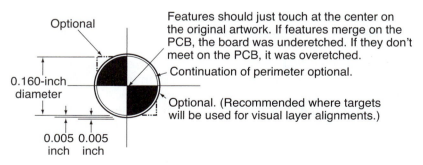

Figure 8-10   Recommended optical target shape.

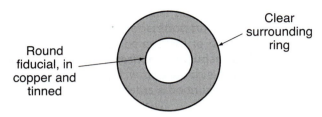

Figure 8-11 Standard round fiducial—size per the equipment manufacturers.

get. However, as mentioned in Section 8.2.1, current SMEMA standards call for a single round dot, as shown in Figure 8-11, with the round fiducial tinned and surrounded by bare substrate— no solder mask, no silkscreen—to maximize contrast. Your equipment manufacturers will tell you the type of fiducial their equipment is optimized for. In either case, it is important to note that the fiducial should always be etched into the copper layer, not added in another layer. Since the features that the assembly equipment has to deal with, such as footprint and pad locations, are etched in copper, any variation such as artwork shrink or stretch from the CAD data will be detectable in the fiducial locations. In this manner, any shifts in copper features can be compensated for by the equipment.

Dispensing inspection systems are used immediately after a dispensing operation to verify the presence, volume, and uniformity of the attachment media, such as solder paste or epoxy. Fiducials may be required for the automated registration of the workpiece prior to inspection. For special clearance and contrast rules around the features to be inspected, consult the inspection equipment vendor.

Vision-guided automatic placement refers to the use of integrated vision inspection of the component leads/metallizations and the actual placement land features. Such systems provide correction for both linear and nonlinear PWB error, plus compensation for component errors. Land-to-part systems are often used to achieve the required accuracies for fine-pitch and high I/O component placement. With fine-pitch parts, the minimal error budget necessary for acceptable placements is generally exceeded by the artwork, board, and component errors, meaning that the placement system must somehow provide less than 0.0 error. In other words, the placement system must compensate for built-in material errors if reasonable yields are to be achieved. This is the reason, as described in Chapter 5, that generally three global fiducials are required.

For design rules regarding clear space around the inspected features, and feature-to-board contrast in land-to-part-guided placement, consult the vision or placement system vendor.

Post-placement inspection involves automated vision verification of the assembly process. Simple systems may do little more than verify that components have been placed in all the required locations. More sophisticated machines will verify the component presence and will detect the correct orientation of polarity and pin-1 registration features. The most powerful machine can even verify component markings and can do a visual inspection of the components for cosmetic defects.

Such scanners are usually located between the placement and soldering operations so that any defects may be touched up easily prior to soldering.

Component spacing requirements, compatible components, and contrast rules vary widely between vendors. So, consult your vision equipment supplier for applicable design rules.

Basic x-ray inspection systems allow go or no-go solder joint analysis, while laminography systems allow viewing of different layers of the parts of the assembly, like a medical CT scanner. These systems' z-axis focal point is adjustable to allow looking at the layers of a solder joint, under components, and even inside boards. The user can additionally view wire bonds and individual layers of a multi-layer PWB. Such systemsBasic x-ray inspection systems allow go or no-go solder joint analysis, while laminography systems allow viewing of different layers of the parts of the assembly, like a medical CT scanner. These systems' z-axis focal point is adjustable to allow looking at the layers of a solder joint, under components, and even inside boards. The user can additionally view wire bonds and individual layers of a multi-layer PWB. Such systems are very useful for locating hidden defects, like broken wire bonds, gas entrapment voids, blowholes, or hidden contaminants inside solder joints.

X-ray systems are slow inspection tools, and a computer interpretation of their data requires a great deal of custom development work and computing power. Their primary use in PCA inspection is to inspect the quality of solder joints under BGA packages and any other packages whose terminations make visual inspection difficult or impossible. Their speed, or lack of it, makes it difficult to perform 100 percent inspection on high-speed lines. Therefore, they may be used for spot checks of suspect locations and also be used in failure analysis and sample testing. However, from the designer's viewpoint, an x-ray machine's ability to see through things makes the machine very tolerant of circuit layout variations. Generally, any design complying with the basic spacing and clearance rules given in Chapter 5 should present no problem in x-ray inspection.

8.2.6   Designing So Rework Is Possible

"Avoid rework, it's as much fun as a heart attack." SMT veterans have stressed this until it has become a cliché. Actually, SMCs are relatively easy to remove and replace, but the job is quite time-consuming. It requires specialized tools and skills, and rework expertise on the part of repair operators. There are a host of otherwise high-tech manufacturers who are out of control in their rework process, and they do not have any idea of what reliability their reworked boards may have. The moral of the rework challenge is twofold.

1. Give heavy weight to those design suggestions that improve manufacturing yields. Even infrequently occurring defects should be eliminated by design rather than by rework.
2. Develop controlled rework processes and procedures. Know how much heat adjacent components will see during rework. There should be a limit on rework time in any area of the board, established from observed time at reflow temperatures, before problems develop. This avoids formation of excessive copper/tin intermetallics, which degrade solder performance. Problems can include damage to the PWB, copper dissolution into solder, damage to adjacent components, and the component termination materials leaching into multiple reflowed joints. Whenever a component is removed, the solder should also be removed.

Documentation

A basic requirement of any repair/rework department is good documentation. However, THT assemblies are usually easy to second guess when documentation is not available. This is not the case with SMT. The components are small. Many look like identical twins even though they come from entirely different parents. Often, there is little or no nomenclature marked on SMCs to allow us to visually distinguish one from the next, and SMCs may be packed on the circuit like sardines in a can. After several engineering changes, the original circuit designer may not be able to figure out "who's on first" without considerable bench time. Clear and rigorously maintained documentation and revision-level markings are mandatory for SMT work.

8.3   Nonmanufacturing Issues in Design Reviews

This entire section of the design review should be under the tutelage of the marketing and sales representatives of the team. Here is where the designers must somehow take their heart off their sleeve and look at the new product like the most critical of reviewers—the customers—might view it. Below, we will develop a few tools to help in this delicate cardiac surgery.

8.3.1   Material Costs

First, how much will it have to cost the customer? If it is cheap enough, customers will be a bit forgiving. But all the patriotism in the world will not protect those U.S. manufacturers of consumer goods that cost twice as much, do not last half as long, and have less features than the goods of our foreign neighbors.

8.3.2   Reliability

Under this heading, step one must be the establishment of a reliability target. Most engineers want to build the most reliable product possible. As proof mounts regarding the cost advantages of good SQC, even the stingiest managers are joining the quality-first ranks.

Today's competitive marketplace viciously enforces attention to quality and reliability in the design of products. Customers expect products that:

- work as they are supposed to work (are quality products)
- continue to work for a reasonable time (are reliable)
- require infrequent and inexpensive repair (maintainable)
- are easy to use (user friendly)
- are safe to use and do not pollute the environment (safe)
- are esthetically pleasing (sexy)[8]

But, as pragmatic thinkers, we realize that these goals may come in conflict with material costs. No one doubts the reliability of a Rolex. But those who can afford Rolexes are few and far between. Within product cost limits, is it as reliable as it can possibly be? More importantly, is it as reliable as it *should* be? These are the questions that the team must answer.

One simple way to determine reliability (in review, the product's ability to work according to specifications for an extended period of time) is to put the product in service and see how long it lasts. Unfortunately, this simplistic approach is often too slow in flagging problems. In the process of a design review, the team should be constantly mindful of the uncompromising and often unsympathetic standards that the customer brings to the judgment of their work. To ensure the design will meet and exceed the customer's expectations, the reliability of electronic products can be predicted using methods presented in MIL-HDBK-217D.[9]

8.3.3   Marketability

The most basic step of marketability is to see that the product really does what it is intended to do. This sounds so obvious as to not need mention. However, with a little thought, we all can recall irritating examples that prove product reliability is sometimes bypassed in the headlong rush to market new products.

After we have established, by testing breadboards, that the product will do what it is supposed to do, we must ask some disturbing questions. Is this what the customer wants done? How does it stack up against the competitor's products?

Benchmarketing is a management term meaning picking the best in a given category and setting it as a benchmark that must be excelled by your design. By benchmarking every facet of a design, you can assure excellence in any product that meets the tests. Table 8-3 lists areas that generally deserve benchmarks. In some, marketing may target them to be best in a niche rather than best in the world. Our Rolex example serves to show why.

In any case, benchmarking is a valuable tool when used to defuse design ego and view a product objectively. And we know from personal experience that after we have lived and breathed and sweated with a design project for 18 months, we may need a powerful tool to separate ego from design. Yet, until this separation occurs, no objective evaluation can be done.

Credibility

Interpreting Table 8-3, by credibility we mean how believable will the device's claims be. How easy will it be for customers to understand the product's plus points? Will you be able to win customer support by simply advertising the demonstrable truth about the product?

Table 8-3   Guide to Marketing Benchmarks.

| Item | Competitive Evaluation | | | |
	System A	System B	System C	Target
Credibility	Claimed	Advertised	Customer test	Customer letters
Delivery	30 days	2 to 6 weeks	3 to 4 weeks	<2 weeks
Features	3 standard, 1 new	3 standard	3 standard, 2 new	3 standard, 3 new
Price	$1,000.00	$1,100.00	$1,500.00	$995.00
Reliability	B-10 of 1 year	B-10 of 1 year	B-10 of 2 years	B-10, 32 months
Service	Slow and costly	Few centers	Factory (slow)	Factory (fast)

*Analysis shown is of a hypothetical product and niche.

A major airline once set a milestone in regrettable advertising with a series of commercials showing travelers being herded into the innards of a plane that was filled wall to wall with cattle. The implication was that the other guy's planes had no more space than a cattle car. The visual impact and attention-grabbing value of the advertisement was undeniable. It hit right at the heart of all travelers who felt like they were being herded to the meat market in tiny plane seats. But it did not achieve sales results for the airline. Why? Because the airline running the commercial had some of the tightest seating in the industry. The clever and catchy advertisement could not cover up for a lack of credibility in the basic message.

For top-flight credibility, develop your product with features that are easily understood. Features should be clearly demonstrated to work in solving customer problems and providing customer needs. To beat the benchmark, your product's features should do this better than any competition in the niche.

Delivery

Here, design for manufacturability is the cornerstone of success. Will you be able to reach the customer before the competition reaches him with one of theirs? Delivery alone has decided many a sales contest when the customer's need was critical. Delivery is also important in obtaining the impulse buyer's nod. These two substantial customer segments are virtually hidden from view for the company that lags competitors in delivery.

Features

Benefits are what sell a product. Features are always related to advantages and benefits. For instance, compare a cellular phone to a radio telephone of the past. The cellular phone features automated cell assignment and central-station switching. This means that the phone automatically switches to the nearest receiving station as you move through a city. The advantages are clear crisp sound rivaling fixed-line telephones and enough frequencies to meet the user demand. Older radio telephones suffered from poor reception in some areas, and there just were not enough frequencies to go around. The benefit to the user is a mobile phone that provides the communications capabilities of an office telephone, regardless of where you take it. The customer likes it because it lets him keep in touch, not because he knows it employs cell-switching technology. In other words, the benefit sells the product, not the feature. When viewed by benefits, does your product exceed the market benchmark?

Reliability

Has the product been tested to establish that it will perform as specified over its full life cycle? Can these tests be demonstrated to those customers concerned with reliability and quality for their money? Is the product more reliable than the competitive benchmark?

Service

Even the most reliable hardware occasionally needs a tuneup. Customer enthusiasm can turn to bitterness if a good product costs too much or if service takes too long. And customers are most unsympathetic when they pay their good money for service, and the product still does not work. Is the product service strategy, implementation, pricing, speed, and testing the best in your market?

Price

There is nothing arcane about this section of the marketability evaluation. If you can develop a product that meets and exceeds competitive benchmarks on the other five fronts and it can be sold profitably for less than all the other market entries, you will have a story customers are likely to find interesting. Advertising and selling such a product is a joy. Sales can look a customer straight in the eye and categorically state, "For the money, we just do not think you can beat us." No sales voodoo or hocus-pocus sells like this sort of honest conviction.

8.3.4 Conclusion

When the team designs a product that fits all the SMT rules of Chapters 1 through 7, meets any required rules in Chapters 9, and then beats all of the above six market benchmarks, get set for a resounding success. You will have it.

8.4 Review

After reading and understanding this chapter, you should be able to answer these questions and note the reference location for the information in the chapter.

1. What is an "NIV" program?
2. Without a NIV program, what resources are required for incoming inspections?
3. What tests should be conducted on bare board material and prepregs?
4. Give four examples of board sizes that could be fabricated without scrap from a 24" × 36" panel.
5. Why are clear areas needed on boards?
6. What is the maximum allowable board coplanarity?
7. What are two issues that are leading to the phaseout of PLCC packages?
8. For an SOW-20 package, assume that the allowable lead co-planarity is 5 mils. If a 7-mil stencil is used to deposit solder paste onto the board, what is the worst-case lead penetration percent?
9. Why should chip resistors be mounted with their resistive element (and markings) facing up?
10. What is "selvage" on a PCB/panel?
11. What circuit uses can selvage have?
12. Why is even heatsink mass distribution important?
13. Why are gas escape holes necessary in the wave/flow soldering process?
14. What is the purpose of solder thieves? Sketch an example of one on a footprint.
15. Explain the 45-degree and 60-degree inspection rules.
16. What is the purpose of global fiducials? Of local fiducials?
17. Explain two uses of an AOI system.
18. What are x-ray systems used for in PCB inspection?

8.5   References

1. Paul, Willard, Jr., *Reader's Digest,* March 1988, p. 47.

2. Barton, Martin, "Implementing Surface Mount Technology (SMT) Into Telecommications Products," *Proceedings of the 1987 SMTA Technical Symposium,* SMTA, Edina, MN, October 28, 1987.

3. Boccia, Louis J., "A Simplified Geometrical Model for Lead Planarity," *Printed Circuit Assembly,* April 1988, p. 12.

4. Haimovich, Joseph, Ph.D., "Intermetallic Compound Growth in Tin and Tin-Lead Platings Over Nickel and Its Effects on Solderability," *12th Annual Electronics Manufacturing Seminar Proceedings,* Naval Weapons Center, China Lake, CA, February 1988, p. 51.

5. van Buul, J.A.N., and Gnant, Russell S., "Advances in Hybrid Chip Placement," *Semiconductor International,* September 1987.

6. Mead, Dr. Donald C., "Machine Vision in SMT," *Assembly Engineering,* September 1987, p. 40.

7. Amick, Christopher G., "Machine Accuracy Requirements for Surface Mounted Assembly," *Proceedings of the Society of Manufacturing Engineers Assembly Conference, Society of Manufacturing Engineers,* Ann Arbor, MI, 1986.

8. Pennucci, Nicholas J., "Estimating Reliability Using MIL-HDBK-217D," *Quality,* September 1986, p. 40.

9. MIL-HDBK-217D, *Reliability Prediction of Electronic Equipment.*

Hybrid Circuits and
Multi-Chip Modules (MCMs)

Objective

The object of this chapter is to introduce the reader to various issues that apply to thick-film and thin-film hybrid circuits, and to three types of multi-chip modules (MCMs).

- define hybrid circuits
- discuss materials for hybrids
- discuss thick- and thin-film hybrids
- discuss embedded components in hybrid circuits
- Identify TCE mismatch issues
- use the Manson-Coffin equation
- discuss types of MCMs (MCM-L, MCM-D, and MCM-C)

Introduction

Together, surface-mount and hybrid technology make a powerful team. In this chapter, we will focus on how they can be allied to miniaturize and improve electronic circuitry. The circuit density of SMT, the power-dissipation ability of a hybrid substrate, and the fact that THTs can be

---

*NOTE: While we will not take up MCMs until the end of this chapter, all the design rules for ceramic and polymer hybrids presented in the first seven sections of the chapter are germane to that discussion.

lead prepped for surface mounting can combine to make a 3-watt DC/DC converter in 6.45 square cm (1 square inch) of board area.

9.1   Overview of Hybrid Technology

9.1.1   The Changing Hybrid Market

Worldwide, the hybrid marketplace is being carried along with the continued market growth of surface-mount components. One report indicates that over the five-year period ending in 2007, Low Temperature Co-fired Ceramic (LTCC) hybrids will have the highest growth rate among hybrid circuit types,[1] and we wil concentrate on LTCCs in this chapter.

SMT, a 50-Year-Old Hybrid Technology?

The dramatic changes that are underway in hybrid devices as a result of SMT should be sufficient evidence to set aside any notion that surface mounting is the same old stuff that hybrid engineers have been doing for the past 50 years. Indeed, components have been surface mounted on hybrids since the late 1950s. The oldest SMT circuit one of the authors (Hollomon) has seen was a military product designed by Centralabs in World War II. But, more recent advances in IC and passive component packaging have brought a host of new SMT benefits to the designer of hybrid circuits. Hybrid assemblies may now contain a variety of silicon, ceramic, and SMT devices that would have seemed like a Jules Verne fantasy back in 1960.

What We Mean by Hybrid

Definitions for "hybrid" are just about as plentiful as hybrid circuits themselves. So let's start our discussion by defining "hybrid." As befits the twenty-first century, let's examine some definitions that can be found on the World Wide Web:

(1) A combination of passive and active electrical devices on an insulating substrate to perform a complete circuit function; or, (2) A combination of one or more integrated circuits with one or more discrete components; or, (3) The combination of more than one type of integrated circuit into a single package.

At: http://www.icinsights.com/links/glossary.html

An insulating base material with various combinations of interconnected film conductors, film components, semiconductor die, passive components, and bonding wire which form an electronic circuit.

At: http://www.eppic-faraday.com/glossary.html

A microcircuit consisting of elements which are a combination of the film circuit type and the semiconductor circuit type, or a combination of one or both of these types and may include discrete add-on components.

At: http://www.eleceng.adelaide.edu.au/Personal/alsarawi/Packaging/node80.html

Mounting technique, where several elements are placed on a substrate (Al_2O_3-ceramics) by special pastes by silk-screen print, followed by a burning process. Semiconductors are bonded "naked" (without package) or soldered in SMD packages. Complete modules are often founded in synthetic resin.

At: http://www.mypage.bluewin.ch/bm98/faq/glossary.htm

As is apparent, there is not universal agreement on the definition of a hybrid circuit.

Note that by some of these definitions, a standard through-hole component FR-4 PWB with one inductor formed by a spiral trace of printed wiring may be a hybrid and an alumina ceramic circuit filled with chip-and-wire ICs but having no substrate integral elements may not be a hybrid. Hmmm. Admittedly, classifying hybrids has its humorous challenges. So, we will do our best with using the term.

9.1.2   An Overview of Thick-Film Processing

Thick film refers to the printing method used to fabricate circuit board features on hybrid circuits named for the process. Thick-film conductors and circuit elements are printed onto circuits using screen or stencil printing techniques. Thick-film printing techniques typically produce film thicknesses from 7.5 microns (~0.0003") to 12.5 microns (~0.0005"). Thin films, as described below, are generally from 200 angstroms (0.0000079") to 5 microns (~0.0002").[2]

Figure 9-1 shows a typical process flow that would be used to build a multi-layer thick-film circuit. As indicated, these techniques allow the construction of resistive, capacitive, and inductive elements as well as circuit conductors and bonding sites. Special techniques also allow the construction of thick-film optoelectronic and electroluminescent elements.

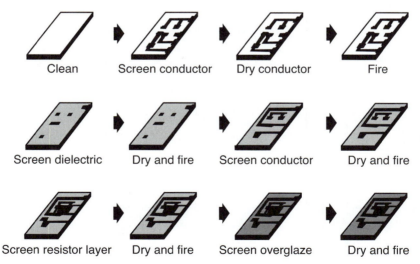

| Clean | Screen conductor | Dry conductor | Fire |

| Screen dielectric | Dry and fire | Screen conductor | Dry and fire |

| Screen resistor layer | Dry and fire | Screen overglaze | Dry and fire |

Figure 9-1   Typical process flow for multi-layer thick-film circuits.

9.1.3   An Overview of Thin-Film Processing

Thin films in the thickness range of 200 to 5000 angstroms are produced by vacuum deposition using sputtering or pyrolytic evaporation techniques. Evaporative deposition through a mask may be used to build up additive circuitry but subtractive processes are more common. Subtractive circuits are produced by sequential evaporative deposition of thin films over the entire substrate surface followed by a photolithographic removal of the undesired portions of the film. In the subtractive process, the resistive coating is generally applied first and then covered by a conductive film. The desired resistors are exposed by conductor etching. A typical thin-film process flow is presented in Figure 9-2.

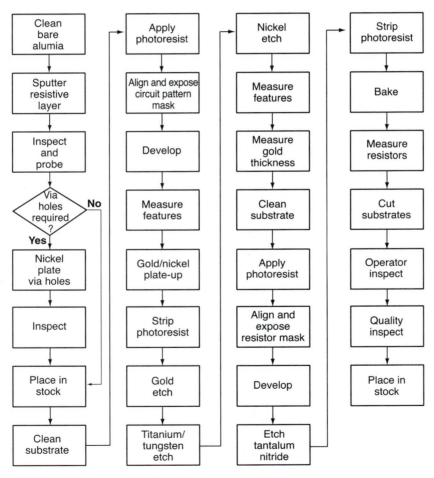

*Thanks to Amplica, Inc., of Newbury Park, California, for technical assistance with this flowchart.

Figure 9-2 Typical process flow for thin-film hybrid substrates.

The preceding discussion is an introduction to what a hybrid is. Since our focus here is on SMT, we will have to leave the in-depth study of this technology to other texts, a number of which are listed in our bibliography.

Turning our attention to SMT, let's explore the benefits and limitations of surface mounting for hybrid circuits. We will look at the rules for SMT use in hybrid circuits today, and we will study examples of actual circuits where these rules have been applied.

9.2   What SMT Offers in Hybrid Circuitry

Surface mounting must clearly provide hybrid engineers some important benefits, otherwise the usage growth that continues would not be happening. What are these benefits? And conversely, what can hybrids offer to SMT? When would SMT limitations speak against its use in hybrids? Just how do SMT methods compare to chip-and-wire density? We will answer these questions in the following pages.

9.2.1   Benefits of SMT for Hybrids

SMT benefits for hybrid circuits are somewhat different than the plus points we listed for SMT over THT. While SMT offers substantial real estate savings when compared to THT, it is seldom possible to shrink a hybrid circuit dramatically just by switching from chip-and-wire ICs to SMC packaged devices. But, SMT does provide hybrid engineers with some powerful new tools. In some cases, it may even save board real estate.

Miniaturization

SMT presents a mixed bag in this category. TAB procedures may allow greater miniaturization than chip-and-wire techniques for complex ICs. Some fine-pitch packaged SMCs also compete very well with chip-and-wire density. But 1.27-mm-pitch (0.050") packaged devices generally do not provide the density available with classical hybrid approaches, so miniaturization of existing hybrid circuits would probably involve a combination of custom VLSI and SMT. We will explore the factors influencing SMT versus chip-and-wire density further in Section 9.2.4.

Component Protection

Protection of chip-and-wire ICs on a hybrid circuit is a major element of hybrid design. For very benign environments, a simple conformal coating may provide all the protection that is needed. More hostile surroundings often dictate that the entire hybrid be hermetically sealed within a protective package. Such protective packaging can form a major element of the cost of a hybrid circuit. In contrast, SMC packages generally offer all the protection needed for a

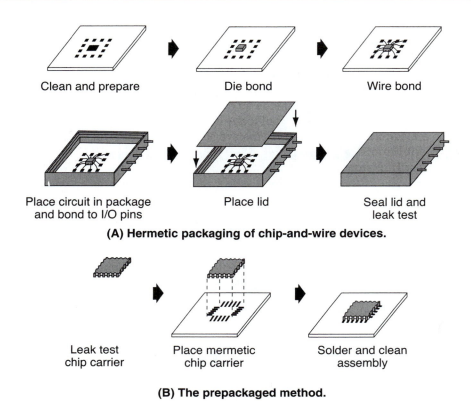

Clean and prepare　　　　　Die bond　　　　　Wire bond

Place circuit in package　　　　　Place lid　　　　　Seal lid and
and bond to I/O pins　　　　　　　　　　　　　　　　leak test

(A) Hermetic packaging of chip-and-wire devices.

Leak test　　　　　Place mermetic　　　　　Solder and clean
chip carrier　　　　　chip carrier　　　　　　assembly

(B) The prepackaged method.

Figure 9-3　Comparing approaches for the hermetic protection of a die.

die. Figure 9-3 compares a hermetically sealed chip-and-wire hybrid with a hybrid using hermetic chip carriers.

Component Testability

Bare dies are from difficult to impossible to fully parametrically and environmentally characterize before they are wire bonded to the hybrid circuit. By substituting SMT ICs, the hybrid manufacturer can fully test devices before assembly. Thus, SMT hybrids tend to provide significantly higher manufacturing first-pass yields. Yield improvements translate directly into lower manufacturing costs.

Ease of Rework

Where chip-and-wire dice are defective, the wires and eutectic or epoxy die bond must be removed. The bonding sites must be cleaned and reconditioned and a new IC must be die-bonded to the substrate, wire-bonded to the circuit, and tested. This process takes time, costs money, and risks irreparable damage to the circuit assembly. So, as you are doing the repair, you pray the replacement IC passes the first test and does not have to go through the whole cycle over again. Using SMCs lets the hybrid manufacturer pretest components and thus reduce the need for rework. Therefore, when rework is needed, it is a simple matter of spot heating leads and pads with special tools.

Reduction in Screenings

Each successive conductive and dielectric screened layer adds both cost and reliability concerns to a thick-film hybrid. Several resistive screenings may be required for different resistor values. By substituting discrete resistors for odd-value resistances, it may be possible to reduce resistive screenings, thus lowering costs, raising substrate yields, and boosting reliability.

Complex circuits requiring multi-layer hybrid interconnect structures are costly. As of this writing, multi-layer hybrids are limited to about one-third the number of layers achievable with polymer MLBs. Using SMT discretes, hybrid designers can replace screened crossovers with discrete components, dramatically lowering the cost of the substrate and the final product. Figure 9-4 compares multi-layer hybrid and crossover techniques that are used to achieve the same circuitry.

Via Fabrication

Vias are costly in all substrate technologies, but particularly so in ceramic boards. Zero-ohm resistors are used as a jumper to perform a crossover without resorting to extra layers and associated vias. The one-cent part compares very favorably with the cost of two vias and a trace on a secondary layer.[3] Figure 9-4 shows the use of just such an SMC to avoid layer interconnect vias.

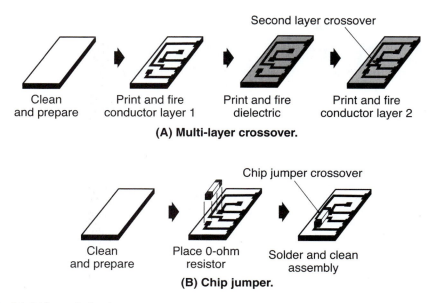

Second layer crossover

Clean and prepare   Print and fire conductor layer 1   Print and fire dielectric   Print and fire conductor layer 2

(A) Multi-layer crossover.

Chip jumper crossover

Clean and prepare   Place 0-ohm resistor   Solder and clean assembly

(B) Chip jumper.

Figure 9-4   Multi-layer hybrid versus component crossover techniques.

Table 9-1 Comparison of Typical Resistive Ink Systems.

Parameter	Thick-Film Technology		Thin-Film Technology		
	Ceramic	*Polymer*	*Chromium Si*	*Ni Chrome*	*Ta Nitride*
Conductor ink (typical)	Ag, Au, Cu, PdAg, PtAg	Ag, Au, Cu		Al, Au, Cu	Al, Au, Cu
Noise (dB)	−35 − +20	0 − +35		<−30	<−10
Power handled (watts/cm²)	≤15.5	3.875		≤15.5	6.2
Resistance range (ohms/sq.)	1 − 1000M	10 − 100K	3.4K − >5K	10 − 300	5 − 500
Resistance stability in %ÅR (1000 hours @ 25°C/50% RH)	≤±0.25	≤±2		≤±0.1	≤±0.1
Resistance stability in %ÅR (rated power 1000 hour/25°C/50% RH)	≤±0.25	≤±3		0.1	≤±0.1
Resistance stability in %ÅR (thermal/150°C 1000 hours)	≤±0.25	≤±2	<0.1	≤±0.1	≤±0.2
TCR tracking (ppm)	10	50		1	1
TCR (ppm/°C)	±50 − 300	±250 − 1000	−5 to −10	±0 to 50	±40
Trim stability, in ÅR	≤±0.25	≤±2		≤±0.01	≤±0.01
Voltage coefficient resistor (ppm/v)	0.5 − 5	25 to 100		0	1

Substrate Limitations

In classic hybrid approaches, polymer substrates are restricted by the limitations of polymer-compatible resistive inks, as shown in Table 9-1, and by die and wirebond site integrity. By allowing use of discrete resistors for high-tolerance requirements and where TCR is critical, SMT reduces concerns about polymer thick-film (PTF) ink systems. And, packaged SMC semiconductors eliminate concerns about die and wire bonds on polymer-based systems.

9.2.2 Benefits of Hybrids for SMT

The benefit equation is far from a one-way street. Hybrid techniques provided the technological seeds for SMT, and hybrid methods have a great deal to offer SMT engineers. Below, we will cover some hybrid engineering solutions to surface-mount problems.

Compatible TCEs

When CLLCCs are required for the hermetic protection of ICs, engineers must find a way to eliminate TCE mismatch problems. Adding leads to the chip carriers can solve the problem, but leads add cost and introduce an additional reliability concern by doubling the number of solder joints. Leads also increase the standoff above the board. For the sake of cleaning, a greater standoff is a benefit, but for close-order stacking of cards, it is a decided disadvantage. These considerations are illustrated in Figure 9-5.

One way to attack the TCE mismatch problem without resorting to leads is to use a substrate matching the TCE of the ceramic chip carrier. Several choices are available. Alumina ceramic is one of these. Alumina is the same material that the chip carrier package is made from, and also the material of choice for most hybrid substrates.

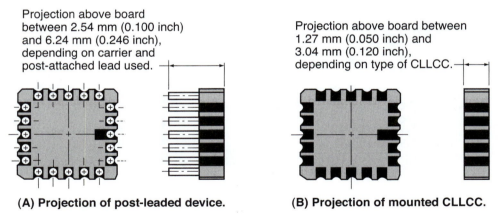

(A) Projection of post-leaded device.   **(B) Projection of mounted CLLCC.**

Figure 9-5 Comparison of post-leaded and leadless CCCs.

Using 96 percent pure alumina substrates, small- to moderate-size CLLCCs may be safely mounted direct to the board, even when high dissipations are involved. Large chip carriers having low internal dissipation may also be direct mounted to compatible substrates. The upward limit is established by the temperature delta between the device under power and the substrate and the overall chip carrier size, as shown by the following formula and the diagram of Figure 9-6.[4]

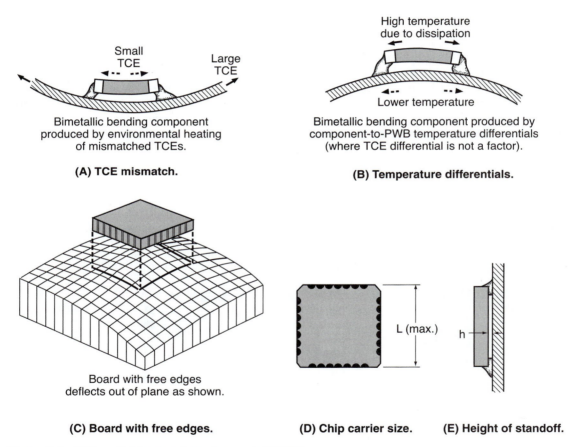

(A) TCE mismatch.

Bimetallic bending component produced by environmental heating of mismatched TCEs.

(B) Temperature differentials.

Bimetallic bending component produced by component-to-PWB temperature differentials (where TCE differential is not a factor).

(C) Board with free edges.

Board with free edges deflects out of plane as shown.

(D) Chip carrier size.

(E) Height of standoff.

Figure 9-6 Differential temperature stresses in CLLCCs.

Power Cycling Produces Complex Stresses

The stresses produced by power on/off cycles are complex, as shown in Figure 9-6. TCE mismatches between ceramic components and polymer boards act like a bimetallic strip, producing a bending stress (Figure 9-6A). Differential heating in component and board tends to oppose this bending. Acting on substances of matching TCE, as shown in Figure 9-6B, differential heating would produce a bimetallic bending effect in the opposite direction. Power dissipation from a mounted chip produces an out-of-plane displacement. This is illustrated in Figure 9-6C.

The Manson-Coffin Equation Modified

Expanding on the original work represented in the generalized Manson-Coffin equation, Norris and Landzberg of IBM modified the formula to define and model the complex bending stresses to account for the effects of maximum temperature and frequency of temperature cycling. Shah and Kelly added consideration of the dwell time at high temperature.

Based on his own work plus research from Hall and his associates, Englemaier of Bell Laboratories modified the original Manson-Coffin equation, producing the following model:

$$N_f = \frac{1}{2} \left(\frac{\Delta\tau}{2\epsilon'_f} \right)^{(1/C)}$$

(Eq. 9-1)

where,

N_f = Mean cycles to failure,

ϵ'_f = Fatigue ductility coefficient (0.325),

C = Fatigue ductility exponent (-0.442 (-6) $\times$ $10^{-4} \bar{T}_s + 1.74 \times 10^{-2} \ln(1 + f)$),

$\Delta\tau$ = Shear strain range $\left(\dfrac{L}{\sqrt{2h}} \ \Delta(\alpha\Delta T)_{ss} \times 10^{-4} \ \{\text{in percent}\} \right)$,

$\bar{T}_s$ = Mean cyclic solder joint temperature (°C),

f = Cyclic frequency ($1 \le f \le 1000$ cycles/day),

L = Chip carrier size (Figure 9-6D),

h = Height of standoff (Figure 9-6E).

and, $\Delta(\alpha\Delta T)_{ss}$ = The in-plane steady-state thermal expansion mismatch $\alpha_s(T_s - T_o) - \alpha_c(T_c - T_o)$, where

α_c = TCE of LCC,

α_s = TCE of substrate,

T_c = Temperature of chip,

T_s = Temperature of substrate,

T_o = Temperature in °C (power off, steady state),

Printed Components

Another advantage of ceramic substrates is that they lend themselves well to the process of printing components directly on the board. In our discussion of components used as crossovers, we mentioned that SMT often allows the elimination of multiple screenings when printing resistors with a variety of ink-resistive values. Still, hybrid engineers prefer to print components where practical for several reasons. First, by in situ fabrication, solder-bond requirements are reduced and reliability is increased. Also, printed resistors can save space. Resistors as small as 1 mm (~0.040") are simple to produce using thick-film processes. With care, resistive tracks of <0.7 mm (0.030") are producible, and thin-film processes reduce the above limits by more than an order of magnitude.

Small inductances and low-value capacitors may also be fabricated directly on the substrate. Electroluminescent elements and photosensitive arrays can be produced using thick- and thin-film processes. Thick-film carbon inks are also being used to replace selective gold plating for keyboard switch contacts and gold contact fingers.

Thermal Management

The thermal conductivity of alumina ceramic is about twenty-three times that of organic polyimide film, and thirteen times greater than epoxy/glass composites. Further, beryllia ceramic is about eighty-eight times as thermally conductive as epoxy/glass. High-dissipation circuitry can use the increased thermal conductivity of ceramic substrates to great advantage.

9.2.3 Limitations of SMT for Hybrids

Above, we studied some of the plus points of mixing SMT with hybrid technology. Fortunately for the engineering profession, there are still factors that must be weighed, and trade-offs that must be analyzed before selecting surface-mount technology for a hybrid circuit. Below, we will look at some hybrid application features that work against SMT's benefits.

Miniaturization

As we have said, some types of SMT will save space on some types of hybrids. The poor miniaturization candidates for SMT are those circuits with few high pin-count devices. Also, the type of SMC under consideration determines the level of real estate that is required, in comparison to chip-and-wire devices. High pin-count TAB devices and fine-pitch packages fare well in this comparison. Low pin-count devices fare very poorly, and 1.27-mm (50-mil) SMCs are generally on the losing end regardless of pin count. We will look more closely at the deciding factors in the real estate wars next, in Section 9.2.4.

Package-Imposed Thermal Limits

Even with cavity-up ceramic chip carriers soldered directly to ceramic boards, there is some standoff height between the carrier and the board. Thus, packaged component thermal resistance is likely to be greater than that of the same die bonded directly to the board. This is particularly true where the primary dissipation path is through the substrate. Thermal-transfer

Table 9-2 Properties of Die-Bonding Materials.

Material	Property	
	Thermal Conductivity (Watts/°C/cm²)	TCE (ppm/°C @ 25°C)
Epoxy, conductive	0.0062	15–16
Epoxy preforms	0.00155	15–16
Gold/silicon eutectic	0.8835	
Gold/tin solder 80/20	0.6975	
Silicon	0.3656	7.38

epoxies can help the package-to-board dissipation, but they still do not approach direct chip mounting. Table 9-2 shows the thermal properties of some typical die-bond materials.

Substrate Cost
It is not accurate to make the unqualified statement that thick-film network boards are more expensive than PWBs with the associated resistors needed to equal the hybrid board. It is common practice to build hybrid circuits on FR-4—or even paper phenolic substrates—using polymer ink systems. Such substrates may be quite cost effective compared to PWB and chip-resistor assemblies (comparing apples and apples). But, the limited resistance range and TCR stability of polymer inks often dictates the use of ceramic systems for hybrids. With ceramic thick-film substrates, costs are generally higher than for analogous polymer board assemblies with chip resistors.

Substrate Size Limits
Just as we cannot categorically rule in favor of polymer PWBs for cost, we cannot give polymer boards the nod in a substrate size contest either. Hybrid circuits can be made with polymer inks printed on very large polymer- or porcelain-coated metal-core boards. However, the ceramic board, which is the foundation of the hybrid industry, does carry some rather restrictive size limitations. Above 100 mm (~4.0") square, substrate costs per square inch begin to climb sharply. The absolute limits are somewhere between 150 mm (~6.0") and 300 mm (~12.0") square depending on the processing parameters and substrate thickness.

Frequency Limits of Ceramic Substrates
Ceramic substrates restrict frequencies due to their relatively high dielectric constant. Alumina has a K ranging between 9 and 10. Microwave circuits operating at 1 GHz are routinely built on beryllia, which has a K about 70 percent that of alumina. But X-Band police radar runs at 24.150 GHz, and such frequencies dictate low-K polymer substrates. Of course, hybrid techniques may still be used on polymer boards.

Higher Capacitance of Ceramics

Ceramic insulation materials result in higher capacitances than polymers typically exhibit. Thus, signal traces of a given geometry will have a lower impedance. For bipolar devices, this lower impedance translates to a higher driving power and to lower termination resistances. Thus, power dissipation will be higher. This relation limits the use of ceramic substrates with power devices.[5]

Via Fabrication on Ceramic Substrates

There are two types of vias in ceramic substrates. One type is formed in the thick-film printing process, and it communicates between layers above the ceramic substrate. The other is a through-feed made by metallizing a hole in the ceramic. This second type provides one method of interconnecting circuitry on opposite sides of a dual-side substrate. However, the feed-through-type via is expensive, and there are a limited number of shops qualified to supply it.

Limits to Number of Layers

Regardless of the substrate material, there are limits to the number of layers that can be reliably produced with the thick- and thin-film processes.

Thin films are usually restricted to single conductive layers. However, Brown-Boveri & Cie, Inc. of Switzerland announced a multi-layer process at the International Society for Hybrid Microelectronics (IMAPS) International Symposium on Microelectronics (1986). By using a spin-coating of polyimide as a dielectric layer, they reported success with dual conductive layers and expected to carry the technology beyond two layers.[6] That same year, Honeywell discussed a four-layer process using thin films of copper and polyimide on ceramic substrates.[7]

As of this writing, most thick-film circuits have less than ten conductive layers. Toshiba reported a new process based on a polymer silver-conductive ink printed on polycarbonate resin film. Local resin flow into prepunched areas under heat and pressure was used to form vias between layers. Resistors were sputtered from NiCr. The authors stated that they had built a sixteen-layer prototype and that the process is not particularly constrained as to the number of layers.[8] None of these technologies, however, have approached the interconnect densities of laminated polymer MLBs.

Technology and Manufacturing Problems

Surface mounting alone is a rather formidable technological hurdle to clear. Hybrid technology, if it is not already in-house, certainly adds to the learning curve. Even with outside contractors helping, taking on SMT and hybrid manufacturing at one sitting is a challenge. Where outside contractors are not used for hybrid manufacturing, the acquisition of hybrid process equipment is also a deterrent to mixing hybrid and surface-mount technologies.

LTCC Design Rules

The bulk of ceramic technologies we have been discussing will be fabricated using Low Temperature Co-fired Ceramic (LTCC) technologies. While we have introduced some design rules here, we cannot cover LTCC design in detail. Each manufacturer has its own rules, and three examples of this can be found at the websites of:

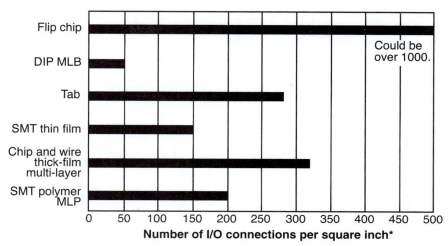

Figure 9-7   Densities of interconnect technologies compared.

CTS Microelectronics: http://www.ctscorp.com/components/ltcc_home.htm
Amkor: http://www.wadsworth-pacific.com/hirai/amkor.pdf
LTCC: http://www.ltcc.de

In addition to the above sources, the International Microelectronics and Packaging Society (IMAPS) has a great deal of information available at their conferences and technical papers. Check for their information at http://www.imapas.org.

9.2.4   The Density Impact of SMT in Hybrid Designs

Actual circuit density is heavily influenced by application, IC to passive-component mix, IC technology level, and a host of other factors. Therefore, our comparison is, at best, an approximation based on typical conditions. With this disclaimer stated, Figure 9-7[9] compares the densities of several interconnect technologies.

9.3   Materials for SMT Hybrids

SMT hybrid materials differ somewhat from those we are familiar with in building SMT assemblies on FR-4 print-and-etch substrates. Let's review the materials, pointing out those areas where differences exist and those areas where our previous discussion of materials apply to hybrids as well.

9.3.1   Discrete Components

We will start our material review with discrete components, where we will find a great deal of similarity with our discussion of components covered in Chapter 2.

Resistors

Chip resistors for SMT hybrids are the same as chip resistors for polymer SMT. The major difference that hybrid technology brings to resistor selection is the possibility of fabricating the resistor directly on the substrate in lieu of placing a resistor chip. In situ components will be discussed below.

Trimming potentiometers are generally added as discrete components and are specified for hybrids in the same way that they are specified for standard SMT assemblies.

Capacitors

Discrete capacitors are selected in the same way that they are specified for standard FR-4 SMT assemblies. However, the use of palladium silver as a metallization on a ceramic board may relieve the need for diffusion barriers to protect the chip leads from silver leaching. Where silver is present in the substrate metallizations, it is common practice to use a solder containing 2 percent silver (such as 62/36/2), which substantially slows leaching.

Tantalum or aluminum electrolytic SM capacitors are available. Single-layer silicon capacitors are also used on hybrids.

Capacitors may also be fabricated in the hybrid substrate, but only small capacitances are practical, using thick-film techniques. Therefore, capacitors are usually added as discrete components and are not fabricated in situ.

Trimmer capacitors are specified for hybrids in the same way that they are for standard SMT assemblies.

Transistors

With transistors, we have several options on a hybrid. The familiar SOT family is commonly used. For high-frequency work, the leadless inverted device (LID) and MO types are also available. In addition, direct chip mounting and wire bonding is another alternative that may be considered with hybrid technology.

Inductors

Inductors may be formed in situ, but such devices are limited to very small inductances. Chip inductors like those presented in Chapter 2 are commonly used where higher values are required. Tunable coils are also available and are selected in the same way they are for non hybrid SMT assemblies.

Thermistors

Chip thermistors for hybrid circuits are the same as we have studied for other SMT applications and should be specified for hybrids per the guidelines given in Chapter 2.

Other Components

The full range of components mentioned in Chapter 2 may also be surface mounted on hybrids. To review, this includes connectors, sockets, switches, transformers, relays, fuses, and jumpers (or zero-ohm resistors).

9.3.2   Integrated Circuits

SOICs

Originally developed for placement on small ceramic circuits for watches, the SOIC has been used in hybrid surface-mount processes since its birth in the late 1960s. Because its dual-in-line I/O allows feed-throughs in one axis under the component, the SOIC is particularly useful for single-layer hybrids. Small SOICs are quite space efficient when compared with quad I/O packages. However, at somewhere between 20 and 28 pins, the real estate advantage begins to shift toward devices having four-sided I/O.

PLCCs

In 28-pin to 68-pin packages, the PLCC provides a low-cost surface-mountable option to chip-and-wire mounting. However, the 1.27-mm (50-mil) PLCC does not rival chip-and-wire density. Above 68 pins, other packaging alternatives are often chosen for space efficiency.

CCCs

Ceramic leaded and leadless chip carriers provide a hermetic package with a TCE suitable for mounting (within our stated design rules) directly to ceramic hybrids. One caution, however. The stated rules do not include the direct mounting of large carriers where high temperature differentials will exist between the package and substrate due to power dissipation during device cycling.

QPFPs

The fine-pitch QPFP will certainly find a receptive market in SMT hybrids. It is much more space efficient than the 1.27-mm (50-mil) PLCC, coming close to chip-and-wire density. And, the quad flat pack allows a full pretest of components before placement.

QCFPs

QCFPs carry the same space efficiencies and pretest capacities noted above, plus providing a hermetic package. For high pin-count devices, the leads provide a flexural element, eliminating the concern about temperature deltas between device and board. The lead standoff also makes cleaning under the package much easier than cleaning under CLLCCs. However, the leads add Z-axis clearance requirements and do not provide the thermal path to the substrate that the leadless carrier boasts. The QCFP will be an interesting package on hybrids, but probably will not replace the CLLCC overnight.

TapePak

Even more space efficient than the QPFP and just as pretestable, the TapePak is expected to become a popular package with hybrid designers.

TAB

TAB rivals—and often eclipses—chip-and-wire densities. The development of flexible equipment for automatically excising and outer-lead bonding TAB devices will spell an increasing use of this technology, particularly for very high I/O devices.

Flip Chip

Flip chips are semiconductor dice specially prepared for face-down bonding by the reflow of tiny solder bumps grown on their bonding pads. Bumping was also described in Chapter 2. The flip-chip technique is flow charted in Figure 9-8[10] and diagrammed in Figure 9-9.[11]

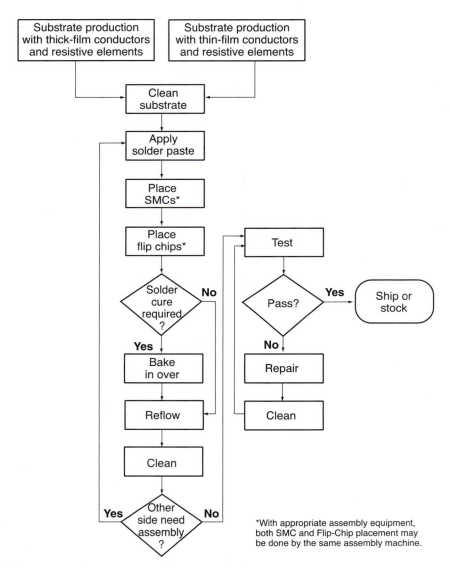

Figure 9-8   Flow chart of a typical Flip Chip process.

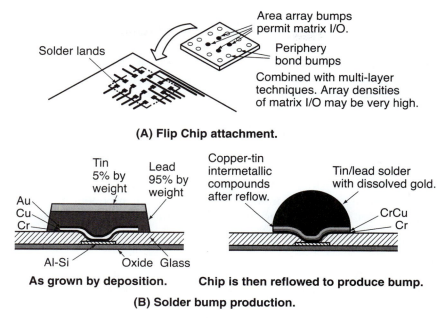

(A) Flip Chip attachment.

As grown by deposition.   Chip is then reflowed to produce bump.

(B) Solder bump production.

Figure 9-9   Diagram of a typical Flip Chip process.

9.3.3   In Situ Component Materials

Thin-film elements are restricted in value by the materials available and by board size constraints. Thick-film elements are limited in value both by the conservation of board area and the peculiarities of the screening process. As Figure 9-10 shows, squeegee pressure causes some thinning of the printed image at the center of the screen opening. For small features, this effect is negligible. But larger features show marked height variations from the edge to the center.[12]

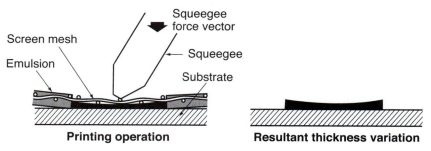

Printing operation   Resultant thickness variation

Figure 9-10   Squeegee pressure produces uneven print height.

Resistors

Thick-film resistor materials are selected to suit the substrate on which they are printed. Ceramic and refractory substrates permit the use of resistor inks with high firing temperatures. These high-firing-temperature inks exhibit better TCR stability than do polymer systems and offer the widest range of resistances of any hybrid resistive materials.

Polymer boards and coated metal-core substrates rule out the use of high firing temperatures for resistor inks. Polymeric binder inks, firing at relatively low temperatures around 220°C, allow the printing of resistors on nonrefractory board materials, including polyimides and even epoxy glass.

Sputtered or evaporated coatings of materials, such as nichrome, tantalum nitride, or chromium silicide, are used to form resistors in the thin-film process.

The layout of hybrid resistors will be detailed in Section 9.5.3. Thermal engineering for in situ resistor dissipation also differs from the familiar discrete component models. Reference 13 presents the methods for calculating and modeling dissipation from a thick- or thin-film resistor and discrete components on a hybrid substrate.[13]

Capacitors

Capacitors are formed by multi-layer thick-film techniques, printing metallized plates that are separated by dielectric layers. Figure 9-11 shows a typical thick-film approach to capacitor fabrication and restates the formula presented earlier (in Chapter 6) for the calculation of the capacitance. Note from the formula that the thin dielectrics and high-dielectric constants inherent in the thick-film process both work in favor of the capacitance that can be achieved with hybrid versus multi-layer polymer boards. Still, there are serious limits on layer construction, and the costs of multiple print and fire operations also limit the use of in situ capacitors to low values.

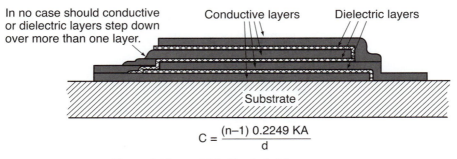

$$C = \frac{(n-1)\ 0.2249\ KA}{d}$$

Figure 9-11   In situ capacitors for multi-layer thick-film hybrids.

Inductors

Inductors may be fabricated using thick- or thin-film techniques. Figure 6-7 (Chapter 6) shows the construction details for designing a required inductance, while the formulae are given in the text. As the formula indicates, only small-value inductors can be practically fabricated in situ. A very large board area would be required to lay out even a moderate inductance. The small line-and-space dimensions that can be manufactured with thin films allow considerably higher values in a given space than do thick films.

Networks

Resistor, and even RC networks, may be fabricated in situ. In fact, a service industry has grown out of providing resistor and resistor-capacitor hybrid networks, sometimes with SMT active elements, on small SIPs. Such SIP supercomponents can move a whole neighborhood of PWB components from urban sprawl to one high-rise area. The result is space and manufacturing cost economy.

9.3.4   Substrates

Ceramic substrates (mainly alumina and some beryllia) so dominate hybrid manufacturing that the terms "ceramic substrate" and "hybrid" almost seem synonymous. However, the entire range of materials discussed in Chapter 6, Section 6.3.2, and even more exotic materials, are used for hybrid substrates. Ceramics are popular because of several properties that fit thick-film processing requirements. The relatively camber-free smooth surface of a ceramic substrate allows close-tolerance screen printing, and the refractory nature of the material permits the use of very stable, high firing-temperature resistive inks. However, by using special inks and care in printing, the thick-film process has even been applied to paper substrates.

9.3.5   Materials

Hybrid circuits use many of the same coating and attachment materials that we have previously studied for print-and-etch boards. Below, we will add to and modify our PWB materials discussion.

Dielectric Layers

Insulative layers of various types are used for crossovers, capacitor buildup, and the construction of multi-layer hybrids. Dielectrics are chosen based both on the electrical purpose they will

Table 9-3   Dielectric Material Characteristics

Property and Unit of Measurement	Dielectric Type		
	Thick-Film Ceramics	Thin-Film Ceramics	Polymer Film
Breakdown voltage (V/mil)	>500	≥100	≤100
Capacitance range (typ., in pF)	≤1000	≤50	≤500
Conductive ink styles	PdAg/PdAu/Au	Au/Cu	Cu/Ag
Dielectric constant (@ 1 kHz)	8 to 14		4 to 6
Dissipation factor range (%)	<3.5		>5
Hermetic properties	Excellent	Fair	Poor
Insulation resistance (typ., in ohms)	>10^{11}	>10^{10}	>10^7
TCE matches with substrate	~96% Al_2O_3	~96% Al_2O_3	Polyimide
Temperature coefficient capacitor (ppm/°C)	<250		
Trimming methods	Abrasive	No	Abrasive

serve and on the substrate material and conductive-ink system on which they will be applied. Table 9-3 lists some commonly used materials and details their physical and electrical properties.

Masks

As described above, hybrid dielectric layers traditionally function like the familiar solder mask of a printed wiring board. Even on single-layer circuits with no crossovers or in situ capacitors, a dielectric layer is often applied to protect circuitry and passivate resistors. Wet- or dry-film polymer mask materials may also be used on ceramic hybrids, and they find wide application on polymer-based boards.

Conformal Coatings

The full range of coating materials discussed in Chapter 5, Table 5-4 are useful in protecting SMT hybrid circuits.

Conductive Adhesives

Conductive adhesives are sometimes used in lieu of solder-to-bond chip passives to circuits where soldering would be difficult because of metallizations or limitations on the processing temperatures. There are two basic types of electrically conductive adhesives (ECAs). There are isotropically conductive adhesives (ICAs) which contain silver as their conductive component carried in epoxy. A controlled quantity of ICA is applied to each pad of a footprint and a component is then placed into the adhesive. The adhesive is cured with heat while pressure is applied to the component. Each pad and component termination are then connected by a conductive layer.

There are also anisotropically conductive adhesives (ACAs), also known as Z-axis adhesives. In these, the conducting material is spaced such that it does not conduct until pressure is applied to it, forcing the conductors together between each pad and lead. Thus, the material may be applied in a continuous line across lands and then aligned to conduct in the Z axis over land areas only. Even small separations, such as 0.25-mm (~0.010") spaces between leads, remain fully dielectric. Dielectric separation can be retained down to about 0.03 mm (~0.001"). The material offers potential relief to solder yields with fine-pitch components.

Packages

External packaging is often required for the protection of chip-and-wire hybrid assemblies. The use of packaged SMT ICs may negate the need for such a protective housing. However, flip chip and TAB technologies, which are planar mounting methods, offer limited protection for the dies and may dictate post-packaging of the assembly.

Metal packages offer both hermeticity and a wide range of form factors, but typically they add considerable cost and weight to the system. Ceramic packages are available in hermetic and nonhermetic designs, with the costs being clearly higher for the hermetic styles. Finally, light and inexpensive plastic packages may be used to shield circuits from impact and environmental contamination. Plastic packages are used where hermeticity is not required.

9.4   Special SMT Design Rules for Thick-Film Hybrids

The design rules presented in our first eight chapters are basically sound rules for SMT hybrid design. Below, we will cover those points where hybrid rules depart from PWB norms.

Please be forewarned that hybrid microelectronics is a major technology. We cannot, in one chapter, begin to catalog the tons of data available and published on the topic. For anything more than a minor use of hybrid techniques in your SMT designs, additional study will be mandatory.

9.4.1   Component Orientation

For reflow attachment of components, the general rules for Type 1 SMT circuits (presented in Chapter 5) apply for hybrid as well as standard SMT PWBs. These rules should be reviewed and might bear revision for conductive epoxy attachment. The adhesive vendor should furnish design rule guidance where Chapter 5 rules do not pertain.

In situ resistors should all be oriented for trimming from the same side if possible. This means that their long axes should be parallel, and there should be a clear area of substrate (no features) on the side that will be trimmed.

9.4.2   Interconnect Layout

Not surprisingly, hybrid thick-film techniques have a dramatic impact on minimum line-and-space rules. Fine geometries at the limit of PWB manufacturing technology are routine in hybrid thick films. Below, we will analyze the impact of thick-film technology on layout.

Traces and Spaces

The shrinking of feature sizes has continued as technologies have advanced. Minimum line separation is generally equal to minimum feature size and for hybrid circuits that can be as small as 1 mil lines and spaces.

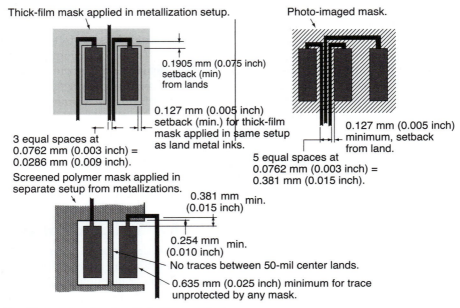

Thick-film mask applied in metallization setup.

Photo-imaged mask.

0.1905 mm (0.075 inch) setback (min) from lands

0.127 mm (0.005 inch) setback (min.) for thick-film mask applied in same setup as land metal inks.

0.127 mm (0.005 inch) minimum, setback from land.

3 equal spaces at 0.0762 mm (0.003 inch) = 0.0286 mm (0.009 inch).

5 equal spaces at 0.0762 mm (0.003 inch) = 0.381 mm (0.015 inch).

Screened polymer mask applied in separate setup from metallizations.

0.381 mm (0.015 inch) min.

0.254 mm (0.010 inch) min.

No traces between 50-mil center lands.

0.635 mm (0.025 inch) minimum for trace unprotected by any mask.

Figure 9-12   Clearance rules for thick-film conductors.

Clearances Around Solder Lands

Clearances around solder lands are determined by the type of mask used to protect the conductive traces from unwanted short circuits. Rules for this are covered in Figure 9-12, but designers should also consult their LTCC fabricators rules, as noted in Section 9.2.3.

Crossovers

Where only a few crossovers are needed, options are to use zero-ohm resistors or print and fire dielectric and conductive layers over the lines that must be crossed. Figure 9-13 illustrates both methods, and presents IMAPS guideline dimensions for screened crossovers.[14]

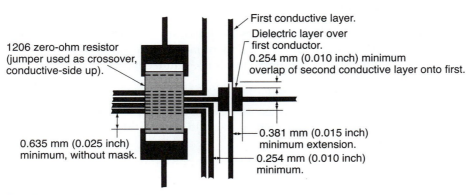

First conductive layer.

Dielectric layer over first conductor.
0.254 mm (0.010 inch) minimum overlap of second conductive layer onto first.

1206 zero-ohm resistor (jumper used as crossover, conductive-side up).

0.635 mm (0.025 inch) minimum, without mask.

0.381 mm (0.015 inch) minimum extension.

0.254 mm (0.010 inch) minimum.

Figure 9-13   Crossover design rules.

Vias

Vias may be constructed by multi-layer thick-film techniques. Very simple multi-level circuits may require no more than a few crossovers, as discussed above. However, denser circuitry often dictates that several successive layers of traces be stacked up with insulating layers of dielectric material between. When this is the case, individual thick-film layers may be interconnected using thick-film vias. Multi-layer thick-film via design rules are presented in Figure 9-14.[15]

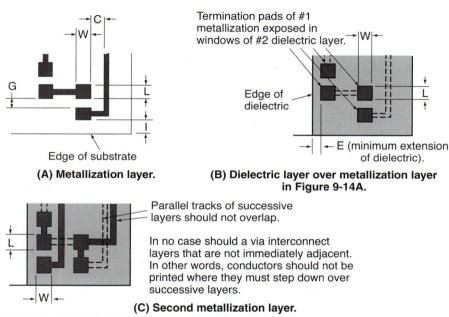

(A) Metallization layer.

(B) Dielectric layer over metallization layer in Figure 9-14A.

(C) Second metallization layer.

Parallel tracks of successive layers should not overlap.

In no case should a via interconnect layers that are not immediately adjacent. In other words, conductors should not be printed where they must step down over successive layers.

Dimension	Modifier	Manufacturing Simplicity		
		Simple (mm/inch)	Average (mm/inch)	Difficult (mm/inch)
C	min.	0.508 (0.020)	0.381 (0.015)	0.254 (0.010)
E	min.	0.508 (0.020)	0.381 (0.015)	0.3048 (0.012)
G	min.	0.508 (0.020)	0.381 (0.015)	0.3048 (0.012)
I	min.	0.508 (0.020)	0.381 (0.015)	0.3048 (0.012)
L	min.	0.508 (0.020)	0.381 (0.015)	0.3048 (0.012)
W	min.	0.508 (0.020)	0.381 (0.015)	0.3048 (0.012)

(D) Table of values.

Figure 9-14   Multi-layer thick-film via design rules.

Mixing SMT with hybrid technology presents the designer with a wealth of options for interconnects. Vias, for instance, may be fabricated by thick-film techniques, as discussed above. They may also be constructed using PTHs on double-sided or multi-layer PWBs. On ceramic boards, double-sided metallizations may be joined by print through-hole or wraparound methods. The various options for layer interconnects are covered in Figure 9-15. These are often mixed and matched on polymer thick-film boards.

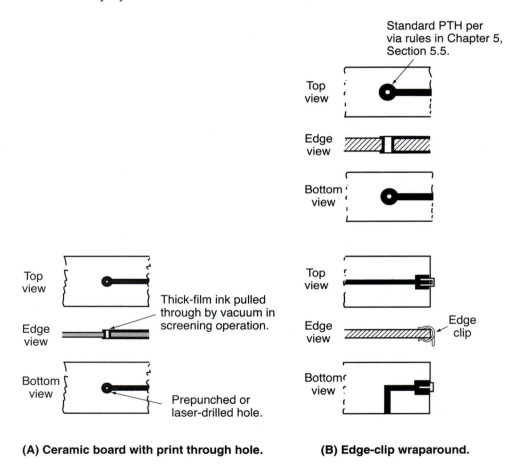

(A) Ceramic board with print through hole.   **(B) Edge-clip wraparound.**

Figure 9-15   Options for layer interconnects in SMT hybrids.

9.4.3   In Situ Component Design

Hybrids take their name from their mixture of in situ components and the components added to the board as separate elements. Therefore, it is clear that a study of SMT hybrid technology will include rules for in situ component production. Below, we will look at some rules for the design of components commonly built as part of hybrid substrates.

Resistors

Step 1 in resistor design is the selection of the resistive paste or pastes required to produce the necessary resistance values in a given circuit. Standard ranges for standard resistive pastes range from 10Ω/square to $1M\Omega$/square. Each paste can be used to design a range of resistors, for example, the 10Ω/square paste can generally be ued for resistors from 3Ω to 100Ω, and the $1M\Omega$/square paste can be used for resistors from $300K\Omega$ to $10M\Omega$. Pastes of intermediate resistances cover the middle ground. Note that "ohms/square" is an abstract term. It defines the resistance obtained from a specified thickness of ink (per paste manufacturer's print-thickness specifications) printed in a square pattern of any dimensions. To double that resistance, we would print a resistor two squares long by one square wide. To halve resistance, we could print our resistor one square long by two squares wide. In theory, as long as we could maintain print thickness, it would not matter whether the unit of measure was square angstroms or square miles, the resistance would remain the same. Of course, the maximum power dissipation of the 1 square-mile resistor would be considerably higher than that of the 1 square-angstrom device. As an example, Figure 9-16 shows a resistor printed 1 square wide by 5 squares long. If the paste used for this resistor is rated at 100Ω/square, the resistance between the two cross-hatched substrate terminations would be $5 \times 100 = 500\Omega$, plus or minus the paste tolerance.

Since some paste resistive values overlap, the designer's job is to choose the minimum number of pastes required for the fabrication of all in situ resistors. Remember, each paste must be applied in a separate print-and-fire cycle.

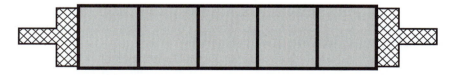

Figure 9-16   Resistive paste example for a 500Ω resistance using 100Ω/square paste.

After selecting the paste to be used, the next step in designing resistors is the calculation of the number of squares required for each individual resistor. This is determined by dividing the resistance required, in ohms, by the paste value in ohms/square. The formula is:

$$S = \frac{R}{P} \qquad \textbf{(Eq. 9-2)}$$

where,
  S = Number of squares,
  R = Resistance required in ohms,
  P = Resistivity of paste in ohms/square.

That being the case, "squares," as calculated above, is not a square area. It is, rather, a definition of the aspect ratio required to obtain the desired resistance. We generally select a resistive paste that will not produce a lopsided aspect ratio.

With the number of squares established, the form factor of the resistor must be determined. Form-factor selection is influenced by the area requirements for power dissipation, the grid of the CAD system, and the minimum values for length and width as stated in Figure 9-17.[16] For additional data on dissipation, see Reference 17.[17]

Where all the required values of resistance fall within the range of one ink, that one ink may suffice. Widely spread values may require several inks. Since each ink requires a separate print-and-fire operation, it is well to limit in situ resistors to one or no more than two inks. If a wide range of values is needed, screenings may be limited by selecting one or two inks that will produce the bulk of the resistors required. Chip resistors may be used to produce the less common values.

The best practice is a single screening unless trimming requirements force more. Resistors not screened are added as SMT discretes.

Where very-close-tolerance resistors (better than ±20 percent) are required, the resistors may be printed with larger areas (= lower values) than required and then trimmed to the required value by removing material in a trimming operation. Passive trimming involves the removal of material, decreasing the area while monitoring the increasing resistance. Active trimming is the removal of resistive material while monitoring some other electrical function that is influenced by the resistance in question. In either case, most trimming today is done with the use of lasers to vaporize patterns starting at the edges of the film until the resulting resistance value is correct. Virtually all SMT resistors are also laser trimmed.

Typical trimming begins with the laser moving in from the edge of the fired paste in a line at 90 degrees to the edge. As the resistance approaches the desired value, the laser turns 90 degrees and moves parallel to the edge. This slows the rate of resistance change per length of cut, resulting in a more precise trim.

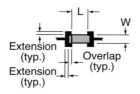

S = number of squares
S = L/W and as defined below.

A, the minimum area required, is
calculated thus: A = D/50
   where D = the maximum power
   dissipation (in watts).
W, the width of resistor, is
calculated thus: W = √A/S
   where S = the number of squares.
L, the length of the resistor, is
calculated thus: L = S · W.

Dimension	Modifier	Manufacturing Simplicity		
		Simple (mm/inch)	Average (mm/inch)	Difficult (mm/inch)
L	min.	0.127 (0.050)	1.016 (0.040)	1.016 (0.040)
W	± 1% tolerance min.	0.016 (0.040)	0.889 (0.035)	0.889 (0.035)
	± 5% tolerance min.	0.762 (0.030)	0.635 (0.025)	0.635 (0.025)
Overlap	min.	0.254 (0.010)	0.2032 (0.008)	0.2032 (0.008)
Extension	min.	0.381 (0.015)	0.254 (0.010)	0.2032 (0.008)
S	min.	0.5 sq.	0.5 sq.	0.3 sq.
S	max.	5.0 sq.	10.0 sq.	10.0 sq.

(A) Resistor form factors.

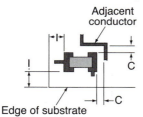

Dimension	Modifier	Manufacturing Simplicity		
		Simple*	Average*	Difficult*
C	min.	0.127 (0.005)	0.127 (0.005)	0.127 (0.005)
I	min.	0.381 (0.015)	0.254 (0.010)	0.254 (0.010)

*Values are in millimeters and inches.

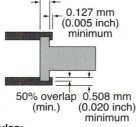

Probe pads on grid for
trimming operation.
Probe located as close as
possible to resistor pads.

Where possible,
layout resistors
to be trimmed
in uniform direction.

Clear area for trimming

(B) Resistor near another conductor.

0.127 mm
(0.005 inch)
minimum

50% overlap   0.508 mm
(min.)   (0.020 inch)
minimum

Top hat rules:
1. Preferred where voltage sensitivity is
   a concern.
2. Noise is reduced by increasing the resistor
   length and/or area.
3. Top hats are not recommended where
   tolerances of less than 1% are required.

(C) Top hat resistor design.

Matched resistors rules:
1. Matched resistors are to be of the
   same ink.
2. Matched resistors should be in close
   proximity to each other.
3. Matched resistors should be located
   on a common axis.

(D) Matched resistor design.

Figure 9-17   Screened resistor form-factor rules.

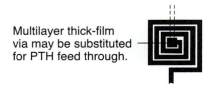

Multilayer thick-film
via may be substituted
for PTH feed through.

Figure 9-18   Thick-film construction of an on-board inductor.

Inductors

Inductors are designed in hybrids in the same way that they are designed on polymer PWBs. Figure 6-7 (Chapter 6) shows some typical geometries and formulae for calculating the inductance of in situ inductors. Figure 9-18 illustrates use of multi-layer thick-film techniques to construct an on-board inductor.

Capacitors

Chapter 6 (Figure 6-6) gave us the following formula for in situ capacitor values:

$$C = \frac{(n-1)\ 0.2249K\ (A)}{d}$$

where,

    C = The capacitance in pF,
    n = The number of conductive plates,
    K = The dielectric constant of the insulative layers,
    A = The area of the electrode plate in square inches, in square inches,
    d = The thickness of the dielectric layer between each electrode, in inches.

We discussed capacitors in Section 9.3. To review, the preceding formula is valid for both thick-film capacitors and standard polymer PWBs. However, several characteristics of thick films have a bearing on the capacitance values achievable with the technology. There are practical limits on the number of conductive layers that we can print with thick films. Thus, "n" may be limited to less layers in thick films than in MLBs. Conversely, the relatively high dielectric constant, "K," of glass-frit-type insulative layers, and the thin "d" dimensions producible in thick films work to the benefit of capacitance values.

9.5   Special SMT Design Rules for Thin-Film Hybrids

The design rules in the first eight chapters, plus the preceding thick-film discussion, form the backbone of thin-film SMT hybrid design. Below, we will cover those points where thin-film technology requires a departure from this.

9.5.1 Component Orientation

As with thick-film circuits, the general rules for Type 1 SMT circuits apply for thin films assembled by reflow soldering. The rules may bear revision for conductive epoxy attachment, and adhesive vendors can furnish design rule guidance as needed in those instances where thin-film circuits are assembled with conductive epoxy.

9.5.2 Wiring Layout

Thin-film technology generally involves the deposition of a conductive film across the full surface of the substrate, followed by masking, photolithographic exposure of the mask, washing away of the unwanted portions of the mask, and etching of the pattern not protected by the remaining mask. Because this process involves the photographic registration of conductive patterns, metallization-feature to mechanical-feature registration is not guaranteed. Therefore, it is a good practice to photolithographically generate a set of orientation features during the conductive pattern-generation process.

Multi-layer thin-film techniques are in practice and may be studied where required. However, currently very few SMT hybrids are built using thin films with more than one conductive layer. Therefore, we will consider only single-layer thin films herein.

Traces and Spaces

Where space permits, line-and-space dimensions per Level 2 density, as shown in Figure 5-11 of Chapter 5, keep manufacturing costs low. However, on good quality ceramic surfaces, thin-film technology will readily permit the production of 0.025-mm (0.001") lines and spaces. On sapphire or other highly polished substrates, lines as small as 5 microns (0.0002") can be fabricated commercially, so thin-film techniques lend themselves to tight single-layer geometries.

Clearances Around Solder Lands

The clearances around solder lands, as specified in Chapter 5 (Section 2 and Section 4), are determined more by circuit-assembly considerations than by metallization techniques. Whenever possible, they should be followed regardless of the substrate manufacturing process.

Crossovers

Crossovers may be accomplished using dielectric covers over small sections of first-layer conductor. Design rules per thick films may be applied to thin-film crossovers.

9.5.3 In Situ Component Design

Resistors

Thin-film resistors are inherently stable and exhibit excellent TCR tracking and extreme trim stability. Resistors may be fabricated to close tolerances using thin-film techniques.

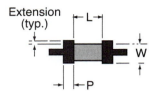

Dimension	Modifier	Manufacturing Simplicity		
		Simple*	Average*	Difficult*
L	min.	0.254 (0.010)	0.127 (0.005)	0.127 (0.005)
W	min.	0.063 (0.0025)	0.025 (0.001)	0.013 (0.0005)
S*	min.	0.2	0.1	0.05
S*	max.	100	1000	2000
Extension	min.	0.254 (0.010)	0.127 (0.005)	0.076 (0.003)
P	min.	0.051 (0.002)	0.025 (0.001)	0.013 (0.0005)

*Values are in millimeters and inches.

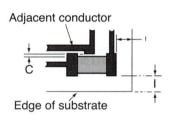

Dimension	Modifier	Manufacturing Simplicity		
		Simple*	Average*	Difficult*
C	min.	0.254 (0.010)	0.127 (0.005)	0.051 (0.002)
I	min.	0.254 (0.010)	0.0254 (0.010)	0.127 (0.005)

*Values are in millimeters and inches.

Figure 9-19   Thin-film etched-resistor form-factor rules.

One of the most constraining factors in thin-film resistor design is the general requirement to abide with a single resistive layer. This means that all thin-film resistors must be constructed using only a one-sheet resistivity value.

The thick-film formulae presented earlier apply also to thin-film resistors. The typical geometry limits for thin-film resistors are covered in Figure 9-19.[18] Some resistive thin films are self-passivating, and others will require a protective overglaze.

Inductors

The small fine-line geometries that can be etched using thin-film printing techniques allow the construction of considerably higher inductances than what can be achieved with thick films or PWBs. The formulae for inductors are the same as those presented for PWBs in Figure 6-7 of Chapter 6.

9.6 SMT Hybrid I/O Interconnects and Packaging

Next, let's look at how a circuit interfaces with, and is protected from, the world around it.

9.6.1 SIP and DIP Lead Frames

Lead frames are often used to make SIP or DIP components from SMT hybrids. Lead frames on 2.54-mm (0.100") or 1.27-mm (0.050") centers are available. Typical solder land configurations for edge-attached lead frames are shown in Figure 9-20.

9.6.2 Clip Leads for Planar Mounting

Using the dimensions shown in Figure 9-20 for solder lands, leads may also be attached to form a planar-mounting DIP or quad component.

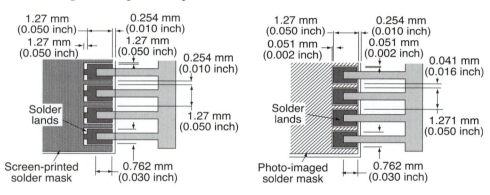

(A) Screened mask—1.27-mm centers.   **(B) Photoimaged mask—1.27-mm centers.**

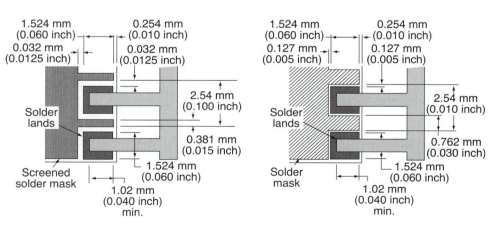

(C) Screened mask—2.54-mm centers.   **(D) Photoimaged mask—2.54-mm centers.**

Figure 9-20   Solder land dimensions for edge-attached lead frames.

9.6.3   Connectors and Headers

Surface-mount connectors and headers may be reflow mounted to ceramic or polymer substrates. However, our previous cautions regarding the weakness in repeated connector cycling may speak against this. Connectors having a mechanical interface to the substrate will provide a far better life under repeated cycling. Mechanical connections generally require holes in the substrate. Thus, widespread use of mechanically fastened connectors may rule against selecting ceramic as a substrate material.

9.6.4   Package-Mounted I/O

As mentioned earlier in this chapter, hybrid circuits having exposed semiconductors are often protected with external packaging. Hermetic metal packages, hermetic and nonhermetic ceramic packages, and nonhermetic plastic-molded housings are used for this purpose. But, whatever the packaging direction, system I/O must somehow be routed through the package. Where hermeticity is required, feedthrough pins may be fired into glass in the wall of the package. The pins are usually located at the same height as the top of the substrate and are connected to the circuit with wire bonds. Packaging vendors can lend assistance in your package and feedthrough design.

9.6.5   Protective Packaging and Coating

Packaging

Where protective packaging must provide a hermetic seal, seal integrity becomes a major quality issue for manufacturing. The protective purpose of the hermetic seal may not be met if there is even a very small leak. During operation, as the circuit heats, gas in the package expands and begins to leak to the outside. This process continues until the internal hot gas reaches equilibrium pressure with the ambient atmosphere or the device is switched off. After the hot internal pressure has equalized with the ambient pressure, switching the circuit off will cause an internal vacuum with respect to ambient atmosphere and can draw damp or otherwise contaminated air into the package.

Leak rates are measured in the cubic centimeters of air that would pass through the leak per second at a differential pressure of 1 atmosphere (Atm.cc/S.). Rates greater than 10^{-5} Atm.cc/S. are called gross leaks. Slower leaks are called fine. To graphically depict what is meant by a fine leak, we have shown in Table 9-4 the time required for 1 cc of air to pass through various leaks.

Table 9-4   Comparison of Leak Rates.*

Leak Rate (Atm.cc/S.)	Time Required
10^{-1}	1 cc of air leaks every 10 seconds
10^{-3}	1 cc of air leaks every 17 minutes
10^{-5}	1 cc of air passes through leak in 28 hours
10^{-8}	1 cc of air passes through leak in 3 years
10^{-11}	1 cc of air leaks in 3,000 years

*Gillespie, T., "Semiconductor Seal Testing," *Journal of IMAPS—Europe,* No. 15, January 1988, p. 46.

Hermetically sealed hybrids are tested for hermetic integrity using several methods. Unfortunately, the methods that will detect gross leaks fail to locate fine leaks, and vice versa. Figure 9-21[19] shows the leak rate ranges of some common leak-test methods. To ensure all leaks are detected, tests covering the full range of concern, with sufficient overlap, should be selected.

Coatings

Hybrid circuits are generally coated with a protective overglaze. Ceramic thick-film overglazes are usually low melting temperature vitreous glass materials. Overglaze protection is required for many resistive inks. It may also be used to cover conductors and act as a solder resist. Screened or photoimaged solder resists may serve these same functions and are commonly used on polymer substrates.

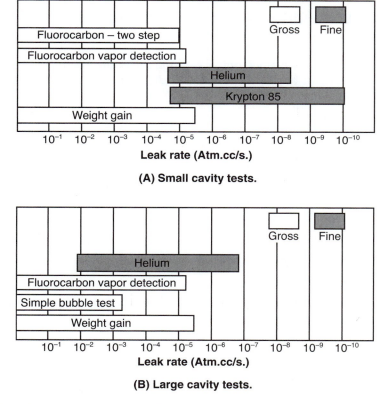

Figure 9-21   Detection ranges of various leak tests.

9.7   SMT Manufacturing Methods for Hybrids

Below, we will provide a brief view of SMT hybrid manufacturing methods and the equipment used in the assembly process. We will not have space to begin to give a complete accounting of various classes of equipment. If you are shopping for a line, please look at more than just this text.

9.7.1   Unique Elements of SMT Manufacturing of Hybrids

Incoming Inspection

Hybrid technology brings new materials to the manufacturing operation. Receiving inspection must be prepared to test or NIV certify resistive, conductive, and dielectric layer materials, substrate materials, and packaging/coating materials for the hybrid process.

Ceramic Processing

If your SMT hybrid operation will make extensive use of ceramic substrates, you may want to process ceramic boards internally rather than buy them from subcontractors.

Thick-Film Printing

Thick-film conductors must be screened onto a ceramic substrate. Then a firing furnace will dry and fire the ink on the circuit board. After firing, additional screenings will add resistive and insulating layers.

Thin-Film Printing

Thin film printing requires the following steps:

- CAD design of the film required
- Vacuum deposition of the thin film by cathode sputtering
- Photolithographic imaging of, for example, a 1-micron pattern
- Etching to remove non photo-masked material
- Plasma etching
- Passivation

Laser Trimming

Laser passive trimming can quickly bring a network of thick-film resistors to ±1 percent of specified value. Active trimming may take longer but is preferred if circuit performance is the more important parameter compared to exact resistor values.

Attachment-Media Application

The printing of solder paste on hybrids is not very different from solder printing on polymer PWBs. Where substrates are ceramic, they are typically referenced from two edges rather than from tooling pins in holes.

SMC Placement

Like solder application, placement of SMCs on hybrids is similar to standard PWB placement. Again, edge referencing may be required for a given substrate material.

Curing and Soldering

The curing and soldering of SMT hybrids is done in the same way that it is for standard PWBs using SMCs. Few hybrids use through-hole connection techniques, so reflow soldering is the typical approach.

Cleaning

SMCs on hybrid substrates present the same challenges as they pose on PWBs. Hybrids are cleaned in essentially the same manner as PWB assemblies. Extra care may be required to ensure that small hybrids are not displaced on the cleaning system belt by strong spray jets.

Test

The tight geometries achievable in hybrid layouts are a boon to miniaturization, but are often a challenge to test engineering. Special probe cards provide test access to closely spaced test points on an SMT hybrid.

Repair and Rework

Repair and rework of SMT hybrids may be done with hot air or with heated nonoxidizing gas, just as is commonly done with SMT PWB assemblies. Single-side-populated hybrids on ceramic boards are also reworked on micro reflow stations. Such a repair station commonly uses a glow-plug-ignited flame on hydrogen gas, which is then directed below the circuit board for reflow of precise spots. Preheat from a resistance element guards against cracking due to thermal shock.

9.7.2   Typical Process Flows

Figure 9-22 shows several ways that both through-hole and surface-mount components may be integrated into a hybrid manufacturing process.

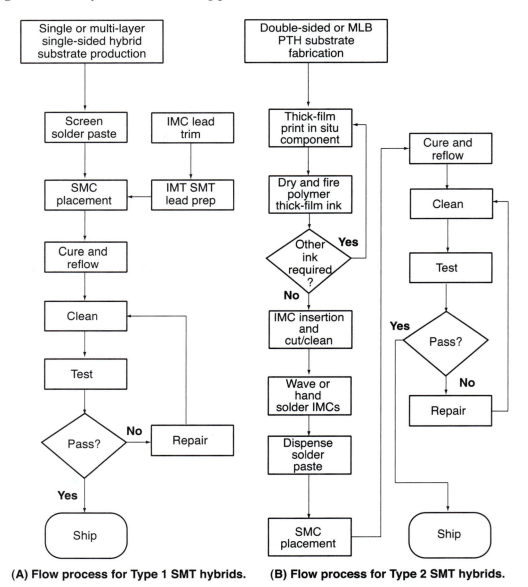

(A) Flow process for Type 1 SMT hybrids.   **(B) Flow process for Type 2 SMT hybrids.**

Figure 9-22   Two typical flow processes for SMT hybrid production.

9.8   Fundamentals of MCM Design

At clock rates of 25 MHz and below, there is very little need for MCMs. As frequencies move to 50 MHz and beyond, the MCM begins to be an attractive packaging solution. They are already widely used in workstations and high-performance computers, cellular telephones, laptop computers, palmtops, and other hand-held electronics. Advanced automotive electronics such as steering control, engine management, suspension controls, collision avoidance, and satellite positioning will provide an expanding market for MCMs for some time to come.

Why are designers so excited about MCMs for high-performance systems? The answer is clear when we compare the densities of various packaging techniques. Let us define density as the percentage of substrate area occupied by silicon IC die. This provides a tool to make such a packaging-density comparison. Thus:

$$e_{IC} = (IC\ die\ area/total\ substrate\ area)*100\%$$

By this measurement, if we completely cover a PC board with PLCCs, we will achieve an e_{IC} of about 6 percent. MCMs routinely deliver e_{IC} ratings in the range of 12 percent to 60 percent. By allowing designers to pack die more closely, interconnects are drastically shortened. For fast-transition signals, shorter interconnects mean fewer problems with false triggering from reflected signals, large reductions in interconnect parasitics, and greatly improved management of EMI/RFI. Both coupled and radiated noises are substantially diminished, thus the interest in this technology.

There are three generic types of MCM circuits based on the technology used to produce the substrate for the circuit. MCM-C substrates are made using hybrid thick-film techniques. The C is drawn from the usual substrate material, ceramic. MCM-D substrates may also be ceramic, but the D refers to deposited, indicating that the circuit is metallized using thin-film deposition. An MCM-L has a substrate made of a polymer laminate, like an ordinary circuit board. We will look at each separately. However, the distinctions between each are not always clear. In the drive for cost and real estate reduction, techniques from hybrid thick-film and thin-film technology often spill over into laminate circuit boards. The intent in presenting three distinct types is to avoid boundaries to engineering creativity.

9.8.1   MCM-C

The thick-film MCM-C may be made using standard print-and-fire techniques or cofired substrate. As of our second edition, substrates having lines as small as 100 microns have been made using the MCM-C approach. Theoretical limits for multiple layers are around 100 layers, but circuits of more than ten or fifteen layers are rarely seen.

Interconnects in all areas have followed a rather steady progression toward ever smaller feature size, both in IC fabrication as well as circuit board and hybrid circuits. Line widths produced by thick-film, thin-film, and laminate-etch techniques shrunk continuously over the past decade. There is no reason to believe this trend will not continue. Thus, the feature sizes mentioned here are provided to help designers select between styles of MCMs. For current limits, check with substrate suppliers.

9.8.2   MCM-D

The MCM-D thin-film deposited interconnect structure uses technology born in the semiconductor industry to build the conductive paths on the substrate. Metallizations are made by vapor deposition or ion sputtering. MCM-D substrate routinely employ interconnects down to 20 microns in width, with similar spacings between adjacent traces. Very high e_{IC} densities are possible. D type MCMs, like Ferraris, are fast and pretty. They are also, like Ferraris, seldom cheap. Both the processes and the materials for the D substrate tend to be costly. Even the dielectric polymers used are precious, costing as much as $200.00 per gram. Dielectric layers are applied by spin coating, so material loss adds to the cost of each successive layer.

9.8.3   MCM-L

The MCM-L is a multi-chip module built on a laminated-polymer substrate. Per square inch, it is the undisputed price leader among MCMs. However, typical laminated-polymer interconnect structures do not come close to ceramics in either interconnect density or the ability to dissipate the heat of a closely packed cluster of fast ICs. Seeing these factors not as limitations but as opportunities for the low-cost laminate approach, innovative packaging engineers have pushed the density envelope of the MCM-L by adapting MCM-C and MCM-D processes. Thus, while space and trace rules usually start at 100 microns, new techniques have yielded densities down to 25 microns. An example of an MCM-L panel is shown in Figure 9-23.

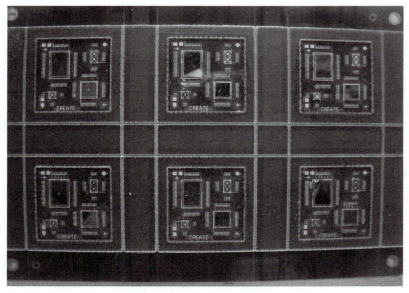

Figure 9-23   Panelized MCM-L with clear coating.

9.8.4   Variants on the Theme

The MCM-LD is one obvious merger of technology. This packaging approach involves the use of a laminated substrate with a surface layer or layers fabricated using thin-film techniques. The substrate may be a multi-layer circuit. Thus, the MCM-LD can provide a very-high-density interconnect structure at a reasonable cost.

Another technology merger, the MCM-LF or laminated-film MCM, rivals the MCM-D in density and adaptability to the needs of high-speed circuitry. It involves fabrication of two or more thin-film-on-polymer circuits which are then laminated onto one high-density interconnect structure. Interstitial connections are formed with microvias as small as 0.1 mm (0.004"). Since it avoids the need for spin-coated dielectrics, the cost of the MCM-LF is much lower than the typical cost for the MCM-D approach.

Technology advances are pushing the cost of very-high-density MCM substrates downward at a rapid rate. Clearly, the approach will gain rapidly in use. We can also expect the substrate fabrication techniques to spread in usage. Why limit high-density interconnect techniques to just a few IC chips? Why not fabricate the entire circuit on a single high-density card? Unfortunately, space does not permit us to give in-depth coverage to the MCM. The subject is complex enough to deserve a book all its own. Let us simply note the importance of the growing technology base known as MCMs and encourage further study for all who find PWB real estate or performance a critical issue.

9.9   Review

After reading and understanding this chapter, you should be able to answer these questions and note the reference location for the information in the chapter.

1. Define hybrid circuits in your own words.
2. Define thick-film circuits, including film thickness.
3. Define thin-film circuits, including film thickness.
4. Describe four benefits of pairing hybrid technology with SMT packaging.
5. Describe issues surrounding TCE differences between materials in SMT and hybrid circuits.
6. For the variables in the modified Manson-Coffin equation, note which ones are unique to hybrid circuits.
7. What kind of printed components can be added to a hybrid circuit?
8. What are the thermal conductivities of alumina and beryllia ceramics relative to each other and to FR-4 substrates?
9. What are the limits to the number of layers in thick-film and thin-film circuits?
10. Describe how an in situ resistor is designed and fabricated using thick-film technology.
11. If the resistivity of a thick-film material is 1000Ω/square and power dissipation issues dictate that the resistor will need to use 1 mm squares, sketch to 10X scale the relative form factor of a 4.5K resistor.
12. What technique is used to trim in situ hybrid resistors?
13. What types of packaging can be used for thick-film hybrids?

9.10 References

1. "The Market for European-Manufactured Hybrid Circuits and MCMs—2003 edition." IMS Research, London, 2003. Available (by subscription) at www.imsresearch.com.

2. Hamill, A. T., Konsowski, Steven G., et al., *Hybrid Microcircuit Design Guide,* IMAPS, Reston, VA, 1982.

3. Balde, John W., Caswell, Greg, et al., *Surface Mount Technology,* International Society for Hybrid Microelectronics, Reston, VA, 1984, p. 3.

4. Lau, John H., and Rice, Donald W., "Solder Joint Fatigue in Surface Mount Technology, State of the Art," *Solid-State Technology,* October 1985, p. 91.

5. Op. cit., ref. 3.

6. Ackerman, Karl-Peter, Hug, Rolf, and Berner, Gianni, "Multilayer Thin Film Technology," *Proceedings of International Symposium on Microelectronics, Atlanta 1986,* IMAPS, Reston, VA, 1986, p. 519.

7. Kompelien, D., Moravec, T. J., and DeFlumere, M., "A New Hybrid Technology: High Density Thin Film Copper/Polyimide Multilayer System," *Proceedings of International Symposium on Microelectronics, Atlanta 1986,* IMAPS, Reston, VA, 1986, p. 749.

8. Ohdaira, Hiroshi, Saito, Masayuki, and Iida, Atsuko, "A New Polymeric Multilayers Substrate," *Proceedings of International Symposium on Microelectronics, Minneapolis 1987,* IMAPS, Reston, VA, 1987, p. 515.

9. Smith-Vargo, Linda, "PCB Makers Spiral Into the Future," *Electronic Packaging & Production,* February 1988, p. 67.

10. Pedder, D. J., "Flip Chip Solder Bonding for Microelectronic Application," *Journal of IMAPS—Europe,* No. 15, January 1988, p. 4.

11. Ibid.

12. Frecska, Tamas, "Theoretical Model for Multilevel Screen Printing of Thick Film Compositions," *Proceedings of International Symposium on Microelectronics, Minneapolis 1987,* IMAPS, Reston, VA, 1987, p. 314.

13. Op. cit., ref. 2, p. 74.

14. Ibid.

15. Ibid.

16. Op. cit., ref. 2, p. 77.

17. De Mey, G., and Van Schoor, L., "Thermal Analysis of Hybrids with Mounted Components," *Journal of IMAPS—Europe,* No. 15, January 1988, p. 28.

18. Ibid.

19. Ibid.

appendix A

Component and Land Geometries

A.1   SMT Component Geometries
A.2   ANSI/IPC-SM-782 Land Patterns
A.3   Land-Pattern Variations

Please read the Introduction before using any of the footprints in this appendix!

Introduction

In the first edition of this book, we compiled working drawings such as the one seen in Figure A, that the reader might need to accompany the information in this book. These drawings and others are being continuously updated by the IPC, JEDEC, and others. Therefore, *there has been no attempt to make this appendix up to date for the second edition.* The best reference as of this writing for the land geometries (footprints) used in SMT PCB design is IPC standard IPC-7351 "Generic Requirements for Surface Mount Design and Land Pattern Standard,"[1] which supersedes IPC-SM-782A and its accompanying calculator. IPC-7351, which is available at the IPC website, http://www.ipc.org, includes many EIA/JEDEC-registered components and guidelines for both wave and reflow soldering. It allows for certain assumptions in the land patterns, including:

- fabrication tolerances
- placement tolerances
- side joint minimums
- toe joint minimums
- heel joint minimums

Surface-mount components covered in IPC-7351 include:

- chip resistors
- chip capacitors
- tantalum capacitors
- inductors
- MELF diodes
- SOD123 diodes
- SOT23 transistors
- SOT89 transistors
- SOT143 transistors

- SOT223 transistors
- TO252 (DPAK) transistors
- SOIC and other variations of two-sided gull-wing ICs
- QFP variations
- J-leaded ICs with leads on two and four sides (for example, PLCCs)
- area arrays, such as BGAs and CGAs
- "no lead" ICs, such as LLPs and SONs

Process variables known to the design team should be factored into the land pattern design, and final design should be determined through experimentation on the designers' own production line.

Developed in conjunction with IPC-7351 were a set of libraries, land pattern calculators and viewers designed in interface with a variety of ECAD software tools, and a freeware version of the IPC-7351 land pattern calculator. The calculator, along with a number of libraries, is available at PCB Libraries' website, http://www.pcblibraries.com. With this calculator available, continuing manual updates of land patterns may no longer be necessary.

It is important for the user to recognize that the primary purpose of a land pattern is not to provide a resting place for a component. It is to provide the proper copper area on the board to accomplish these two goals:

- provide for the creation of a strong and easily inspected solder joint
- minimize the risk of solder bridging between joints

As noted in Chapter 5, these two criteria have conflicting dimensional requirements. Providing for a strong and easily inspected solder joint demands that we have a large copper area, and minimizing the risk of bridging demands as much space as possible between the copper features. Good footprint design must balance these two demands.

For viewing purposes and as basic references, we will present in this appendix standard SMT component outlines, standard land geometries for the common SMT components, and recommendations for custom land patterns for specific applications. Except as specifically indicated, referenced EIA and JEDEC specifications are the sources for all component outline and dimensional data given. Unless otherwise noted, metric dimensions control. Inch conversions are direct and are rounded to three places. Where such conversions do not agree with the conversions shown in cited specifications, the correct conversion has been substituted for the specification conversion. Where metric control results in cumbersome inch conversions, round the inch dimensions to the nearest 0.005 inch for inch control. Conversely, where inch control produces inconvenient metric dimensions, the metric numbers should be rounded to the nearest 0.1 mm, unless otherwise noted.

A.1   SMT Component Geometries

The outline drawings for standard SMT components are shown in the following text. This material is meant to be used in conjunction with Chapter 2 in the specification of and engineering with SMCs. Some of this material may be superseded by that shown in IPC-7351, "Generic Requirements for Surface-Mount Land Patter and Design Standard."

A.1.1   SMT Passives

We will start with the components with the simpler form factors, the *passives*. Following are the outlines for capacitors of various types, typical resistors including trimmers and networks, thermistors, and various coils, transformers, and inductors.

Ceramic Chip

Ceramic chip capacitors are available in five standard sizes (four preferred, plus one additional). Figure A-1 shows the dimensional information for standard-size chip capacitors.

There is also an infinite variety of nonstandard-dimension parts that are made for special applications. They follow similar form factors and are rarely larger than 7.62 mm (0.300") in length or width. For sake of handling and mechanical orientation, special sizes should have an aspect ratio that is distinctly apart from 1:1. Cubic parts are particularly troublesome to handle mechanically. In the same way, parts that are nearly the same height as they are in width are troublesome in mechanical sorters.

Chip capacitor end terminations are generally coated with thick-film mixtures of silver, palladium, and glass frit. Where soldering and/or rework processes will require exposure to typical soldering temperatures (240°C) for a total of over 10 seconds, a diffusion barrier should be specified.

For a time, diffusion barriers became so common that there was serious talk of deleting the unprotected palladium/silver termination from manufacturers' catalogs. However, several major automotive and telecommunication users objected to the nickel-barrier part feeling that it was only a solution to poor process control and might introduce problems of its own. Where nickel terminations contribute to component cracking or solder-joint stress failures, the soldering process should be adjusted to permit the use of plain palladium/silver parts.

Nickel-barrier specifications are covered in the Chapter 2 discussion of chip capacitors.

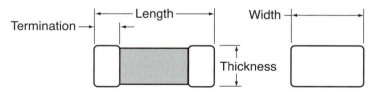

NOTES
1. For discussion of end terminations, see Chapter 2,
   Figure 2-4, and related text, plus the discussion here.
2. For land patterns, see Figures A-29, A-46, and A-48.
   For vias, see Figures A-47 and A-49.
3. For dimensions, see Table A-1.

Figure A-1   Chip-capacitor outline drawing.

Tantalum and Aluminum Electrolytic Capacitors

Tantalum and other electrolytic capacitors are generally packaged per the outlines shown in Figure A-2 and Figure A-3. Dimensions for these drawings are given in Table A-1, Table A-2, and Table A-3, respectively. Note that the inset terminations of the extended-range brick allow some parts to fit the standard brick lands.[2]

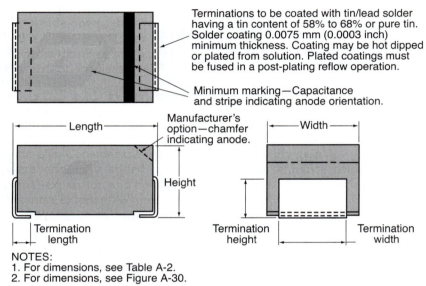

Figure A-2 Electrolytic "brick" capacitor outlines.

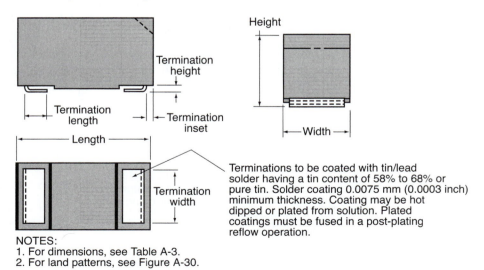

Figure A-3 Entended-range electrolytic "brick" capacitors.

Table A-1   EIA Chip Capacitor Standard Sizes.

EIA Size Designation	Dimensions (mm/in)*			
	Length	*Width*	*Thickness (Max)*	*Termination*
0805	1.8–2.2 0.070–0.087	1.0–1.4 0.040–0.055	1.3 0.050	0.3–0.6 0.012–0.024
1206	3.0–3.4 0.118–0.134	1.4–1.8 0.055–0.070	1.5 0.060	0.4–0.7 0.016–0.028
1210	3.0–3.4 0.118–0.134	2.3–2.7 0.090–0.106	1.7 0.067	0.4–0.7 0.016–0.028
1812	4.2–4.8 0.166–0.190	3.0–3.4 0.118–0.134	1.7 0.067	0.4–0.7 0.016–0.028
1825**	4.2–4.8 0.166–0.190	6.0–6.8 0.236–0.268	1.7 0.067	0.4–0.7 0.016–0.028

* Inch dimensions control, and metric dimensions are shown rounded to the nearest 0.1 mm.

** Not included in SMTA Component Standards Group recommended sizes. *Surface Mount Technology Component Standardization Proposals—Commercial-Grade Components,* SMTA, Edina, MN, 1986.

Table A-2   Standard Capacitance Electrolytic "Brick" Capacitors.

EIA RS-288 Size	Size Code/Standard Capacitor Range (mm/in)			
	A Case	*B Case*	*C Case*	*D Case*
Length	3.0–3.4 0.118–0.134	3.3–3.7 0.130–0.146	5.7–6.3 0.224–0.248	6.8–7.6 0.268–0.299
Width	1.3–1.8 0.050–0.070	2.6–3.0 0.102–0.118	2.9–3.5 0.114–0.138	4.0–4.6 0.157–0.181
Height	1.4–1.8 0.055–0.070	1.7–2.1 0.067–0.083	2.2–2.8 0.087–0.110	2.5–3.1 0.098–0.122
Termination length	0.5–1.1 0.020–0.043	0.5–1.1 0.020–0.043	0.5–1.1 0.020–0.043	0.5–1.1 0.020–0.043
Termination width	1.1–1.3 0.043–0.051	2.1–2.3 0.083–0.090	2.1–2.3 0.083–0.090	2.3–2.5 0.090–0.098
Termination height	0.7 (min) 0.028	0.7 (min) 0.028	1.0 (min) 0.040	1.0 (min) 0.040

* Inch dimensions control, and metric dimensions are rounded to the nearest 0.1 mm.

Table A-3   Extended Range Electrolytic "Brick" Capacitors.

EIA RS-228 Size	Size Code/Extended-Range Capacitor Range (mm/in)			
	3518 Case	*3527 Case*	*7227 Case*	*7257 Case*
Length	3.3–3.7 0.130–0.146	3.3–3.7 0.130–0.146	6.9–7.5 0.272–0.295	6.9–7.5 0.272–0.295
Width	1.6–2.0 0.063–0.079	2.4–3.0 0.095–0.118	2.4–3.0 0.095–0.118	5.2–6.2 0.205–0.244
Height	1.7–2.1 0.067–0.083	1.7–2.1 0.067–0.083	2.5–3.1 0.098–0.122	3.0–3.7 0.118–0.146
Termination length	0.6–1.0 0.024–0.040	0.6–1.0 0.024–0.040	0.8–1.2 0.031–0.047	0.8–1.2 0.031–0.047
Termination width	1.6–1.8 0.063–0.071	2.4–2.6 0.095–0.102	2.4–2.6 0.095–0.102	5.4–5.8 0.213–0.228
Termination height	0.7 0.028	0.7 0.028	1.0 0.040	1.2 0.047
Termination inset	0.4–0.6 0.016–0.024	0.4–0.6 0.016–0.024	0.6–0.8 0.024–0.032	0.6–0.8 0.024–0.032

* Inch dimensions control, and metric dimensions are shown rounded to the nearest 0.1 mm.

Variable Capacitors

The current multiplicity of SMT package outlines offered in variable capacitors might lead us to speculate that "variable" applies to the shape of the device rather than to the capacitance value. Work is underway to set standards. However, no standard was available as of this writing. The outlines shown in Figure A-4,[3,4] illustrate some "typical" packages that are widely available in the commercial market.

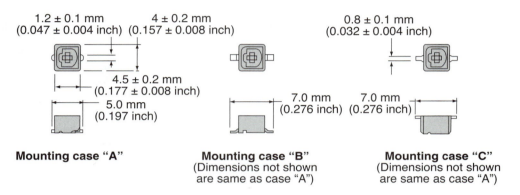

(A) MuRata Erie components available in single plate and multilayer capacitance ranges. Shipping containers are 12-mm reel or stick.

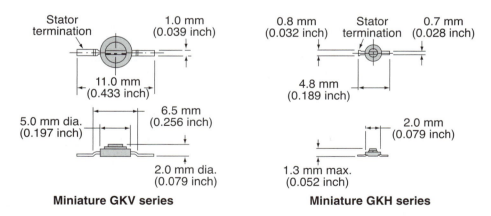

(B) Two of the many SMT cases available from Sprague. Capacitance ranges from 2.5 to 40.0 pF. These parts are typical of hybrid-style trimming capacitors.

Figure A-4 Typical SMT outlines for variable capacitors.

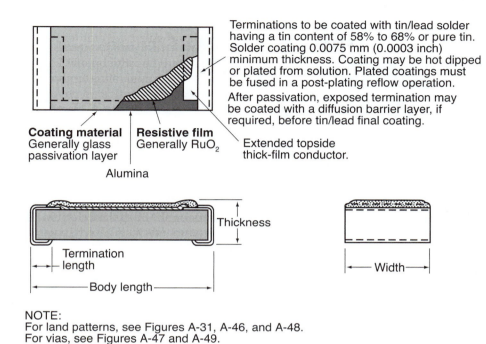

Terminations to be coated with tin/lead solder having a tin content of 58% to 68% or pure tin. Solder coating 0.0075 mm (0.0003 inch) minimum thickness. Coating may be hot dipped or plated from solution. Plated coatings must be fused in a post-plating reflow operation.

After passivation, exposed termination may be coated with a diffusion barrier layer, if required, before tin/lead final coating.

Coating material | **Resistive film**
Generally glass | Generally RuO$_2$
passivation layer |

Extended topside
thick-film conductor.

Alumina

Thickness

Termination
length

Body length

Width

NOTE:
For land patterns, see Figures A-31, A-46, and A-48.
For vias, see Figures A-47 and A-49.

Figure A-5   Standard SMT chip resistor outline.

Chip Resistors

Figure A-5 covers standard EIA/IS 30 chip resistor outlines; Table A-4 gives the dimension values. In addition to the standard parts, there are a great variety of custom forms available for specific application needs.

Table A-4   Chip Resistor Dimensions.

| Size Code | Dimensions (mm/in)* | | | |
	Body Length	Width	Thickness	Termination Length
RC0805	1.8–2.2	1.0–1.4	0.3–0.7	0.3–0.6
	0.070–0.087	0.040–0.055	0.012–0.028	0.012–0.024
RC1206	3.0–3.4	1.4–1.8	0.4–0.7	0.4–0.7
	0.118–0.134	0.055–0.070	0.016–0.028	0.016–0.028
RC1210	3.0–3.4	2.3–2.7	0.4–0.7	0.4–0.7
	0.118–0.134	0.090–0.106	0.016–0.028	0.016–0.028

* Inch dimensions control and metric conversions are shown rounded to the nearest 0.1 mm.

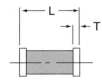

NOTES:
1. Terminations to be coated
   with tin/lead solder having a tin
   content of 58% to 68% or pure tin.
   Solder coating 0.0075 mm
   (0.0003 inch) minimum
   thickness. Coating may be
   hot dipped or plated from
   solution. Plated coatings must
   be fused in a post-plating
   reflow operation.
2. For land patterns, see
   Figure A-32.
3. This outline also used for
   cylindrical diodes, etc.

| Case | Dimensions (mm/inch) | | |
Style	L	D	T
MLL34	3.30–3.70 (0.130–0.146)	1.50–1.70 (0.059–0.067)	0.29–0.55 (0.011–0.022)
MLL41	4.80–5.20 (0.189–0.205)	2.44–2.54 (0.096–0.100)	0.35–0.51 (0.014–0.020)

Figure A-6   Cylindrical leadless resistor outline.

Cylindrical Resistors
MELF, or other cylindrical resistor forms, are generally specified per the outline shown in Figure A-6. This outline conforms to EIAJ standards and is essentially a JEDEC DO-35 without the leads.

Trimmer Potentiometers
Trimmer potentiometers, like their capacitor cousins, are as variable in their package outlines as in their electrical parameters. Figure A-7[5,6,7] shows several typical designs.

Resistor Networks
Resistor networks are generally in either a leadless or leaded ship carrier and in SOIC-style packaging. The chip carriers are typically the same as the various standard 1.27-mm (0.050-inch) JEDEC outlines, as covered under *Actives* in Section A.1.2. Some SOIC styles are also in standard JEDEC-outline packages. However, many are wider than the narrow JEDEC part but not as wide as the SOLICs. Most special resistor packages conform to the outline drawing

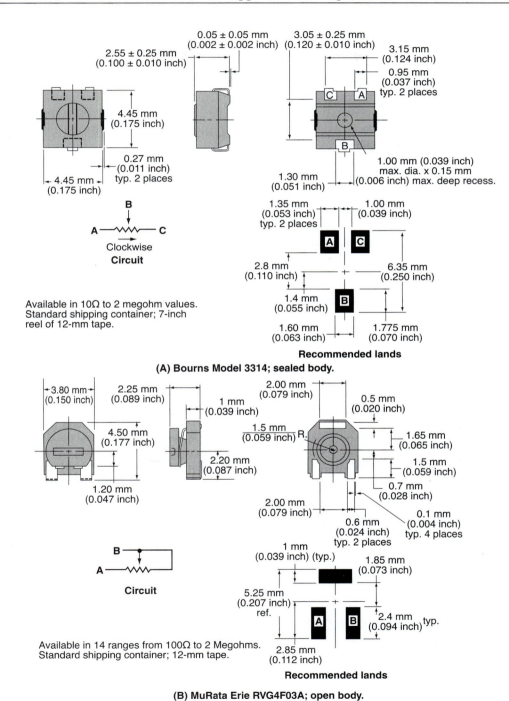

Figure A-7   Typical trimmer potentiometer outlines.

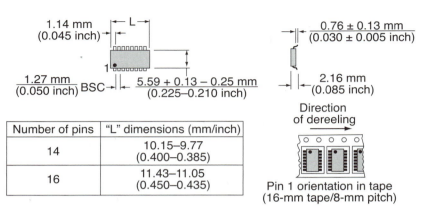

Number of pins	"L" dimensions (mm/inch)
14	10.15–9.77 (0.400–0.385)
16	11.43–11.05 (0.450–0.435)

NOTES:
1. Terminations to be coated with tin/lead solder having a tin content of 58% to 68% or pure tin. Solder coating 0.0075 mm (0.0003 inch) minimum thickness. Coating may be hot dipped or plated from solution. Plated coatings must be fused in a post-plating reflow operation.
2. For land recommendations, see Figure A-50.

Figure A-8   Resistor networks in nonstandard SOIC packaging.

shown in Figure A-8. But, a word of warning is in order. These special resistor network SOs may not be governed by any standard. Without standards control, manufacturers are free to choose package dimensions to suit their requirements. Before proceeding with a design for nonstandard parts, consult the vendor for the exact specifications and outline.

Thermistors
SMT thermistors are generally supplied in chip format with dimensions taken from the ceramic chip-capacitor outlines or with custom chip dimensions. Standard EIA monolithic capacitor dimensions are shown in Figure A-1 and Table A-1. For nonstandard parts, consult the device supplier for dimensions.

Tuning Coils and Transformers
In the absence of standards, a wide variety of alternatives are on the market. Consult the vendors for dimensional data.

Inductors
SMT inductors are supplied in open and protected cases. In the absence of a standard, many protected case parts are modeled after tantalum brick-style components, as shown in Figure A-2 and Figure A-3. Some of these "brick" inductors fit the tantalum brick dimension scheme and others are dimensioned differently, but they follow the same basic form factor as the brick. Other protected parts follow their own unique molded construction. In addition, a wide variety of nonstandard parts, particularly of the open body construction, are offered for special applications. Consult the manufacturers for details, and use the SMTA component recommendations as a guideline for the specifications.

A.1.2   Actives

Diodes

Diodes may be packaged in MLL-34, SOD-80 or MLL-41 packaging, as shown earlier under MELF-style resistors. Larger diodes may be in the ¼-watt MELF resistor format. Both single and dual diodes are offered in SOT packages, such as shown in Figure A-9.

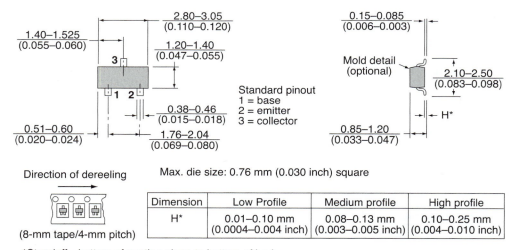

*Standoff—bottom of seating plane to bottom of body

(A) SOT-23 (EIA TO-236) outline.

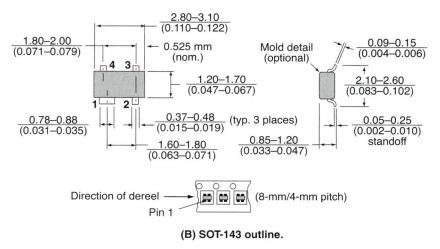

(B) SOT-143 outline.

Figure A-9   Typical SMT transistor package outlines. (*Continues on next page*)

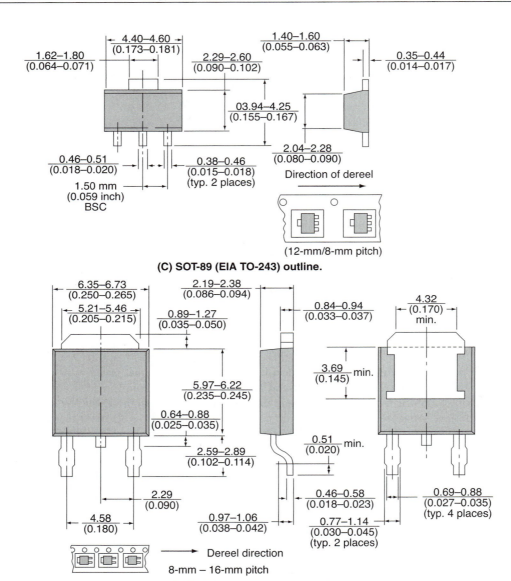

(C) SOT-89 (EIA TO-243) outline.

(D) DPAK power transistor package.

NOTES:
1. Values are in millimeters and inches.
2. For SOT-23 lands, see Figures A-34, A-51, and A-52.
3. For SOT-143 lands, see Figure A-35.
4. For SOT-89 lands, see Figures A-36, A-54, and A-55.
5. For DPAK lands, see Figure A-56.

6. Terminations to be coated with tin/lead solder having a tin content of 58% to 68% or pure tin. Solder coating 0.0075 mm (0.0003 inch) minimum thickness. Coating may be hot dipped or plated from solution. Plated coatings must be fused in a post-plating reflow operation.

Figure A-9   *(Continued)* Typical SMT transistor package outlines.

Transistors

Figure A-9A, Figure 9B, and Figure A-9C detail the package outlines for common SMT transistor packages. Power transistors and diodes are often housed in lead-formed variants of the TO-220 through-hole package. Figure A-9D shows a typical SMT approach to packaging a power device.

SOICs

The JEDEC standard gull-wing-leaded SOIC and SOLIC are covered in Figure A-10 and Table A-5.

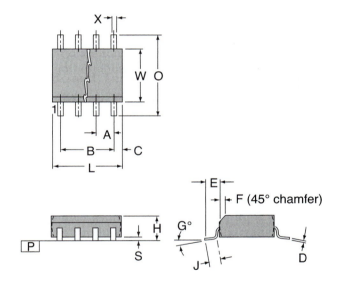

Direction of dereeling

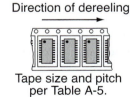

Tape size and pitch per Table A-5.

NOTES:
1. Terminations to be coated with tin/lead solder having a tin content of 58% to 68% or pure tin. Solder coating 0.0075 mm (0.0003 inch) minimum thickness. Coating may be hot dipped or plated from solution. Plated coatings must be fused in a post-plating reflow operation.
2. For land patterns, see Figures A-37 and A-57.

Figure A-10   Gull-wing-leaded SO and SOL IC package outlines.

Table A-5   Dimensions for Gull-Wing-Leaded SO and SOLIC Packages.

Symbol	SO Package Dimensions (mm/inch)						
	SOIC 8	SOIC 14	SOIC 16	SOLIC 16	SOLIC 20	SOLIC 24	SOLIC 28
A	1.27 BSC 0.050	1.27 BSC 0.050	1.27 BSC 0.050	1.27 BSC 0.050	1.27 BSC 0.050	1.27 BSC 0.050	1.27 BSC 0.050
B	3.81 (REF.) 0.150	7.62 (REF.) 0.300	8.89 (REF.) 0.350	8.89 (REF.) 0.350	11.43 (REF.) 0.450	13.97 (REF.) 0.550	16.51 (REF.) 0.650
C (Nominal)	0.53 0.021	0.51 0.020	0.050 0.020	0.69 0.027	0.69 0.027	0.72 0.028	0.70 0.027
D	0.18–0.23 0.007–0.009	0.18–0.23 0.007–0.009	0.18–0.23 0.007–0.009	0.23–0.32 0.009–0.013	0.23–0.32 0.009–0.013	0.23–0.32 0.009–0.013	0.23–0.32 0.009–0.013
E (Nominal)	1.05 0.041	1.05 0.041	1.05 0.041	1.40 0.055	1.40 0.055	1.40 0.055	1.40 0.055
F	0.25–0.50 0.01–0.02	0.25–0.50 0.01–0.02	0.25–0.50 0.01–0.02	0.25–0.50 0.01–0.02	0.25–0.50 0.01–0.02	0.25–0.50 0.01–0.02	0.25–0.50 0.01–0.02
G	0–8"	0–8"	0–8"	0–8"	0–8"	0–8"	0–8"
H	1.3–1.75 0.053–0.069	1.3–1.75 0.053–0.069	1.3–1.75 0.053–0.069	2.4–2.6 0.094–0.102	2.4–2.6 0.094–0.102	2.4–2.6 0.094–0.102	2.4–2.6 0.094–0.102
J	0.5–1.15 0.02–0.045	0.5–1.15 0.02–0.045	0.5–1.15 0.02–0.045	0.58–1.25 0.023–0.049	0.58–1.25 0.023–0.049	0.58–1.25 0.023–0.049	0.58–1.25 0.023–0.049
L	4.75–5.0 0.188–0.197	8.53–8.74 0.336–0.344	9.78–10.00 0.385–0.394	10.11–10.41 0.398–0.413	12.6–13.0 0.496–0.512	15.21–15.6 0.599–0.614	17.7–18.11 0.697–0.713
O	5.8–6.2 0.228–0.244	5.8–6.2 0.228–0.244	5.8–6.2 0.228–0.244	10.0–10.6 0.394–0.419	10.0–10.6 0.394–0.419	10.0–10.6 0.394–0.419	10.0–10.6 0.394–0.419
P (Planarity)	±0.05 ±0.002	±0.05 ±0.002	±0.05 ±0.002	±0.05 ±0.002	±0.05 ±0.002	±0.05 ±0.002	±0.05 ±0.002
S	0.01–0.02 0.004–0.008	0.01–0.02 0.004–0.008	0.01–0.02 0.004–0.008	0.01–0.02 0.004–0.008	0.01–0.02 0.004–0.008	0.01–0.02 0.004–0.008	0.01–0.02 0.004–0.008
W	3.8–4.0 0.150–0.158	3.8–4.0 0.150–0.158	3.8–4.0 0.150–0.158	7.4–7.6 0.291–0.299	7.4–7.6 0.291–0.299	7.4–7.6 0.291–0.299	7.4–7.6 0.291–0.299
X	0.36–0.45 0.014–0.018	0.36–0.45 0.014–0.018	0.36–0.45 0.014–0.018	0.36–0.48 0.014–0.019	0.36–0.48 0.014–0.019	0.36–0.48 0.014–0.019	0.36–0.48 0.014–0.019
Tape size/ pitch	12 mm/ 8 mm	16 mm/ 8 mm	16 mm/ 8 mm	16 mm/ 12 mm	24 mm/ 12 mm	24 mm/ 12 mm	24 mm/ 12 mm

SOJICs

The JEDEC standard "J"-leaded SO (SOJIC) package outline is shown in Figure A-11.

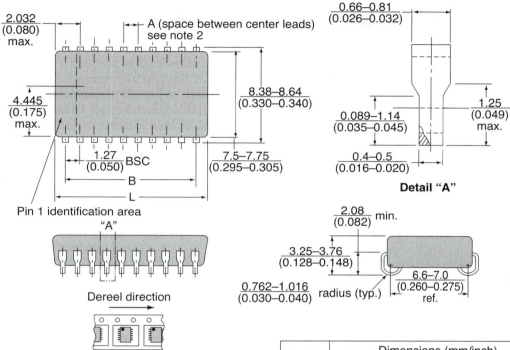

NOTES:

1. Terminations to be coated with tin/lead solder having a tin content of 58% to 68% or pure tin. Solder coating 0.0075 mm (0.0003 inch) minimum thickness. Coating may be hot dipped or plated from solution. Plated coatings must be fused in a post-plating reflow operation.

2. For some applications requiring a large power bus, "A" is increased to 5.08 mm/0.200 inch or 3.81 mm/0.150 inch by omitting 3 pins for parts with an even number of 50-mil spaces per side, or 2 pins for parts with an odd number of 50-mil spaces per side.

3. For land patterns, see Figure A-37.

4. Dimensions are in millimeter and inches.

Number of pins	Dimensions (mm/inch)		
	A (see Note 2)	B	L
14	1.27 (0.050)	7.62 (0.300)	9.40–9.65 (0.370–0.380)
16	1.27 (0.050)	8.89 (0.350)	10.67–10.92 (0.420–0.430)
18	1.27 (0.050)	10.16 (0.400)	11.94–12.19 (0.470–0.480)
20	1.27 (0.050)	11.43 (0.450)	13.21–13.46 (0.520–0.530)
22	1.27 (0.050)	12.70 (0.500)	14.48–14.73 (0.570–0.580)
24	1.27 (0.050)	13.97 (0.550)	15.75–16.00 (0.620–0.630)
26	1.27 (0.050)	15.24 (0.600)	17.02–17.27 (0.670–0.680)
28	1.27 (0.050)	16.89 (0.650)	18.29–18.54 (0.720–0.730)

Figure A-11   "J"-leaded SOJIC package outline.

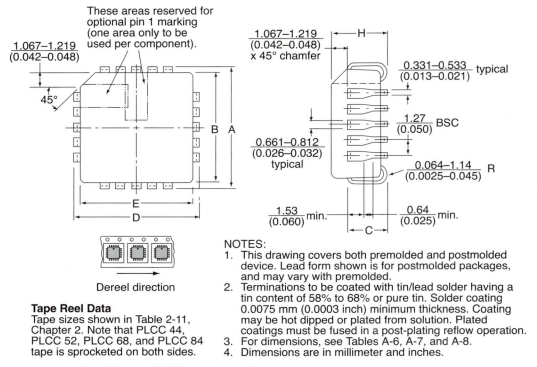

NOTES:
1. This drawing covers both premolded and postmolded device. Lead form shown is for postmolded packages, and may vary with premolded.
2. Terminations to be coated with tin/lead solder having a tin content of 58% to 68% or pure tin. Solder coating 0.0075 mm (0.0003 inch) minimum thickness. Coating may be hot dipped or plated from solution. Plated coatings must be fused in a post-plating reflow operation.
3. For dimensions, see Tables A-6, A-7, and A-8.
4. Dimensions are in millimeter and inches.

Tape Reel Data
Tape sizes shown in Table 2-11, Chapter 2. Note that PLCC 44, PLCC 52, PLCC 68, and PLCC 84 tape is sprocketed on both sides.

Figure A-12   A 50-mil Type A PLCC package outline.

PLCCs

The JEDEC postmolded and premolded 1.27-mm-pitch (0.050") PLCC package outlines are shown in Figure A-12. The dimensions for the drawing are given in Table A-6.

Table A-6   Square Postmolded PLCC Dimensions (Type A).

| Package Pin Count | Dimensions* (mm/inch) | | | | | | Lands |
	A	B	C	D	E	H	
28	11.94–12.95 0.470–0.510	11.43–11.582 0.450–0.456	2.29–3.05 0.090–0.120	See "A"	See "B"	4.19–4.57 0.165–0.180	For IPC lands specially designed for this part, see Figure A-38. For universal lands, see Figure A-58.
44	17.02–18.03 0.670–0.710	16.51–16.662 0.650–0.656	2.29–3.05 0.090–0.120	See "A"	See "B"	4.19–4.57 0.165–0.180	
52	19.56–20.57 0.770–0.810	19.05–19.202 0.750–0.756	2.29–3.30 0.090–0.130	See "A"	See "B"	4.19–5.08 0.165–0.200	
68	24.64–25.65 0.970–1.010	24.13–24.333 0.950–0.958	2.29–3.30 0.090–0.130	See "A"	See "B"	4.19–5.08 0.165–0.200	

Table A-7 Square Premolded PLCC (JEDEC MO-047) Dimensions (Type A).

Package Pin Count	Dimensions* (mm/inch)						Lands
	A	B	C	D	E	H	
20	9.78–10.03 0.385–0.395	8.89–9.042 0.350–0.356	2.29–3.05 0.090–0.120	See "A"	See "B"	4.19–4.57 0.165–0.180	For IPC lands specially designed for this part, see Figure A-39. For universal Type A PLCC lands, see Figure A-58.
28	12.32–12.57 0.485–0.495	11.43–11.582 0.450–0.456	2.29–3.05 0.090–0.120	See "A"	See "B"	4.19–4.57 0.165–0.180	
44	17.40–17.65 0.685–0.695	16.51–16.662 0.650–0.656	2.29–3.05 0.090–0.120	See "A"	See "B"	4.19–4.57 0.165–0.180	
52**	19.94–20.19 0.785–0.795	19.05–19.202 0.750–0.756	2.29–3.30 0.090–0.130	See "A"	See "B"	4.19–5.08 0.165–0.200	
68	25.02–25.27 0.985–0.995	24.13–24.333 0.950–0.958	2.29–3.30 0.090–0.130	See "A"	See "B"	4.19–5.08 0.165–0.200	
84	30.10–30.35 1.185–1.195	29.21–29.413 1.150–1.158	2.29–3.30 0.090–0.130	See "A"	See "B"	4.19–5.08 0.165–0.200	
100**	35.18–35.43 1.385–1.395	34.29–34.493 1.350–1.358	2.29–3.30 0.090–0.130	See "A"	See "B"	4.19–5.08 0.165–0.200	
124**	42.80–43.05 1.685–1.695	41.91–42.113 1.650–1.658	2.29–3.30 0.090–0.130	See "A"	See "B"	4.19–5.08 0.165–0.200	

* Dimensions in this table apply to Figure A-12.
** JEDEC-qualified packages but are not widely available.

Most PLCCs consumed today are of the postmolded variety. However, original JEDEC Type A PLCC specifications were unclear on construction. Today, the MO-047 premolded part specification is applied to both premolded and postmolded packages. Note that the table of dimensions, Table A-6, lists specifications for only four postmolded-package pin counts. Some manufacturers needing a postmolded package for other pin counts use a dimensioning scheme like the table shows but adjust the dimensions to suit the desired pin count. Others commonly follow the premolded dimension charts (Table A-7 and Table A-8) even though they postmold

Table A-8 Rectangular† Premolded PLCC (JEDEC MO-052) Dimensions (Type A).

Package Pin Count	Dimensions* (mm/inch)						Lands
	A	B	C	D	E	H	
18	8.05–8.31 0.317–0.327	7.14–7.315 0.281–0.288	2.29–3.05 0.090–0.120	11.61–11.86 0.457–0.467	10.72–10.87 0.422–0.428	4.19–4.57 0.165–0.180	For IPC lands specially designed for this part, see Figure A-39. For universal Type A PLCC lands, see Figure A-58.
18L	8.13–8.51 0.320–0.335	7.24–7.518 0.285–0.296	2.29–3.05 0.090–0.120	13.21–13.59 0.520–0.535	12.32–12.60 0.485–0.496	4.19–4.57 0.165–0.180	
22	8.13–8.51 0.320–0.335	7.24–7.518 0.285–0.296	2.29–3.05 0.090–0.120	13.21–13.59 0.520–0.535	12.32–12.60 0.485–0.496	4.19–4.57 0.165–0.180	
28	9.78–10.03 0.385–0.395	8.89–9.042 0.350–0.356	2.29–3.05 0.090–0.120	14.86–15.11 0.585–0.595	13.97–14.12 0.550–0.556	4.19–4.57 0.165–0.180	
32	12.32–12.57 0.485–0.495	11.43–11.582 0.450–0.456	2.29–3.05 0.090–0.120	14.86–15.11 0.585–0.595	13.97–14.12 0.550–0.595	4.19–4.57 0.165–0.180	

† Rectangular parts are not included in the SMT Association-recommended components.
* Dimensions in this table apply to Figure A-12.

the parts. The IPC-recommended land patterns for premolded and postmolded parts are not identical twins. Therefore, users should be sure to verify the dimension scheme followed on PLCCs before selecting land patterns, or they should resort to the alternate pattern given in Section A.3, which is suitable for either type.

TapePaks

The TapePak package is unique in that it is shipped in a protective ring. After testing the device, the end user excises it from the ring and forms the leads for reflow attachment, as shown in Figure A-13. The dimensions for a TapePak package are given in Table A-9.

There are three main JEDEC-registered TapePak body sizes. The part shown in the graphics of Figure A-13 has a body that is 7.264 mm (0.286") square and is available with 32 pins on a 25-mil (0.635-mm) pitch, or 40 pins spaced at a 20-mil (0.508-mm) pitch. The medium-lead-count part has a body that is 12.83 mm (0.505") square. It comes with 52 leads at 25 mil or 84 pins at 20 mil. The high pin-count part offers 100-, 132-, and 180-pin counts with pitches of 25, 20, and 15 mils (0.381 mm), respectively.

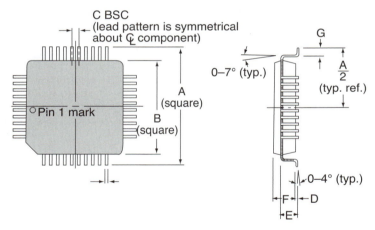

NOTES:
1. Dimensions are in millimeter and inches.
2. Recommended outline for formed and excised packages.
3. Refer to Table A-9 for package dimensions.
4. Terminations to be coated with tin/lead solder having a tin content of 58% to 68% or pure tin. Solder coating 0.0075 mm (0.0003 inch) minimum thickness. Coating may be hot dipped or plated from solution. Plated coatings must be fused in a post-plating reflow operation.
5. For land patterns, see Figure A-63 for 20-mil lead pitch parts, and Figure A-65 for 10-mil pitch.

6. JEDEC MO-071 registration covers 32-, 40-, 52-, 84-, 100-, 132-, and 180-pin packages. There is a proposal to create a very high pin count family having a 30.6-mm (1.205-inch) square body and 168, 220, 284, 360, and 400 pins at 25-mil, 20-mil, 15-mil, 12-mil, and 10-mil lead pitches, respectively.
7. For data and dimensions on unexcised part dimensions, test point dimensions, etc., consult the manufacturer. The unexcised part is shown in Figure 2-17 of Chapter 2.

Figure A-13   TapePak package outline.

Table A-9 Dimensions for TapePak Packages.

Package Description		Dimensions*						
Body Size	Pin Count	A	B (Body Size)	C (Pitch)	D	E	F	G
7.25 mm 0.286 inch square	32	8.99–9.30 / 0.354–0.366	7.26 / 0.286	0.635 / 0.025 BSC	0.05–0.25 / 0.002–0.010	1.14–1.35 / 0.045–0.053	1.55–1.75 / 0.061–0.069	0.048–0.635 / 0.019–0.025
	40	8.99–9.30 / 0.354–0.366	7.26 / 0.286	0.508 / 0.020 BSC	0.05–0.25 / 0.002–0.010	1.14–1.35 / 0.045–0.053	1.55–1.75 / 0.061–0.069	0.048–0.625 / 0.019–0.025
12.83 mm 0.505 inch square	52	14.58–14.88 / 0.574–0.586	12.83 / 0.505	0.635 / 0.025 BSC	0.05–0.25 / 0.002–0.010	1.50–1.65 / 0.059–0.065	1.93–2.13 / 0.076–0.084	0.048–0.635 / 0.019–0.025
	84	14.58–14.88 / 0.574–0.586	12.83 / 0.505	0.058 / 0.020 BSC	0.05–0.25 / 0.002–0.010	1.50–1.65 / 0.059–0.065	1.93–2.13 / 0.076–0.084	0.048–0.635 / 0.019–0.025
20.45 mm 0.805 inch square	100	22.76–23.06 / 0.896–0.908	20.45 / 0.805	0.635 / 0.025 BSC	0.28–0.48 / 0.011–0.019	1.73–1.88 / 0.068–0.074	1.93–2.13 / 0.076–0.084	0.76–0.91 / 0.030–0.036
	132	22.76–23.06 / 0.896–0.908	20.45 / 0.805	0.508 / 0.020 BSC	0.28–0.48 / 0.011–0.019	1.73–1.88 / 0.068–0.074	1.93–2.13 / 0.076–0.084	0.76–0.91 / 0.030–0.036
	180	22.76–23.06 / 0.896–0.908	20.45 / 0.805	0.381 / 0.015 BSC	0.28–0.48 / 0.011–0.019	1.73–1.88 / 0.068–0.074	1.93–2.13 / 0.076–0.084	0.76–0.91 / 0.030–0.036
	T.B.D. (In Design)		20.45 / 0.805	0.284 / 0.012 BSC				
	T.B.D. (In Design)		20.45 / 0.805	0.284 / 0.012 BSC				
30.6 mm 1.205 inch square (Proposed)	168		30.61 / 1.205	0.635 / 0.025 BSC				
	220		30.61 / 1.205	0.508 / 0.020 BSC				
	284		30.61 / 1.205	0.381 / 0.015 BSC				
	360		30.61 / 1.205	0.284 / 0.012 BSC				
	400		30.61 / 1.205	0.254 / 0.010 BSC				

*Dimensions are millimeter/inch; inch dimensions control.

PQFPs and CQFPs

Figure A-14 and Table A-10 cover the package outline for the Plastic Quad Flat Pack (PQFP) per JEDEC MO-069. Note 2 (in Figure A-14) details the variations of ceramic quad flat packs (CQFPs) from the plastic part outline shown. Basically, the ceramic package provides hermeticity. It is slightly smaller in body than the plastic device to provide additional space for the lead-form anvil without risk to the hermetic seals of the leads. Further, the corner protrusions, or bumpers, are omitted because ceramic bumpers would be too fragile to serve any useful purpose. Otherwise, the two packages are identical.

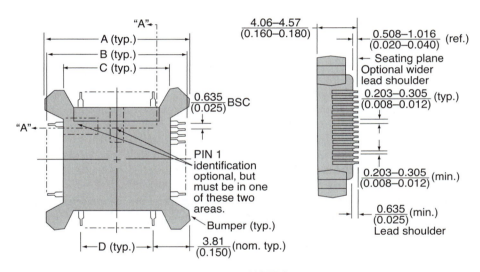

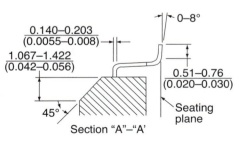

NOTES:
1. Terminations to be coated with tin/lead solder having a tin content of 58% to 68% or pure tin. Solder coating 0.0075 mm (0.0003 inch) minimum thickness. Coating may be hot dipped or plated from solution. Plated coatings must be fused in a post-plating reflow operation.
2. Ceramic quad flat pack per this outline, except body width and length 1.27 mm (0.050 inch) smaller to allow room for lead-form anvil without risk of damage to glass seals, and except corner bumpers not included as ceramic bumpers would chip easily.
3. For land recommendations, see Figure A-40.
4. Values are in millimeter/inch. Inch dimensions control.

Figure A-14 Plastic and ceramic quad flat pack outlines.

Table A-10   Plastic and Ceramic Quad Flat Pack Dimensions.

Pin Count	Dimensions (mm/inch)			
	A	B	C	D
52	15.16–15.32 0.597–0.603	14.61–14.86 0.575–0.585	11.35–11.51 0.447–0.453	7.62 BSC 0.300
68	17.70–17.86 0.697–0.703	17.15–17.40 0.675–0.685	13.89–14.05 0.547–0.553	10.16 BSC 0.400
84	20.24–20.40 0.797–0.803	19.69–19.94 0.775–0.785	16.43–16.59 0.647–0.653	12.70 BSC 0.500
100	22.78–22.94 0.897–0.903	22.23–22.48 0.875–0.885	18.97–19.13 0.747–0.753	15.24 BSC 0.600
132	27.86–28.02 1.097–1.103	27.31–27.56 1.075–1.085	24.05–24.21 0.947–0.953	20.32 BSC 0.800
164	32.94–33.10 1.297–1.303	32.39–32.64 1.275–1.285	29.13–29.29 1.147–1.153	25.40 BSC 1.000
196	38.02–38.18 1.497–1.503	37.47–37.72 1.475–1.485	34.21–34.37 1.347–1.353	30.48 BSC 1.200
244	45.64–45.80 1.797–1.803	45.09–45.34 1.775–1.785	41.83–41.99 1.647–1.653	38.10 BSC 1.500

CLLCCs

The ceramic leadless chip carrier was a very early entry in the SMT component arena. Thus, several variations developed before standards activity provided a clear guidance to package design.

Figure A-15 and Table A-11 (page 432)cover some common features of the four most standard 50-mil lead-pitch packages (JEDEC Types A, B, C, and D). Figure A-16 (page 432) details how these four packages vary. The dimensions for the device types shown in Figure A-16 are listed in Table A-12 (page 433). Several additional specifications, mentioned in the notes of Figure A-16, cover JEDEC Types E and F rectangular parts.

1. Metallization on sides in termination area, including castellation (where used).
2. Cavity down (Types A and D) metallized on top, with bottom optional.
3. Cavity up (Types B and C) metallized on bottom, with top optional.
4. Nontermination metallizations for heat transfer only, and must be clear of terminations.

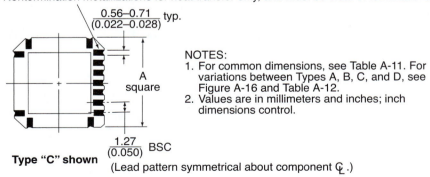

NOTES:
1. For common dimensions, see Table A-11. For variations between Types A, B, C, and D, see Figure A-16 and Table A-12.
2. Values are in millimeters and inches; inch dimensions control.

Type "C" shown

(Lead pattern symmetrical about component ₵.)

Figure A-15   Common features of 1.27-mm-center (0.050-inch) leadless chip carriers under JEDEC MS-002 (Type A), JEDEC MS-003 (Type B), JEDEC MS-004 (Type C), and JEDEC MS-005 (Type D).

Table A-11   Dimensions for Standard 50-mil Lead-Pitch CLLCC Packages.

Number of Terminals	"A" Dimensions	
	Millimeters	*Inches*
16†*	7.4–7.8	0.291–0.307
20†	8.7–9.1	0.342–0.358
24†*	10.0–10.4	0.390–0.410
28	11.2–11.7	0.441–0.460
44	16.3–16.8	0.640–0.660
52*	18.8–19.3	0.740–0.760
68	23.9–24.4	0.940–0.960
84	28.8–29.6	1.135–1.165
100*	33.9–34.7	1.335–1.365
124*	41.3–42.0	1.625–1.655
156*	51.7–52.5	2.035–2.065

* Package registered by JEDEC but are not widely used as of this writing.

† Packages available in Type C only.

Note: Inch dimensions control. Metric dimensions are rounded to nearest 0.1 mm.

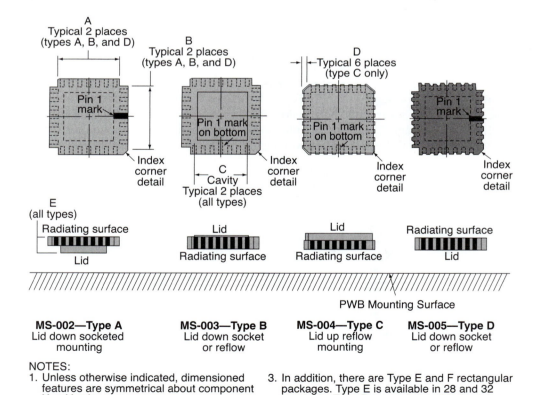

Figure A-16   Variations among common 50-mil leadless chip-carrier outlines.

Table A-12 JEDEC Leadless Chip-Carrier Dimensions for 50-Mil Center Parts.

Package	Dimensional Symbol	Number of Terminations (Maximum)										
		16	20	24	28	44	52	68	84	100	124	156
MS-002	A	N/A	N/A	N/A	8.76–9.01 / 0.345–0.355	13.82–14.12 / 0.544–0.556	16.34–16.69 / 0.643–0.657	21.39–21.79 / 0.842–0.858	26.47–26.87 / 1.042–1.058	31.47–32.03 / 1.239–1.261	39.09–39.65 / 1.539–1.561	49.25–49.81 / 1.939–1.961
MS-003 MS-005	B	N/A	N/A	N/A	9.27 Ref. / 0.365	14.38 Ref. / 0.566	16.94 Ref. / 0.667	22.05 Ref. / 0.868	27.15 Ref. / 1.069	32.13 Ref. / 1.265	39.75 Ref. / 1.565	49.91 Ref. / 1.965
MS-002	C	N/A	N/A	N/A	7.06–7.37 / 0.278–0.290	8.9–12.4 / 0.350–0.490	9.8–15.0 / 0.386–0.590	11.6–20.1 / 0.458–0.790	13.5–25.1 / 0.530–0.990	15.3–30.2 / 0.602–1.190	18.0–37.8 / 0.710–1.490	20.9–48.0 / 0.822–1.890
MS-002	E	N/A	N/A	N/A	2.29–3.05 / 0.090–0.120	2.29–3.05 / 0.900–0.120	2.29–3.05 / 0.090–0.120	2.29–3.05 / 0.090–0.120	2.29–3.05 / 0.090–0.120	1.91–4.06 / 0.075–0.160	1.91–4.06 / 0.075–0.160	1.91–4.06 / 0.075–0.160
MS-003	C	N/A	N/A	N/A	7.06–8.12 / 0.278–0.320	8.9–13.2 / 0.350–0.520	9.8–15.7 / 0.386–0.620	11.6–20.8 / 0.458–0.820	13.5–25.9 / 0.530–1.020	15.3–31.0 / 0.602–1.220	18.0–38.6 / 0.710–1.520	20.9–48.8 / 0.822–1.920
MS-003	E	N/A	N/A	N/A	1.40–3.05 / 0.055–0.120	1.40–3.05 / 0.055–0.120	1.40–3.05 / 0.055–0.120	1.40–3.05 / 0.055–0.120	1.40–3.05 / 0.055–0.120	1.40–3.05 / 0.055–0.120	1.40–3.05 / 0.055–0.120	1.40–3.05 / 0.055–0.120
MS-004	C	6.53–7.82 / 0.257–0.308	7.80–9.09 / 0.307–0.358	9.02–10.41 / 0.355–0.410	10.31–11.63 / 0.406–0.458	12.57–14.22 / 0.495–0.560	12.57–14.22 / 0.495–0.560	12.6–21.9 / 0.495–0.862	12.6–27.1 / 0.495–1.066			
MS-004	D	1.02 / 0.040	1.02 / 0.040	1.02 / 0.040	1.02 / 0.040	1.02 / 0.040	1.02 / 0.040	1.02 / 0.040	1.02 / 0.040			
MS-004	E	1.63–2.54 / 0.064–0.100	1.63–2.54 / 0.064–0.100	1.63–2.54 / 0.064–0.100	1.63–2.54 / 0.064–0.100	1.75–3.05 / 0.069–0.120	2.08–3.05 / 0.082–0.120	2.08–3.05 / 0.082–0.120	2.08–3.05 / 0.082–0.120			
MS-005	C	N/A	N/A	N/A	7.07–7.37 / 0.278–0.290	8.9–12.4 / 0.350–0.490	9.8–15.0 / 0.386–0.590	11.6–20.1 / 0.458–0.790	13.5–25.1 / 0.530–0.990	15.3–30.2 / 0.602–1.190	18.0–37.8 / 0.710–1.490	20.9–48.0 / 0.822–1.890
MS-005	E	N/A	N/A	N/A	1.27–2.03 / 0.050–0.080	1.27–2.03 / 0.050–0.080	1.27–2.03 / 0.050–0.080	1.27–2.03 / 0.050–0.080	1.27–2.03 / 0.050–0.080	1.27–2.41 / 0.050–0.095	1.27–2.41 / 0.050–0.095	1.27–2.41 / 0.050–0.095

Notes:
1. Dimensions in this table apply to Figure A-16.
2. Values are in millimeters and inches. Inch dimensions control.
3. N/A indicates that this is not a standard package.
4. For IPC standard lands for MS-003 and MS-004, see Figure A-41. For universal lands, see Figure A-58.

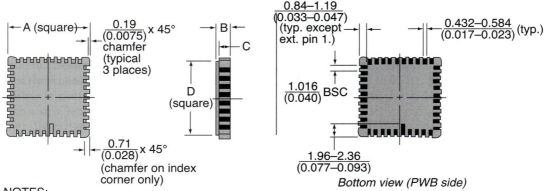

NOTES:
1. Unless otherwise indicated, dimensions shown centered are centered on the component axis.
2. For dimensions, see Table A-13.
3. For land patterns, see Figure A-62.
4. Dimensions are in millimeters and inches.

Figure A-17 JEDEC MS-009 40-mil leadless chip carrier devices.

Figure A-17 covers the 40-mil CLLCCs under JEDEC MS-009. Tabular dimensions for the 40-mil part are shown in Table A-13.

CLDCCs

Figure A-18 (page 436) covers the package outline for postleaded ceramic chip carriers under JEDEC MS-008. Table A-14 presents the tabular dimensions for the MS-008 part shown in Figure A-18.

As the drawing in Figure A-18 indicates, these specifications afford the manufacturer considerable latitude in selecting each detail of the lead configurations, an area critical to manufacturing yields and long-term reliability of the component. Some component manufacturers have exercised this freedom in favor of the end user's needs, providing top-brazed leads with high compliancy. Other manufacturers have taken advantage of the liberal specifications to cut their own production costs at the expense of device reliability. They have used stiff leads, often side brazed, to simplify meeting the coplanarity requirements. Such leads have too little compliance to absorb TCE stresses. The result is a higher field-failure rate of the solder joints. The moral, when selecting CLDCCs, is make sure that the leads will both meet coplanarity specifications and offer sufficient compliance to suit the design requirements.

JEDEC MS-007 specifically covers postleaded CCCs with side-brazed leads. Because side attachment reduces the available flexural element of the lead, we have not included this specification herein.

Preleaded ceramic chip carriers are also available. These have lead frames with hermetic glass-to-metal seals around the lead exit from the body. Preleaded CCCs in 50-mil centers are

Table A-13 JEDEC MS-009 40-Mil Leadless Chip-Carrier Dimensions.

Dimensional Symbol	Number of Terminations (Maximum)								
	16	20	24	32	40	48	64	84	96
A	5.84–6.22 0.230–0.245	8.26–8.64 0.325–0.340	8.76–9.14 0.345–0.360	10.54–10.92 0.415–0.430	12.07–12.50 0.475–0.492	14.07–14.53 0.554–0.572	18.08–18.62 0.712–0.733	23.14–23.75 0.911–0.935	26.19–26.80 1.031–1.055
B	1.02–2.41 0.040–0.095	1.40–2.79 0.055–0.110	1.40–2.79 0.055–0.110	1.40–2.79 0.055–0.110	1.40–2.79 0.055–0.110	1.40–3.18 0.055–0.125	1.40–3.18 0.055–0.125	1.40–3.18 0.055–0.125	1.78–3.43 0.070–0.135
C	0.76–1.52 0.030–0.060	1.14–1.91 0.045–0.075	1.14–1.91 0.045–0.075	1.14–1.91 0.045–0.075	1.14–1.91 0.045–0.075	1.14–2.29 0.045–0.090	1.14–2.29 0.045–0.090	1.14–2.29 0.045–0.090	1.52–2.54 0.060–0.100
D	5.33–5.72 0.210–0.225	7.49–7.87 0.295–0.310	7.87–8.38 0.310–0.330	9.40–10.16 0.370–0.400	11.05–11.68 0.435–0.460	12.70–13.59 0.500–0.535	15.24–15.88 0.600–0.625	16.38–17.91 0.645–0.705	16.38–17.91 0.645–0.705

Values are mm/inch. Inch dimensions control.

Table A-14 Dimensions for JEDEC MS-008 Postleaded Ceramic Chip Carriers.

Dimensional Symbol	Number of Terminations (Maximum)							
	28	44	52	68	84	100	124	156
A	11.81–12.83 0.465–0.505	16.84–17.96 0.663–0.707	19.35–20.52 0.762–0.808	24.38–25.65 0.960–1.010	29.41–30.78 1.158–1.212	34.52–35.84 1.359–1.411	42.14–43.46 1.659–1.711	52.30–53.62 2.059–2.111
B	8.76–9.02 0.345–0.355	13.82–14.12 0.544–0.556	16.33–16.69 0.643–0.657	21.39–21.79 0.842–0.858	26.47–26.87 1.042–1.058	31.47–32.03 1.239–1.261	39.09–39.65 1.539–1.561	49.25–49.81 1.939–1.961
C	9.27 0.365	14.38 0.566	16.94 0.667	22.05 0.868	27.15 1.069	32.13 1.265	39.75 1.565	49.91 1.965
D	2.67–4.95 0.105–0.195	2.67–4.95 0.105–0.195	2.67–4.95 0.105–0.195	2.67–4.95 0.105–0.195	2.67–4.95 0.105–0.195	2.29–5.46 0.090–0.215	2.29–5.46 0.090–0.215	2.29–5.46 0.090–0.215
E	0.89–1.27 0.035–0.050	0.89–1.27 0.035–0.050	0.89–1.27 0.035–0.050	0.89–1.27 0.035–0.050	0.89–1.27 0.035–0.050	0.89–1.27 0.035–0.050	0.89–1.27 0.035–0.050	0.89–1.27 0.035–0.050

Values are mm/inch. Inch dimensions control.

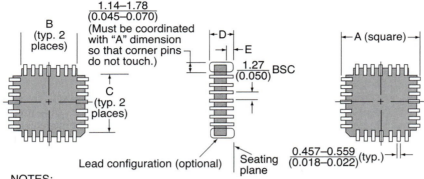

NOTES:
1. Lead pattern and A and B dimensions are centered about component center.
2. MS-007 also covers ceramic leaded carriers. We do not recommend these parts because their side-brazed leads offer less compliancy than top-brazed MS-008 parts.
3. Terminations to be coated with tin/lead solder having a tin content of 58% to 68%. Solder coating 0.0075 mm (0.0003 inch) minimum thickness. Coating may be hot dipped or plated from solution. Plated coatings must be fused in a post-plating reflow operation.
4. For dimensions, see Table A-14. For land, see Figure A-58.
5. Values are in millimeter/inch.

Figure A-18   JEDEC MS-008 50-mil-center postleaded ceramic chip carrier.

covered in JEDEC MS-044, as shown in Figure A-19. While this standard covers only 68- and 84-pin devices, many preleaded variants have been developed. Some are higher pin counts in 50-mil parts, and others have finer pitches. Fine pitches are offered in 0.8-mm (0.031"), 0.64-mm (0.025"), 0.5-mm (0.020"), and even 0.46-mm (0.018") varieties.

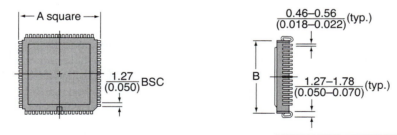

NOTES:
1. Terminations to be coated with tin/lead solder having a tin content of 58% to 68% or pure tin. Solder coating 0.0075 mm (0.0003 inch) minimum thickness. Coating may be hot dipped or plated from solution. Plated coatings must be fused in a post-plating reflow operation.
2. Dimensions are in millimeter/inch.
3. For land patterns, see Figure A-42.

Dimensional symbol	Dimensions (mm/inch)	
	68 Pin	84 Pin
A	25.02–25.27 (0.985–0.995)	30.10–30.35 (1.185–1.195)
B	23.11–24.13 (0.910–0.950)	28.2–29.2 (1.110–1.150)

Figure A-19   Preleaded ceramic chip carrier per JEDEC MS-044.

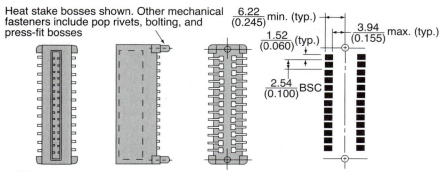

NOTES:
1. Typically available in up to 60 rows (120 contacts).
2. Many varieties on the market. Dimensions here are for reference only. Check with vendor for outline and land recommendations.
3. Values are in millimeter/inch. Inch dimensions control.
4. Terminations to be coated with tin/lead solder having a tin content of 58% to 68% or pure tin. Solder coating 0.0075 mm (0.0003 inch) minimum thickness. Coating may be hot dipped or plated from solution. Plated coatings must be fused in a post-plating reflow operation.

Figure A-20   Details of a typical surface-mount card edge connector.

A.1.3   Other Components

Connectors

Figure A-20 covers the package outline of a typical SMT-style card-edge connector. Figure A-21 shows an approach to an SMT header. There are a multitude of application demands that impact the design of connectors. Because of this, standards are difficult to develop, and thus have

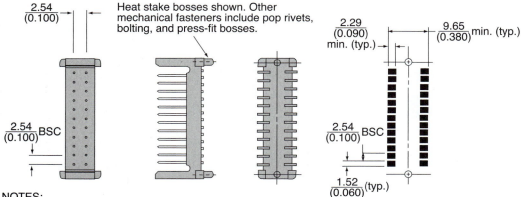

NOTES:
1. Available in 10- to 100-pin configurations.
2. Many varieties are on the market. Dimensions here are for reference only. Check with vendor for outline and land recommendations.
3. Values are in millimeter/inch. Inch dimensions control.
4. Terminations to be coated with tin/lead solder having a tin content of 58% to 68% or pure tin. Solder coating 0.0075 mm (0.0003 inch) minimum thickness. Coating may be hot dipped or plated from solution. Plated coatings must be fused in a post-plating reflow operation.

Figure A-21   Details of a typical SMT header.

moved slowly. Without standards, a wide range of outlines have been brought to the market. The configurations shown in Figure A-20 and Figure A-21 are offered for reference only, but the general approach to lead form which is shown is quite common among SMT connectors. As these drawings suggest, SMT connectors are often modeled after IMC versions and are given special lead forms for surface mounting. Generally, the housing material is also different from IMC types in order to accommodate high-temperature exposure during reflow soldering.

Sockets

Like connectors, sockets are application-driven items that have proven difficult to standardize. Like connectors, many variations of sockets have filled the standards vacuum. To bring some sense to the many offerings, we can divide them roughly by application into three basic families of SMT sockets.

First, there are sockets that allow the surface mounting of through-hole components. Figure A-22 shows a typical example of such a part. Some such sockets are gull winged as shown, while others have J or I leads. Some provide mechanical attachment bosses or have provisions for screw attachment. The socket vendors can supply details of the outlines and the solder lands for a given device.

Second, there are sockets for reflow-solderable SMCs. The intent of these parts is primarily to facilitate device removal and reinstallation. A typical PLCC type is shown in Figure A-23. Note the two variations of lead depicted in Figure A-23. Most SMT sockets use one or the other. Surface-mountable SMC sockets are available for PLCCs, SOICs, CLLCCs, and other surface-mount components.

Third, there are sockets designed for nonreflowable CLLCCs. These are similar to the PLCC socket shown in Figure A-23 except that they generally have separate hold-down covers that snap in over the chip carrier, securing it firmly into the socket.

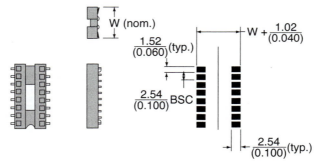

NOTES:
1. Terminations to be coated with tin/lead solder having a tin content of 58% to 68% or pure tin. Solder coating 0.0075 mm (0.0003 inch) minimum thickness. Coating may be hot dipped or plated from solution. Plated coatings must be fused in a post-plating reflow operation.
2. Typical offerings include sockets for standard 300-mil DIP in 8, 14, 16, 18, and 20 pins; 400-mil DIP in 22 pins; and 600-mil DIP in 24, 28, and 40 pins.
3. Many varieties are on the market. Dimensions here are for reference only. Check with vendor for outline and land recommendations.
4. Some sockets are offered with features for mechanical anchoring to PWB.
5. Values are in millimeter/inch. Inch dimensions control.

Figure A-22   Detail of a typical SMT DIP socket.

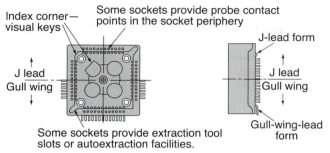

Index corner—visual keys

Some sockets provide probe contact points in the socket periphery

J lead
Gull wing

Some sockets provide extraction tool slots or autoextraction facilities.

J-lead form

J lead
Gull wing

Gull-wing-lead form

NOTES:
1. Terminations to be coated with tin/lead solder having a tin content of 58% to 68% or pure tin. Solder coating 0.0075 mm (0.0003 inch) minimum thickness. Coating may be hot dipped or plated from solution. Plated coatings must be fused in a post-plating reflow operation.
2. A wide variety of desings are available for the socketing of PLCC and CLLCCs, in all standard pin counts.

Figure A-23 Typical PLCC socket facilitates device removal/ replacement.

Because there are so many types of sockets and no firm standards to guide us in deciding which to cover herein, we have not included dimensional data on sockets; vendors can provide detailed information. Some sockets fit the same land pattern as the device they carry, and these provide a welcome measure of flexibility worth considering in the selection process.

Switches

As with the devices mentioned above, standards development has lagged far behind device-outline development in switches. Outwardly, most of today's SMT switches look like lead-prepped versions of through-hole parts. Figure A-24 shows a typical package outline for a DIP

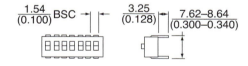

$\frac{1.54}{(0.100)}$ BSC $\frac{3.25}{(0.128)}$ $\frac{7.62\text{--}8.64}{(0.300\text{--}0.340)}$

(A) Typical DIP switch.

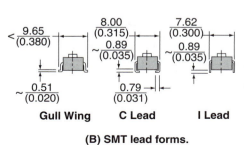

$<\frac{9.65}{(0.380)}$

$\frac{8.00}{(0.315)}$

$\frac{0.89}{(0.035)}$

$\frac{0.51}{(0.020)}$

Gull Wing

$\frac{7.62}{(0.300)}$

$\sim\frac{0.89}{(0.035)}$

$\frac{0.79}{(0.031)}$

C Lead   **I Lead**

(B) SMT lead forms.

NOTES:
1. There are a wide variety of designs on the market. Dimensions presented here are for reference only. Check with vendor for the actual outline and land recommendations.
2. Values are in millimeter/inch. Inch dimensions control.
3. Terminations to be coated with tin/lead solder having a tin content of 58% to 68% or pure tin. Solder coating 0.0075 mm (0.0003 inch) minimum thickness. Coating may be hot dipped or plated from solution. Plated coatings must be fused in a post-plating reflow operation.
4. For lands, see Figure A-43.

Figure A-24 SMT lead forms for switches.

Table A-15   Advantages and Disadvantages of Lead Forms.

Lead Form	Trade-Off Discussion	
	Advantages	*Disadvantages*
Gull wing	Strong solder joints. Full joint is visible.	Uses extra PWB area. Inspect only with close look.
C or J	Strong solder joints. Quick visual solder inspection. Minimizes PWB area.	Part of joint hidden from detailed visual inspection.
I	Quick visual solder inspection. Minimizes PWB area.	Weak solder joints.

switch and details several ways it may be lead prepped for surface mounting. As we mentioned in Chapter 2, though SMC switches look like their IMC cousins, they often differ substantially in materials and sealing details so as to withstand the SMT manufacturing environment.

Since switches present the designer with options of lead form, this is a good place to discuss the trade-offs involved in lead-form selection. Table A-15 lists the advantages and disadvantages for each. The comments regarding the visual inspection of J leads may seem suprising. However, referring to Figure A-25, one can quickly and visually spot an open on a J-leaded part by the absence of reflected light. Gull wings will require a much closer look. However, the full solder joint and under-the-heel wetting is difficult to see on the J-lead part. Here, the gull-wing part has a distinct advantage.

Whatever lead form is used, if the parts are to be stored for any time, the lead ends as well as the lead sides should have a protective solder overplate. Untreated beryllium-copper or Kovar surfaces quickly become unsolderable, and worse, the open end of a lead that was trimmed after solder coating provides a starting point for corrosion and surface blistering that may interfere with solderability elsewhere on the lead. This comment is doubly important for I-lead parts because the I lead inherently forms the weakest solder joint of the three types mentioned.

Crystals
Crystal standards work is proceeding. Surface-mount crystals are generally packaged in two alternatives. Many are in metal cans similar to IMC crystals but with SMT lead forms. Others are in LLCC packages but use only designated pins on two sides of the package. Consult the device manufacturers for the exact outlines and recommended land patterns.

Relays
Relays are another class of devices that suffer from a paucity of standards and an abundance of varied packaging. Currently, most SMT relays, outwardly, look like IMC versions that are lead prepped for surface mounting. However, like switches, SMT relays often differ from IMC

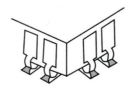

Figure A-25   Visual inspection of "J" and gull-wing lead forms.

devices in both construction details and materials. Such differences are dictated by the harsh treatment of the components in the SMT assembly processes.

Jumpers

Chip jumpers are actually zero-ohm resistors. They are controlled by the specifications governing chip resistors (see Chapter 1). The major difference between chip resistors and jumpers is that the zero-ohm parts will carry substantially higher power levels than a resistor of the same body size.

DIPs

DIP packages may be lead prepped for surface mounting. Figure A-26 covers the package outline for a 16-pin DIP and shows several ways it can be customized for surface mounting. Other package pin counts follow the same customizing scheme.

IPC-SM-782 lists the I lead as the preferred form for lead-prepped DIPs. The I form is preferred because the gull wing is more difficult to form within coplanarity specifications.[8] Also, the I lead is more space efficient than the gull wing. However, the gull-wing configuration, when properly formed per the specifications shown in Figure A-26, provides a stronger, more stress-resistant solder bond than does the I lead. Also, both parts fit the same land pattern, discounting concerns about real estate. Given appropriate tooling and process control to produce consistently coplanar feet, the gull wing is the better choice where high shear and tensile forces will be encountered. Note that "appropriate" includes the tooling that will protect the plastic body of the device from lead forming and trimming stress.

Whichever lead form is chosen, the parts should not be lead prepped and then stored. Lead trim-and-form operations should occur immediately before component assembly and soldering because of concerns about solderability, as discussed earlier in the section on switches.

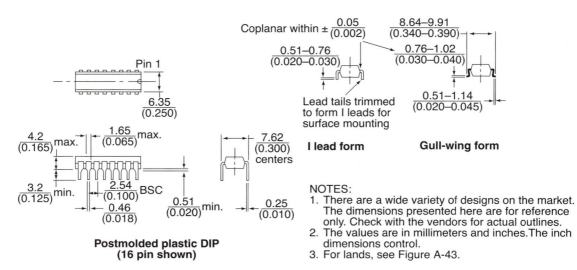

Figure A-26   Lead prepping DIPs for surface mounting.

A.2 Land Patterns

The following lands are suggested by the IPC in SM-782A for surface-mount components.[9] For reflow-soldered boards, the screened pattern of solder paste should be the same as the land patterns.

A.2.1   SMT Passives

Chip component lands are determined by the size of the chip component. IPC-SM-782 states that the following formulae are to be used for land geometry determination for nonstandard components and are the basis of the geometries in the standard.

To calculate the land geometry of the device shown in Figure A-27, we need to determine the land width (W), the land length (L), and the gap between the lands (S). To do this, we calculate the land width by

$$W = W_c \max - K \tag{Eq. A-1}$$

the land length by

$$L = H_c \max + T_c \max + K (\text{for resistors}) \tag{Eq. A-2}$$

and

$$L = H_c \max + T_c \max - K (\text{for chip caps})$$

and the gap between the lands by

$$S = L_c \max - 2T_c \max - K \tag{Eq. A-3}$$

where,

   W_c is the component width,
   H_c is the component height,
   L_c is the component length,
   T_c is the termination distance,
   K is a constant (0.25 mm / 0.010").

For tantalum bricks, L = metallization height (the termination height shown in Figure A-2).

Round all dimensions to a convenient number directed at accommodating expected board fabrication tolerances.

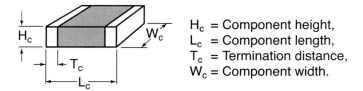

H_c = Component height,
L_c = Component length,
T_c = Termination distance,
W_c = Component width.

Figure A-27   IPC-SM-782 Standard land geometry nomenclature.

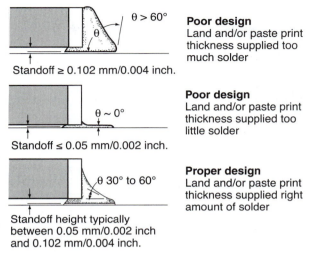

Poor design
Land and/or paste print thickness supplied too much solder

θ > 60°

Standoff ≥ 0.102 mm/0.004 inch.

Poor design
Land and/or paste print thickness supplied too little solder

θ ~ 0°

Standoff ≤ 0.05 mm/0.002 inch.

Proper design
Land and/or paste print thickness supplied right amount of solder

θ 30° to 60°

Standoff height typically between 0.05 mm/0.002 inch and 0.102 mm/0.004 inch.

Figure A-28 Solder-joint quality determination.

After the calculations, the result of each equation is rounded to the nearest convenient measurement; i.e., S = 0.062 (Equation A-3) would be rounded to 0.060" if the inch dimensions are to control, or converted to 1.57 mm and rounded up to 1.6 mm if metric dimensions are to control.

When using the formulae to determine a land for a nonstandard part, testing should be conducted to determine that the land geometry produced by the formulae yields good solder fillets and proper standoff height. "Good" solder fillets are determined by the solder wetting angle, as shown in Figure A-28.

Ceramic Chips

Figure A-29 shows the IPC-recommended land pattern for EIA standard-size ceramic chip capacitors. Alternate reflow lands are shown in Figure A-46 and Figure A-47. Optional flow-soldering optimized lands are covered in Figure A-48 and Figure A-49.

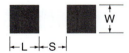

W

├─L─┼─S─┤

NOTES:
1. For component end terminations (per Figure A-1).
2. Component center at center of geometric pattern.
3. Where IPC-SM-782 inch conversions of metric dimensions were not exact, correct conversions, rounded to three places, have been substituted.
4. For alternate patterns, see Figure A-46 and Figure A-47 for reflow soldering, and Figure A-48 and Figure A-49 for flow soldering.

Component Type	Dimensions (mm/inch)		
	Width (W)	Length (L)	Space (S)
0805 Capacitor	1.40 (0.055)	1.50 (0.059)	0.80 (0.031)
1206 Capacitor	1.60 (0.063)	1.60 (0.063)	1.80 (0.071)
1210 Capacitor	2.60 (0.102)	1.80 (0.071)	1.80 (0.071)
1812 Capacitor	3.20 (0.126)	1.80 (0.071)	3.20 (0.126)
1825 Capacitor	6.60 (0.260)	1.80 (0.071)	3.20 (0.126)

Figure A-29 Lands for ceramic-chip capacitors per IPC-SM-782.

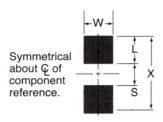

Component Designation	Dimensions (mm/inch)			
	L	S	W	X
Standard capacitance (A) 3216	2.0 (0.079)	0.8 (0.031)	1.4 (0.055)	4.8 (0.189)
(B) 3528	2.0 (0.079)	1.2 (0.047)	2.4 (0.094)	5.2 (0.205)
(C) 6032	2.4 (0.094)	3.2 (0.126)	2.4 (0.094)	8.0 (0.315)
(D) 7243	2.4 (0.094)	4.2 (0.165)	2.6 (0.102)	9.0 (0.354)
Extended range 3518	2.0 (0.079)	0.8 (0.031)	2.0 (0.079)	4.8 (0.189)
3527	2.0 (0.079)	0.8 (0.031)	2.8 (0.110)	4.8 (0.189)
7227	3.0 (0.118)	3.2 (0.126)	2.8 (0.110)	9.2 (0.362)
7257	3.0 (0.118)	3.2 (0.126)	6.0 (0.236)	9.2 (0.362)

Symmetrical about C̸ of component reference.

NOTES:
1. Values are in millimeters and inches.
2. All tolerances are ±0.1 mm/± 0.004 inch.
3. For inch control, round inch dimensions to nearest 0.005 inch.
4. For components, per Figures A-2 and A-3.

Figure A-30   Electrolytic capacitor land dimensions.

Tantalum and Aluminum Electrolytic Capacitors
Figure A-30 shows the IPC-recommended land pattern for EIA standard "brick" style electrolytic capacitors.

Variable Capacitors
Several of the many and varied outlines of trimming capacitors on today's market are shown in Figure A-4. For land pattern recommendations for a specific part, consult the component manufacturer.

Chip Resistors
Figure A-31 shows the IPC-recommended land pattern for chip resistors. Alternate reflow-process optimized lands are shown in Figure A-46 and Figure A-47. Lands optimized for flow soldering are covered in Figure A-48 and Figure A-49.

Cylindrical Components
Figure A-32 shows the IPC-recommended land pattern for cylindrical parts under the 1/4-watt MELF and the MLL34, SOD80, and MLL41 designations. The notches shown on the inner faces of the lands are intended to help stabilize cylindrical parts during reflow. In some applications, the notches have been omitted. Without notches, the parts may be secured by adhesives or fixturing.

To notch or not to notch is discussed further under the MELF discussion in Section A.3.1.

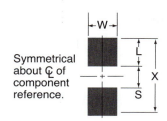

Symmetrical about ℄ of component reference.

Component Designation	Dimensions (mm/inch)			
	L	S	W	X
RC0805	1.5 (0.059)	0.8 (0.031)	1.4 (0.055)	3.8 (0.150)
RC1206	1.6 (0.063)	1.8 (0.071)	1.6 (0.063)	5.0 (0.197)
RC1210	1.6 (0.063)	1.8 (0.071)	2.6 (0.102)	5.0 (0.197)

NOTES:
1. Values are in millimeters and inches.
2. All tolerances are ±0.1 mm/± 0.004 inch.
3. For inch control, round inch dimensions to nearest 0.005 inch.

4. Use for components, per Figure A-5.
5. For alternate lands, reflow optimized, see Figure A-46 and Figure A-47; for flow-solder optimized, see Figure A-48 and Figure A-49.

Figure A-31   Chip-resistor land patterns.

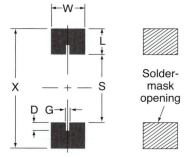

Symmetrical about ℄ of component reference.

Solder-mask opening

NOTES:
1. Notch (dimensions D and G) is optional to reduce swimming during reflow soldering. (See discussion of notch under MELFs in Section A.3.1)
2. Values are in millimeters and inches.
3. Tolerances are ±0.1 mm/± 0.004 inch.
4. For inch control, round inch dimensions to nearest 0.005 inch.
5. Use for components outlined in Figure A-6.

Component Designation	Dimensions (mm/inch)					
	D	G	L	S	W	X
MLL-34	0.7 (0.028)	0.3 (0.012)	1.4 (0.055)	2.2 (0.087)	2.0 (0.079)	5.0 (0.197)
SOD-80	0.8 (0.031)	0.3 (0.012)	1.6 (0.063)	2.2 (0.087)	2.0 (0.079)	5.4 (0.213)
MLL-41	1.0 (0.039)	0.3 (0.012)	1.8 (0.071)	3.4 (0.134)	2.8 (0.110)	7.0 (0.276)
$\frac{1}{4}$-watt MELF	0.8 (0.031)	0.3 (0.012)	2.0 (0.079)	4.4 (0.173)	2.5 (0.098)	8.4 (0.331)

Figure A-32   Cylindrical component land patterns.

Trimmer Potentiometers

Figure A-7 shows the manufacturer's recommended land patterns for the two outlines covered herein. For other outlines (again, there are many), consult the component vendors.

Resistor Networks

Where resistor networks are in standardized JEDEC component outlines such as the SOIC or CLLCC, standard SOIC or CLLCC land patterns are used. For nonstandard package outlines, consult the component vendor for land recommendations. Lands for one widely used nonstandard package are shown in Figure A-50.

Thermistors

Thermistors in EIA-standard package sizes fit capacitor IPC land patterns, such as shown in Figure A-29. Or, use the process-tailored capacitor lands from Figure A-46 and Figure A-47, or from Figure A-48 and Figure A-49. Where nonstandard body sizes are required, the IPC chip component formula in Figure A-27 or the customized land formulae in Figure A-46 or Figure A-48 may be used to determine the land geometry.

Tuning Coils and Transformers

The many and varied outlines on today's market prevent us from presenting a standard pattern herein for tuning coils and transformers. For land-pattern recommendations for a specific part, consult the component manufacturer.

Inductors

Figure A-33 shows the IPC-recommended land pattern for inductors in tantalum brick-style packaging (parts similar to tantalum bricks in form factor but having a specialized inductor rather than the tantalum package dimensions). For open core and other packaging formats not covered herein, consult the manufacturer for package outline and land geometry information.

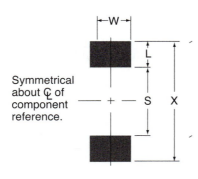

Component Designation	Dimensions (mm/inch)			
	L	S	W	X
Size A	1.4 (0.055)	1.4 (0.055)	2.4 (0.094)	4.2 (0.165)
Size B	1.6 (0.063)	2.4 (0.094)	3.0 (0.118)	5.6 (0.220)

NOTES:
1. Values are in millimeters and inches.
2. All tolerances are ±0.1 mm/± 0.004 inch.
3. For inch control, round inch dimensions to nearest 0.005 inch.
4. Use for component designations listed in chart.

Figure A-33   Chip inductor lands for tantalum brick-style packages.

A.2.2   Actives

Diodes

Diodes are packaged in both chip-style and MELF packaging using the lands per Figure A-32. Single and dual diodes are also available in SOT packages using the lands patterns shown below. LEDs follow these same packaging directions.

Transistors

Figure A-34 shows the IPC-recommended land pattern for transistors in SOT-23 packaging while Figure A-35 covers SOT-143 lands, and Figure A-36 (page 448) presents land patterns for the SOT-89. Alternate SOT-23 lands are shown in Figure A-51, Figure A-52, and Figure A-53. SOT-89 CAD vias are shown in Figure A-53, and SOT-89 alternates are covered in Figure A-54 and Figure A-55. An additional land pattern for power transistors per the DPAK outline is presented in Figure A-56.

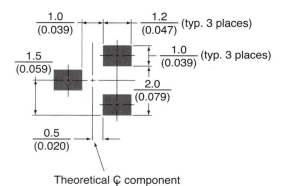

NOTES:
1. Values are in millimeters and inches.
2. Tolerances not stated in IPC-SM-782.
3. For inch control, round inch dimensions to nearest 0.005 inch.
4. Use for components covered under SOT-23 (Figure A-9A).
5. For lands optimized for reflow soldering, see Figure A-51 and Figure A-52. Lands shown here are excellent for flow soldering. For CAD vias suitable for lands shown here, see Figure A-53.

Figure A-34   SOT-23 land pattern dimensions.

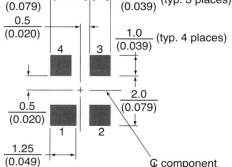

NOTES:
1. Values are in millimeters and inches.
2. Tolerances not stated in IPC-SM-782.
3. For inch control, round inch dimensions to nearest 0.005 inch.
4. Use for components covered under the SOT-143 designation in Figure A-9B.

Figure A-35   SOT-143 land pattern dimensions.

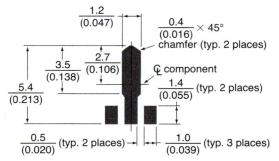

NOTES:
1. Values are in millimeters and inches.
2. Tolerances not stated in IPC-SM-782.
3. For inch control, round inch dimensions to nearest 0.005 inch.
4. Use for components covered under the SOT-89 designation in Figure A-9C.
5. For application-optimized alternate, see Figure A-54 and Figure A-55.

Figure A-36 SOT-89 land pattern dimensions.

SOICs

Figure A-37 shows the IPC-recommended land patterns for JEDEC standards SO, SOL, and SOJICs. CAD via recommendations for these land patterns are covered in Figure A-57. Figure

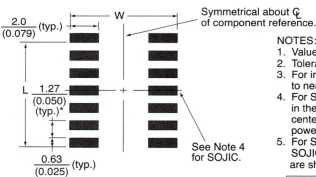

*See Note 4 for SOJICs.

NOTES:
1. Values are in millimeters and inches.
2. Tolerances are ±0.1 mm/± 0.004 inch.
3. For inch control, round inch dimensions to nearest 0.005 inch.
4. For SOJICs with an open area, missing pin(s), in the center, leave out appropriate number of center lands. Open area is clearance for large power bus trace.
5. For SO and SOLICs per Figure A-10, and SOJICs per Figure A-11. Via recommendations are shown in Figure A-57.

SO and SOLIC			
Component Designation	Dimensions (mm/inch)		Package Drawing
	L	W	
SO-8	3.8 (0.150)	7.6 (0.299)	See Figure A-10.
SO-14	7.62 (0.300)	7.6 (0.299)	
SO-16	8.9 (0.350)	7.6 (0.299)	
SOL-16	8.9 (0.350)	11.6 (0.457)	
SOL-20	11.43 (0.450)	11.6 (0.457)	
SOL-24	13.97 (0.550)	11.6 (0.457)	
SOL-28	16.5 (0.650)	11.6 (0.457)	

SOJIC			
Component Designation	Dimensions (mm/inch)		Package Drawing
	L (Note 4)	W	
SOJ-14	7.62 (0.300)	8.89 (0.350)	See Figure A-11.
SOJ-16	8.89 (0.350)	8.89 (0.350)	
SOJ-18	10.16 (0.400)	8.89 (0.350)	
SOJ-20	11.43 (0.450)	8.89 (0.350)	
SOJ-22	12.70 (0.500)	8.89 (0.350)	
SOJ-24	13.97 (0.550)	8.89 (0.350)	
SOJ-26	15.24 (0.600)	8.89 (0.350)	
SOJ-28	16.51 (0.650)	8.89 (0.350)	

Figure A-37 SO, SOL, and SOJIC land pattern dimensions.

A-50 displays lands for the nonstandard SOM IC (per the component outline given in Figure A-8). CAD vias for the SOM are also exhibited in Figure A-57.

PLCCs

Figure A-38 shows IPC-recommended land patterns for JEDEC postmolded PLCCs. This figure covers lands specifically designed for components dimensioned in Table A-6 and illustrated in Figure A-12. Postmolded parts per this outline may also be placed on universal Type A lands as illustrated in Figure A-58 with vias per Figure A-59, Figure A-60, and Figure A-61.

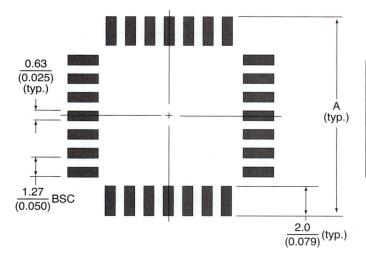

Pin Count	Dimension "A"
28	14.0 (0.551)
44	19.0 (0.748)
52	21.6 (0.850)
68	26.6 (1.047)

NOTES:
1. Values are in millimeters and inches.
2. Tolerances are ±0.1 mm/± 0.004 inch.
3. For inch control, round inch dimensions to nearest 0.005 inch.
4. For CAD vias, see Figure A-59, Figure A-60, and Figure A-61.
5. For lands recommended by IPC for premolded standard parts, see Figure A-39. For universal Type A PLCC lands, see Figure A-58.
6. For components, such as shown in Figure A-12 and dimensioned in Table A-6.

Figure A-38   JEDEC Standard postmolded PLCC lands.

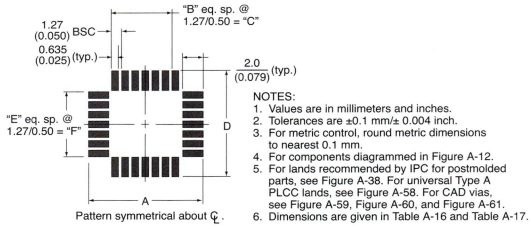

NOTES:
1. Values are in millimeters and inches.
2. Tolerances are ±0.1 mm/± 0.004 inch.
3. For metric control, round metric dimensions to nearest 0.1 mm.
4. For components diagrammed in Figure A-12.
5. For lands recommended by IPC for postmolded parts, see Figure A-38. For universal Type A PLCC lands, see Figure A-58. For CAD vias, see Figure A-59, Figure A-60, and Figure A-61.
6. Dimensions are given in Table A-16 and Table A-17.

Figure A-39   JEDEC MO-047 premolded PLCC square and rectangular lands.

Figure A-39 covers premolded packages in the square and rectangular formats. Parts fitting these lands are illustrated in Figure A-12 and are dimensioned in Table A-7. Universal Type A lands, per Figure A-58, may also be used for parts dimensioned per the premolded standards. Vias are shown in Figure A-59, Figure A-60, and Figure A-61.

Table A-16 and Table A-17 provide the land dimensions for square packages per JEDEC MO-047 and rectangular packages per JEDEC MO-052, respectively.

Table A-16   Premolded Square Packages Per JEDEC MO-047.

Pin Count	Dimensions*					
	A	B	C	D	E	F
20	$\dfrac{10.80}{0.425}$	4	$\dfrac{5.08}{0.200}$	Same as A	Same as B	Same as C
28	$\dfrac{13.34}{0.525}$	6	$\dfrac{7.62}{0.300}$	Same as A	Same as B	Same as C
44	$\dfrac{18.42}{0.725}$	10	$\dfrac{12.7}{0.500}$	Same as A	Same as B	Same as C
52	$\dfrac{20.96}{0.825}$	12	$\dfrac{15.24}{0.600}$	Same as A	Same as B	Same as C
68	$\dfrac{26.04}{1.025}$	16	$\dfrac{20.32}{0.800}$	Same as A	Same as B	Same as C
84	$\dfrac{31.12}{1.225}$	20	$\dfrac{25.40}{1.000}$	Same as A	Same as B	Same as C
100	$\dfrac{36.20}{1.425}$	24	$\dfrac{30.48}{1.200}$	Same as A	Same as B	Same as C
124	$\dfrac{43.82}{1.725}$	30	$\dfrac{38.10}{1.500}$	Same as A	Same as B	Same as C

* Values are millimeters/inches; tolerances are ±0.1 mm/± 0.004 inch. For metric control, round metric dimensions to nearest 0.1 mm.

Table A-17   Premolded Rectangular Packages Per JEDEC MO-052.

Package Type	Dimensions*					
	A	B	C	D	E	F
18 pin	$\dfrac{9.02}{0.355}$	3	$\dfrac{3.81}{0.150}$	$\dfrac{12.57}{0.495}$	4	$\dfrac{5.08}{0.200}$
18L pin	$\dfrac{9.40}{0.370}$	3	$\dfrac{3.81}{0.150}$	$\dfrac{14.61}{0.575}$	4	$\dfrac{5.08}{0.200}$
22 pin	$\dfrac{9.40}{0.370}$	3	$\dfrac{3.81}{0.150}$	$\dfrac{14.61}{0.575}$	6	$\dfrac{7.62}{0.300}$
28 pin	$\dfrac{10.80}{0.425}$	4	$\dfrac{5.08}{0.200}$	$\dfrac{16.00}{0.630}$	8	$\dfrac{10.16}{0.400}$
32 pin	$\dfrac{13.46}{0.530}$	6	$\dfrac{7.62}{0.300}$	$\dfrac{16.00}{0.630}$	8	$\dfrac{10.16}{0.400}$

* Values are millimeters/inches; tolerances are ±0.1 mm/± 0.004 inch. For metric control, round metric dimensions to nearest 0.1 mm.

PQFPs and CQFPs
Figure A-40 shows the IPC-recommended land patterns for plastic and ceramic quad flat packs.

TapePaks
TapePak lands were not covered in the March, 1987, revision of IPC-SM-782. Lands for TapePak packages are covered under alternate patterns in Figure A-65.

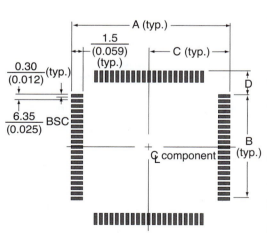

Pin Count	Dimensions (mm/inch)			
	A	B	C	D
84	20.5 (0.807)	12.7 (0.500)	10.25 (0.404)	3.75 (0.148)
100	23.0 (0.906)	15.24 (0.600)	11.50 (0.453)	3.73 (0.147)
132	28.0 (1.102)	20.32 (0.800)	14.00 (0.551)	3.69 (0.145)
164	33.2 (1.307)	25.4 (1.000)	16.60 (0.654)	3.75 (0.148)
196	38.2 (1.504)	30.48 (1.200)	19.10 (0.752)	3.71 (0.146)
244	42.5 (1.673)	38.10 (1.500)	21.25 (0.837)	2.05 (0.081)

NOTES:
1. Values are in millimeters and inches.
2. Tolerances are ±0.1 mm/± 0.004 inch.
3. For inch control, round inch dimensions to nearest 0.005 inch.
4. For component, such as shown in Figure A-14.

Figure A-40   Land patterns for JEDEC PQFPs and CQFPs.

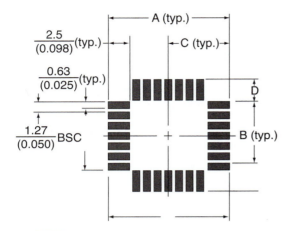

NOTES:
1. Values are in millimeters and inches.
2. Tolerances are ±0.1 mm/± 0.004 inch.
3. For inch control, round inch dimensions
   to nearest 0.005 inch.
4. For those components covered in Figure A-15
   and Figure A-16, and Table A-11 and Table A-12.
5. For universal lands, see Figure A-58; for vias,
   see Figure A-59, Figure A-60, and Figure A-61.

Pin Count	Dimensions (mm/inch)			
	A	B	C	D
16	9.8 (0.386)	3.8 (0.150)	4.90 (0.193)	2.68 (0.106)
20	11.1 (0.437)	5.08 (0.200)	5.55 (0.219)	2.69 (0.106)
24	12.4 (0.488)	6.35 (0.250)	6.20 (0.244)	2.71 (0.107)
28	13.6 (0.535)	7.62 (0.300)	6.80 (0.268)	2.67 (0.105)
44	18.8 (0.740)	12.70 (0.500)	9.40 (0.370)	2.73 (0.107)
52	21.3 (0.839)	15.24 (0.600)	10.65 (0.419)	2.71 (0.107)
68	26.4 (1.039)	20.32 (0.800)	13.20 (0.520)	2.73 (0.107)
84	31.5 (1.240)	25.4 (1.000)	15.75 (0.620)	2.73 (0.107)
100	36.5 (1.437)	30.48 (1.200)	18.25 (0.719)	2.70 (0.106)
124	44.2 (1.740)	38.10 (1.500)	22.10 (0.870)	2.73 (0.107)
156	54.3 (2.138)	48.26 (1.900)	27.15 (1.069)	2.71 (0.107)

Figure A-41   Lands for JEDEC 50-mil CLLCCs.

CLLCCs

Figure A-41 shows the IPC-recommended land patterns for 50-mil-center square and rectangular reflow-mountable ceramic leadless chip carriers. Universal lands, per Figure A-58, may also be used for these CLLCCs; however, they require a slightly greater board real estate. Vias for CLLCCs are shown in Figure A-59, Figure A-60, and Figure A-61. CLLCCs with 40-mil lead pitch are covered in Figure A-62.

CLDCCs

Figure A-42 shows IPC-recommended land patterns for JEDEC MS-044 ceramic postleaded chip carriers. Universal Type A lands, per Figure A-58, may also be used for parts dimensioned per MS-044. Vias for these CLDCCs are shown in Figure A-59, Figure A-60 and Figure A-61.

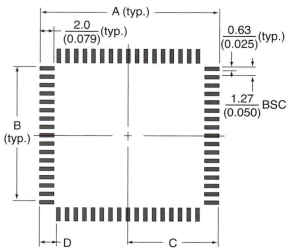

Pin Count	Dimensions (mm/inch)			
	A	B	C	D
68	26.4 (1.039)	20.32 (0.800)	13.2 (0.520)	2.73 (0.107)
84	31.0 (1.220)	25.4 (1.000)	15.5 (0.610)	2.48 (0.098)

NOTES:
1. Values are in millimeters and inches.
2. Tolerances are ±0.1 mm/± 0.004 inch.
3. For inch control, round inch dimensions to nearest 0.005 inch.
4. For components per Figure A-19.
5. For universal Type A lands, see Figure A-58; for vias, see Figure A-59, Figure A-60, and Figure A-61.

Figure A-42   JEDEC MS-044 Cerdip-type CLDCC lands.

A.2.3   Other Components

Connectors

Figure A-20 and Figure A-21 include the recommended land patterns for the typical SMT-style connectors they depict. Since nonstandardized connectors may be found in such a variety of outlines, consult the connector manufacturers for exact dimensional data on any given part.

Sockets

Figure A-22 shows a land pattern for a typical DIP socket. Again, see the socket vendor for the exact outline and land geometry information.

Switches

Standard DIP switches with custom surface-mount lead forms are mounted on a land pattern for the appropriate DIP lead form, as shown below. Lands for other package formats should be specified by the switch manufacturer.

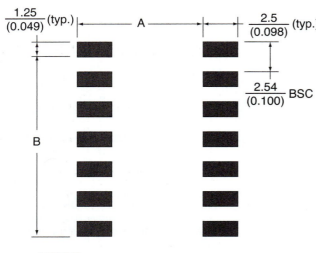

| DIP | Dimensions (mm/inch) | |
Package	A	B
300 mil. 8 pin	7.7 (0.303)	7.62 (0.300)
300 mil. 14 pin	7.7 (0.303)	15.24 (0.600)
300 mil. 16 pin	7.7 (0.303)	17.78 (0.700)
300 mil. 18 pin	7.7 (0.303)	20.32 (0.800)
300 mil. 20 pin	7.7 (0.303)	22.86 (0.900)
400 mil. 22 pin	10.2 (0.402)	25.40 (1.000)
600 mil. 24 pin	15.2 (0.598)	27.94 (1.100)
600 mil. 28 pin	15.2 (0.598)	33.02 (1.300)

NOTES:
1. Values are in millimeters and inches.
2. Tolerances are ±0.1 mm/± 0.004 inch.
3. For inch control, round inch dimensions to nearest 0.005 inch.
4. For components per Figure A-26. Also used for lead prepped parts, such as shown in Figure A-24.

Figure A-43   Land patterns for SMT lead-prepped DIPs.

DIPs
Figure A-43 shows the IPC-recommended land pattern for DIPs when they are lead prepped for surface mounting. The pattern is intended for the I lead form, but it will also work without modification for the gull-wing lead form. Vias for DIP land patterns are presented in Figure A-65.

Crystals
Consult the crystal manufacturers for the land recommendations for crystals.

Relays
Consult the manufacturers for the land recommendations for relays.

Jumpers
Chip jumpers, packaged per standard chip-resistor sizes, will fit the same land patterns that the corresponding resistor size would require. See Figure A-31 for IPC standard chip-resistor lands, and Figure A-46, Figure A-47, Figure A-48, and Figure A-49 for process customized patterns.

A.3 Land Pattern Variations

The following lands differ from IPC recommendations in order to meet specific application objectives. For reflow-soldered boards, the screened pattern of solder paste should be the same as the land patterns.

A.3.1   SMT Passives

There are several potential difficulties with the formulae for standardized land geometries shown in Section A.2.1.

First, the formulae tie the extension of the lands past the ends of the components to the component thickness. This could result in a very large number of land patterns being required for a number of component thickness variations. We could easily exceed the size of the aperture library available on typical photoplotting equipment. Of greater consequence, very tall components would require a very long land extension. Excessive extensions are undesirable because they contribute to component misalignment and because they take up valuable board real estate. Experience has shown that a fillet of sufficient strength is formed with a land extension of only about 0.65 mm (0.025"). Greater extension has little impact on solder joint strength. In fact, too much solder can actually detract from solder joint resistance to thermal cycling or board flexure-induced stress, as shown in Figure A-44 and Figure A-45. Also, past a certain point, greater extension has little impact on fillet formation.

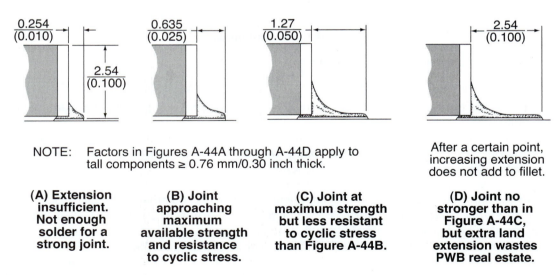

NOTE: Factors in Figures A-44A through A-44D apply to tall components ≥ 0.76 mm/0.30 inch thick.

After a certain point, increasing extension does not add to fillet.

(A) Extension insufficient. Not enough solder for a strong joint.

(B) Joint approaching maximum available strength and resistance to cyclic stress.

(C) Joint at maximum strength but less resistant to cyclic stress than Figure A-44B.

(D) Joint no stronger than in Figure A-44C, but extra land extension wastes PWB real estate.

Figure A-44   View of fillet formation vs. land extension.

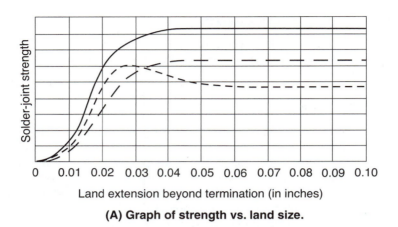

(A) Graph of strength vs. land size.

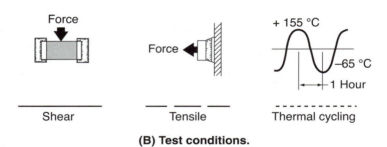

(B) Test conditions.

Figure A-45   Land extension vs. solder-joint strength.

Note that the data illustrated in Figure A-44 and Figure A-45 apply to chip components, not leaded parts.

Leadless chip carriers derive the bulk of their solder joint reliability from the fillet outside the component body. Thus, a minimum pad extension of 1.27 mm (0.050") is necessary to produce the maximum stress-resistant CLLCC joint.

For J-leaded parts, both the heel and toe of the joint contribute about equally. A pad extension of 0.635 mm (0.025") past the edges of the lead, inside and outside, is sufficient. SOICs and other gull-wing parts derive their solder joint reliability primarily from the heel of the lead. Thus, the inner extension is critical for gull-wing forms.

Second, as discussed under Chip Capacitors and Resistors later, reflow and wave-soldering process requirements differ. A land optimized for one process will not perform well in the other. A land that compromises between the needs of both processes will not give the consistent high yields and joint reliability that can be delivered by process-optimized lands.

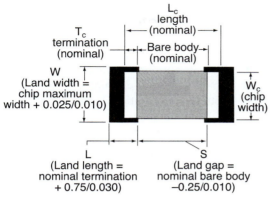

NOTES:
1. Values are in millimeters and inches.
2. All land dimensions calculated from English component dimensions (maximum or nominal as noted), rounded to the nearest 0.005 inch for inch control (or 0.05 mm for metric control).
3. For dimensions, see Table A-18. For component outline, see Figure A-1 for capacitors, and Figure A-5 for chip resistors.

Figure A-46 Chip capacitor and resistor lands tailored for reflow-soldering manufacturing yields.

Ceramic Chip Capacitors and Resistors

The standardized lands shown above are compromises between lands dimensioned specifically for reflow soldering and lands optimized for wave soldering.

For reflow soldering, lands should be designed to center the parts using solder surface-tension forces. Also, it is vital that the lands be sized to minimize tombstoning.

In a study of land size versus tombstoning defects, solder paste manufacturer, Indium Corporation of America, found that the gap between the lands should essentially match the space between the terminations (bare body) on the component. Either reducing or increasing the gap from this ideal will increase tombstoning.[10] The reflow-optimized lands shown in Figure A-46 and dimensioned in Table A-18 are constructed along this model, with an allowance for basic manufacturing tolerances.

Table A-18 Standard Chip Lands Tailored for Reflow Soldering.

Component Type	Dimensions (mm/inch)		
	Land Width	Land Length	Land Gap
0805 Capacitor	1.65 / 0.065	1.27 / 0.050	0.89 / 0.035
1206 Capacitor	2.03 / 0.080	1.27 / 0.050	1.78 / 0.070
1210 Capacitor	2.92 / 0.115	1.27 / 0.050	1.78 / 0.070
1812 Capacitor	3.68 / 0.145	1.27 / 0.050	3.18 / 0.125
1825 Capacitor	7.11 / 0.280	1.27 / 0.050	3.18 / 0.125
0805 Resistor	1.65 / 0.065	1.27 / 0.050	0.89 / 0.035
1206 Resistor	2.03 / 0.080	1.27 / 0.050	1.78 / 0.070
1210 Resistor	2.92 / 0.115	1.27 / 0.050	1.78 / 0.070

Note: Inch dimensions control.

Another custom land feature of interest is prelocated vias. The idea here is to include design-rule-conforming vias on each land. These vias are located as close as permitted to the land while still remaining on grid. We arbitrarily place the component center on the grid, or centered between grids, as required to ensure grid conformance of the predesigned vias. This artifact is intended for CAD libraries. When a particular component is called from the library, its lands and associated vias are placed on the grid for testability. Vias may then be moved to a more remote location where necessary. Figure A-47 shows automatic 50- and 100-mil grid vias for reflow-soldered chips.

Flow-soldering defects differ from reflow soldering in several ways. Because parts are glued to the board, they could not tombstone if their life depended on it. Instead, flow defects generally involve the amount of solder applied to the joint. Sometimes, it's a feast. Excessive solder on a joint can short circuit to adjacent lands or terminations. This defect is called bridging. At other times, it is a famine. Too little solder (or none at all) results in an insufficient or even absent solder connection between the solder land and the device termination. Defects from insufficient solder are called "solder starved" joints. Missing connections are called skips, opens, or voids.

Using solder lands narrower than the component body will reduce bridging. Such lands would cause serious component misregistration (swimming) in reflow, but glued-down parts stay put just fine.

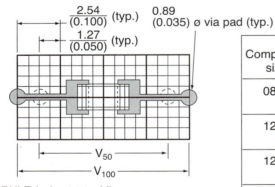

$\dfrac{2.54}{(0.100)}$ (typ.) $\dfrac{0.89}{(0.035)}$ ø via pad (typ.)

$\dfrac{1.27}{(0.050)}$ (typ.)

Component size	Dimensions (mm/inch)	
	V_{50} = 50-Mil Grid	V_{100} = 100-Mil Grid
0805	5.08 (0.200)	5.08 (0.200)
1206	6.35 (0.250)	7.62 (0.300)
1210	6.35 (0.25)	7.62 (0.300)
1812	7.62 (0.300)	7.62 (0.300)
1825	7.62 (0.300)	7.62 (0.300)

RULE (using test grid):
Place component center on grid or half grid. Locate via at minimum clearance from land, and then snap out to first available test probe grid.

NOTES:
1. Values are in millimeters and inches; tolerances are ±0.05 mm/ ± 0.002 inch.
2. Assumes photo-imaged SMOBC or bare copper trace-pads-only surface layer; 0.035-inch via pad.
3. For metric control, round metric dimensions to correspond with metric test grid and comply with rules.
4. For lands per Figure A-46, and components per Figure A-1 or Figure A-5.

Figure A-47   Auto vias for reflow-soldered chip components.

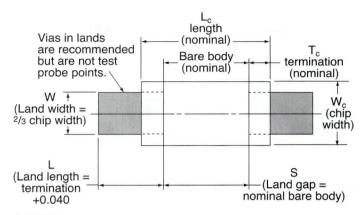

NOTES:
1. All land dimensions calculated from nominal English
   component dimensions and rounded to the nearest 0.005 inch.
2. For component per Figure A-1 or Figure A-5.

Figure A-48   Chip capacitor and resistor lands tailored for flow-soldering manufacturing yields.

To limit opens, solder lands are extended well past the ends of the component. Thus, even if eddy currents in the molten solder cause a void around the device termination, solder will still contact the extended land. Solder-wetting forces will draw it in to the device termination, forming an acceptable joint. Figure A-48 shows lands dimensioned to suit this flow-soldering model. Tabular dimensions for the drawing are in Table A-19.

Table A-19   Standard Chip Lands Tailored for Flow Soldering.

	Dimensions (mm/inch)		
Component Type	*Land Width*	*Land Length*	*Land Gap*
0805 Capacitor	$\dfrac{0.76}{0.030}$	$\dfrac{1.52}{0.060}$	$\dfrac{1.14}{0.040}$
1206 Capacitor	$\dfrac{1.02}{0.040}$	$\dfrac{1.52}{0.060}$	$\dfrac{2.03}{0.080}$
1210 Capacitor	$\dfrac{1.65}{0.065}$	$\dfrac{1.52}{0.060}$	$\dfrac{2.03}{0.080}$
1812 Capacitor	$\dfrac{2.16}{0.085}$	$\dfrac{1.52}{0.060}$	$\dfrac{3.43}{0.135}$
1825 Capacitor	$\dfrac{4.32}{0.170}$	$\dfrac{1.52}{0.060}$	$\dfrac{3.43}{0.135}$
0805 Resistor	$\dfrac{0.76}{0.030}$	$\dfrac{1.52}{0.060}$	$\dfrac{1.14}{0.045}$
1206 Resistor	$\dfrac{1.02}{0.040}$	$\dfrac{1.52}{0.060}$	$\dfrac{2.03}{0.080}$
1210 Resistor	$\dfrac{1.65}{0.065}$	$\dfrac{1.27}{0.060}$	$\dfrac{2.03}{0.080}$

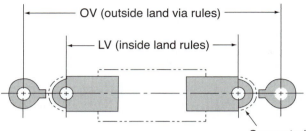

RULES (using test grid):
OV—Locate component center on grid or half
grid. Locate via at minimum clearance
from land, and then snap via out to first
available test grid.
LV— Locate component as above. Locate via
on test grid. At least 0.46/0.18 from end
of maximum length component.

Suggested option
for in-land pad:
0.89/0.035 ø via pad
(typ.)

NOTES:
1. Values are in millimeters and inches;
   tolerances are ±0.05 mm/ ± 0.002 inch.
2. Where in-land vias are used, all other
   design rules re: vias in-lands, must be met.
3. Assumes photo-imaged solder mask or
   no mask.
4. For metric control, round metric dimensions
   to correspond with metric test grid and
   comply with rules.
5. For lands, per Figure A-48, and components,
   per Figure A-1 or Figure A-5.

Figure A-49   Auto vias for flow-soldered chips. (For standard chip dimensions, see Figure A-20.)

Figure A-49 shows the automatic CAD-located vias for flow-soldered chip capacitors and resistors. Note that the flow-solder process allows, and even encourages, the location of vias much closer than would be usual in reflow soldering. This is discussed under via design in Chapter 5, Section 5. Table A-20 gives the necessary dimensions.

Table A-20   Dimensions for Automatic CAD-Located VIAs.

Component Size	Dimensions* (50-mil Test Grid)		Dimensions* (100-mil Test Grid)	
	OV	LV	OV	LV
0805	5.08 / 0.200	3.81 / 0.150	5.08 / 0.200	—
1206	6.35 / 0.250	5.08 / 0.200	7.62 / 0.300	5.08 / 0.200
1210	6.35 / 0.250	5.08 / 0.200	7.62 / 0.300	5.08 / 0.200
1812	7.62 / 0.300	6.35 / 0.250	7.62 / 0.300	—
1825	7.62 / 0.300	6.35 / 0.250	7.62 / 0.300	—

*Values are in millimeters and inches.

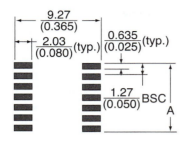

Package designation	Dimension "A"
SOM-8	3.81 (0.150)
SOM-14	7.62 (0.300)
SOM-16	8.89 (0.350)

NOTES:
1. For SOIC-like nonstandard packages, per Figure A-8.
2. Values are in millimeters and inches. Tolerances are ± 0.1 mm/ ± 0.004 inch.
3. For metric control, round the metric dimensions to nearest 0.05 mm.

Figure A-50   Lands for common nonstandard SOIC resistor network.

Cylindrical Resistors

The IPC-782A specifications for MELFs show a locating notch centered on the inside of the land. The purpose of the notch is to reduce swimming and roll-off of cylindrical parts during soldering. The specification states that the notch is optional and that without the notch, parts may be secured in some other way.

Some SMT manufacturers do glue or fixture MELFs for reflow. Some find that unrestrained reflow with the notches works well. Other users have reported no motion problem in the unrestrained reflow of parts without a notch. Several of these no-notch advocates have large-volume manufacturing operations where, if reflow problems were going to occur, they would have been noticed. Finally, there is also a battery of manufacturers who have had trouble with round parts, with or without notches in the lands. The difference is in the process control regarding the parts and board solderability, the surface contour of the board lands, warpage of boards, and the reflow process. Regarding process, vapor-phase reflow is more likely to upset cylindrical parts than is IR reflow.

Resistor Networks

A common nonstandard SOIC-like package used for resistor networks was illustrated in Figure A-8. Figure A-50 is a customized SOIC land pattern for this medium body SO.

A.3.2   Actives

Transistors

The SOT-23 land pattern recommended in IPC-SM-782 was intended as a compromise between lands optimized for reflow soldering and those tailored for wave soldering. In fact, it is close to an optimum for wave soldering and may be a poor choice for zero-defect reflow. The following discussion of the derivation of a reflow-optimized land pattern may prove useful in guiding nonstandard component land derivation as well.

For components where coplanarity problems may exist, we size lands to present enough solder to bridge the gaps produced by nonplanar leads. (Remember, in reflow soldering, your only control of the solder to the joint is in the solder paste print area width, length, and solder deposit thickness. Most chip components dictate a wet-paste print thickness of ≤0.010".) Since the SOT-23 is a three-lead device, it always exhibits perfect coplanarity as long as its leads extend past the bottom of its body. Thus, we are free to reduce the land size to the minimums required by the component. Smaller lands conserve board space and also produce very stress-resistant joints. (Remember, overly stiff joints fracture or break the components when stressed, and too much solder on leaded components stiffens the leads, reducing their flexural ability.)

Figure A-51 shows a tolerance analysis of the JEDEC dimensions for the SOT-23 at both maximum and minimum material conditions. Since the lead of the SOT-23 is relatively thin, only 0.085 to 0.152 mm (0.003" to 0.006") thick, we know that a land extending at least 0.13 mm (0.005") beyond a component lead will roughly equal the lead height and will produce a strong solder fillet with minimum excess extension. Therefore, we can predict a stress-resistant optimum reflow joint by setting a 0.005" extension around the limits of the tolerance pattern in Figure A-51. However, the SOT-23 inherently sees unequal wetting forces during reflow soldering, so we have extended the outer end of the single leg land in Figure A-51 an extra 0.010" to help balance forces. In practice, this modification reduces reflow defects common with SOT-23s.

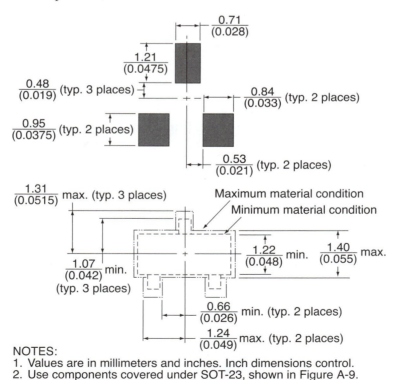

NOTES:
1. Values are in millimeters and inches. Inch dimensions control.
2. Use components covered under SOT-23, shown in Figure A-9.

Figure A-51   SOT-23 tolerance analysis—a land design challenge.

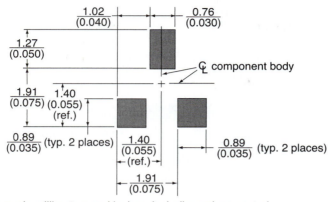

NOTES:
1. Values are in millimeters and inches. Inch dimensions control.
2. Use for components covered under SOT-23 in Figure A-9A. See Figure A-53 for standard vias.

Figure A-52 SOT-23 reflow-optimized lands.

Figure A-52 shows a conversion of the theoretical numbers we established in Figure A-51, changing them to more manageable dimensions. The rounding has been in favor of the convenient inch dimensions, but could just as easily be slanted to favor the convenient metrics where metric dimensions will control.

Figure A-53 shows CAD auto vias for SOTs. These can be used with either the lands shown in Figure A-52 or the IPC-recommended (flow) lands from Figure A-34.

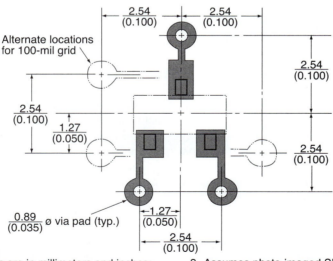

NOTES:
1. Values are in millimeters and inches; tolerances are ± 0.05 mm/ ± 0.002 inch.
2. For metric control, round the metric dimensions to correspond with metric test grid and comply with design rules.
3. Assumes photo-imaged SMOBC or bare copper and pads-only surface layer.
4. Used for lands per Figure A-52.
5. For SOT-143, use 2-via pattern on both sides.

Figure A-53 CAD auto vias for SOT-23 and SOT-143 devices.

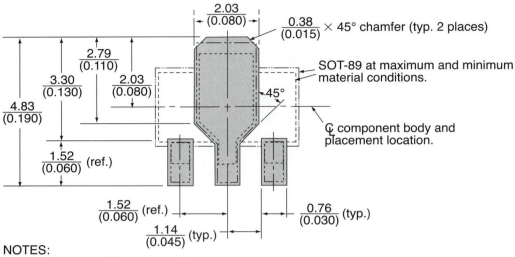

Figure A-54   Lands to maximize the dissipation from SOT-89s.

NOTES:
1. Values are in millimeters and inches; tolerances are ± 0.05 mm/± 0.002 inch.
2. For metric control, round the metric dimensions to correspond with metric test grid and comply with design rules.
3. Used for components covered under SOT-89 in Figure A-9C. See Figure A-55 for standard via locations.
4. Not recommended for flow soldering because of difficulties in assuring full solder bonding between the center tab of the component and the heak-sink solder land.

SOT-89s

The SOT-89 center tab is the die flag as well as part of the device termination. The SOT-89 relies on heat-sinking through its center metal tab for a substantial portion of its dissipation. The standard land, with a width of 1.2 mm (0.048"), leaves up to 33 percent of this thermal path unsoldered, thus limiting its effectiveness in dissipation. The graphics of a tolerance analysis was used to provide a full thermal path, and the resulting land pattern is shown in Figure A-54. This alternate pattern is recommended for any high-dissipation applications of SOT-89s. Figure A-55 shows CAD auto vias for SOT-89 lands.

DPAKs

Figure A-56 shows a land recommendation for the DPAK power transistor and diode package. Similar land patterns, amended for variations in device dimensions, may be used for other SMT parts that are made by lead prepping TO-220 devices and other IMC packages.

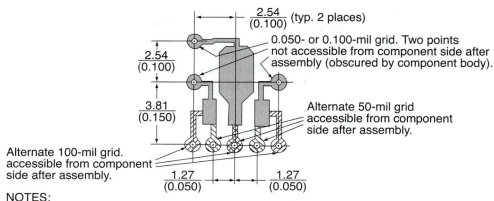

NOTES:
1. Values are in millimeters and inches; tolerances are ± 0.05 mm/ ± 0.002 inch.
2. Assumes photo-imaged SMOBC or bare copper and pads-only surface layer.
3. For metric control, round the metric dimensions to correspond with metric test grid and comply with design rules.
4. Use vias for lands, per Figure A-54.

Figure A-55   CAD auto vias for SOT-89 transistors.

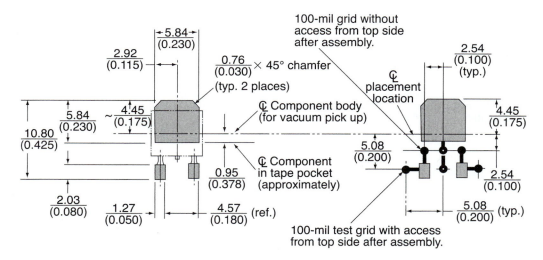

NOTES:
1. Values are in millimeters and inches; tolerances are ± 0.05 mm/ ±0.002 inch.
2. Assumes photo-imaged SMOBC or bare copper and pads-only surface layer.
3. For metric control, round the metric dimensions to correspond with metric test grid and comply with design rules.
4. Used for components, per Figure A-90.

Figure A-56   Lands and CAD auto vias for DPAK power transistors.

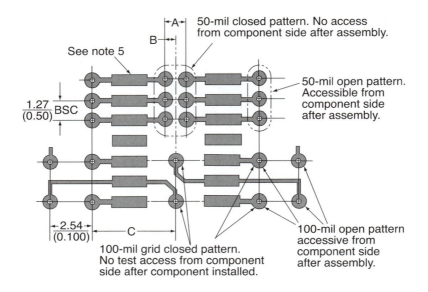

NOTES:
1. Values are in millimeters and inches; tolerances are ± 0.05 mm/ ± 0.002 inch.
2. Assumes photo-imaged SMOBC or bare copper and pads-only surface layer.
3. For metric control, round the metric dimensions to correspond with metric test grid and comply with design rules.
4. Use for SO and SOL ICs per Figure A-10, and on land patterns per Figure A-37. For SOM ICs per Figure A-8, and on lands per Figure A-50.
5. For very-high-density design, lands of 0.5 mm (0.20 inch) wide x 1.9 mm (0.075 inch) long, with rounded ends, may be substituted for IPC standard lands.

Package type body size	Dimensions (mm/inch)		
	A	B	C
SO IC, 150 mil	1.27 (0.050)	0.635 (0.025)	5.08 (0.200)
SOM IC, 220 mil	2.54 (0.100)	1.27 (0.050)	7.62 (0.300)
SOL IC, 300 Mil	5.08 (0.200)	2.54 (0.100)	7.62 (0.300)

Figure A-57   CAD auto vias for SO, SOL, and SOM ICs.

SOICs

Figure A-57 shows 50- and 100-mil grid CAD auto vias for SO and SOL ICs and gives via recommendations for the nonstandard SOM (medium body) device.

PLCCs

Figure A-58 shows a universal land pattern for Type A PLCCs. Figure A-59 covers 50- and 100-mil grid CAD auto vias for the JEDEC 50-mil PLCCs and CLLCCs on all appropriate land patterns herein. The dimensions are listed in Table A-21 (page 468) and Table A-22 (page 469), respectively.

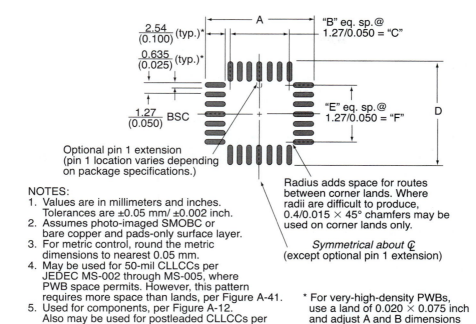

$\dfrac{2.54}{(0.100)}$ (typ.)*

$\dfrac{0.635}{(0.025)}$ (typ.)*

$\dfrac{1.27}{(0.050)}$ BSC

"B" eq. sp.@
1.27/0.050 = "C"

"E" eq. sp.@
1.27/0.050 = "F"

Optional pin 1 extension
(pin 1 location varies depending
on package specifications.)

Radius adds space for routes
between corner lands. Where
radii are difficult to produce,
0.4/0.015 × 45° chamfers may be
used on corner lands only.

Symmetrical about ℄
(except optional pin 1 extension)

NOTES:
1. Values are in millimeters and inches.
   Tolerances are ±0.05 mm/ ±0.002 inch.
2. Assumes photo-imaged SMOBC or
   bare copper and pads-only surface layer.
3. For metric control, round the metric
   dimensions to nearest 0.05 mm.
4. May be used for 50-mil CLLCCs per
   JEDEC MS-002 through MS-005, where
   PWB space permits. However, this pattern
   requires more space than lands, per Figure A-41.
5. Used for components, per Figure A-12.
   Also may be used for postleaded CLLCCs per
   Figure A-18. For CAD auto vias, see Figure A-59.

* For very-high-density PWBs,
  use a land of 0.020 × 0.075 inch
  and adjust A and B dimensions
  to suit component.

Figure A-58   Universal lands for Type A PLCCs (premolded and postmolded).

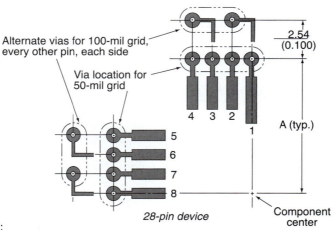

Alternate vias for 100-mil grid,
every other pin, each side

Via location for
50-mil grid

$\dfrac{2.54}{(0.100)}$

A (typ.)

28-pin device

Component
center

NOTES:
1. Values are in millimeters and inches;
   tolerances are ±0.05 mm/ ± 0.002 inch
2. For metric control, round the metric
   dimensions to correspond with metric
   test grid and comply with design rules.
3. Assumes photo-imaged SMOBC or
   bare copper and pads-only surface layer.

4. Use for components listed in Table A-21,
   per appropriate IPC recommendation.
5. Use for Type A universal lands, per
   Figure A-58. Dimensions A to suit A or D
   dimension for rectangular parts.

Figure A-59   CAD auto vias for PLCCs.

Table A-21   Dimensions for Figure A-58.

Package Type(s)	Dimensional Data					
	A	B	C	D	E	F
20 Square (MO-047)	$\dfrac{11.43}{0.450}$	4	$\dfrac{5.08}{0.200}$	Same as A	Same as B	Same as C
28 Square (all Type A)	$\dfrac{13.97}{0.550}$	6	$\dfrac{7.62}{0.300}$	Same as A	Same as B	Same as C
44 Square (all Type A)	$\dfrac{19.05}{0.750}$	10	$\dfrac{12.70}{0.500}$	Same as A	Same as B	Same as C
52 Square (all Type A)	$\dfrac{21.59}{0.850}$	12	$\dfrac{15.24}{0.600}$	Same as A	Same as B	Same as C
68 Square (all Type A)	$\dfrac{26.67}{1.050}$	16	$\dfrac{20.32}{0.800}$	Same as A	Same as B	Same as C
84 Square (MO-047 and MS-008)	$\dfrac{31.75}{1.250}$	20	$\dfrac{25.40}{1.000}$	Same as A	Same as B	Same as C
100 Square (MO-047 and MS-008)	$\dfrac{36.83}{1.450}$	24	$\dfrac{30.48}{1.200}$	Same as A	Same as B	Same as C
124 Square (MO-047 and MS-008)	$\dfrac{44.45}{1.750}$	30	$\dfrac{38.10}{1.500}$	Same as A	Same as B	Same as C
156 Square (MS-008)	$\dfrac{54.61}{2.150}$	38	$\dfrac{48.26}{1.900}$	Same as A	Same as B	Same as C
18 Rectangular (MO-052)	$\dfrac{10.16}{0.400}$	3	$\dfrac{3.81}{0.150}$	$\dfrac{12.70}{0.500}$	4	$\dfrac{5.08}{0.200}$
18L Rectangular (MO-052)	$\dfrac{10.16}{0.400}$	3	$\dfrac{3.81}{0.150}$	$\dfrac{14.86}{0.585}$	4	$\dfrac{5.08}{0.200}$
22 Rectangular (MO-052)	$\dfrac{10.16}{0.400}$	3	$\dfrac{3.81}{0.150}$	$\dfrac{14.86}{0.585}$	4	$\dfrac{7.62}{0.300}$
28 Rectangular (MO-052)	$\dfrac{11.43}{0.450}$	4	$\dfrac{5.08}{0.200}$	$\dfrac{16.51}{0.650}$	8	$\dfrac{10.16}{0.400}$
32 Rectangular (MO-052)	$\dfrac{13.97}{0.550}$	6	$\dfrac{7.62}{0.300}$	$\dfrac{16.51}{0.650}$	8	$\dfrac{10.16}{0.400}$

Table A-22   Dimensions for Figure A-59.

Pin Count	Dimensional "A"—Various Carrier Outlines				
	Type A PLCC (Postmolded)	Premolded PLCC	50-mil CLLCC	CLDCC (JEDEC MS-044)	Type A Univeral (See Note 5)
16	N/A	N/A	$\dfrac{6.35}{0.250}$	N/A	N/A
20	N/A	$\dfrac{6.35}{0.250}$	$\dfrac{6.35}{0.260}$	N/A	$\dfrac{7.62}{0.300}$
24	N/A	N/A	$\dfrac{7.62}{0.300}$	N/A	N/A
28	$\dfrac{8.89}{0.350}$	$\dfrac{7.62}{0.300}$	$\dfrac{7.62}{0.300}$	N/A	$\dfrac{8.89}{0.350}$
44	$\dfrac{11.43}{0.450}$	$\dfrac{10.16}{0.400}$	$\dfrac{10.16}{0.400}$	N/A	$\dfrac{11.43}{0.450}$
52	$\dfrac{12.70}{0.500}$	$\dfrac{11.43}{0.450}$	$\dfrac{11.43}{0.450}$	N/A	$\dfrac{12.70}{0.500}$
68	$\dfrac{15.24}{0.600}$	$\dfrac{13.97}{0.550}$	$\dfrac{13.97}{0.550}$	$\dfrac{13.97}{0.550}$	$\dfrac{15.24}{0.600}$
84	N/A	$\dfrac{16.51}{0.650}$	$\dfrac{16.51}{0.650}$	$\dfrac{16.51}{0.650}$	$\dfrac{17.78}{0.700}$
100	N/A	$\dfrac{19.05}{0.750}$	$\dfrac{19.05}{0.750}$	N/A	$\dfrac{20.32}{0.800}$
124	N/A	$\dfrac{22.86}{0.900}$	$\dfrac{22.86}{0.900}$	N/A	$\dfrac{24.13}{0.950}$
156	N/A	N/A	$\dfrac{27.94}{1.100}$	N/A	$\dfrac{29.21}{1.150}$

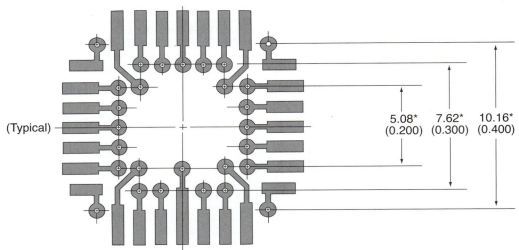

* Dimensions shown for reference. Adjust to suit component pin count and land pattern used.

28-pin PLCC lands

NOTES:
1. Vias are not accessible for in-circuit test from component side.
2. Used for components, per Figure A-12.

Figure A-60   Modified via layout for dense applications. (28-pin PLCC lands are shown—premolded or Type A postmolded.)

Figure A-60 presents a high-density 50-mil test grid alternative to the via arrangement shown in Figure A-59. The vias shown in Figure A-60 are not accessible from the component side after assembly.

Figure A-61 shows another alternative, using filled or capped vias to allow via location within the solder lands. Filled and capped vias are discussed further in Chapter 5, Section 5. This design may be used for trace-length control in high-frequency design or for space considerations. However, using standard test fixtures, these vias are not component-side accessible after assembly.

CLLCCs

Figure A-62 presents land patterns and 50- and 100-mil grid CAD auto vias for 40-mil center CLLCCs. The pertinent dimensions are given in Table A-23 (page 472).

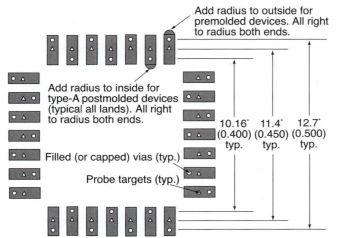

Add radius to outside for premolded devices. All right to radius both ends.

Add radius to inside for type-A postmolded devices (typical all lands). All right to radius both ends.

Filled (or capped) vias (typ.)

Probe targets (typ.)

10.16*   11.4*   12.7*
(0.400) (0.450) (0.500)
typ.   typ.   typ.

NOTES:
1. With appropriate lands, may be used with all styles of chip carriers having 50-mil lead pitch, per Figure A-12.
2. Vias are not accessible from component side after assembly.

* Dimensions shown for reference. Adjust to suit component pin count and land pattern used.

Figure A-61 Modified via layout for VHF/UHF design using filled or capped vias.

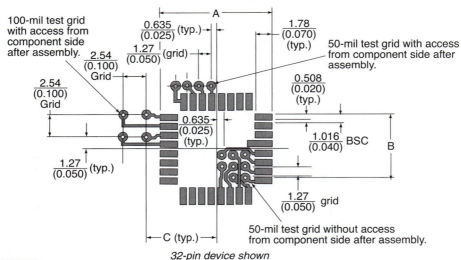

100-mil test grid with access from component side after assembly.

$\frac{2.54}{(0.100)}$ Grid

$\frac{2.54}{(0.100)}$ Grid

$\frac{1.27}{(0.050)}$ (typ.)

$\frac{0.635}{(0.025)}$ (typ.)

$\frac{1.27}{(0.050)}$ (grid)

$\frac{0.635}{(0.025)}$ (typ.)

A

$\frac{1.78}{(0.070)}$ (typ.)

50-mil test grid with access from component side after assembly.

$\frac{0.508}{(0.020)}$ (typ.)

$\frac{1.016}{(0.040)}$ BSC

B

$\frac{1.27}{(0.050)}$ grid

C (typ.)

50-mil test grid without access from component side after assembly.

32-pin device shown

NOTES:
1. Values are in millimeters and inches; tolerances are ±0.05 mm/ ±0.002 inch.
2. Assumes photo-imaged SMOBC or bare copper and pads-only surface layer.
3. For metric control, round the metric dimensions to correspond with metric test grid and comply with design rules.
4. Test via pattern varies dependent on pin count. Component X and Y center lines may be on grid or may be centered between grid lines, whichever yields most space-efficient pattern (if CAD permits).
5. Features are symmetrical about ₵, except Pin 1 feature and vias.
6. Use for components per Figure A-17, and dimensions per Table A-6 through A-8.

Figure A-62 Land patterns and CAD auto vias for 40-mil center CLLCCs.

Table A-23   Dimensional Data for Figure A-62.

Pin Count	Dimensions		
	A	*B*	*C*
16	$\dfrac{7.24}{0.285}$	$\dfrac{3.05}{0.120}$	$\dfrac{5.08}{0.200}$
20	$\dfrac{9.78}{0.385}$	$\dfrac{4.06}{0.160}$	$\dfrac{6.35}{0.250}$
24	$\dfrac{10.29}{0.405}$	$\dfrac{5.08}{0.200}$	$\dfrac{6.35}{0.250}$
32	$\dfrac{12.07}{0.475}$	$\dfrac{7.11}{0.280}$	$\dfrac{7.62}{0.300}$
40	$\dfrac{13.59}{0.535}$	$\dfrac{9.14}{0.360}$	$\dfrac{7.62}{0.300}$
48	$\dfrac{15.62}{0.615}$	$\dfrac{11.18}{0.440}$	$\dfrac{8.89}{0.350}$
64	$\dfrac{19.69}{0.775}$	$\dfrac{15.24}{0.600}$	$\dfrac{11.43}{0.450}$
84	$\dfrac{24.77}{0.975}$	$\dfrac{20.32}{0.800}$	$\dfrac{13.97}{0.550}$
96	$\dfrac{27.81}{1.095}$	$\dfrac{23.37}{0.920}$	$\dfrac{15.24}{0.600}$

TapePaks

Lands and vias for the 20-mil and 25-mil pitch TapePak are shown in Figure A-63 and the corresponding dimensions in Table A-24. Lands and vias for 15-mil 180-pin devices are shown

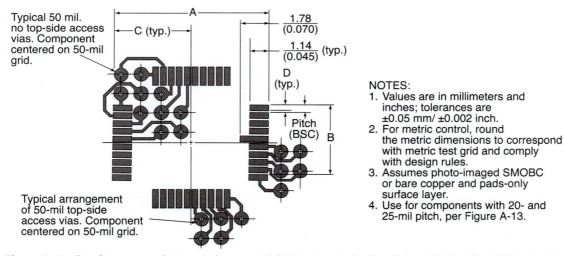

Figure A-63   Land recommendations and suggested CAD auto vias for TapePaks with 20-mil and 25-mil pitch.

Table A-24 Dimensional Data for Figure A-63.

Pitch	Pin Count	Dimensions (mm/inch)			
		A	B	C	D
0.653 0.025	32	9.78 / 0.385	4.57 / 0.180	4.89 / 0.1925	0.38 / 0.015
	52	15.24 / 0.600	7.62 / 0.300	7.62 / 0.300	0.38 / 0.015
	100	23.50 / 0.925	15.24 / 0.600	11.75 / 0.4625	0.38 / 0.015
	168*	—*	26.16* / 1.030	—*	0.38* / 0.015
0.508 0.020	40	9.78 / 0.385	4.57 / 0.180	4.89 / 0.1925	0.30 / 0.012
	84	15.24 / 0.600	10.16 / 0.400	7.62 / 0.300	0.30 / 0.012
	132	23.50 / 0.925	16.26 / 0.640	11.75 / 0.4625	0.30 / 0.012
	220*	—*	27.43* / 1.080	—*	0.30* / 0.012

*Proposed future package design. Final configuration and lead count may change.

in Figure A-64 (page 474). These are not variants of the IPC-SM-782 rules since these packages were not covered in the March, 1987, revision of IPC land-pattern specifications, which were available to us as of the publication of this work. The dimensions shown are the recommendations of National Semiconductor, the developer of the package. These dimensions are the result of both in-house and user research on mounting methods and solder joint integrity.

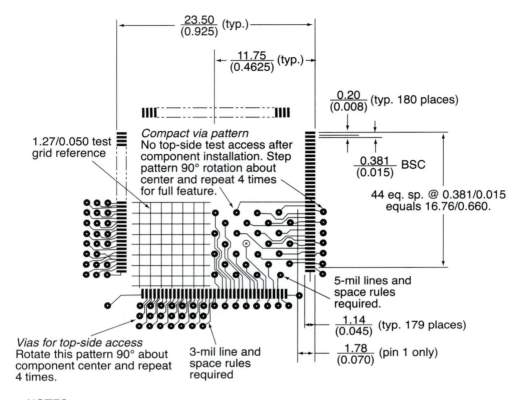

Figure A-64   Land recommendations and CAD auto vias for 180-pin TapePaks.

A.3.3   Other Components

Switches

CAD auto vias for SMT lead-formed DIP switches are shown in Figure A-65. The same pattern is applicable to I-lead, gull-wing, and J-lead forms.

DIPs

Figure A-65 shows 50- and 100-mil grid CAD auto vias for SMT lead-formed DIP lands. Table A-25 gives the dimensional data.

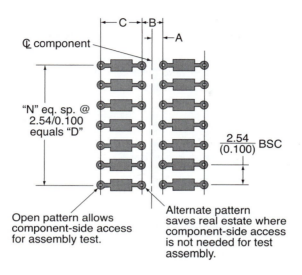

NOTES:
1. Values are in millimeters and inches. Tolerances are ±0.05 mm/±0.002 inch. Inch dimensions control. For metric control, round the metric dimensions to nearest 0.005 inch.
2. May be used with 50- or 100-mil test grids.
3. Use for components per Figure A-26, and similar lead forms, such as Figure A-24. Used for land patterns, per Figure A-43.

Figure A-65   CAD auto vias for SMT lead-formed DIPs.

Table A-25   Dimensional Data for Figure A-65.

	Dimensional Data (mm/inch)				
Dip Package	*A*	*B*	*C*	*D*	*N*
300 mil, 8 pin	$\dfrac{1.27}{0.050}$	$\dfrac{2.54}{0.100}$	$\dfrac{5.08}{0.200}$	$\dfrac{7.62}{0.300}$	3
300 mil, 14 pin	$\dfrac{1.27}{0.050}$	$\dfrac{2.54}{0.100}$	$\dfrac{5.08}{0.200}$	$\dfrac{15.24}{0.600}$	6
300 mil, 16 pin	$\dfrac{1.27}{0.050}$	$\dfrac{2.54}{0.100}$	$\dfrac{5.08}{0.200}$	$\dfrac{17.78}{0.700}$	7
300 mil, 18 pin	$\dfrac{1.27}{0.050}$	$\dfrac{2.54}{0.100}$	$\dfrac{5.08}{0.200}$	$\dfrac{20.32}{0.800}$	8
300 mil, 20 pin	$\dfrac{1.27}{0.050}$	$\dfrac{2.54}{0.100}$	$\dfrac{5.08}{0.200}$	$\dfrac{22.86}{0.900}$	9
400 mil, 22 pin	$\dfrac{2.54}{0.100}$	$\dfrac{5.08}{0.200}$	$\dfrac{5.08}{0.200}$	$\dfrac{25.40}{1.000}$	10
600 mil, 24 pin	$\dfrac{5.08}{0.200}$	$\dfrac{10.16}{0.400}$	$\dfrac{5.08}{0.200}$	$\dfrac{27.95}{1.100}$	11
600 mil, 28 pin	$\dfrac{5.08}{0.200}$	$\dfrac{10.16}{0.400}$	$\dfrac{5.08}{0.200}$	$\dfrac{33.02}{1.300}$	13

A.4 References

1. *IPC-7351-Generic Requirements for Surface Mount Land Pattern and Design Standard.* IPC Association Connecting Electronic Industries, Bannockburn, IL, 2005.

2. *EIA Specification RS-228, Fixed Electrolytic Tantalum Capacitors,* EIA, Washington, DC.

3. *Surface Mounted Components Catalog and Applications Manual,* Catalog No. 61-07A, MuRata Erie North America, Smyrna, GA, 1987, p. 7.

4. "Ceramic Dielectric Trimmer Capacitors," *Sprague-Goodman Engineering Bulletin SG-305A,* Sprague-Goodman Electronics, Garden City Park, NY, 1987.

5. *The Trimmer Primer IV—A Look Below the Surface of SMT,* Bourns Trimpot, Riverside, CA, 1986.

6. *Bourns Model 3314 SMD 4mm Sealed Single-Turn Product Bulletin,* Bourns Trimpot, Riverside, CA, 1987.

7. *Trimming Potentiometers,* Catalog No. 61-09, MuRata Erie North America, Smyrna, GA, 1987, p. 15.

8. Gray, Foster; Prasad, Ray; et al., "Surface Mount Land Patterns (Configurations and Design Rules)," *ANSI/IPC-SM-782,* IPC, Lincolnwood, IL, 1987, p. 13.

9. *Ibid.*

10. Dr. Lee, Ning-Cheng, and Evans, Gregory, "Solder Paste: Meeting the SMT Challenge," *Screen Image Technology for Electronics,* June 1987, p. 36.

appendix B

SMT Standards

B.1 American Society of Testing and Materials (ASTM)
B.2 Electrostatic Discharge Association
B.3 Electronics Industries Alliance (EIA)
B.4 Japan Electronics and Information Technologies Association (JEITA)
B.5 International Electrotechnical Committee (IEC)
B.6 Association Connecting Electronic Industries (IPC)
B.7 Joint Electronic Device Engineering Council (JEDEC)
B.8 Military and DoD Specifications and Standards
B.9 Surface-Mount Equipment Manufacturer's Association (SMEMA)
B.10 Surface-Mount Technology Association (SMTA)

We have cataloged some of the critical standards in this appendix, but in this industry, standards are continuously being changed and updated. For up-to-the-minute information, contact the organizations shown in Table B-l.

Table B-1 Sources of Information on Standards.

Organization	Address	Phone
American National Standards Institute (ANSI), http://www.ansi.org	1430 Broadway, New York, NY 10018	(212) 354-3300
ASTM, http://www.astm.org/	1916 Race St., Philadelphia, PA 19103	(215) 299-5599
British Standards Institute (BSI)	2 Park St., London W1A 12BS, UK	(441) 629-9000
Canadian Standards Association, http://www.csa.ca	178 Rexsdale Blvd., Rexsdale, Ontario, Canada	(416) 747-4000
Defense General Supply Center (Department of Defense)	1. DGSC-SSC, Richmond, VA 23297-5000	(804) 275-3900
	2. Naval Publications and Forms Center, 5801 Tabor Avenue, Philadelphia, PA 19120	(215) 697-2000
Electrostatic Discharge Association, http://www.esda.org	7900 Tunin Rd., Rome, NY 13440	
EIA, http://www.eia.org	2500 Wilson Blvd., Arlington, VA 22201	(703) 907-7500
Japan Electronics and Information Technologies Association (formerly EIAJ) http://www.jeita.or.jp/english	2-2, Marunouchi 3 Chome, Chiyoda-Ku, Tokyo 100, JAPAN	(03) 213-1071
IEC, http://www.iec.ch	P.O. Box 131, 1211 Geneva 20, Switzerland	(022) 340150
IEEE, http://www.ieee.org	345 East 47th St., New York, NY 10017	(212) 705-7900
IPC, http://www.ipc.org	3000 Lakeside Dr., Suite 309S, Bannockburn, IL 60015	(312) 677-2850
JEDEC, http://www.jedec.org	2500 Wilson Blvd., Arlington, VA 22201	(703) 907-7559
SMEMA, the IPC SMEMA Council, http://www.smema.org	71 West St., Medfield, MA 02052	(617) 359-7928
SMTA, http://www.smta.org	5200 Willson Rd., Suite 107, Edina, MN 55424	(612) 920-SMTA

B.1   American Society of Testing and Materials (ASTM)

ASTM has over 125 standards that apply to electronic devices. Included are:

- F72-95(2001) Standard Specification for Gold Wire for Semiconductor Lead Bonding
- F357-78(2002) Standard Practice for Determining Solderability of Thick-Film Conductors
- F375-89(2005) Standard Specification for Integrated Circuit Lead Frame Material
- F458-84(2001) Standard Practice for Nondestructive Pull Testing of Wire Bonds
- F459-84(2001) Standard Test Methods for Measuring Pull Strength of Microelectronic Wire Bonds
- F508-77(2002) Standard Practice for Specifying Thick-Film Pastes
- F584-87(1999) Standard Practice for Visual Inspection of Semiconductor Lead Bonding Wire
- F637-85(2001) Standard Specification for Format, Physical Properties, and Test Methods for 19 and 35 mm Testable Tape Carrier for Perimeter Tape Carrier-Bonded Semiconductor Devices
- F1661-96(2002) Standard Test Method for Determining the Contact Bounce Time of a Membrane Switch

B.2   Electrostatic Discharge Association

The ESDA publishes a number of standards and protection procedures for dealing with electrostatic energy and electrical overstress. Examples include:

- ANSI/ESD S1.1-1998: Wrist Straps
- ANSI/ESD-SP5.4-2004: Transient Latch-Up Testing—Component Level Supply Transient Stimulation

B.3   Electronics Industries Alliance (EIA)

The EIA is made up of these constituent groups:
- CEA, The Consumer Electronics Association
- ECA, The Electronic Components, Assemblies, and Materials Association
- GEIA, The Government Electronics and Information Technology Association
- JEDEC, see below
- TIA, The Telecommunications Industry Association

The EIA and JEITA (Japan Electronics and Information Technologies Association) are working to bring the U.S. and Japanese component packaging programs together under a single standard. The two groups have an agreement to circulate new standards proposals between groups for comment before any adoption. This activity should lead to world standards for new packaging, but the resistance to change of existing packages makes any retroactive action less likely.[1]

Three major EIA component standard works contain the bulk of data needed by SMT component users. These are on Components, Assembly, and Quality. Examples of other standards are shown below, but for a detailed listing of the many individual offerings, contact EIA.

B.3.1   Components

Specification Number	Subject Matter
EIA-JEDEC-95	Outlines for Semiconductor Devices
EIA-PDP-100	Outlines for Passive Devices
EIA-RS-481	8 mm through 200 mm Embossed Carrier Taping and 8 mm and 12 mm Punched Carrier Taping of SMCs for Automatic Handling

B.3.2   Assembly

Specification Number	Subject Matter
EIA-CB-11	Guidelines for the Surface Mounting of MLC Capacitors

B.3.3   Quality

Specification Number	Subject Matter
EIA-IS-46	Test Procedure for SMC Resistance to VPS

B.4   Japan Electronics and Information Technologies Association (JEITA)

Please note the comments about any future joint activity, as listed under the EIA above. JEITA, formerly EIAJ, has published hundreds of technical standards and, through cooperation with IEC, is working on international standards.

B.5   International Electrotechnical Committee (IEC)

The IEC has a wide range of published standards covering all facets of electronic assembly—from components through processes and QC/test issues. The organization also publishes the *IEC Bulletin*, which reports on current activity. Contact IEC for further information on either the standards or the *IEC Bulletin*. The IEC is also the secretariat for the IECQ-CECC, or the International Electrotechnical Commission Quality Assessment System for Electronic Components.

B.6   IPC Association Connecting Electronic Industries

The IPC is the association that writes standards relating to printed board design, assembly and testing, as well as electronic components' standards as they relate to assembly—e.g., standards for lead formations. We will not try to list all the IPC standards, since they are in a constant state of flux given the nature of the industry. Here are the categories of standards and publications listed on the IPC's website, http://www.ipc.org:

- Electronics Assembly
  - Acceptance
  - Advanced
  - Cleaning
  - Components
  - General
  - Material
  - Process Support
  - Quality/Test
  - Rework/Repair
  - Solderability
- Circuit Board Fabrication
  - Acceptance
  - Flex
  - HDI
  - HS/HF
  - Materials
- Quality/Test
- Design
- Lead Free
- Market Research/Management

In each of these categories the IPC has their standards and a number of other timely publications. For example, given the state of confusion surrounding the lead-free issues, it is appropriate that they list over 100 publications under "Lead Free." Examples of standards under other categories include:

B.6.1   Electronics Assembly—Acceptance

- A-610: Acceptability of Electronic Assemblies
- CM-770E: Component Mounting Guidelines for Printed Boards
- J-STD-001D: Requirements for Soldered Electrical and Electronic Assemblies
- J-STD-002B: Solderability Tests for Printed Boards
- J-STD-020C: IPC/JEDEC Moisture/Reflow Sensitivity Classification for Nonhermetic Solid State Surface Mount Device
- 7525: Stencil Design Guidelines
- 9191: General Guidelines for Implementation of SPC

B.6.2   Electronics Assembly—Cleaning

- SA-61A: Post Solder Semi-Aqueous Cleaning Handbook
- TP-104-K: Cleaning and Cleanliness Test Program, Water Soluble Fluxes
- TR-468: Factors Affecting Insulation Resistance Performance of Printed Boards
- TR-476A: Electrochemical Migration

B.6.3   Circuit Board Fabrication—Acceptance

- A-600: Acceptability of Printed Boards
- DW-426: Specification for Assembly of Discrete Wiring
- 6011: Generic Performance Specification for Printed Boards
- 6015: Qualification & Performance Specification for Organic Multichip Module Mounting and Interconnecting Structures
- 6016 Qualification & Performance Specification for High Density Interconnect (HDI) Layers or Boards
- 6016A: Microwave End Product Board Inspection and Test

B.6.4   Circuit Board Fabrication—Quality

- 2524: PWB Fabrication Data Quality Rating System
- 6012B: Qualification and Performance Specification for Rigid Printed Boards
- 9252: Guidelines and Requirements for Electrical Testing of Unpopulated Printed Boards
- MS-810: Guidelines for High Volume Microsection
- OI-645: Standard for Visual Optical Inspection Aids
- QL-653A: Certification of Facilities that Inspect/Test Printed Boards, Components & Materials

Design

- 1902: IPC/IEC Grid Systems for Printed Circuits
- 2141A: Design Guide for High-Speed Controlled Impedance Circuit Boards
- 2220: Design Standards Series, for Rigid and Flexible Boards
- 2226: Sectional Design Standard for High Density Interconnect (HDI) Boards
- 2252: Design Guide for RF/Microwave Circuit Boards
- 2501: Definition for Web-Based Exchange of XML Data
- 2615: Printed Board Dimensions and Tolerances
- D-330: Design Guide Manual

Lead Free

- 1065: Material Declaration Handbook
- 1066: Marking, Symbols, and Labels for Identification of Lead-Free and Other Reportable Materials

- 7095A: Design and Assembly Process Implementation for BGAs
- Over 100 other publications, by the IPC and others, relating to lead free electronics processing

Again, the reader is cautioned to note that the above listing only shows examples of the full listing of documentation available on IPC website.

B.7   Joint Electronic Device Engineering Council (JEDEC)

JEDEC set device standards. They exist for every individual device package and are too numerous to list in this book. The EIA/JEDEC components' blanket standards, which are listed, are two important starting points in component outline data. They are also listed above under EIA, the parent organization of JEDEC. Two general standards are listed below.

B.7.1   Components

Specification Number	Subject Matter
JEP-95	Registered and Standard Outlines for Solid-State and Related Products
EIA-PDP-100	Outlines for Passive Devices

B.8   Military and DoD Specifications and Standards

While the first edition of this book presented many of the military publications related to electronic components, design, and testing, we have chosen not to do that for this second edition. Like so many of the other standards that are applicable to this field, they are in a constant state of flux, and any listing would likely be out of date.

The three types of documentation that apply to issues that must be considered when designing a document to military procurement standards are:

- Military Handbooks: MIL-HDBK refers to a guidance document that contains standard procedures and or standard technical and/or engineering information. Examples of Handbooks are MIL-HDBK-103, "List of Standard Microcircuit Drawings," and MIL-HDBK-198, "Capacitors, Selection and Use of."

- Military Specifications: MIL-SPEC refers to documents that specify technical requirements for purchased components or materials that is either a modified commercial item or an item unique to military products. Some commercial components are also used as part of the COTS program—commercial off the shelf. Examples of Specifications include MIL-C-18832, "Connector, Breakaway, Electro-Mechanical," and MIL-D-87157, "Displays, Diode, Light Emitting, Solid State, General Specification For."

- Military Standards: MIL-STD refers to documents that establish requirements for engineering and technical aspects of processes, procedures, and methods. Examples of Standards include MIL-STD-1285, "Marking of Electrical and Electronic Parts," and

MIL-STD-1329, "Switches, RF Coaxial, Selection of."

All of these types of documents are available either to the general public or on a limited-access basis. There are several sources for military documents:

- The Acquisition Streamlining and Standardization Information System (ASSIST) program at http://assist.daps.dla.mil. You may be required to register and login with the ASSIST program to obtain information and/or documents from ASSIST, depending on the nature of your request.

- The Defense Supply Center Columbus (DDSC) at http://www.dscc.dla.mil/Programs/MilSpec. Most of the MIL documents at Columbus are readily available, and there are links to other government sites with further documentation.

Other documentation besides the three formal document types listed above are also available at some of the sites. For example, the DSCC has available a "Standardized Microcircuit Cross-Reference" that can be used to search for a standard MIL part number, a commercial manufacturer's part number, or by part description, such as "FET switch drivers."

The authors would note that MIL-STD-2000 and other DOD soldering and material standards have been cancelled. Check carefully if you have reason to use older documentation to be certain that the document is still in force and that you have the latest revision.

B.9   Surface-Mount Equipment Manufacturer's Association (SMEMA) Council of the IPC

The SMEMA Council has issued standard primarily for interfaceability of manufacturing equipment. These standards have been adopted by equipment manufacturers who are member companies of SMEMA. Thus, SMEMA equipment will interface electrically and mechanically in a straightforward manner for a pass through the manufacturing line. The following SMEMA standards and specifications are available in .pdf format on their website:

SMEMA 1.2: Mechanical Equipment Interface Standard
SMEMA 3.1: Fiducial Mark Standard, calling for a round fiducial
SMEMA 4: Reflow Terms and Definition
SMEMA 5: Screen Printing Terms and Definitions
SMEMA 6: Electronics Cleaning Terms and Definitions
SMEMA 7: Fluid Dispensing Terms and Definitions
SMEMA Standard Recipe File Format Specification

B.10   Surface-Mount Technology Association (SMTA)

The SMTA does not set standards but has an extensive database of .pdf files relating to surface-mount activities, as well as a book store with many publications. These files can be accessed on their website. One set of guidelines that does find widespread use is the following:

SMTA Testability Guidelines TP-101C (Hard Copy)

appendix C

SMT Bibliography

This appendix is intended to supplement the references listed in the body of the book. In Section C.1, we will list published books that provide information on surface mount. Section C.2 will cover the more readily available periodical sources of information.

C.1   Book Resources

We have not made an effort to provide a comprehensive list, and, at the rate that new material is appearing, there would be little point. Check R. R. Bowker's *Books in Print* for additional titles. This listing is available through most public libraries and includes all books currently in print that have an International Standard Book Number (ISBN) assigned. Also, on-line searches will provide many more resources.

The following listing is largely available from the "Bookstore" at the Surface-Mount Technology Association (SMTA) website, http://www.smta.org, and gives an indication of the volume of printed material available on SMT-related topics.

Assembly

SMART Group; "0201 Chip Component Design, Assembly and Soldering Guide"
- Yacobi, B. G, Hubert, M.; "Adhesive Bonding in Photonics Assembly and Packaging"
- Rowland, R.; "Applied Surface-Mount Assembly: A Guide to Surface-Mount Materials and Processes"
- Licari, J. J.; "Coating Materials for Electronic Applications: Polymers, Processes, Reliability, and Testing"
- Scheel, W.; "Handbook for Critical Cleaning"
- Cala, F.; "Handbook of Aqueous Cleaning Technology for Electronic Assemblies"
- Klein-Wassink, R.; "Manufacturing Techniques for Surface-Mounted Assemblies"
- Hutchins, C.; "Understanding and Using SMT & Fine Pitch Technology"
- Prasad, R.; "Surface-Mount Technology: Principles and Practices, 2nd Edition"

Design

- Solberg, V.; "Design Guidelines for Surface Mount and Fine-Pitch Technology, 2nd Edition
- Blankenhorn, J.; "Designing Using BGA and Flip Chip Technology"
- Harper, C. A.; "High-Performance Printed Circuit Boards"
- Johnson, H.; "High-Speed Digital Design: A Handbook of Black Magic"
- Jawitz, M.; "Printed Circuit Board Materials Handbook"

General

- Christiansen, D.; "Electronics Engineers' Handbook, 4th Edition"
- Conner, G.; "Lean Manufacturing for the Small Shop"

Packaging

- Blackwell, G. R., ed.; "Electronic Packaging Handbook"
- Lau, J.; "Flip Chip Technologies"
- Lau, J.; "Ball Grid Array Technology"
- Fjelstad, J.; "An Engineer's Guide to Flexible Printed Circuit Technology"
- Liu, J.; "Conductive Adhesives for Electronics Packaging"
- Harper, C.; "Electronic Packaging and Interconnect Handbook, 4th Edition"
- Hwang, J. S.; "Ball Grid Array and Fine-Pitch Peripheral Interconnections"
- Lau, J. and Lee, R.; "Microvias for Low-Cost, High-Density Interconnects"
- Sergent, J. E. and Krum, A.; "Thermal Management Handbook for Electronic Assemblies"

Soldering

- Woodgate, R.; "A Guide to Defect-Free Soldering"
- Hwang, J. S.; "Environment-Friendly Electronics: Lead-Free Technology"
- Puttlitz, K. J. and Stalter, K. A., eds.; "Handbook of Lead-Free Solder Technology for Microelectronic Assemblies"
- Willis, B.; "Lead-Free Assembly and Soldering" (CD)
- Manko, H.; "Soldering Handbook for PC and SM"
- Klein-Wassink, R.; "Soldering in Electronics, 2nd Edition"

Test/Reliability/Quality

- Hobbs, G. K.; "Accelerated Reliability Engineering: HALT & HASS"
- Scheiber, S.; "Building a Successful Board-Test Strategy, 2nd Edition"
- Condra, L.; "Reliability Improvement with Design of Experiments, 2nd Edition"
- Bhotel, K. and Bhote, A.; "World-Class Quality: Using Design of Experiments to Make It Happen"
- Shina, S. G.; "Six Sigma for Electronics Design and Manufacturing"
- Plumbridge, W. J., Matela, R. J., and Westwater, A.; "Structural Integrity and Reliability in Electronics in a Lead-Free Environment"
- Hutchins, C.; "Troubleshooting the SMT/FPT Process"

C.2   Periodical Resources

The following is a list of general-interest magazines that regularly devote space to the coverage of SMT. We have not performed an exhaustive search for magazines and apologize to any publishers who may feel they should be included in this listing. Again, please write to us through the publisher and ask to be added to the next update. The reader is encouraged to contact associates in related job classifications and industry segments for additional periodicals that would be of value.

- *Advanced Packaging,* http://www.appennnet.com. Contains articles and advertising of interest to semiconductor packaging engineers. Published monthly by PennWell Corp, Nashua, NH. Distributed free to qualified subscribers.

- *Circuits Assembly,* http://www.circuitsasembly.com. Contains articles and advertising of interest to engineers and managers in the electronics assembly businesses. Published monthly by UP Media Group, Atlanta, GA. Distributed free to qualified subscribers.

- *Connector Specifier,* http://www.csmag.com. Consists of articles and advertising applicable to connector engineering and the applications of interconnect technology. Published monthly by PennWell Corp, Nashua, NH. Distributed free to qualified subscribers.

- *EDN,* http://www.edn.com. Articles and advertising of interest to electronics system designers. Published 48 times a year (biweekly with additional monthly issues). Published by Reed Electronics Group, Oak Brook, IL. Distributed free to qualified subscribers.

- *Electronics Cooling,* http://www.electronics-cooling.com. Contains articles and advertising applicable to engineers and managers concerned with thermal issues in electronics design. Published quarterly by Flomerics, Inc, Marlborough, MA. Distributed free to qualified subscribers.

- *Semiconductor Packaging,* http://www.semiconductor.net/packaging. Includes articles and advertising of interest to electronics manufacturing and process engineers and system designers. Published monthly by Reed Electronics Group, Oak Brook, IL. Distributed free to qualified subscribers.

- *Evaluation Engineering,* http://www.evaluationengineering.com. Has articles and advertising that is of interest to QC, QA, and test engineers and managers. Published monthly by A. Verner Nelson Associates, Nokomis, FL. Distributed free to qualified subscribers.

- *Global SMT & Packaging,* http://www.globalsmt.net. Has articles and advertising of interest to engineers in electronics design and manufacturing. Published monthly by Trafalgar Publications, Ltd., Glastonbury, UK. Distributed free to qualified subscribers.

- *Journal of Microelectronics and Electronic Packaging,* http://www.imaps.org. Has articles and advertising applicable to engineers and managers of hybrid-circuit-manufacturing businesses. Includes SMT data. Published quarterly for members of the International Microelectronics and Packaging Society (IMAPS), Washington, DC.

- *IEEE Transactions on Components and Packaging Technologies,* http://www.ieee.org. The journal of IEEE Components, Hybrid, and Manufacturing Technology Proceedings. Published by the Institute of Electrical and Electronic Engineering, New York, NY.

- *IEEE Transactions on Electronics Packaging Manufacturing,* http://www.ieee.org. The journal of IEEE Components, Hybrid, and Manufacturing Technology Proceedings. Published by the Institute of Electrical and Electronic Engineering, New York, NY.

- *Journal of Surface-Mount Technology,* http://www.smta.org. Contains articles of interest to SMT engineers. Distributed to members of the Surface Mount Technology Association (SMTA). Published quarterly by the SMTA, Edina, MN.

- *Printed Circuit Design and Manufacturing,* http://www.pcdandm.com. Contains articles and advertising of interest to PWB designers. Distributed free to qualified subscribers. Published monthly by UP Media Group, Atlanta, GA.

- *SAMPE Journal,* http://www.sampe.org. Contains articles, SAMPE papers, and advertising of interest to materials engineers and electronics designers. Published bimonthly by the Society for the Advancement of Materials and Process Engineering, Covina, CA. Distributed free to SAMPE members, with paid subscriptions for others.

- *Semiconductor International,* http://www.semiconductor.net. Has articles and advertising that is of interest to engineers and managers in semiconductor manufacturing. Distributed free to qualified subscribers. Published monthly by Cahners Publishing Co., Newton, MA.

- *Solid-State Technology,* sst.pennnet.com. Furnishes articles and advertising of interest to those engineers and managers involved in designing or applying semiconductors or semiconductor packaging. Distributed free to qualified subscribers. Published monthly by PennWell Publishing Co., Nashua, NH.

- *SMT Magazine,* http://www.smtmag.com. Contains articles and advertising that is applicable to engineers and managers in SMT businesses. Published monthly by PennWell Corp., Nashus, NH. Distributed free to qualified subscribers.

- *Test & Measurement World,* http://www.reed-electronics.com/tmworld. Has articles and advertising of interest to QC, QA, and test engineers and managers in electronics manufacturing. Published monthly by Reed Electronics Group, Oak Brook, IL. Distributed free to qualified subscribers.

- *A Scientific Guide to SMT,* C. Lea. A highly recommended comprehensive guide to the science of the SMT process. There are fourteen chapters covering an SMT overview, components, substrates, design and assembly, wave soldering, reflow soldering, non-solder attachment, and solderability. Particularly commendable is the scientific discussion of the physical mechanisms of the soldering process. Contains 569 pages, well illustrated. Published by Electro Chemical Publications Ltd., 8 Barns St., Ayr, KA7 1XA, Scotland. Not available through U.S. distributors.

- *Surface-Mount Directory.* An exhaustive listing of components, including 1,200 outline drawings and a listing of contract services and consultants, equipment, materials, supplies, and CAD/CAM hardware and software. A required item for libraries where SMT component and supply specifications are an issue. Contains over 4,000 pages, illustrated. Published by Info-Mation, Inc., 904 Town Center, New Britain, PA 18901.

- *Surface-Mounted Assemblies,* J. Pawling, ed. A concise discussion of all aspects of SMT. Includes SMT beginnings, substrates, passive and active devices, circuit layout, footprints, component placement and attachment, and post-attachment processes. Contains 227 pages, illustrated. Published by Electro Chemical Publications Ltd., 8 Barns St., Ayr, KA7 1XA, Scotland. Not available through U.S. distributors.

- *Surface-Mount Technology.* Charles Henry Mangin and S. McCleland. This book provides an overview of SMT told primarily through six case studies which cover SMT start-ups, including Iomega, Hewlett-Packard, Philips, Rank Zerox, and Texas Instruments. Contains ten chapters, 219 pages, illustrated. Published by CEERIS International, POB 939, Old Lyme, CT 06371-0939.

appendix D

SMT Test Board Patterns

Test patterns on coupons can yield valuable insight into the accuracy and degree of the process control of PWB manufacturing and assembly processes. Such patterns will testify to PWB under-/over-etching, trace adhesion, cleaning, insulation resistance, solderability, etc. Coupons are particularly valuable when standard patterns are used because a large bank of heuristic data has already been assembled, allowing a prompt accurate interpretation of the test observations.

IPC-recommended test coupons are shown in Figure D-1 (pages 493 and 494). By incorporating coupons on all board or panel perimeters, you can develop meaningful test data on the board and assembly quality. That is, test data can be correlated to previous data because the data pertain to exactly the same test pattern(s). To evaluate the test results, obtain copies of IPC-A-600C, *Acceptability of Printed Boards—Guidelines* and IPC-A-610, *Acceptability of Printed Board Assemblies.*

Additional IPC offerings for PWB quality control and assurance are listed below:

Specification Number	Subject Matter
IPC-D-2223	Sectional Design Standard for Flexible Printed Boards
IPC-D-2221A	Sectional Design Standard on Rigid Organic Printed Boards
IPC-HM-860	Specification for Multi-layer Hybrid Circuits
IPC-L-4101	Specifications for Base Materials for Rigid and Multi-layer Printed Boards
IPC-L-4103	Specifications for Base Materials for High Speed/High Frequency Applications
IPC-RF-6013A	Quality and Perform Specifications for Flexible Printed Boards
IPC-D-352	Electronic Design Data Description for Printed Boards in Digital Form
IPC-CC-830	Quality and Performance of Electrical Insulating Compound for Printed Wiring Assemblies
IPC-D-351	Printed Board Drawings in Digital Form
IPC-D-325	Documentation Requirements for Printed Boards
IPC-MC-6011	General Performance Specifications for Printed Boards
IPC-NC-349	CNC Formatting for Drillers and Routers
IPC-D-322	Guidelines for Selecting Printed Wiring Board Sizes Using Standard Panel Sizes

In addition, IPC-D-330 is a compendium of design guidelines and specifications, and IPC-TM-650 and IPC-MI-660 gather test methods and incoming inspection techniques, respectively.

All IPC specifications and compendiums may be obtained from the IPC Association Connecting Electronic Industries, 7380 Lincoln Avenue, Lincoln Avenue, Lincolnwood, IL 60646, USA. Their telephone number is (312) 677-2850, and their FAX number is (312) 677-9570.

We would like to close this section with a special note of thanks to the IPC and the International Electronic Packaging Society (IEPS) for their generous support of this work through the provision of data and specifications.

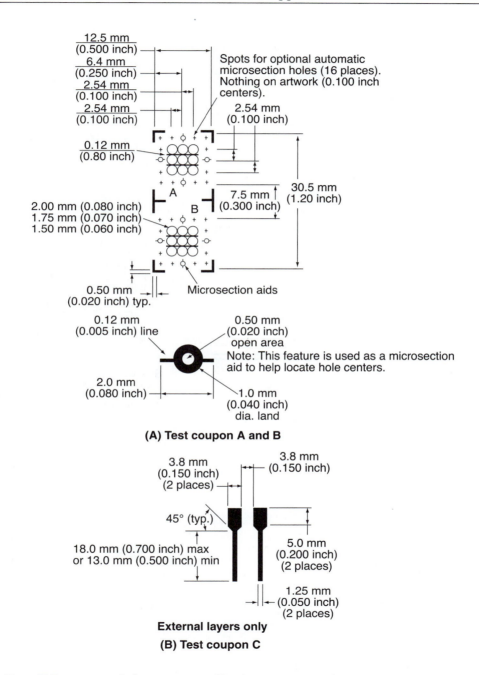

12.5 mm
(0.500 inch)

6.4 mm
(0.250 inch)

2.54 mm
(0.100 inch)

2.54 mm
(0.100 inch)

Spots for optional automatic
microsection holes (16 places).
Nothing on artwork (0.100 inch
centers).

2.54 mm
(0.100 inch)

0.12 mm
(0.80 inch)

A

30.5 mm
(1.20 inch)

7.5 mm
(0.300 inch)

2.00 mm (0.080 inch)
1.75 mm (0.070 inch)
1.50 mm (0.060 inch)

B

0.50 mm
(0.020 inch) typ.

Microsection aids

0.12 mm
(0.005 inch) line

0.50 mm
(0.020 inch)
open area

Note: This feature is used as a microsection
aid to help locate hole centers.

2.0 mm
(0.080 inch)

1.0 mm
(0.040 inch)
dia. land

(A) Test coupon A and B

3.8 mm
(0.150 inch)
(2 places)

3.8 mm
(0.150 inch)

45° (typ.)

18.0 mm (0.700 inch) max
or 13.0 mm (0.500 inch) min

5.0 mm
(0.200 inch)
(2 places)

1.25 mm
(0.050 inch)
(2 places)

External layers only

(B) Test coupon C

Figure D-1   IPC recommended test coupons. *(Continues on next page)*

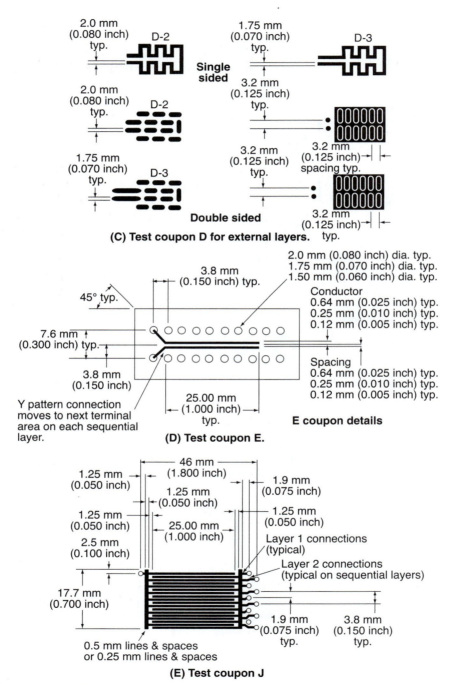

(C) Test coupon D for external layers.

(D) Test coupon E.

(E) Test coupon J

Figure D-1 *(Continued)* IPC recommended test coupons.

appendix E

Glossary of SMT Terms

In this glossary, we will not attempt to define common dictionary words except where they carry a unique connotation for surface mounting. Also, because of space limitations, we will not define words in common usage in through-hole assembly or general electronic engineering. Such words are amply defined elsewhere. We will concentrate on defining words peculiar to SMT and SMT-related areas of hybrid technology.

Abrasive Trimming—The trimming of material from a component, such as a resistor, to bring it to a specified value, where such trimming is performed by means of abrasive cutting. Abrasive dust is generally directed through a nozzle and at the part in a miniature air blast.

ACPI—An acronym for Automated Component Placement and Insertion. Also, the name of a committee dealing with related issues in the Joint Electron Device Engineering Council (JEDEC).

Active—A component that changes its basic character under applied voltage. For example, semiconductors, which are conductive only when certain conditions of applied voltage are met, are active devices, as are integrated circuits (ICs) which are primarily composed of transistors.

Active Trim—The trimming of a component while monitoring a remote electrical parameter that the trimmed component controls.

Additive Process—A process for manufacturing PWB conductive patterns by the deposition of conductive material on a substrate.

Adhesion Failure—The rupture of an adhesive bond within the adhesive material or at the adhesive-component interface.

Air Gap—The minimum space between conductors on a PWB.

Air Knife—A high-pressure airflow directed through an elongated narrow slot-style nozzle and used to sweep excess solder and solder bridges off flow-soldered boards. Particularly useful in removing bridges and icicles during the flow soldering of SMCs.

Alumina—Polycrystalline or amorphous aluminum oxide, Al_2O_3. Commonly used as a hybrid substrate and material for leadless chip carriers. *See also* Sapphire.

Ambient—The environment immediately surrounding a component or assembly.

Annular Ring—The portion of conductive material immediately surrounding a through hole or via.

Angle of Attack—The angle that the squeegee of a screen printer forms with the screen.

Annular Ring—That portion of conductive material completely surrounding a hole in a PWB. *See also* Pad.

Aqueous Flux—An organic soldering flux that is soluble in distilled water.

ANSI—An acronym for the American National Standards Institute located at 1430 Broadway, New York, NY 10018 USA.

Artwork—An accurately scaled representation of the desired pattern for a set of features on a PWB. Examples would be the conductor layer(s) and the solder mask layer(s).

ASIC—An acronym for Application Specific Integrated Circuit, a custom integrated circuit (IC) developed for a specific user application.

Aspect Ratio—The ratio of the length to the width of a feature, component, etc. For a via, the ratio of the length of the hole to its diameter.

"A" Stage—The low molecular weight stage of a "resin" in which it is readily soluble and fusible.

ASTM—An acronym for the American Society of Testing and Materials, which is located at 1916 Race Street, Philadelphia, PA 19103 USA. Available at http://www.astm.org.

ATE—An acronym for "automated test equipment."

Automatic Placement—The automated assembly of SMT boards using pick-and-place machinery. An adjective phrase to describe equipment that performs this process.

Auto Route—A computer-aided design (CAD) term referring to the capability of the CAD system to automatically interconnect components given inputs of a net list and a component placement database.

Axial—A term describing a discrete device having leads coming out of its ends along the longitudinal axis; an example would be a 1/4 watt axial carbon resistor.

Azeotropic—An adjective used to describe a liquid made up of two or more distinct fluid compounds but having a constant boiling point at a given composition.

Back Bonding—The bonding of dies with their noncircuitry side toward the substrate, as in chip-and-wire techniques. Antonym: Face Bonding.

Backward Crosstalk—Noise induced into a quiet line, seen at the end of the quiet line that is closest to the signal source. *See also* "Forward Crosstalk."

Backplane—A printed wiring board (PWB) assembly fitted with multiple connectors and conductors for interconnecting several other PWB assemblies.

Bake—1. A baking process used in SMT to dry volatiles from solder paste, activate flux, and to firmly fix components prior to reflow soldering. 2. To heat for the purpose of performing a bake. Sometimes called Bake Out.

Ball Bond—The thermocompression termination of a (typically) gold wire interconnect to a land.

Bare Board—An unpopulated (no components mounted) printed board.

Base Material—The rigid or flexible insulating material upon which a conductive pattern may be formed.

Bathtub Curve—A plot of typical failures versus time.

Bed of Nails—A test-probing device that has spring-loaded test pins in an equidistant-spaced matrix. Boards or assemblies that are to be tested will have test points spaced on exact multiples of the matrix spacing.

Beryllia—A ceramic of beryllium oxide, BeO. Beryllia is highly thermally conductive. It is used as a substrate where its thermal conductivity is useful. The dust of BeO is toxic when inhaled.

Beryllium Copper—An alloy of beryllium and copper noted for good electrical and thermal conductivity and high wear resistance. It is used in lead frames of surface-mount components (SMCs) to promote dissipation.

BIT—An acronym for *Binary digIT*. A bit is considered the indivisible limit of smallness in binary information. It has the value of either 1 or 0.

Bleed—In screen or stencil printing, the act of the substance being printed diffusing into adjacent areas beyond the limits of the desired print.

Blind Via—a via that starts at either the top or bottom layer of a printed board and ends with a connection to an inner layer. Like a blind alley, it does not go completely through the board.

Bond—1. The interconnect formed on the surface of a bare semiconductor chip, or the interconnect between the wire leading from such a chip and a bonding pad on a board. This generally refers to the thermal compression or ultrasonic bonding process in chip and wire hybrids. 2. The connection between the bottom side of a semiconductor die and a circuit board. Chips may be attached by a eutectic, epoxy, or solder die bonds.

Bonding Pad—A metallized site for the attachment of a wire through a wire bond.

Bonding Wire Crossover—A potential defect where the routing of bond wires from I/O bonding pads on an IC to its external I/O bonding pads forces bond wires to crisscross, raising a risk of short circuiting.

Breakaway—*See also* Snapoff.

Breakout—Portion(s) of a substrate scribed, notched, or otherwise mechanically processed so that it may be broken apart after assembly operations.

Bridge—1. A soldering process defect. An unwanted solder connection formed between closely spaced leads or metallized features on a PWB assembly. 2. The act of forming such a defect. *Antonym:* Open.

BSI—An acronym for the British Standards Institute, which is located at 2 Park Street, London W1A-12BS, UK.

"B" Stage—The medium molecular weight condition of a "resin" where it is less soluble than in the "A" Stage, but is plastic and fusible. "Prepregs" are resin/reinforcement compositions with B-Stage resin.

Bump—1. In tape automated bonding (TAB) technology, a thermocompression or reflow-bondable hemispherical metal deposit. Bumps may be grown on TAB tape (as in bumped-tape technology) or on the IC die (as in bumped-chip technology). Bumped-die technology, with gold bumps grown on the wafer, is the common approach as of this writing. 2. In flip-chip technology, the name applied to the solder hemispheres grown on the bonding sites of IC dies to facilitate their reflow attachment to substrates.

Buried Via—A via not opening to either outer surface of a multi-layer board (MLB).

Burn-In—The process of electrically stressing a device or assembly (usually at a high ambient temperature) for a sufficient time to eliminate failure due to component infant mortality. A large percentage of device failures generally occur during early life. Devices that do not succumb to infant mortality tend to live to their predicted life span. Thus, by eliminating infant mortalities from a population, the apparent reliability of the population may be significantly raised. Sophisticated burn-in tests log real-time anomalous event information to the nodal level during test.

Butt Lead—An SMT lead form; also called an I lead. The lead extends from the component center and turns toward the component base like a standard dual in-line package (DIP) lead but is cut to extend just below the device body.

CAD—An acronym for "computer-aided design."

CAE—An acronym for "computer-aided engineering."

CAM—An acronym for "computer-aided manufacturing."

Castellation—A semicircular cutout centered on the metallization on the periphery of a leadless chip carrier intended for a solder connection.

CAT—An acronym for "computer-aided testing."

CCC—An acronym for "ceramic chip carrier." May be leaded or leadless. *See also* Ceramic Chip Carrier.

Centering—Describes a device used for precisely registering the center of a component so that it may be accurately placed or handled by an automated machine.

Centipoise—A unit of measure of the coefficient of dynamic viscosity. One centipoise equals one one-hundredth of 1 dyne sec/cm^2.

Ceramic Chip Carrier (CCC)—A ceramic IC package with leads on four sides. May be leadless or leaded.

Ceramic Quad Flat Pack (CQFP)—A fine pitch hermetic surface-mount semiconductor package with a gull-wing lead form and peripheral I/O on all four sides.

Chase—The frame that supports a screen or stencil.

Chip—1. A single substrate on which all the active and passive circuit elements have been fabricated in situ using one or all of the semiconductor techniques of diffusion, passivation, masking, photoresist, and epitaxial growth. *See also* IC. 2. An adjective used to describe a small uncased discrete component, i.e., *Replace a large radial capacitor with a tiny disc leaded capacitor with an SMT chip capacitor.*

Chip and Wire—A hybrid interconnect method employing faceup-bonded IC devices that are interconnected to the substrate by wire bonding.

Chip Capacitor—A small uncased rectangular or square discrete device that introduces capacitance into an electronic circuit.

Chip Carrier—A high-density packaging technique often used for ICs. Its input and output terminals are around the perimeter of the device. Chip carriers are available in leadless and leaded formats.

Chip-on-Board—Chip-and-wire technique applied to traditional PWB materials in lieu of hybrid substrates.

Chip Resistor—A small ceramic substrate-based resistor, generally from 3.2 mm (125 mils) to as little as 0.5 mm (20 mils) in length. The ceramic chip is an inert substrate with the resistor formed (generally from ruthenium oxide) on the surface. Chip resistors are small, have extremely low shunt or parasitic capacitance, no inductance, and are stable.

Chip Scale Package—An active device for which the overall package size is no greater than 20 percent larger than the silicon die contained in the package.

Chip Shooter—A slang term for automatic placement equipment that is optimized for the very-high-speed placement of passive chips.

CIM—An acronym for "computer-integrated manufacturing."

Clam Shell Fixture—A test fixture that closes like a clam's shell to allow the probing of two sides of a PWB assembly simultaneously.

CLDCC—The preferred acronym for "ceramic leaded chip carrier."

Clinch—In electronics assembly, to bend the inserted leads in order to mechanically secure the leaded components in their holes before soldering.

CLLCC—The preferred acronym for "ceramic leadless chip carrier."

COB—An acronym for "chip on board."

Code 39—The most commonly used bar-code standard for the identification of components, assemblies, etc. It provides adequate field area and the alphanumeric data capability required for the application.

Coefficient of Thermal Expansion (CTE)—A variant of "Thermal Coefficient of Expansion."

Cold—When applied to solder joints in electronics, this denotes the condition also called "upset joint" where two or more pieces being joined by solder have moved during the solder solidification process. Such joints are reliability concerns. They are recognized by a grainy rough surface finish.

Cold-Crack—A metallurgical term for flaws that develop in certain alloys and metals upon low-temperature exposure.

Cold Work—The embrittlement resulting from repeated flexural stresses to a metal or alloy.

Component—The preferred term for a discrete electrical element.

Component Side—For THT/IMC boards, the side of the board bearing the components and the opposite of the "solder side" which travels through the molten solder during the wave-soldering process.

Computer Integrated Manufacturing (CIM)—The high-level computer intergration of a manufacturing operation. CIM generally is taken to include the computer integration of computer-aided engineering (CAE), computer-aided design (CAD), computer-aided manufacturing (CAM), computer-aided test (CAT), and includes test program generation from CAE/CAD and CAE/CAD feedback and upgrading from CAM/CAT.

Condensation Soldering—*See also* Vapor-Phase Soldering.

Conduction Reflow—A reflow soldering process whereby assemblies are conductively heated for soldering by being passed over heated platen(s) with an intimate thermal contact to the heated surface(s).

Conformal Coating—A protective coating, normally insulative, that is sprayed on or applied by dipping to protect a finished assembly against certain environmental effects.

Contact Printing—Screen-printing terminology for printing with no snapoff. This is commonly done in stencil printing.

Controlled Collapse—In flip-chip technology, controlling the reduction in standoff of a chip during reflow. This may be done by the growth of high melting-point pedestals within the solder bumps, etc.

Convection Reflow—Reflow soldering in a convection oven.

Coordinatograph—A high-precision drafting machine used to cut scribe-coat films in the preparation of IC or microcircuit original artwork.

Coplanar—Lying in a single plane. In SMCs, the coplanarity specification defines the maximum variance of any lead from a single plane at the points where the leads contact the substrate for mounting.

Copper/Invar/Copper—Also called Copper-Clad Invar. A laminated metallic material composed of two outer layers of copper over an inner layer of Invar. Used to form heat-sink, TCE-restraint, stiffening, and power/ground layers in organic laminate MLBs. In proper proportions, Copper/Invar/Copper can match the TCE of an MLB to the TCE of the leadless ceramic components mounted on it.

Copper/Molybdenum/Copper—Also called Copper-Clad Moly. A laminated metallic material composed of two outer layers of copper over an inner layer of Molybdenum, which is noted for its high strength-to-weight ratio. Used to form heat-sink, TCE-restraint, stiffening, and power/ground layers in organic laminate MLBs. In proper proportions, copper-clad molybdenum can match the TCE of an MLB to the TCE of the leadless ceramic components mounted on it.

Core—In PWBs, a central layer or layers of a different material than that used elsewhere in the board laminate. May serve as power, ground, heat-sink, TCE-restraint, and stiffening layer(s), or as a combination of these functions.

Coupon—A small test board fabricated with a production run of boards under matching manufacturing and control conditions. Test coupons generally have an intricate pattern and/or small holes suitable for testing the PWB pattern and/or the via quality.

CQFP—An acronym for "Ceramic Quad Flat Pack."

Cream—A solder formulation of tiny solder spheres or powder, a liquid flux, and a thixatropic vehicle, plus, in some instances, a flux activator. Solder creams may be designed for application by screening, stencil printing, or dispensing. *Synonymous with* Ink, Paste.

Crosstalk—An unwanted signal produced in an interconnect line by inductive and/or capacitive coupling with an adjacent line(s). *See also* "Backward Crosstalk" and "Forward Crosstalk."

CSA—An acronym for the Canadian Standards Associations, located at 178 Rexsdale Blvd., Rexsdale, Ontario, CANADA.

CSP—See "Chip Scale Package."

"C" Stage—The high molecular weight, fully cured state of a resin. In the C Stage, the resin is low in solubility and is infusible.

CTE—An acronym for "Coefficient of Thermal Expansion." *See also* Thermal Coefficient of Expansion.

Dam—A dielectric barrier printed or fabricated over a conductor trace to prevent unwanted solder migration along the conductor and away from the intended reflow sites.

Decoupling—To prevent noise in signal lines from appearing on power supply and ground lines.

Definition—1. Screening terminology for the sharpness of the demarcation line between a printed pattern and the adjacent substrate area. 2. Photolithographic terminology for the sharpness of demarcation between patterns and background in film masters, masks, or etched patterns.

Delid—The action of removing the hermetically sealed lid from a hybrid circuit package.

Dendrite—A galvanic growth of circuit metallization between conductors of opposite charges. Often, the dendrite takes on an appearance like the branches of a tree.

Dendritic—Of, or pertaining to, dendrites or their growth.

Desmear—The removal of (primarily) melted resin smear and (secondarily) drilling debris following the drilling process that generates a via or through hole.

Detritus—Loose material produced by spatter from a trimming process that adheres to a substrate or assembly after trimming.

Dewet—In soldering, the action of molten solder pulling back from an area where it has been applied because of insufficient wetting forces in the application area.

Dewetting—A solder defect. The condition where solder lands, metallizations, or leads have dewetted.

Die—The bare silicon form of an electronic component. Plural is "dice."

Differential Scanning Calorimetry (DSC)—A quantitative test for the cure state of a resin by determining the glass transition temperature (Tg) of a sample.

DIP—An acronym for "dual in-line package."

Direct Chip Assembly—Assembly methods involving the direct mounting of semiconductor chips on substrates. Some examples are COB, Flip Chip, and Chip and Wire.

Direct Emulsion—In screening terminology, an emulsion coating applied directly to screen material as a liquid as opposed to solid emulsions transferred to the screen from a film backing.

Direct-Metal Mask—A mask produced by etching a desired positive pattern into a metal-foil sheet. One application for such masks is stencil printing.

Discrete—1. A component, such as a resistor, capacitor, transistor, etc., that has an individual identity, as opposed to an in situ component of a hybrid. 2. A component forming a single circuit element as opposed to an integrated circuit composed of multiple circuit elements. 3. Used as an adjective, it means of, or pertaining to, such a component.

DMA—An acronym for "dynamic mechanical analysis."

DoD—An acronym for the U.S. Department of Defense. Also in common usage as DOD.

Double-Side Populated—An adjective that describes a board (usually primarily SMT) that has components located on both sides.

DPAK—Also known generically as a TO-252AA package, this is an SM transistor package with a large metal heat-sink area on its bottom, similar in format to the TO-220 package with a surface-mount lead form. Used primarily for high-power transistors and fast-switching rectifiers.

DRAM—An acronym for "dynamic RAM." A read/write memory that must be clocked or refreshed at regular intervals to maintain its memory integrity.

Drawbridge—A term for a solder defect in which a small chip component raises one termination producing an open during reflow soldering. The defect is caused by unbalanced forces of solder wetting and surface tension acting on the two ends of the chip. When this imbalance is great enough to overcome gravity, the component begins to stand on one end like a drawbridge opening. The term is used for parts in any stage of erection great enough to produce a solder open on the raising end. Also, the act of drawbridging. *Synonymous with* Pop Wheelies, Tombstone (preferred), Manhattan Effect.

DSC—An acronym for "differential scanning calorimetry."

Dual In-Line Package (DIP)—An insertion-mountable package having a row of extended I/O pins on two opposing sides. The extended pins are typically mounted through hole in a PWB. The format is generally used to package ICs.

Dual Wave—An SMT flow-soldering approach using two solder waves. The first is a high- pressure turbulent wave to throw solder on all metallizations, reducing the likelihood of solder skips. The second is a low-pressure, laminar flow wave designed to remelt and remove any bridges and icicles left by the turbulent wave.

Dynamic Mechanical Analysis (DMA)—A quantitative test method for resin materials involving the measurement of the change in stiffness of laminates and prepregs with changes in temperature.

Dynamic Printing Force—Screening terminology for the fluid force that operates on a thixatropic paste causing it to act as a liquid and flow through a screen or stencil, wetting the workpiece to be printed.

Edge Rate—Alternate term for "Rise Time." The rate of change of voltage, typically in a digital circuit, with respect to time, measured from 10 percent of maximum voltage to 90 percent of maximum voltage.

EIA—An acronym for the Electronics Industries Association, located at 2001 Eye Street, Washington, DC 20006 USA.

Electromagnetic Interference (EMI)—Unwanted signals induced in a circuit by magnetic fields.

Electromagnetic Pulse (EMP)—A brief high-power electromagnetic radiation. EMPs are generated over a brief period during a nuclear or thermonuclear explosion.

Electroplating—The deposition of conductive material, typically copper or solder, from a plating solution to base material by the passage of electrical current.

Electroless Plating—The deposition of conductive material, typically copper, from a plating solution to base material without the use of electrical current.

EMI—An acronym for "electromagnetic interference."

EMP—An acronym for "electromagnetic pulse."

Eutectic—A metallurgical term for an alloy composition that has the lowest melting point of that alloy series. Alternately, where the liquidus and solidus temperatures are equal. Unlike other alloy compositions in the series, the eutectic has no plastic or mushy phase. Its liquidus and solidus phases coincide, making eutectic solders useful for soldering to reduce the likelihood of upset solder joints. In the tin/lead series, the eutectic alloy is approximately 63 percent Sn, 37 percent Pb.

Exponential Failures—Failures that occur at an exponentially increasing rate. Also called "wear-out" failures.

Extender—A material added to a plastic, paste, etc., to reduce the amount of the primary material required.

Face Bonding—The bonding of dies with their circuitry side toward the substrate, as in flip-chip mounting. *Antonym of* Back Bonding.

Fiducial—A vision target in the conductive layer of a printed board used by automated assembly equipment to locate features on the board.

Fillet—The concave-shaped interfacial connection formed by solder between a device termination and the solder land on a PWB.

Film—Hybrid terminology for a coating deposited by thick-film (screen print), or thin-film (evaporation, or sputtering) processes. *See also* Thick Film, Thin Film.

Final Seal—In hybrid processing, the lid-sealing operation that covers the circuit assembly such that further processing cannot be done to the circuit without disassembling the package (delidding).

Fine Leak—In hermetically sealed packages, a leak passing less than 10^{-5} cm^3/sec. at a pressure differential of one atmosphere.

Fire—To bake, at high temperatures, a green ceramic or printed thick-film pattern in order to alter its basic properties and bring it to a condition suitable for use.

First-Pass Yield—The percentage of the product conforming to specifications as it leaves the manufacturing line and arrives at final test. This percentage is generally expressed as failures in parts per million (PPM) for high-yield processes and a percentage of conforming parts for low-yield lines (greater than 1000 PPM nonconforming assemblies, or less than 99.9 percent conforming).

Fissure—In ceramics, a crack in a dielectric in the area of metallizations which is usually due to stresses produced in an incorrect firing cycle. Also, to crack the dielectric in such a fashion.

Flag—An area of a component lead frame used for support of the semiconductor die.

Flat Pack—1. A semiconductor package having leads on two or four sides. The lead-frame material is very thin and fragile. Flat packs are typically shipped with the lead frame untrimmed and unformed. The parts are commonly trimmed and formed into gull-wing format for surface mounting. 2. A style of hybrid package with a thin Z-axis dimension and two rows of ribbon-wire leads, one extending from each long side. Before forming, the leads extend in a single plane. Such packages are designed for gull-wing lead forming to make a surface-mountable hybrid circuit.

Flexible—When applied to machines or production lines, this indicates that the system can readily accommodate a range of assembly specifications. In the most extreme usage, "flexible automation," it implies that equipment has the capability to automatically recognize and adapt to a wide range of assembly tasks and changes in workpiece specifications. However, when used in advertising or sales claims, this word appears to have lost any precise meaning.

Flip Chip—An IC die prepared for reflow face bonding by the growth of solder hemispheres on the bond pads of the chip. More broadly, the word is used to refer to the full technology of mounting such devices.

Flood—Screening terminology for the nonprinting stroke of a screen printer. This stroke is used to drag paste over the screen before the print stroke. In some printers, the flood stroke is accomplished by a separate flood bar. In others, the flooding is provided by a stroke of the print squeegee where the normal printing pressure is not applied.

Flow—A soldering method involving the immersion of metallizations in molten solder to produce the desired solder joints. (Wave soldering in the dominant flow-soldering method used in THT/IMC PWB assembly.)

Footprint—1. The land pattern for an SMC. 2. The outline that a component body or chip will occupy when placed on an assembly.

Forward Crosstalk—Noise induced into a quiet line, seen at the end of the quiet line that is farthest from the signal source. *See also* "Backward Crosstalk."

FPT—Fine Pitch Technology, the term typically used for ICs with a lead pitch of 0.025" = 25 mils = 0.635 mm.

Frit—Powered glass used as a component in thick-film pastes to give adhesion to the composition after firing, and also used singly as an overglaze for the protection of underlying materials.

GaAs—An acronym for Gallium Arsenide. A composite semiconducting material used in lieu of silicon in very high-speed discrete devices and IC applications.

Gang Placement Equipment—Automatic pick-and-place machinery that places a large number of components in each single machine cycle. *Also called simultaneous placement equipment.*

Glassivation—Semiconductor protection caused by the firing of a pyrolytic glass-frit deposition over the semiconductor face. A die passivation method.

Glass Phase—In hybrid firing, the part of the cycle where the glass-frit binder is in its molten phase.

Glass-to-Metal Seal—A hermetic dielectric seal such as that formed between a lead that passes through a package wall and the wall of the package. The seal is made by filling the space between the lead and package with glass frit and firing it to oxide layers on the metals.

Glass Transistion Temperature (Tg)—In polymer technology, the temperature above which the thermal expansion rate of the polymer increases dramatically and changes from a linear to an asymptotic curve.

Glaze—1. To produce a glassy dielectric layer by firing a thick-film dielectic ink on a substrate. 2. In PWB drilling, to cause a glassy coating to form on the barrel of the drilled hole due to improper drilling rotational or feed speed or due to a dull bit. Glazed holes must be etched to produce a rougher surface for plating adhesion prior to through plating. 3. The glassy surface produced in either of these two processes.

Glue Chip—A slang term for simple merchant market logic ICs. The term derives from the fact that such logic is often used to attach VLSI parts together to form a system.

Green—1. Ceramic technology, of or pertaining to a ceramic material in the unfired state. 2. The color of early epoxy-glass compositions that were green due to the presence of impurities. Now applied as a name for the coloring material added to pure epoxy (which is milky pink) to produce the customary green hue.

Gross Leak—A leak in a hermetically sealed package that passes more than 10^{-5} cm^3/sec. at a pressure differential of one atmosphere.

Gull Wing—A lead form where the leads exit a package body from a plane near or on the Z-axis center of the part, turn downward to just below the body of the device, and then turn outward to form feet for reflow-solder mounting. So named for the lead appearance, which resembles a gull's wing bent downward in flight.

Hairpin Mounting—A high-density through-hole style of mounting axial components where one lead is left straight and the other lead is bent 180 degrees to conserve square mounting area. This mounting method is not allowed for high-reliability assemblies because the hairpin leads are prone to damage from impact, shock, and vibration.

Halo—In ceramic technology, the term for a defect in which glassy halos form around conductors during the firing process. Generally avoided by proper firing profiles and correct material combinations.

HDI—An acronym for High Density Interconnect(ion).

Hidden Via—A via in an MLB that exits through one surface layer but does not come through the opposite surface layer. Also a blind via (preferred).

Hot-Crack—A metallurgical term for the cracking of a metal or alloy upon freezing. It may occur because of stress developed in the alloy due to unequal cooling. The phenomenon is sometimes accompanied by "hotshortness."

Hotshortness—The brittleness exhibited by some alloys and metals at elevated temperatures.

Hybrid—A type of microcircuit that consists of elements that are fabricated in situ directly on the substrate material in combination with discrete add-on components.

IC—An acronym for integrated circuit.

Icicle—A solder stalactite that forms on circuit metallizations or component terminations. Icicles are generally produced by an improperly controlled or designed immersion soldering operation. Also, the formation of such a defect.

IEC—An acronym for the International Electrotechnical Commission.

IEEE—An acronym for the Institute of Electrical and Electronics Engineers.

I²MT—An acronym for International Institute of Manufacturing Technology, an organization conducting SMT and advanced technology continuing education and research. Headquarters are at Box 549, MCV Station, Richmond, VA 23296-0001.

ILB—An acronym for "inner-lead bond."

IMAPS—An acronym for "International Microelectronics and Packaging Society."

IMC—An acronym for "insertion-mounted component." Also known as "THT."

Immersion Soldering—Soldering effected by the dipping of parts to be joined in a molten solder bath, wave, or jet.

Indirect Emulsion Screen—In screening terminology, a screen on which the emulsion is applied as a separate sheet or film pressed into the screen mesh, as opposed to a direct emulsion which is applied as a liquid.

Infant Mortality—The failure of an electronic device or system very early in its expected lifetime. Devices that do not succumb to infant mortality tend to live for their predicted lifespan, but a large percentage of device failures generally occur during their early life. Thus, by eliminating infant mortalities from a population, the apparent reliability of the population may be significantly raised.

Infrared—Of, pertaining to, or being in the band of radiation that has wavelengths from just longer than visible light to just shorter than microwaves. Generally used in SMT to describe a furnace or soldering process that uses such radiation as one of its major heating mechanisms.

Ink—A screen-printable thick-film paste.

In-Line Placement—Placement equipment that passes boards through a number of assembly stations in line with each other. At each station, one component is added in simple in-line systems. More complex systems may add several or many components at each station.

Inner-Lead Bond—A TAB bond to a bonding pad on an IC (an inner lead), as opposed to bonds on the circuit or interconnect structure (outer lead bonds).

Insertion-Mount Component (IMC)—A leaded component designed for mounting to a printed wiring board by insertion of its leads through holes in the board and then the soldering of those leads to circuitry on the substrate surface(s).

Integrated Circuit—A monolithic microcircuit consisting of elements inseparably associated and formed in situ on or within a single substrate (such as silicon).

Interfacial Bond—An interconnection made between conductors on two separate conducting layers of a substrate.

Intermetallic—Of, or pertaining to, a compound formed by two or more metals where the compound has a definite composition, its own unique crystal structure, and properties widely divergent from its constituent materials. Intermetallics are often refractory and brittle. Intermetallic formation in solder joints raises concern about the predictability of the solder joint's reliability.

Invar—A trademark of the Carpenter Technology Corporation. A metal alloy consisting of 64 percent iron and 36 percent nickel. Its notable properties are high strength, good thermal conductivity, and a low TCE.

I/O—The abbreviation for input/output. Generally applied to the outer interconnects of an IC or a PWB assembly, it refers to those interconnects through which a given device or assembly communicates with related devices or assemblies in an electronic system.

IPC—An acronym for the Institute for Interconnecting and Packaging Electronic Circuits (IPC), located at 7380 North Lincoln Avenue, Lincolnwood, IL 60646 USA.

IR—An acronym for "infrared."

ISHM—An acronym for the International Society for Hybrid Microelectronics, whose address is P.O. Box 2698, Reston, VA 22090 USA.

I/SMT—An acronym for the Interconnect/Surface-Mount Technology division of ISHM. *See also* ISHM.

ISO—An acronym for the International Standards Organization.

JEDEC—An acronym for the Joint Electron Device Engineering Council, which is the component standardization arm of the EIA, Washington, DC. *See also* EIA.

J Lead—A lead form commonly used on PLCCs. The lead departs from the package body at about the Z-axis center of the body, turns immediately toward the bottom of the component, and is formed in a radius (resembling the base of the letter J), turning under the package body. It then terminates in a protective pocket under the body.

K—Symbol for the dielectric constant of an insulating material. *See also* K factor.

Kevlar—A trademark of New England Ropes, Inc. Also, a tradename for a high-strength polymer. In SMT, Kevlar is used as a reinforcing and TCE restraining fiber in polymer PWB materials. Kevlar's TCE restraint is the result of the fiber's property of becoming shorter and fatter when heated. Thus, the warp and woof of woven Kevlar cloth restrains a PWB's X and Y expansion but amplifies its Z-axis expansion.

K Factor—A term for the value of thermal conductivity of a material. *See also* K.

Kirkendall Voids—These are voids that occur along the interface between two dissimilar materials that have different diffusion rates. The higher diffusion-rate material develops these voids as it diffuses into the adjacent substance faster than it is replaced by that substance.

Known Good Die—Also KGD, an unpackaged IC die known to meet all performance and electrical specifications for a particular application.

Land—The preferred term for a metallized area used for the attachment of a surface-mounted component termination to a substrate. Also called a *pad*, which is not preferred because of a potential confusion with via pads.

Laser Reflow—Refers to soldering using a laser with the laser generally emitting in the infrared wavelengths to impart a short burst of precisely located heat to reflow a solder joint.

LCC—An early acronym for "Leadless Chip Carrier." Not favored now because of the confusion with "Leaded Chip Carrier."

L Cut—An L-shaped notch made in a resistive material in a resistor trimming operation. The first part of the trim is made perpendicular to the resistor longitudinal axis to rapidly increase the resistance. As the desired value is approached, the trimmer turns the cut 90 degrees so that the rate of resistance change slows and the final value can be more precisely controlled.

LDCC—The preferred acronym for "Leaded Chip Carrier."

Leach—The action of one material diffusing into another. In soldering, leaching applies to the diffusion of metals from the device and/or PWB metallizations into solution in molten solder.

Lead—A length of metallic conductor used for component electrical interconnection. Not present on all components

Leaded Chip Carrier (LDCC)—A chip carrier with leads. The leads may be in the form of an integral lead frame or may be post attached to a leadless chip carrier.

Lead Frame—A metal interconnect structure used to provide I/O terminals for an electronic package or substrate.

Leadless Chip Carrier (LLCC)—A chip carrier, usually ceramic, with no external lead wires. I/O connections are through metallizations around the periphery at the bottom and/or the sides, and/or the top of the device. Also uses the less precise acronym of LCC.

Leadless Pad Grid Array (LPGA)—A package format used primarily for high pin-count ICs. The leadless SMT package provides I/O connections through metallized pads arrayed in a matrix on its base.

Lid—A cover for a hermetic package used with hybrid circuits and chip carriers for mechanical and/or hermetic protection of uncased dice. Also means "To install such a cover."

LID—An acronym for "Leadless Inverted Device." An open ceramic carrier for small semiconductor dice. The LID is designed for surface mounting by reflow attachment of four bottom-side I/O metallizations.

Line Definition—A term for the precision of the line width of a screen printing process. The value is obtained by dividing twice the line edge variation measurement by the line width. The value is typically expressed as a percent.

LLCC—The preferred acronym for "Leadless Chip Carrier."

LPGA—An acronym for "Leadless Pad Grid Array."

Machine Vision—Optical pattern recognition used as a sensor to facilitate the computer guidance of automated equipment.

Manhattan Effect—A term for a solder defect in which a small chip component raises one termination producing an open during reflow soldering. The defect is caused by unbalanced forces of solder wetting and surface tension acting on the two ends of the chip. When this imbalance is great enough to overcome gravity, the component begins to stand on one end. The term is best used for parts standing straight up, but is sometimes applied to parts in any stage of erection which produces a solder open on the raising end. *Synonymous with* Tombstone (preferred), Drawbridge, Pop Wheelies.

Margin—The distance from the edge of a substrate to the beginning of the features of the substrate.

Mask—A thin metal sheet with etched features. Masks are used in lieu of screens for stencil printing and in vapor deposition to define the thin-film deposition areas. Also, means "To cover or protect with a mask."

Matrix Tray—A shipping container specifically designed to protect relatively fragile components, such as bare IC dies. The matrix tray is named for its egg-crate matrix of pockets for part storage. Also called a *Waffle Pack.*

MCM—An acronym for Multi Chip Module wherein more than one IC die (chip) and passive chip components are placed on a substrate.

MCM-C—An acronym for Multi Chip Module—Ceramic.

MCM-L—An acronym for Multi Chip Module—Laminate.

Megahertz (MHz)—A term meaning one million hertz (one million cycles per second).

MELF—An acronym for "Metal Electrode Face (Bonded)." A cylindrical package for two-pin devices. Similar to a DO-35 package but with end metallizations instead of lead wires. The part takes its name from the common practice of face bonding metal caps over the ends to formterminations.

Mesh Size—A measurement of the size of the openings in a screen's cloth; expressed as the number of openings per linear inch.

Metal-to-Glass Seal—*See also* Glass-to-Metal Seal.

Microstrip—A conductive pattern on a PCB that has the conductors on the surface parallel with internal power and/or ground planes with dielectric insulation for separation between the conductors and the planes

Migration—The motion of ions from a material into an adjacent material where the concentration of ions is lower. Often, migration is an undesirable condition, such as in dendritic growth, leaching, or the migration of plasticizers from a polymer.

Misregistration—The offset error between the actual and theoretical X, Y, and theta dimensions of the tooling features to the working features of a substrate.

MLB—An acronym for "multi-layer board."

MLC—An acronym for "monolithic ceramic capacitor."

MLCC—An acronym for "multi-layer chip capacitor."

MLL-34—A cylindrical MELF-type package often used for diodes, usually about 1.5 mm (0.059") in diameter by 3.5 mm (0.138") long. Very similar to, and sometimes interchanged with, the SOD-80 package.

MLL-41—A cylindrical MELF-type package often used for diodes, usually about 2.5 mm (0.098") in diameter by 5 mm (0.197") long.

Monolithic Ceramic Capacitor—A term sometimes used for a chip capacitor laminated from a ceramic multi-layer construction.

Morphology—As applied in engineering, the science and study of the form, size, and shape of things. For example, in SMT, Quality Control is concerned with the morphology of solder pastes.

Multi Chip Module (MCM)—See MCM.

Multi-layer Board (MLB)—A composite wiring board generally laminated from multiple layers with each having a dielectric and at least one conductive layer. Successive MLB layers are generally interconnected by through-plated vias.

Multiple-Circuit Layout—A circuit layout or CAD design of an array of identical circuits used to fill a substrate that is greater than twice the size of a single circuit.

NASA—An acronym for the National Aeronautics and Space Administration, the U.S. governmental space agency. Its address is: NASA Scientific and Technical Information Facility, Technology Utilization Office, P.O. Box 8757, Baltimore, MD 21240-0757 USA.

NASA-JPL—An acronym for the NASA Jet Propulsion Laboratory at the California Institute of Technology, 4800 Oak Grove Drive, Pasadena, CA 91109 USA.

Neck Down—A term for the reduction of the width of conductors only in those areas where they closely approach adjacent features. Using relatively large lines and necking down the traces only where necessary can improve PWB manufacturing yields, thus reducing cost without sacrificing high density.

Noise—Any signal, particularly a random persistent disturbance, that obscures the intended circuit signals.

Ohms/Square—The unit of measure of sheet resistance. It is the resistance exhibited by a square pattern of the resistive material when printed at a given thickness. The area of the square pattern has no bearing on the measured resistance, but it controls the dissipation ability of the film resistor.

OLB—An acronym for "outer lead bond."

Open—A solder defect in either the reflow or flow-soldering process that occurs when insufficient solder is applied to a joint in effecting a connection (*see also Skip*), when poor wetting causes solder to fail to make a connection (*see also Tombstone*), or when a previously formed electrical connection fails and approaches infinite resistance. An *antonym* is: Bridge.

Orange Peeling—A defect of solder masks printed over tin or tin/lead-plated traces on a PWB. The defect occurs when the high temperatures of the PWB surface during SMT soldering operations reflow the metals under the solder mask. When this occurs, the mask takes on a rippled surface texture. In extreme cases, the mask may crack or open. The presence of "orange peeling" indicates the reflow of underlying metals, which points to another serious potential defect, the capillary migration of solder from reflow lands through traces. Such capillary migration may remove enough solder to cause solder-starved joints or open connections.

Outer Lead Bond (OLB)—A TAB bond connecting the TAB lead frame to a substrate, or interconnect structure as opposed to the "Inner Lead Bonds," which connect the TAB lead frame to the IC.

Outer Lead Bonder—A machine for the outer lead bonding of TAB devices.

Overglaze—A glassy coating applied over circuitry or over components to provide mechanical and electrical protection. Also, "To apply such a coating."

P&P—An acronym for Pick & Place. *See also* Placement Machine.

Package—1. A housing or protective covering for an electronic component. 2. A housing for a circuit assembly (usually a hybrid circuit). 3. A PWB assembly. 4. A housing for circuit board(s) or an electronic system. 5. To design or assemble a circuit assembly.

Packaging—1. The act of putting an item(s) in a housing or protective covering. 2. The act of designing an electronic circuit-board assembly—a package. 3. The material forming a protective package.

Pad—1. An annular ring around a via. Sometimes imprecisely used to mean the land for a surface-mount device. 2. A bonding site.

Pad Cap Layer—A surface layer of an MLB having only SMC lands, short or nonexistent connecting traces, and vias to internal routing layers.

Pads-Only Layer—*See also* Pad Cap Layer.

Parasitic—Refers to capacitance, inductance, or resistance not designed into a circuit but produced by circuit interconnections. Parasitics cause losses extraneous to those designed into circuit components.

Parasitics—Parasitic losses in a circuit.

Passive—1. A circuit element that does not change its basic characteristics when an electrical signal is applied to it. Resistors, capacitors, inductors, and transformers are some of the passives. 2. Of, pertaining to, or describing a circuit element that performs its intended electrical function without changing its basic character to do so.

Pass Through—A type of assembly line equipment where the workpieces are automatically transported from an entry point on one end of the machine to an exit point on the other end. Also, a line made up of such machines and linked to automatically pass workpieces through the line.

Paste—1. A formulation of tiny solder spheres or powder, a liquid flux, a thixatropic vehicle, and, in some instances, a flux activator. Solder pastes may be designed for application by screening, stencil printing, or dispensing. *Synonymous with* Cream. 2. Any screenable material for thick-film processing. *Synonymous with* Ink.

Paste Blending—The mixing of two or more inks to obtain a paste with properties between those of the original two. Resistor pastes are sometimes blended to achieve an ohms/square value between the values of the two original materials.

PCB—An acronym for "printed circuit board."

PFIB—An acronym for "Perfluoro Isobutelyne," a highly toxic decomposition product of perfluoronated hydrocarbon fluids (such as those used for vaporphase soldering). PFIB is generated when perfluoronated hydrocarbon fluids are subjected to extremes of high-temperature stress.

PGA—An acronym for a "pin grid array." *See also* LPGA.

P/IA—An acronym for "printed interconnect assembly." An IPC-recommended term for a PWB assembly.

Pick & Place Machine—*See also* Placement Machine.

Pin—1. A lead of an electronic component. 2. A small post forming an interconnect facility when inserted in a PWB.

Pin Grid Array (PGA)—An IMC package used for high pin-count devices. The pin grid array has a matrix of I/O pins extending from its base. *See also* LPGA.

Pin Transfer—A printing method employing pin(s) to pick up and transfer to a printing site a paste or ink. Used in SMT processing for adhesive application and occasionally for solder pastes.

P/IS—An acronym for "printed interconnect structure." The IPC-recommended term for a bare PWB.

Pit—A solder defect. An opening or cavity in the surface of a solidified solder joint. It is usually caused by entrapped gas or volatiles bubbling up through the solder as it is solidifying. It may indicate the hidden presence of voids. *See also* Void.

Pitch—The center-to-center distance between terminations/leads on a component.

Placement—The act of locating an SMC on its correct solder lands on a PWB.

Placement Machine—An automated or semi-automated machine used for the assembly of SMT components onto PWBs. Often called a "Pick & Place" machine because it functions by picking components from specified delivery mechanisms and placing them at programmed locations on a PWB assembly.

Plastic Leaded Chip Carrier (PLDCC or PLCC)—A plastic IC chip carrier package having J-leaded SMT I/O on four sides. Both premolded and postmolded parts are available in this package format.

Plastic Quad Flat Pack (PQFP)—A fine-pitch postmolded plastic surface-mount semiconductor package with a gull-wing lead form, peripheral I/O on all four sides, and is distinguished by extended corners that form bumpers for the protection of the leads during part handling and transport.

Plated-Through Hole (PTH)—A hole in a PWB with a plated metallization on its inner walls that allows the interconnection of circuit metallizations on the opposite sides of two-sided boards and/or an interconnection with the innerlayer metallizations of an MLB.

PLCC—An acronym for "Plastic Leaded Chip Carrier."

Poise—The unit of measure of the coefficient of dynamic viscosity. One poise equals 1 dyne-sec/cm^2.

Pop Wheelies—The physical action that creates a solder defect in which a small chip component raises one termination, thereby producing an open, during reflow soldering. The defect is caused by the unbalanced forces of solder wetting and surface tension acting on the two ends of the chip. When this imbalance is great enough to overcome gravity, the component begins to stand on one end like a drawbridge opening. The term is used for parts in any stage of erection that is sufficient to produce a solder open on the raising end. *Synonymous with* Tombstone (preferred), Drawbridge, Manhattan Effect.

PPM—An acronym for "parts per million."

PQFP—An acronym for "Plastic Quad Flat Pack."

Preform—A small preshaped piece of material, such as solder, for placement at a soldering site. Some solder preforms also have a layer of flux preapplied. Epoxy and hotmelt adhesives may also be applied as preforms.

Printed Circuit Board (PCB)—The precise term for a substrate containing both interconnect wiring traces and in situ components (circuitry). Often used when referring to printed wiring boards as well, but printed wiring boards are actually printed interconnect structures only, having no in situ circuitry.

Printed Wiring Board (PWB)—The preferred term for a wiring board fabricated by photolithographic print-and-etch techniques, and which contains only printed interconnect structure(s) with no components (circuits) formed in situ. Also, sometimes called a Printed Circuit Board (PCB), which is not preferred since the board is not actually an electronic circuit but merely the wiring for a circuit assembly.

Process Control—1. The science of improving manufacturing yields by bringing the variability of each critical element of a manufacturing process under tight control. 2. The measurement and control of a given variable in a manufacturing process.

Process Quality—The percentage of product that conforms to specifications as it leaves the manufacturing line and arrives at final test.

Profile—1. The time-temperature relationship experienced by a workpiece in a process such as linear furnace soldering, with given furnace settings. The X–Y plot of such a time-temperature relationship. 2. To adjust and correctly set the time-temperature relationship of a piece of equipment.

Propagation Delay—The time delay between the input and output of a signal.

PTH—An acronym for "plated through hole."

Pull Test—A test for the force required to pull a component from its mounting lands with the force acting only in the +Z vector.

Push-Off Test—A test for the force required to remove a component from its mounting lands by a single axis push that is acting only in the X or the Y axis.

PWB—An acronym for "printed wiring board."

Quad Flat Pack—Any flat pack with a peripheral I/O on all four sides.

Quad Pack—Literally, a semiconductor package having I/O on all four sides at its periphery. Generally, usage does not include leadless components as quad packs. Also, packages with a distinct name, such as PLCCs, PQFPs, CQPFs, etc., are not called quad packs. It is the generic name for four-sided I/O-leaded semiconductor packages that do not have some other specific name.

Quality—The conformance to specifications by a new device or system.

Radial—Of or pertaining to an electronic component lead form in which the leads extend from the outer radius of the device tangentially. Also, such a device.

Radio-Frequency Interference (RFI)—Unwanted (interference) signals received in a circuit from external or internal radio frequency noise sources.

RAM—An acronym for "random access memory."

Random Access Memory (RAM)—A memory in which data may be read or written in any selected binary address location.

Real Estate—Refers to the space on a printed wiring board; its square area.

Reference Edge—The term for the edge or edges of a substrate that are to locate it for manufacturing processes.

Reflow—A type of soldering that involves the melting of previously applied solder to form device solder bonds on a circuit assembly.

Reliability—The continued conformance to specifications by a device or system over a specified extended period of time.

Repair—The work done to a circuit assembly to cause it to operate, but not necessarily to bring it into complete conformance with its original specifications. Also, the act of performing such work. *See also* Rework.

Resist—A dielectric coating used to mask selected areas of a PWB and prevent solder from adhering to metallizations. Resists may be temporary, allowing the selective solder coating of PWB features during board manufacture, or may be permanent, as in the case of solder masks.

Resistor Drift—A measure of the change in resistance of a resistor over an extended period. It is generally expressed as a percentage of change in resistance over 1000 hours.

Resistor Overlap—In hybrid circuitry, the overlap distance of the resistor print and its conductor trace.

Reverse Reading—In photolithography, a photo tool in which the desired pattern is seen as a mirror image when viewed from the emulsion side of the film.

Rework—Work done to a circuit assembly to bring it into complete conformance with its original specifications. Also, the act of performing such work.

RFI—An acronym for "radio frequency interference."

Right Reading—In photolithography, a photo tool in which the pattern is seen as the required circuit image when viewed from the emulsion side of the film.

Riser—In hybrid MLBs, a via formed by thick-film print-and-fire techniques.

Runner—In a PWB layout, an individual conductor. *Synonymous with* Trace (preferred), Track.

Sapphire—The monocrystalline form of aluminum oxide (Al_2O_3). Used as a substrate material where its highly polishable surface finish, low camber, radiation hardness, and/or uniform lattice structure is needed. Also used for high-speed silicon-on-sapphire MOS devices. *See also* Alumina.

Scallop Marks—Screening terminology for a defect where the edges of a printed pattern are jagged. The defect is seen when printing pressures are incorrect or the screen emulsion is too thin.

Scavenge—Also called "leaching." An action that causes one material to diffuse into another. In soldering, leaching applies to the diffusion of metals from the device and/or PWB metallizations into solution in molten solder.

Score—Hybrid terminology for the process of scribing break lines on ceramic, silicon, glass, or sapphire snapstrates.

Screen—The image-defining mask used in screen printing. Also, the act of using such a mask to print.

Screenability—A measure (often subjective) of the quality of print definition in a screen printing operation.

Scribe Coat—A two-layer polymer film with a clear base and a semiopaque cover that is used for preparation of semiconductor and microcircuit artwork. Circuit patterns, in an enlarged scale, are defined on the film by cutting and stripping away unwanted portions of the opaque cover using a coordinator graph.

Self-Passivating—A term used to describe materials that inherently form a passivating overglaze when fired in a thick-film process. Some resistor inks are self-passivating.

Semiconductor—A solid material like silicon or gallium that has a resistivity midway between that of an insulator and that of a conductor.

Sequencer—A machine designed for extracting leaded components from shipping container carrier tape reels and placing them on an assembly tape reel in the sequence in which they will be inserted in a PWB assembly.

Sequential Placement—A class of placement equipment that programmably places components sequentially, one at a time.

Sequential Simultaneous Placement—A class of placement equipment that programmably places components sequentially, and that places more than one component during each machine cycle (sequence).

Serpentine Cut—A serpentine-shaped or sinusoidal trim cut made in a resistor.

Shadowing—A term for reduced heating in infrared ovens on areas of a workpiece when the adjacent objects on the assembly shield those areas from IR radiation. Now largely eliminated by oven design improvements whereby a greater percentage of the total energy transfer is by conduction and convection than it was in earlier furnace designs.

Shear Rate—The relative rate of flow of a viscous fluid.

Sheet Resistance—The measured resistance of a square pattern of controlled thickness film of resistive material, expressed in ohms/square. *See also* ohms/square.

Silicon Monoxide—A material that is sometimes vapor deposited as a passivating layer for thin-film circuits.

Silver Chromate Paper Test—A qualitative test for the presence of organic halides. Used in SMT soldering to test RMA and other mildly activated fluxes to determine that no organic halides are present in the activators. *See also* Silver Nitrate Test.

Silver Nitrate Test—A qualitative test for the presence of organic halides. Used in SMT soldering to test RMA and other mildly activated fluxes to determine that no organic halides are present in the activators. *See also* Silver Chromate Paper Test.

Simultaneous Placement—The placement of a full complement of SMCs on an assembly, or the placement of a large number of SMCs simultaneously. Also, a class of placement machines that operates in this fashion. Generally, "simultaneous placement" refers to inflexible dedicated automation. Also called "Gang Placement."

Single In-Line Package (SIP)—A package format that is often used for resistor networks or ICs. The package is named for its single line of I/O pins extending from one side. SIP leads are designed for insertion mounting in PWBs.

Single-Layer TAB Tape—In TAB technology, a TAB tape consisting only of a metal foil which is etched to produce repetitive lead-frame patterns and sprocket holes for handling.

SIP—An acronym for "single in-line package."

Skip—An open-circuit defect in flow soldering where insufficient solder to make a desired solder connection is applied to a solder joint. *Antonym:* Bridge. *See also* Open.

Skin Effect—The tendency of high-frequency electrical currents to concentrate at the surface (skin) of a conductor resulting in an increase in apparent resistivity.

Slump—To spread after being deposited. This applies to viscous fluids and thixatropic pastes. The slumping of screened ink before or during firing is a defect that blurs screened image definition.

SM—An acronym for Surface Mount, Surface Mounted, and Surface Mounting. See definition under *Surface-Mount Technology.*

SMA—An acronym for "surface-mount assembly."

SMC—An acronym for "surface-mount(ed) component."

SMD—An acronym for "surface-mount device," SMD is a trademark of Philips. See definition under *Surface-Mounted Component.*

SMOBC—An acronym for "Solder Mask Over Bare Copper."

SMT—An acronym for "surface-mount technology."

Snapback—The return of the screen to its at-rest position after a squeegee stroke.

Snap-Off—A measure of the distance between the top of the substrate and the bottom of the undeflected screen when they are mounted on a screen printer. In printing, the squeegee deflects the resilient screen causing it to contact the board. After the squeegee stroke, the screen snapback lifts the screen from the substrate and returns it to its original at-rest clearance.

Snapstrate—A multiple-circuit substrate that is scribed so that it may be broken (snapped) into individual assemblies after batch assembly processing.

SO—An acronym for "small outline." Refer to SOB, SOD, SOT, SOIC, and SOJ IC.

SOB—1. An acronym for "small-outline bridge." A four-leg SOT-143-like package with an oversized body for housing a full-wave bridge rectifier. 2. The suppressed crying sound made by engineers in reaction to the proliferation of electronics acronyms.

SOD—An acronym for "small-outline diode"; a MELF-type package.

SOIC—An acronym for "small-outline IC," a package format having I/O on two sides. The leads exit the component body around its center line and are formed into a gull-wing shape for surface mounting. The JEDEC standard SOIC body is nominally 3.8 mm (0.150") wide, and the package length varies dependent on the number of 1.27-mm (0.050") center leads per side.

SOJIC—An acronym for "small-outline J-leaded IC." An IC package similar to the SOIC, except that its body is wider and its leads turn under the body in J-lead form rather than turning out in the "gull-wing" fashion.

Solder—1. A relatively low melting point metal or alloy used to join metal(s) having a higher melting point. For SMT, the tin/lead alloys are the most common solders, although lead-free solders are in growing use. 2. To join several metals using solder.

Solder Ball—1. A solder defect where solder forms in small spheres outside the solder-joint area. Solder balls are difficult or impossible to remove from under low-clearance components, but can move while in service, causing short circuits and catastrophic failures. Therefore, solder balling is considered a serious defect that is to be completely avoided. Solder balls may result from the use of overly oxidized solder paste or the violent outgassing of volatiles from improperly baked solder paste during rapid reflow heating. 2. The solder sphere at each interconnect of a ball grid array. 3. The microscopic metal balls that make up the metallic portion of solder paste.

Solder Bridge—A short-circuit defect produced when solder crosses a dielectric space and electrically joins two leads or metallizations that are supposed to be electrically isolated.

Solder Bumps—Solder hemispheres grown on or added to a semiconductor die to facilitate interconnection of the chip. *See also* Flip Chip.

Solder Mask—An insulating material applied using wet or dry techniques that is intended to minimize the adherence of molten solder to the areas the mask covers.

Solder Mask Over Bare Copper—A PWB manufacturing technique that involves the application of the solder mask over bare copper traces with subsequent operations to apply a tin/lead overplate to the solder lands left exposed by the mask. The SMOBC technique eliminates solder-filled capillaries under the mask, thus improving reflow solder yields.

Solder Paste—A solder formulation of tiny solder spheres or powder, a liquid flux, and a thixatropic vehicle, plus, in some instances, a flux activator. Solder pastes may be designed for application by screening, stencil printing, or dispensing.

Solder Resist—*See also* "Solder Mask."

Solder Side—In IMC boards, the side of the board opposite the components. This is the side of the PWB that is immersed in solder during the flow soldering of through-hole component leads.

SOLIC—An oxymoronic acronym for Small-Outline Large IC, a package format similar to the SOIC, but with a wider body that is capable of housing a larger die. The JEDEC standard SOL IC body is nominally 7.6 mm (0.300") wide, and package length varies dependent on the number of 1.27-mm (0.050") center leads per side.

SOMIC—An acronym for Small-Outline Medium IC. A nonstandard package that is sometimes used to house resistor networks with substrates too large to fit in an SOIC. The package is like the SOIC, but it has a nominal body width of 5.6 mm (0.220").

SOS—1. An acronym for "silicon-on-sapphire," a high-speed device in MOS IC technology. 2. An acronym for "silicon-on-silicon," a technology that makes use of a silicon substrate.

SOT-23—An acronym for Small-Outline Transistor (style 23), it uses a JEDEC TO-236 package. A small-signal transistor package. Package is also used for diodes, dual diodes, LEDs, etc. The package has three leads and can house dies up to about 0.6 mm (0.025") square. Its maximum dissipation in free air at 25°C is 200 milliwatts.

SOT-89—An acronym for Small-Outline Transistor (style 89), it uses a JEDEC TO-243 package. A small-outline transistor package capable of handling higher-power applications than the SOT-23. Suitable for dies up to about 1.5 mm (0.060") square. The package can dissipate a maximum of 500 milliwatts in 25°C free air.

SOT-143—An acronym for Small-Outline Transistor (style 143). A small-signal transistor package that is also used for dual diodes, GaAs FETs, RF transistors, etc. The package has four leads and can house dies up to about 0.6 mm (0.025") square. It dissipates a maximum of 200 milliwatts in free air at 25°C.

Space—In PWB layouts, the minimum distance between the adjacent edges of two conductors (not the center-to-center distance).

SRAM—An acronym for "static RAM," a memory device that does require periodic clocking or refresh to maintain memory integrity. *Contrast with* DRAM.

Stair-Step—Term used to describe a screen printing defect where the print edges show the distinct pattern of the screen mesh. This defect is produced by inadequate printing pressure or by the use of screens with too thin an emulsion.

Standoff—In SMT, the distance between the top of the substrate and the bottom of an SMC mounted on it. An adequate standoff height is critical to the cleaning of surface-mount assemblies.

Stencil—Screening terminology for a metal mask that is used in lieu of a screen for printing. For solder printing, stencils are more expensive and require longer lead times than screens. However, stencils last considerably longer, are easier to clean and maintain, and are capable of reproducing finer printed patterns than screens.

Step and Repeat—A process wherein a pattern, or a sequence of automated machine motions, is repeated one or more times at equal increments in the X and/or Y axis. For instance, a small circuit pattern may be stepped and repeated multiple times on a large substrate.

Stick—A plastic shipping container for components. Sticks also may be used as a component delivery mechanism for many types of automated processing equipment.

String—1. The name for the phenomenon evidenced when internal cohesive forces of a dispensed fluid or gel cause a part of the dispensed droplet to draw up as a tail following the withdrawal of the dispensing tool. Excessive stringing must be avoided because such tails after breaking their cohesive connection with the departing tool may fall into areas of the assembly where they will be a quality or reliability problem. 2. The action of stringing.

Stripline—A conductive pattern inside a PCB that has the conductors spaced equidistant between power and/or ground planes with dielectric insulation for separation between the conductors and the planes.

Substrate—In an electronics assembly, the supporting material, with or without its printed interconnections, upon which components are mounted or fabricated to produce a circuit assembly. A multi-layer board is spoken of as a single substrate, even though it is laminated from several layers of composite materials.

Surface-Mounted Component (SMC)—A component designed for planar mounting as opposed to through-hole or chip-and-wire mounting. Variants are standard SMT components, flip chips, and TAB formats.

Surface-Mount Technology (SMT)—The science of planar mounting components on a PWB or on hybrid-printed interconnect substrates.

Surfactant—A contraction of *SURFace ACTive AgeNT*. A material that reduces the surface tension of liquids.

Swim—The act of X–Y motion of components during reflow soldering. Swimming is the phenomenon seen when a part displaces more than its own weight in solder, thus floating during the reflow process. Floating parts tend to move to equalize the forces of wetting and solder surface tension. With good wettability and proper design, these forces cause parts to center in the lands.

TAB—An acronym for "tape automated bonding."

Tape—1. A carrier mechanism for shipping, storing, and automatically feeding SMCs to assembly processes. Carrier tape has pockets cut or embossed in it. Components are retained in the pockets by means of cover tape. The carrier tape has peripheral sprocket holes to facilitate the precise feeding of pocketed components to an assembly machine. Cover tape is peeled back to expose one component after another as they are required for assembly. 2. The act of installing components in carrier tape. NOTE: There are numerous other electronics usages of the word *Tape* derived from its standard dictionary definitions. The above definitions and that shown under *Tape Automated Bonding* are listed herein as they are unique to the SMT process.

Tape and Reel—A term to distinguish carrier tape from other forms of tape used in electronics (axial and radial components are commonly taped for automatic insertion, but such tape is not called *Tape and Reel*, or *Carrier Tape*). The phrase derives from the reel storage system used for the shipping and handling of carrier tapes.

Tape Automated Bonding—A high-density approach to interconnecting IC dies to substrates or leadframes. A tape of either metal foil or a foil/polymer composite is used to carry and provide an interconnect to the die. Sprocket holes are formed in the periphery of the tape to facilitate precise automated handling. The metal foil is etched to form a leadframe. Inner leads of the tape leadframe are thermocompression bonded to the IC die. At final assembly, the chip and leadframe portion are excised from the outer carrier portion of the tape. To interconnect the TAB chip to its higher-level assembly, the outer leads are thermocompression or solder bonded to the lands of the higher-level interconnect structure. *See also* TAB.

TCE—An acronym for "Thermal Coefficient of Expansion."

Termination—A lead or metallization area of a component that provides a point to interconnect an I/O line of the component to other circuitry.

Tg—An acronym for "Transition Temperature, Glass." In polymer technology, the temperature above which the thermal expansion rate of the polymer increases dramatically and changes from a linear curve to an asymptotic curve.

TGA—An acronym for "thermogravimetric analysis."

THC—An acronym for "through-hole component."

THT—An acronym for "through-hole technology."

Theoretical Origin—In an artwork or pattern generated from an artwork, that point from which all features are measured in the X and Y offset.

Thermal Coefficient of Expansion (TCE)—Also called *Coefficient of Thermal Expansion (CTE)*. A measure of the coefficient of expansion of a material under temperature increases. Expressed in parts per million (PPM)/°C. This is critical in SMT where leadless parts have no flexural element to absorb the stress generated by the disparate expansion of substrate and component during temperature cycling. *See also* TCE.

Thermal Stress—1. Stress generated by disparate expansion rates between separate bonded bodies in an assembly during temperature cycling. 2. Stress developed within a homogeneous body due to temperature differentials within the body during temperature cycling. (This sense of the phrase is also referred to as "thermal shock.") 3. Electrical and mechanical stress produced by operation at elevated ambient temperatures.

Thermocompression—A type of bonding process in which pressure and elevated temperatures are used to join two materials by interdiffusion across their boundary.

Thermogravimetric Analysis—A quantitative analytical test used to determine the cure state of a polymer resin by observing the effect of temperature changes on the sample's weight. *See also* TGA.

Thermomechanical Analysis—A quantitative analytical test used on polymer PWBs to determine their dimensional expansion curve in relationship to temperature. TMA can be used to pinpoint the Tg of a polymer material. *See also* TMA.

Thick Film—A film generated on a substrate surface by a screening or stencil printing process. Such films are generally greater than 100 microns (0.004") thick. Typical SMT thick films include conductive, resistive, and dielectric films.

Thin Film—A film accreted on a substrate surface, generally by a vapor-deposition or sputtering process. Such films are generally less than 5 microns (0.0002") thick and, more usually, under 0.3 microns (0.00001"). Typical SMT thin films include conductive, resistive, and dielectric materials.

Thixatropic—A term used to describe the property that certain gels and liquids have to liquefy when stirred or placed under dynamic pressure but to return to the previous viscous state when static. Thixatropic agents allow solder paste to flow through a screen under squeegee pressure, but stand without slumping on the substrate after printing.

Three-Layer TAB Tape—In TAB technology, a tape comprised of a base polymer layer (generally polyimide) with necessary handling features, such as sprocket holes, an adhesive reinforcing layer running the full width of the tape, and a metal-foil layer etched to form the TAB leadframes. Generally, the metal foil of three-layer tapes covers only the center portion of the tape and is just wide enough to encompass the needed leadframe features.

Through Hole—1. A via hole extending from one side to the other in a PWB. 2. The technology and practice of mounting leaded components by insertion in holes through PWBs.

Through-Hole Component—A leaded component designed for mounting to a printed wiring board by the insertion of its leads through holes in the board and then the soldering of those leads to circuitry on the substrate surface(s). *See also* Insertion-Mount Component (IMC).

TMA—An acronym for "thermomechanical analysis."

Tombstone—The preferred term for a solder defect in which a small chip component raises one termination, producing an open, during reflow soldering. The defect is caused by unbalanced forces of solder wetting and surface tension acting on the two ends of the chip. When this imbalance is great enough to overcome gravity, the component begins to stand on one end. The term is best used for parts standing straight up, but is sometimes applied to parts in any stage of erection that produces a solder open on the raising end. Also, the act of tombstoning. *Synonymous with* Drawbridge, Pop Wheelies, Manhattan Effect.

Top Hat—A descriptive term for a type of thick-film resistor geometry resembling a top hat. Top-hat resistors have a projection from one or both sides (like the brim of a top hat). The projection(s) allow a trim notch to be cut into the center of the resistor, creating a serpentine geometry and thereby increasing resistance. Top hats are recommended where high aspect ratios are required.

Trace—In PWB layout, an individual conductor. Also, the width of an individual conductor. *Synonymous with* Runner (not preferred), and Track.

Trace and Space Rules—In PWB layout, the minimum allowable trace and space dimensions.

Track—In PWB layout, an individual conductor. *Synonymous with* Runner (not preferred), and Trace (preferred).

Trim—To remove material (cut a notch) in a film resistor or other component to adjust the value of the component. Laser and abrasive trimming are two typical processes.

Tube—A plastic shipping container for components. Tubes also may be used as a component delivery mechanism for many types of automated processing equipment. *Synonymous with* Stick.

Two-Layer TAB Tape—In TAB technology, a tape made up of a base layer of polymer film (such as polyimide) and a metal-foil layer etched to form the TAB leadframe. Sprocket holes and handling features are fabricated in the periphery of the composite tape to permit the precise indexing of the material in assembly operations.

Type 1 SMT—A surface-mount assembly style that employs surface mounting exclusively and is characterized by the reflow soldering of all solder connections. Type 1 SMT may be used on either a single side or both sides of a board.

Type 2 SMT—A surface-mount assembly style that employs SMCs mixed with through-hole components on the same side of a board assembly and is characterized by the reflow soldering of SMC solder connections and the flow soldering of IMCs where they are mixed on the same side. Type 2 SMT may involve double-sided population using any style of assembly, but is still considered Type 2 as long as one side mixes SMCs and IMCs.

Type 3 SMT—A surface-mount assembly method characterized by having IMCs on the classical component side of the board and SMCs glued to the bottom side of the board. After assembly, the board is positioned with the SMCs down and is flow soldered, simultaneously forming through-hole and SMC solder bonds.

UBM—Under Bump Metallurgy, used to create the first layers in the bumping process for TAB and flip-chip devices.

Ultraviolet Cure—A curing process used to activate the cross linking of certain epoxy adhesives by exposing the material to strong ultraviolet radiation. Also used to cure certain silk-screened inks, such as those used on some PWB silk-screen legends.

Underfill—Encapsulant material typically deposited between a flip-chip device and the substrate on which it is mounted to reduce the effects of the CTE mismatch between the silicon device and the substrate materials.

Underglaze—1. A glass or ceramic glaze that is fired onto a substrate under the area where a resistor will be screened and fired. 2. To apply such a glaze.

Upset—When applied to solder joints in electronics, this term denotes the condition that is also called a "cold joint," where two or more pieces being joined by solder moved during the plastic phase of the solder solidification process. Such joints are reliability concerns. They are recognized by a grainy, rough, surface finish.

Vapor-Phase Soldering (VPS)—A reflow-soldering method that uses a vapor blanket above a boiling fluid to conductively heat a workpiece. The heating is rapid since vapor condensing on the relatively cool workpiece yields its latent heat of vaporization to the workpiece. For typical SMT solder requirements, high-boiling-temperature perfluorinated fluids are used. The most commonly used material boils at 215°C.

VCD—An acronym for "variable center distance," a style of insertion machine capable of programmably adjusting to form-and-insert axials at a variety of lead-center distances.

Via—1. In PWBs, an interlayer connection in a two-sided or multi-layered board provided by a through-plated hole. 2. In hybrids, an interlayer connection provided by printing an upper-layer conductive trace over an opening in the separating dielectric layer. The opening in the dielectric is arranged to expose the lower-layer conductor that is to be interconnected. Also called a *Riser*.

Void—A type of solder defect. A cavity inside a solidified solder joint. Generally caused by entrapped gases or volatiles that are unable to escape as the solder solidifies. *See also* Pit.

VPS—An acronym for "vapor-phase soldering."

Waffle Pack—A matrix tray with "egg crate"-style pockets for the protection of delicate components during shipment, storage, and handling.

Warp and Woof—In screen technology, a description of the lay of the fibers that make up the screen mesh. The warp fibers run parallel to the length of the fabric and are warped alternatively upward and then downward in the weaving process. The woof fibers run crosswise to the fabric's length (or parallel to the length of the weaving machine).

Wave Soldering—An immersion soldering method. Workpieces are moved across a wave or waves of molten solder at a height such that the bottom of the board is momentarily immersed. Leaded components are on the board's top side with their leads inserted through metallized holes in the board. Any SMCs that are to be wave soldered are glued to the bottom (solder) side. Both the leads extending through the PTHs and the bottom-side SMCs are simultaneously soldered in passing through the solder immersion(s). SMT wave soldering generally requires equipment with special features in order to reduce solder defects on the attachment of surface-mount components.

Wetting Agent—A material that reduces the surface tension of liquids. In SMT, a wetting agent improves a cleaning fluid's ability to penetrate under low standoff components.

Wick—A term used to describe flow through a capillary, as when solder flows away from lands through capillaries produced by reflowed metal on masked conductor traces. Wicking is likely to produce defects due to the solder starving of SMC joints.

WIP—An acronym for "Work In Process."

Wire Bond—1. An interconnection between components or between a component and leads of a circuit. A wire bond is made by means of a fine wire bonded to the points to be interconnected. 2. The act of making such interconnections.

Work In Process—Defines the goods involved as raw material, subassemblies, and assemblies in a manufacturing line. A large amount of WIP increases inventory amounts, reduces inventory turns, and generally reduces manufacturing yield while driving costs upward.

"X" Axis—For drafting, normally taken as the up-and-down length of the paper as it is viewed. For automated equipment, convention places the X axis parallel to the flow of workpieces through the machine.

"Y" Axis—For drafting, normally taken as the length across the paper as it is viewed. For automated equipment, convention places the Y axis perpendicular to the flow of workpieces through the machine.

"Z" Axis—For automated equipment, the Z axis is an axis perpendicular to the plane of the workpieces.

Zero Zero—The theoretical origin point of an artwork or the pattern generated from an artwork. That point from which all features are measured in the X and Y offset.

ZIP—An acronym for "zigzag in-line package." An IMC package format in which leads extend from only one side of the package body and alternate (zigzag) between two parallel rows, which are generally spaced 2.54 mm (0.100") apart.

Index